Modern Physical Chemistry

Other books by George H. Duffey

Quantum States and Processes

Applied Group Theory for Physicists and Chemists

Theoretical Physics

A Development of Quantum Mechanics Based on Symmetry Considerations

Modern Physical Chemistry

A Molecular Approach

George H. Duffey
South Dakota State University
Brookings, South Dakota

A Solomon Press Book
Kluwer Academic / Plenum Publishers
New York, Boston, Dordrecht, London, Moscow

Library of Congress Cataloging-in-Publication Data

Duffey, George H.
Modern physical chemistry: a molecular approach/George H. Duffey
p. cm.
Includes bibliographical references.
ISBN 0-306-46395-4
1. Chemistry, Physical and theoretical. I. Title.

QD453-2 .D84 2000
541.3—dc21

00-035702

Book designed by Sidney and Raymond Solomon

ISBN: 0-306-46395-4

233 Spring Street, New York, N.Y. 10013

http://www.wkap.nl/

10 9 8 7 6 5 4 3 2 1

A C.I.P. record for this book is available from the Library of Congress

Printed in the United States of America

To Henry Eyring,
an inspiring original thinker

Preface and Acknowledgments

This book has been designed for chemistry and chemical physics majors in their junior or senior years. However, advanced students and students in related fields may also find it useful as a reference. A student using the text should be familiar with basic general chemistry, some analytic chemistry, elementary calculus, and calculus-level general physics.

Its style is simple, concise, and straightforward. Concepts are *not* hidden in a blizzard of words. Mathematics is employed slowly and carefully as a tool.

General principles are shown to rest on certain key experimental results. In expressing relationships mathematically, the text is as detailed as is expedient. With respect to problems, the text includes many worked-out examples. And where helpful, the explanations are couched in terms of what atoms and molecules are present and what they are doing.

Topics are arranged in a progressive pattern. Simpler theory and examples are considered first; more complicated theory and applications later. More information on the various topics and, in many cases, alternate developments may be found in the references.

I wish to acknowledge my debt to my predecessors and to my students. Our present knowledge of physical chemistry is the product of many hands and minds. Various articles in the *Journal of Chemical Education* and *The American Journal of Physics* have been particularly helpful. These are cited in the reference sections. I want to thank Prof. Orin Quist for his encouragement and help. Also, Corey Halstead, for keyboarding the equations. And finally, the staffs at The Solomon Press and Jonah Shaw have been most helpful.

— George H. Duffey
April 1999

Contents

1

Structure in Solids

1.1 Limits on Homogeneity

TO AN OBSERVER, THE UNIVERSE APPEARS to be made up of objects and beings that act on each other. With an appropriate divider, each object may be broken down into smaller parts. When the object is homogeneous, each part is like any other part of the same size and shape.

But can such subdivision be carried out indefinitely? Or does a person finally reach units whose rupture would destroy the properties of the material making up the object?

Thinkers in ancient times considered both possibilities. However, they were not able to arrive at any conclusion. Only in modern times has the submicroscopic nature of objects been investigated and determined. The simplest evidence involves how the constituent material can be arranged in space, particularly at low temperatures.

1.2 States of Matter

When enough heat is removed from a given material, it becomes frozen; it tends to maintain both a definite volume and shape. Such material is said to be a *solid*. Then heating it above a certain temperature gives it the ability to flow. It assumes a shape determined by the lower surfaces of the containing vessel. The material is then said to be a *liquid*. Above a higher temperature, determined by the applied pressure, it boils. The resulting phase then will fill any small containing vessel, regardless of the vessel's shape or size. The material is then said to be a *gas*. A gas of the given material over the corresponding liquid or solid is called a *vapor*.

We will study the nature of temperature, pressure, heat, and work later. The resulting relationships will enable us to analyze the behavior of transitions among the states. But here our primary concern is geometric, the arrangement of materials in space at low temperatures and what one can deduce from such simple considerations.

In the absence of polarizing forces, a gas or liquid is *isotropic*, with no macroscopic property that varies with direction. An appreciable impressed force field does destroy some of this symmetry and make the system *anisotropic*. The solids called *glasses* are also isotropic in the macroscopic sense. Other solids exhibit some anisotropy. Such a solid exists either as a single unit called a crystal or quasicrystal, or as a conglomerate of such units.

A single *crystal*, placed in a uniform supersaturated solution, grows faster in some directions than in others. Under shear strain, slippage tends to occur along planes parallel to those developed at the surface and along planes intersecting these at regular angles. Under shock, cleavage occurs along such planes. In addition, the compressibility, coefficient of linear expansion, electrical and thermal conductivity, refractive index, and dielectric constant may vary with direction.

The planes along which slippage and cleavage occur form parallel sets intersecting in a regular fashion. Apparently, something is arranged in order in a crystal. Furthermore, the isotropic states seem to arise when the basic parts shift to break down this long-range order. Let us seek information on the bases and their arrangements in space.

A *quasicrystal* does not have the long-range order of a true crystal. It does, however, possess enough short-range order to exhibit crystal-like diffraction patterns. A glass, liquid, or gas does not possess such short-range order.

1.3 Diffraction

In everyday life, one gathers information about the external world with one's senses. The most important of these involve sight and touch.

Some of the light scattered by neighboring objects is collected by one's eyes, then refracted and projected to form images on the retinas. Signals from these are transmitted to appropriate regions of the brain. The resulting mental impressions are supported by tactile evidence. For, each visible body may be touched, felt, squeezed, and possibly moved. Sounds can be traced to macroscopic sources. Similarity, smells and tastes can be.

These procedures tell us nothing about the *minute* internal structure of materials, however. A finer probe is needed. Shorter wavelength radiation, in the X-ray region, is suitable. But since refractors do not exist for such radiation, one has to analyze each scattered pattern directly.

In the open where no scattering occurs, the radiation appears to propagate along definite lines called *rays*. Where scattering does occur, the pertinent rays are bent or forked. Only in certain directions do neighboring rays reinforce each other. We say that *diffraction* has occurred.

At a typical point along a given ray, the pertinent physical properties vary in concert. Within a given medium, variations in the different components are scaled and rotated (and/or translated) versions of each other. Away from its source, electromagnetic radiation consists of a changing electric field strength E perpendicular to a corresponding magnetic field strength H, both perpendicular to the direction of motion.

When a component varies sinusoidally, as figure 1.1 shows, we say the variation is monochromatic, with a definite frequency and wavelength. The *frequency* ν is given by the number of crests passing a fixed point in unit time, while the *wavelength* λ equals the distance between crests. The *phase velocity* w is the velocity at which a given crest moves.

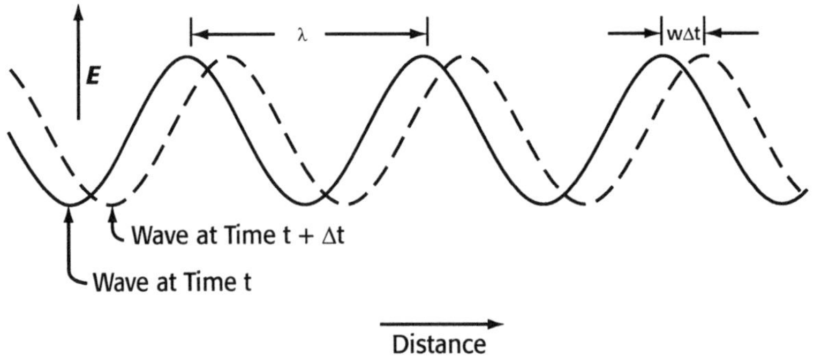

FIGURE 1.1 Electric field strength **E** in a monochromatic electromagnetic wave.

In the simple diffraction setup of figure 1.2, monochromatic (sinusoidal) radiation strikes a plane containing equispaced parallel scattering lines at angle θ and is observed at angle ϕ above the plane. Maxima in the radiation are found at certain angles, which we number, taking 0 as the *order* when ϕ equals θ, 1 as the order for the first larger angle, 2 for the second larger one, and so on.

Each maximum occurs where neighboring rays reinforce each other as much as possible. That is, where a crest of one ray reaches the observer at the same time as a crest of each of its neighbors, a trough at the same time as a trough of each of its neighbors. But at other angles, the diffracted ray labeled 2 cancels part of the one labeled 1, and so on. Thus, the intensity drops very fast on each side of a maximum.

Since the rays labeled 1 and 2 start in phase at their source, they are reflected in phase, with the crests matching at the observation point, when the extra path length along 1 equals an integral number of wavelengths. But from figure 1.2, this extra length equals the projection of the distance between scattering lines on the incident ray 1 minus its projection on the reflected ray 2:

$$n\lambda = d(\cos\theta - \cos\phi). \qquad [1.1]$$

Note that n is the order of the reflection, λ the wavelength of the radiation, d the distance between lines, θ the glancing angle at which the rays strike the grating, and ϕ the angle at which the diffracted rays leave the grating.

A person can determine d directly by counting the number of lines per millimeter under a microscope, or indirectly by measuring how radiation of known wavelength is diffracted. With the short wavelength radiation needed to study crystals and with available gratings, angles θ and ϕ are quite small. Nevertheless, they can be very accurately measured, so λ can be found to about six significant figures.

Example 1.1

X rays of wavelength 0.710 Å strike a grating at a glancing angle θ equal to 5' 0". If the grating contains 200 lines cm^{-1} evenly spaced, at what angle ϕ would the first-order diffraction peak be found?

This diffraction is governed by formula (1.1) with

$$n = 1, \qquad \lambda = 0.710 \times 10^{-8}\ \text{cm}, \qquad d = \frac{1}{200\ \text{cm}^{-1}},$$

$$\theta = \frac{(5\ \text{min})(3.1416\ \text{radians})}{(60\ \text{min deg}^{-1})(180\ \text{deg})} = 0.001454\ \text{radians}.$$

Because the angles are small, higher terms in the expansions

$$\cos\theta = 1 - \frac{\theta^2}{2} + \ldots,$$

$$\cos\phi = 1 - \frac{\phi^2}{2} + \ldots,$$

which are valid when the angles are in radians, can be neglected.

Then

$$\cos\theta - \cos\phi \cong \left(1 - \frac{\theta^2}{2}\right) - \left(1 - \frac{\phi^2}{2}\right) = \frac{1}{2}\left(\phi^2 - \theta^2\right)$$

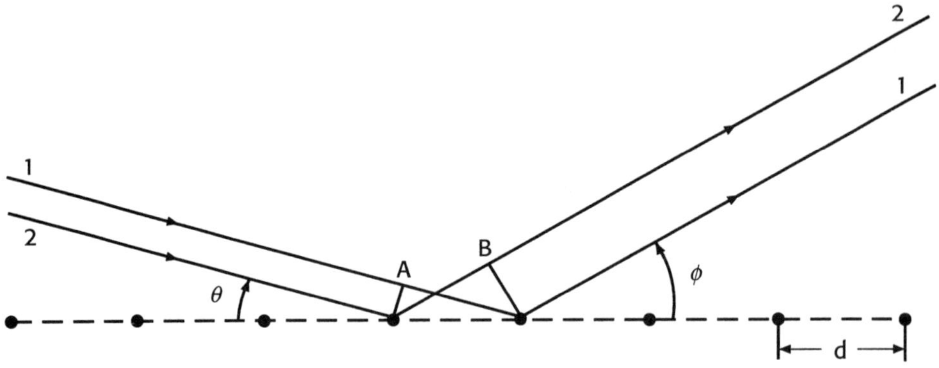

FIGURE 1.2 Diffraction of neighboring rays by a ruled grating.

and equation (1.1) reduces to

$$n\lambda = \frac{1}{2}d\left(\phi^2 - \theta^2\right).$$

So we have

$$\left(1\right)\left(0.710 \times 10^{-8}\ \text{cm}\right) = \frac{1}{\left(2\right)\left(200\ \text{cm}^{-1}\right)}\left[\phi^2 - \left(0.001454\right)^2\right],$$

whence

$$\phi = 2.23 \times 10^{-3}\ \text{radians}$$

or

$$\phi = 7'\ 39''.$$

1.4 Producing Short Wavelength Radiation

The fact that electromagnetic radiation is diffracted by a ruled grating shows that it is composed of waves, as we have assumed. These waves can distinguish between points which scatter appreciably different signals.

But the signals can be appreciably different only if the points are a fair fraction of a wavelength apart. With visible light, in the 7200 Å to 4000 Å range, the limit of resolution in refractive work (with a microscope) is about 1300 Å. Only gross imperfections and grains in solids are larger than this and can be distinguished. The finer structures, the small distances between the parallel planes in a nearly perfect crystal, for instance, can only be revealed by much shorter wavelength radiation. Before discussing the pertinent diffraction, let us consider how such radiation is produced.

A common method involves boiling electrons out of a metal in a vacuum, accelerating them in a certain direction with a high voltage, and causing them to strike a dense target. Associated with the acceleration and deceleration of the negative charge is the emission of electromagnetic radiation of all wavelengths down to a certain limit determined by the applied voltage. From the German word for braking radiation, this radiation is called *bremsstrahlung.*

On penetrating the target, the high energy electrons knock various electrons out of atoms along their paths. Electrons from higher energy levels then fall into the vacated positions and emit radiation. The resulting line spectrum is characteristic of the atoms present in the target because the pertinent energy levels are.

For nearly monochromatic X rays, a person adjusts the tube voltage to obtain as much contrast as possible between the desired line and the background bremsstrahlung. Also, one interposes some material that absorbs the unwanted lines more effectively than the desired line. This filter may be used as a window in the X-ray tube.

1.5 The Action of Crystals on the Radiation

Max von Laue and his students Friedrich and Knipping were the first to find that X rays are suitable for studying the internal structure of crystals. In 1912, they reported having put photographic plates (1) between the crystal and the X-ray source, (2) on either side of the crystal, and (3) behind the crystal. They obtained positive results with the third arrangement, finding a definite diffractive pattern of spots. Laus's method will not be analyzed here; instead, we will consider the simpler Bragg method.

W. H. Bragg and others early found that the radiation scattered by crystals possessed the same ionizing power and was absorbed like the incident radiation. When polychromatic X rays were employed, the scattered radiation consisted of X rays varying in intensity and composition with direction.

W. L. Bragg (son of W. H. Bragg) found that each set of diffracted rays converged as though the incident beam was being reflected from parallel planes within the crystal. See figure 1.3. Turning the crystal altered the reflection. In fact, the emerging beam moved as light does from a similarly rotating optical mirror. The diffracting substance was acting as a pack of parallel mirrors.

But what are these X-ray mirrors in a crystal? W. L. Bragg suggested that they are planes of high atomic density, since he found that reflection occurred from planes parallel to cleavage surfaces. Those planes parallel to good cleavage surfaces were very good X-ray reflectors. Evidence that the mirrors were within the crystal was obtained by Hupka when he showed that roughening the surfaces had no effect on the reflection of X rays although it did cause ordinary light to be scattered diffusely.

1.6 Bragg Reflection

The younger Bragg went on to treat the reflection of monochromatic X rays from a set of planes. The following discussion is based on his work.

Let X rays of given wavelength λ strike parallel partially reflecting planes spaced distance d apart as figure 1.4 shows. In the approximation that specular reflection occurs, the angle between a reflected ray and the pertinent plane equals the angle between the corresponding incident ray and the plane, θ.

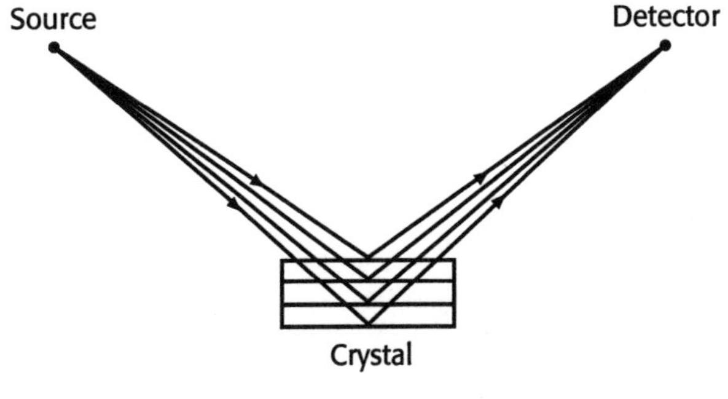

FIGURE 1.3 Diverging incident beam producing a converging reflected beam.

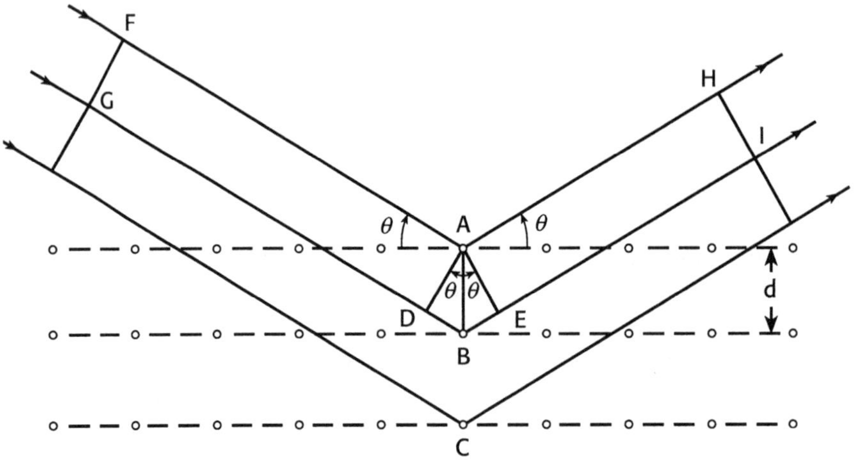

FIGURE 1.4 Reflection of X rays from parallel atomic planes in a crystal.

Since the neighboring rays, *FA* and *GB*, originate at the same distant source, they are nearly parallel. And, equiphase surfaces intersect the two rays at right angles, as line *AD* does. Since the detector is so far away, the reflected rays, *AH* and *BI*, are also nearly parallel. A maximum in intensity is observed where these arrive in phase. Then the wavelets at points A and E, on a perpendicular to the two rays, are in phase. Such occurs when length $DB + BE$ is an integral number of wavelengths λ .

Each side of acute angle *DAB* is perpendicular to a side of the acute angle between *FA* and the first parallel plane. Therefore, the two angles are equal. Similarly, angle *EAB* equals the reflection angle θ. In addition, hypotenuse *AB* of the two small right triangles equals the interplanar spacing d. So from the definition of the sine, we have

$$\mathrm{DB} = d\sin\theta \qquad [1.2]$$

and

$$\mathrm{BE} = d\sin\theta. \qquad [1.3]$$

Setting the sum of these lengths equal to an integral number of wavelengths λ yields the *Bragg equation*

$$n\lambda = 2d\sin\theta \qquad \text{where} \qquad n = \text{ an integer.} \qquad [1.4]$$

When n is 1, the reflection is said to be first order, when n is 2, second order, and so on.

Example 1.2

A crystal reflects X rays of wavelength 1.539 Å most strongly when it is oriented so the angle between the incident rays and the reflecting planes is 37° 15' . If the order of the reflection is 1, what is the interplanar spacing?

Put the given data into formula (1.4),

$$(1)(1.539\ \text{Å}) = 2d\sin(37^\circ\ 15'),$$

and solve for d:

$$d = \frac{1.539\ \text{Å}}{2(0.6053)} = 1.271\ \text{Å}.$$

1.7 Standard Crystal Lattices

The symmetry that a crystal displays has long been known and its cause discussed. In 1690, Huygens suggested that the external form was determined by how particles composing the given crystal were arranged. In 1813, Wollaston wrote that the particles could be replaced by points having suitable properties. Bravais in 1848 adopted this view and discussed the 14 different ways in which the points could be arranged in space.

Since 1912, the Braggs and others have used diffraction methods to study the various parallel planes within crystals. They could thus find internal planes not parallel to external faces and determine all pertinent interplanar spacings. It was found that three sets of prominent planes intersect each other to divide a crystal into small parts. Such a part that can serve as a representative unit of a crystal is called a *unit cell.*

Only a few kinds of geometric solids can be unit cells in 3- dimensional space. Solids having cross sections with 5-fold, 7-fold, 8- fold, or higher-fold axes leave empty regions, as figure 1.5 illustrates. The possibilities that are observed appear in figure 1.6.

For each, we let the lower front left corner be the origin and the three edges that intersect at this point be the x, y, and z axes. The distance that the unit cell extends along the x axis is labeled a, the distance along the y axis b, and the distance along the z axis c. We designate the angle between the y and z axes α, the angle between the z and x axes β , and the angle between the x and y axes γ. The cells in figure 1.6 are then described by the conditions in table 1.1.

Because the X-ray mirrors are planes of high average atomic density, the highest densities appear where the planes intersect. So one expects to find an atom or molecule at each corner of a unit cell, where three prominent planes intersect. Bravais noted that particles may lie at other positions also, as table 1.2 indicates. He considered only *point lattices*, lattice structures in which all the structural units (atoms or molecules) are alike and equivalent to points. More general lattices are described in terms of these, however.

1.8 Miller Indices

Each set of parallel reflecting planes is characterized by the distance between successive planes and by the orientation of any one of them. The former is given by the number d. The latter, on the other hand, is described by the intercepts on the x, y, z axes, or by their reciprocals.

Consider a typical plane of the given set. Suppose that the axes are drawn as in the appropriate figure 1.6 and that the plane cuts the axes at points A, B, and C, as in figure 1.7. If the unit cell edges are taken to be the unit lengths, then the intercepts are $OA/a, OB/b, OC/c$ units from the origin, while the reciprocals of these measures are a/OA, b/OB, c/OC.

As long as the crystal is formed only of equivalent unit cells, these reciprocals are rational. This result is called the *law of rational indices*. On multiplying the reciprocals by an appropriate factor, one obtains three small whole numbers. The smallest of these are the *Miller indices* for the set of planes.

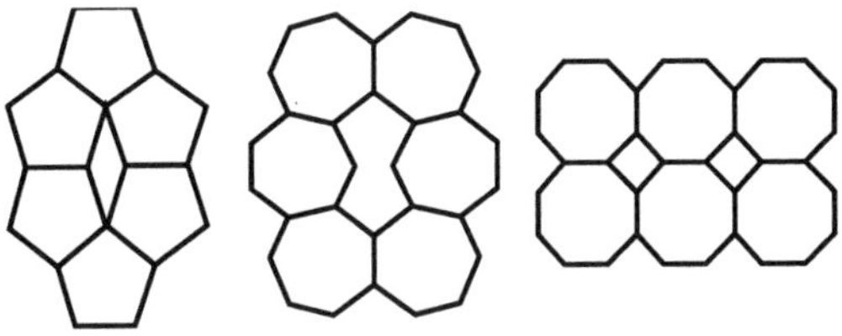

FIGURE 1.5 Failure of 5-, 7-, and 8-sided figures to fill space

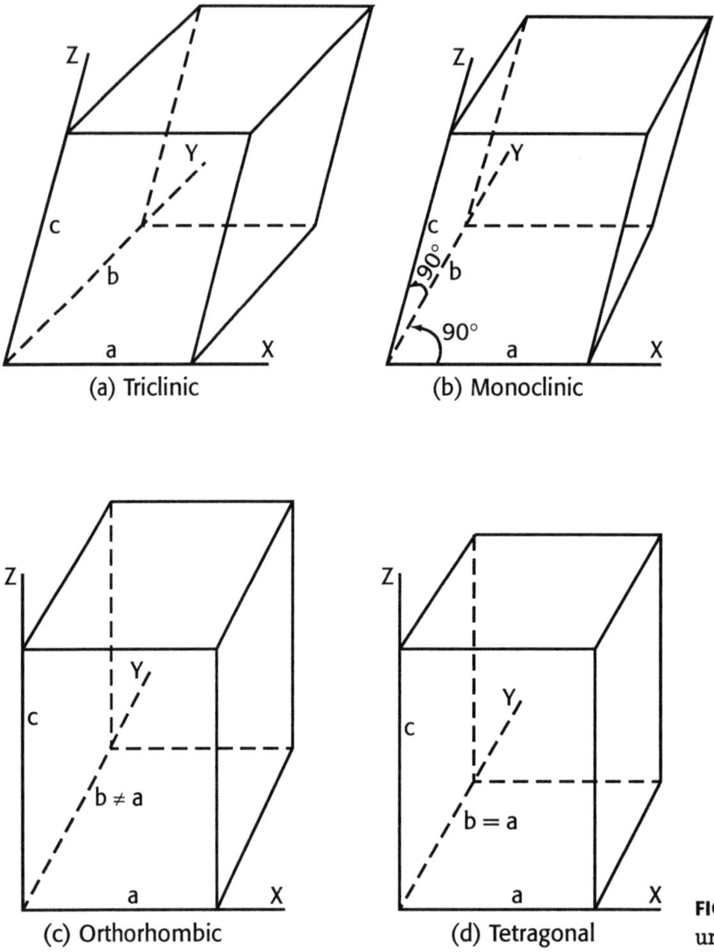

FIGURE 1.6a–d Possible unit cells for filling space.

When the Miller indices for a set of planes are h, k, and l, the planes are called (hkl) planes. A negative Miller index is indicated by a minus sign over the number. Thus, a $(1\bar{1}0)$ plane is parallel to a plane intersecting the x axis at a, the y axis at - b, and the z axis at ∞.

Example 1.3

If a (123) plane intersects the x axis at 6 a, what are its intercepts on the y and z axes?

The Miller indices h, k, l are proportional to reciprocals of the intercepts measured in terms of a, b, and c, respectively. Here

$$1 = f\frac{a}{\mathrm{OA}}, \qquad 2 = f\frac{b}{\mathrm{OB}}, \qquad 3 = f\frac{c}{\mathrm{OC}},$$

with

$$\mathrm{OA} = 6a.$$

Combining the first and fourth equations yields the constant of proportionality

$$f = 6.$$

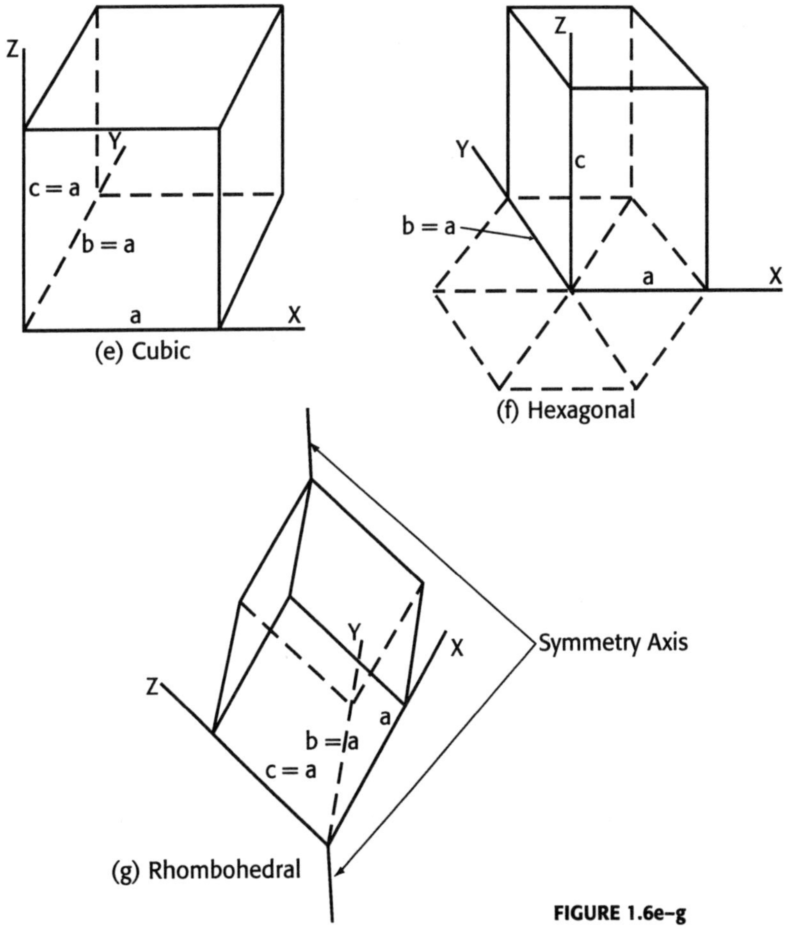

FIGURE 1.6e–g

TABLE 1.1 Three-Dimensional Crystal Systems

Symmetry	*Restrictions on*	
	Unit-Cell Edges	*Angles*
Triclinic	$a \neq b \neq c$	$\alpha \neq \beta \neq \gamma \neq 90°$
Monoclinic	$a \neq b \neq c$	$\alpha = \gamma = 90°$, $\beta \neq 90°$
Orthorhombic	$a \neq b \neq c$	$\alpha = \beta = \gamma = 90°$
Tetragonal	$a = b \neq c$	$\alpha = \beta = \gamma = 90°$
Cubic	$a = b = c$	$\alpha = \beta = \gamma = 90°$
Hexagonal	$a = b \neq c$	$\alpha = \beta = 90°$, $\gamma = 120°$
Rhombohedral	$a = b = c$	$\alpha = \beta = \gamma \neq 90°$

Substituting this into rearranged forms of the second and third equations gives us

$$\mathrm{OB} = 6\frac{b}{2} = 3b, \qquad \mathrm{OC} = 6\frac{c}{3} = 2c.$$

Other (123) planes are parallel to this plane at equidistant intervals.

TABLE 1.2 Bravais Lattices

Positions of Particles	*Symbol*	*Where Found*
Corners only	P	All lattices
Corners and center of one face	C	Monoclinic and orthorhombic
Corners and cnters of all faces	F	Orthorhombic and cubic
Corners and center of body	I	Orthorhombic, tetragonal, and cubic

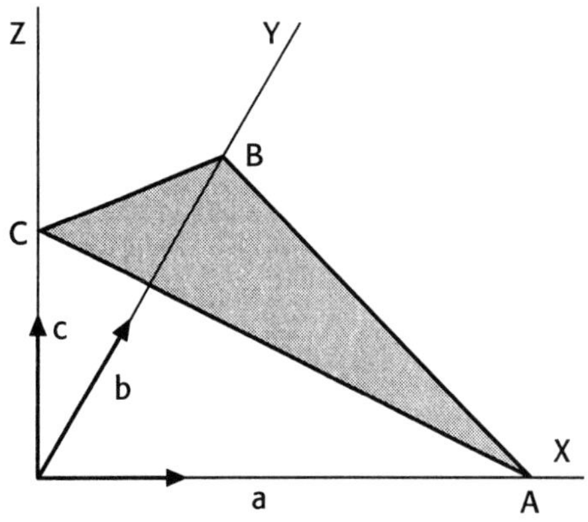

FIGURE 1.7 Intersection of a reflecting plane with the coordinate planes and axes for a crystal lattice.

1.9 Cubic Point Lattices

Crystals constructed from cubic unit cells may favor cubic, octahedral, or intermediate (cuboctahedral) external shapes. Under shear stresses, such a crystal tends to fail along planes parallel to those that develop at the surface on growth, and along planes intersecting these at regular angles. The new surfaces that may appear lie on high-particle-density planes, Now, the highest densities occur where these planes intersect. We expect these places to be occupied by atoms, molecules, or ions.

The pertinent points may lie (a) only at the corners of each unit cell, or (b) at the middle of each face as well as at the corners, or (c) at the center as well as at the corners. The different lattices are said to be simple or primitive (P), face- cantered (F), and bodycentered (I), respectively. These can be distinguished by their three principal interplanar spacings.

In a cubic lattice, angles α, β, and γ all equal 90°; so the (100), (010), and (001) planes are mutually perpendicular. Furthermore, distances a, b, and c are equal; so the spacings between the (100), (010), and (001) planes are equal. Indeed, since a cubic crystal varies in the same way along the x, y and z axes, the (100) set of planes is equivalent to the (010) set and the (001) set.

Likewise, the (110) set of planes is equivalent to the (101), (011), $(\bar{1}10)$, $(\bar{1}01)$, and $(0\bar{1}1)$ sets. Also, the (111) set of planes is equivalent to the $(\bar{1}11)$, $(1\bar{1}1)$, and $(11\bar{1})$ sets. Consequently, a person can study all spacings among the prominent planes in a cubic lattice by determining d_{100}, the distance between successive (100) planes, and d_{110}, the distance between successive (110) planes, and d_{111}, the distance between successive (111) planes, only.

With figures 1.8, 1.9, and 1.10, one can determine how these distances are related to the length of an edge. Indeed, we see that the (100) planes either leave a undivided or cut it into two equal parts. The (110) planes divide the diagonal across the base into equal parts. But the Pythagorean theorem tells us that this diagonal is $\sqrt{2}a$ long. The (111)

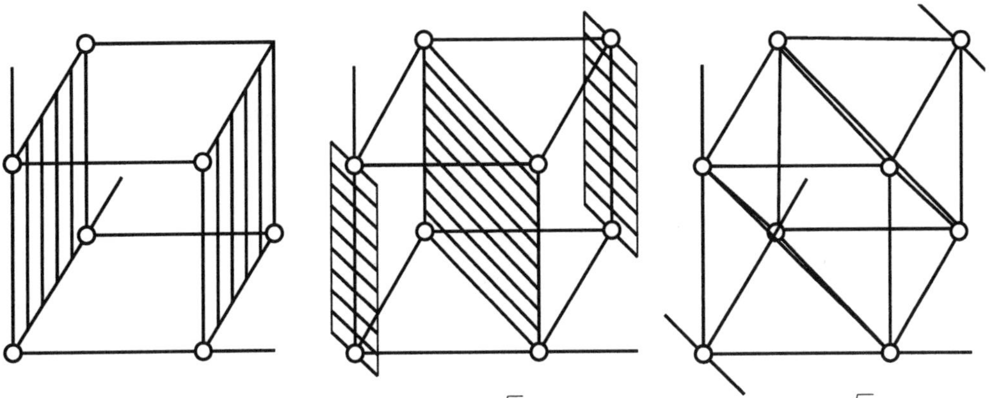

$d_{100} = a$ $d_{110} = \frac{\sqrt{2}}{2}a$ $d_{111} = \frac{\sqrt{3}}{3}a$

FIGURE 1.8 Spacings in a simple cubic lattice.

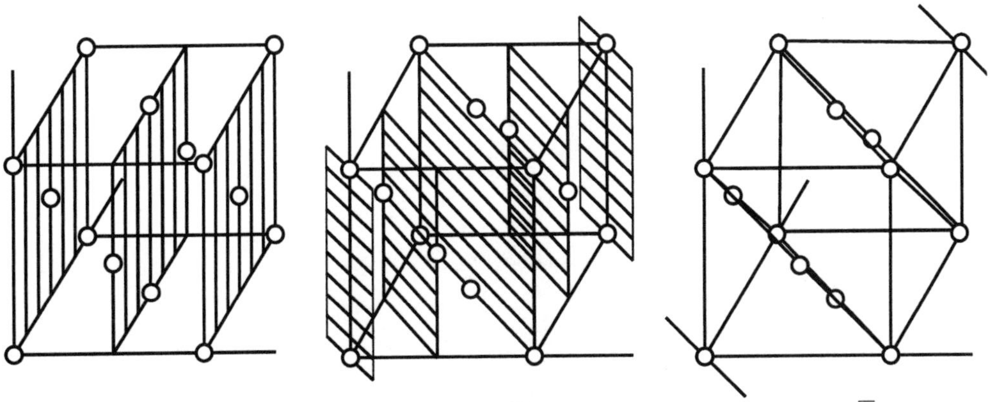

$d_{100} = \frac{1}{2}a$ $d_{110} = \frac{\sqrt{2}}{4}a$ $d_{111} = \frac{\sqrt{3}}{3}a$

FIGURE 1.9 Spacings in a face-cantered cubic lattice.

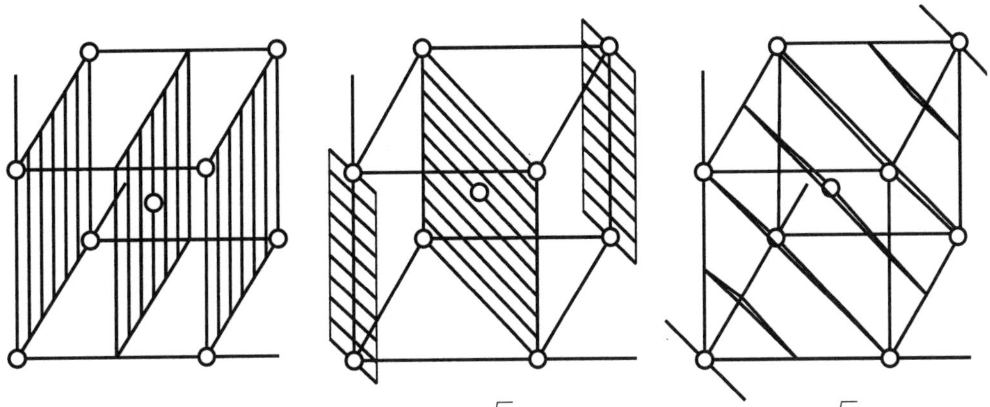

$d_{100} = \frac{1}{2}a$ $d_{110} = \frac{\sqrt{2}}{2}a$ $d_{111} = \frac{\sqrt{3}}{6}a$

FIGURE 1.10 Spacings in a body-cantered cubic lattice.

planes cut the diagonal through the cube into equal parts. This line is the hypotenuse of a right triangle whose sides are $\sqrt{2}a$ and a long; so it is $\sqrt{3}a$ in length.

In a unit cube of the simple cubic lattice, edge a is left undivided, the diagonal $\sqrt{2}a$ is bisected, the diagonal $\sqrt{3}a$ is trisected. So the ratios of the spacings are given by

$$d_{100} : d_{110} : d_{111} = 1 : \frac{\sqrt{2}}{2} : \frac{\sqrt{3}}{3}. \quad [1.5]$$

See figure 1.8

For the face-cantered cubic lattice, figure 1.9 shows that

$$d_{100} : d_{110} : d_{111} = 1 : \frac{\sqrt{2}}{2} : \frac{2\sqrt{3}}{3}. \quad [1.6]$$

Spacings for the body-cantered cubic lattice appear in figure 1.10, whence

$$d_{100} : d_{110} : d_{111} = 1 : \sqrt{2} : \frac{\sqrt{3}}{3}. \quad [1.7]$$

Example 1.4

X-ray studies of a cubic crystal by the Bragg method yielded the ratios

$$d_{100} : d_{110} : d_{111} = 1.000 : 0.709 :1.164.$$

What lattice does the crystal possess?

These numbers are very close to those in (1.6):

$$1 : \frac{\sqrt{2}}{2} : \frac{2\sqrt{3}}{3} = 1.000 : 0.707 : 1.155.$$

So the crystal has a face-cantered cubic lattice.

1.10 Some Simple Crystal Types

The physical entity, the *basis*, associated with a lattice point may consist of one or more atoms or ions, spread out over space. The lattice point may be located at the mean centroid of the entity, or at the equilibrium position for one of the nuclei, or at some other constant position with respect to the equilibrium point for the basis.

A formula unit of a strong electrolyte consists of positive and negative ions, cations and anions. The whole unit may be associated with a lattice point. Or, one may consider the equilibrium position for the center of each cation a lattice point, while the similar position for each anion is a point on another interpenetrating lattice.

Sodium chloride and other alkali halides generally come out of solution as small imperfect cubes. In the presence of impurities, the corners may not develop, causing octahedral planes to appear. These external shapes indicate that the internal lattice is cubic. Accordingly, crystallographers early surmised that the alkali ions and the halide ions are arranged alternately in the three-dimensional checkerboard pattern of figure 1.11.

This was first tested by W. L. Bragg. His X-ray measurements showed that NaCl, KBr, and KI are face centered, while KCL seemed to be simple cubic. Now, a given ion interacts with X rays largely through its electrons. In the first three salts, each anion contains many more electrons than the accompanying cation; so the anion is much more effective

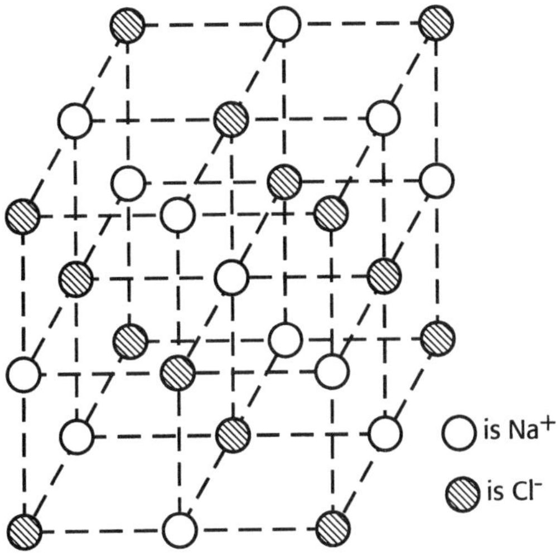

FIGURE 1.11 Unit cube from an NaCl crystal.

as a scatterer. The X rays then react primarily to the face-cantered cubic lattice on which the anions are located. In KCl, on the other hand, the cation and the anion are isoelectronic. The X rays are scattered strongly by both and each K^+ is seen as a Cl^-, each Cl^- as a K^+. Although the K^+ and Cl^- ions are arranged as the cations and anions are in NaCl, the crystal appears to be simple cubic.

The NaCl lattice has been found in other alkali halides except CsCl, CsBr, and CsI. In each of these cesium halide crystals, the center of a unit cube is occupied by one of the ions and the corners by the other, as figure 1.12 shows. The center may be associated with any one of the corners by convention to form the basis, the point lattice being simple cubic.

In the zinc blende form for ZnS, one species of ion forms a face-cantered cubic lattice, while the other species appears at the centers of alternate cubelets, as figure 1.13 illustrates. Four anions are arranged tetrahedrally around each cation; four cations are arranged tetrahedrally about each anion. The same configuration exists in the diamond crystal, with all particles carbon atoms.

The wurtzite form for ZnS also employs tetrahedral arrangements about each ion. However, the longer range order is hexagonal, as figure 1.14 shows.

In graphite, the carbon atoms are bound covalently in condensed benzene-ring layers. Each layer is offset with respect to its neighbors as figure 1.15 indicates; the net result is a hexagonal lattice. The fourth valence electron of each carbon atom is delocalized over its layer and to a slight extent from layer to layer. As a consequence, the bonding between layers is very weak, compared to that within a layer.

1.11 Stability Conditions

The atoms, molecules, oppositely charged ions, or cations and valence electrons of a substance attract each other. The resulting interaction causes the material to condense and freeze if the temperature is low enough and the pressure adequate.

The consequent array tends to be most stable when it is as compact as properties of the constituent particles allow. When the particles are equivalent and spherically symmetric, they pack as marbles do. Six units then surround a given one in a plane, while three contact it from above and three from below. If the centers of the three upper ones lie over the centers of those below, the arrangement is a hexagonal close-packed one; if

O is Cs⁺   ⬡ is Cl⁻

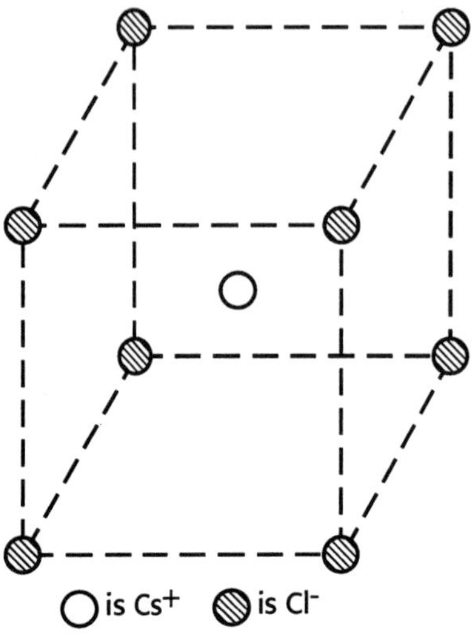

FIGURE 1.12   Unit cube from a
CsCl crystal

(a) In the diamond   ◯ and ⬡ are C

(b) In zinc blende   ◯ is Zn^{++} and ⬡ is $S^=$

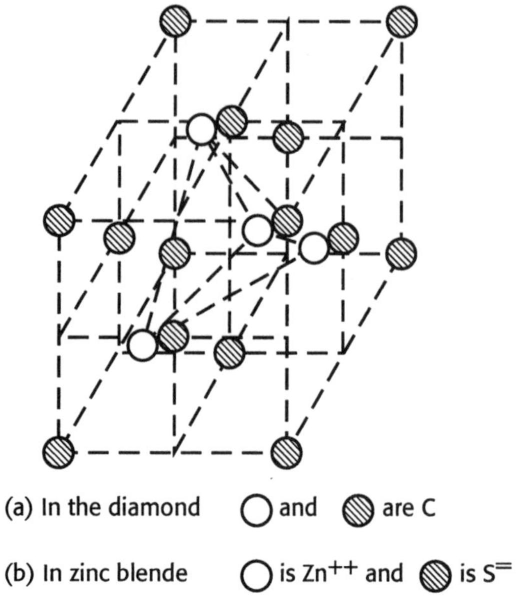

FIGURE 1.13   Arrangement of ions in zinc blende
and of atoms in diamond.

◯ is Zn^{++} ⬡ is $S^=$

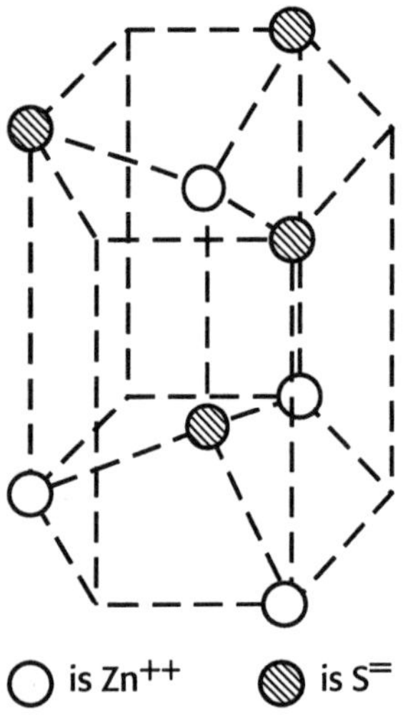

FIGURE 1.14   Unit cell of a
wurtzite crystal.

O is C

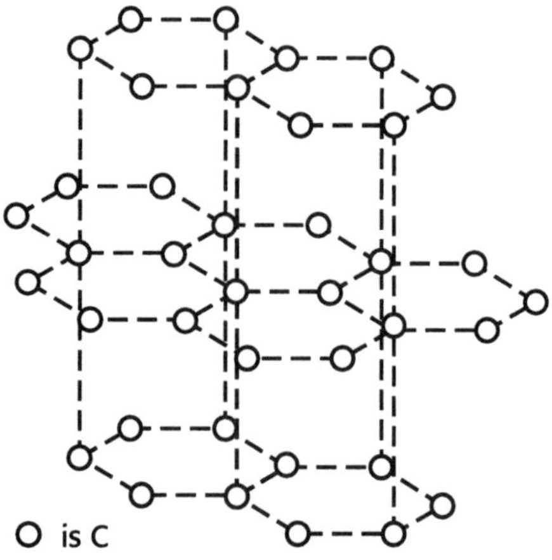

FIGURE 1.15   The graphite lattice.

the triangle through the upper centers is rotated by 60° from that through the lower centers, the arrangement is cubic close packed (a face centered cubic structure). Many metals crystallize in one of these two forms.

When the bases are bound covalently to each other, the structure depends on valence properties of the pertinent atoms. In diamond, for instance, each carbon is bound tetrahedrally to its neighbors. In graphite, the bonding about each carbon is trigonal, as in benzene.

When a strong electrolyte crystallizes, the like-charged ions try to remain apart, while oppositely charged ions move together into contact. The densest array meeting this condition tends to be assumed. When each ion contains only one nucleus, the limits listed in table 1.3 apply. By the way, because like charges repel, oppositely charged ones attract, cations are much smaller than, while anions are much larger than, the corresponding isoelectronic atoms.

When the radius of the smaller ion divided by the radius of the larger ion is between 0.9021 and 0.7321, each small ion can hold eight of the larger ones around itself without squeezing them against each other and the cesium chloride lattice tends to form. When the radius ratio is between 0.7321 and 0.4142, each small ion can hold six large ions around itself without squeezing them against each other and the sodium chloride lattice tends to form. When the radius ratio is between 0.4142 and 0.2247, each small ion can only bond four large ions without squeezing them against each other and either the zinc blende or the wurtzite lattice tends to form. The linear arrangement is observed in the hydrogen bond.

Example 1.5

At what radius ratio do the anions begin to touch each other in the cesium chloride structure?

Consider an electrolyte AB in which the ions are spherical, with

$$r_{A^+} = \text{radius of } A^+,$$
$$r_{B^-} = \text{radius of } B^-.$$

Let r_{A^+} be large enough so that eight B^- ions can be packed around it without touching each other. Keep the oppositely charged ions in contact and decrease the radius of the cation until the anions meet each other along each edge of the unit cube. If a is the length of an edge of the cube, then

$$a = 2r_{B^-}$$

But each diagonal through the cube passes through one half of a B^-, the A^+, and one half of another B^-; consequently,

$$\sqrt{3}a = 2r_{A^+} + 2r_{B^-}.$$

Now, eliminate a from these two equations,

$$\sqrt{3}\left(2r_{B^-}\right) = 2r_{A^+} + 2r_{B^-},$$

TABLE 1.3 Radius Ratios at which an Anion Touches Neighboring Anions as well as Neighboring Cations

Lattice Structure	*Number of Anions around One Cation*	*Arrangement about Cation*	*Cation Radius / Anion Radius when Anions Touch*
Quasicrystal	12	Icosahedral	0.9021
Cesium chloride	8	Cubic	0.7321
Sodium chloride	6	Octahedral	0.4142
Zinc blende or wurtzite	4	Tetrahedral	0.2247
—	3	Triangular	0.1547
—	2	Linear	0.0000

rearrange

$$r_{A^+} = \left(\sqrt{3} - 1\right) r_{B^-},$$

and solve for the desired radius ratio

$$\frac{r_{A^+}}{r_{B^-}} = \sqrt{3} - 1 = 0.732.$$

1.12 Avogadro's Number

From the density of a well formed crystal and from its internal structure, one can determine the mass of a formula unit. Knowing the mass of a mole of the substance, one can then find the number of units in a mole.

By chemical analysis, a person can determine the amount of a pure substance that reacts with, or is chemically equivalent to, a given amount of another substance. The *equivalent mass* is defined as the mass that is equivalent to 3.0000 g tetravalent ^{12}C. Multiplying this mass by the pertinent valence, or the pertinent valence change, yields the mass of one mole. Thus, one mole of ^{12}C weighs 12.0000 g.

In a given discussion, the elementary unit may be taken as an atom, or a molecule, an ion, an electron, a proton,..., or even a specified group of such particles. By definition, a *mole* is the amount of substance that contains the same number of elementary units as there are atoms in 12.0000 g pure ^{12}C. The number of elementary units in a mole is called the *Avogadro number* N_A.

By X-ray methods, a person can determine the nature and the size of a unit cell in a given crystal. From how each particle is shared with neighboring unit cells, one determines the net number n of molecules or formula units in the cell. If m is the mass of a formula unit, we have

$$m = \frac{\rho V}{n}, \qquad [1.8]$$

where ρ is the density and V the volume of the cell. If there are only negligible defects in the crystal, the conventionally determined whole-crystal density can be used for ρ. The volume of the cell is obtainable from the X-ray data.

Avogadro's number N_A equals the mass per mole M divided by the mass per formula unit m.

$$N_A = \frac{M}{m}. \qquad [1.9]$$

In practice, very few crystalline materials yield good data; for, even a crystal that appears to be perfect to the eye may contain many faults. There may be lattice defects, particles missing from lattice points, and particles that are not associated with lattice points, but with interstitial positions. Extensive dislocations of various kinds may occur with respect to inner surfaces. The surfaces may encompass small volumes so that the crystal is broken up into a mosaic of crystallites. The defects tend to decrease the sharpness of diffraction patterns and decrease the over-all density of the crystal.

A solid metal usually exhibits a mosaic structure. An alkali chloride prepared by freezing its melt is generally defective. But when it is slowly crystallized out of a solution at low temperatures, suitable crystals may form. Some naturally occurring crystals of calcite, quartz, and diamond are also suitable.

The best results from X-ray determinations have been compared with those found in other ways. Critical evaluations have led to the value

$$N_A = 6.0221 \times 10^{23}\ \text{mol}^{-1}. \qquad [1.10]$$

Example 1.6

At room temperature, the density of a good lithium fluoride crystal is 2.640 g cm^{-3}. X-ray studies show that the LiF has a face-cantered cubic lattice with a unit-cell edge 4.016 Å long. From these data, calculate Avogadro's number.

Each F^- at a corner of a unit cube is shared with seven other unit cells, while each F^- at the middle of a face is shared with one other unit cell. Since there are eight corners to the cube and six faces, the number of F^- ions effectively in the unit cube is

$$n = \frac{8}{8} + \frac{6}{2} = 4.$$

Because the cube is electrically neutral, the unit cell contains a like number of cations.

Substituting into formula (1.8),

$$m = \frac{\rho V}{n} = \frac{\left(2.640\ \text{g cm}^{-3}\right)\left(4.016 \times 10^{-8}\ \text{cm}\right)^3}{4\ \text{formula units}} = 4.275 \times 10^{-23}\ \text{g (formula units)}^{-1}$$

and then into (1.9) gives us

$$N_A = \frac{M}{m} = \frac{25.939\ \text{g mol}^{-1}}{4.275 \times 10^{-23}\ \text{g (formula units)}^{-1}} = 6.068 \times 10^{23}\ \text{formula units mol}^{-1}.$$

1.13 Surface Structure

At the surface of a condensed phase, the coordination number of a unit (atom, ion, or molecule) is lower than it is in a homogeneous region of the phase. As a result, the units there may adopt a geometric arrangement different from that in the interior.

When the condensed phase is a solid conductor, the surface structure may be determined directly by moving a relatively charged sharp metal tip over the surface, keeping the resulting current low and constant. How the tip moves may be analyzed and screened by a computer. The current is said to tunnel between the tip and the closest point on the surface. So the apparatus is called a scanning tunneling microscope. See figure 1.16. The technique is referred to as *scanning tunneling microscopy* (STM).

To move the tip over the solid with atomic-sized resolution, one employs orthogonal piezoelectric positioners. In figure 1.16, these are labeled x, y, and z. A bias voltage is applied between the tip and the sample to be studied. The resulting current is held constant by controlling the vertical position of the tip, using the z positioner, as the tip is moved over the surface, using the x and y positioners. The computer is programmed so that an image of the surface appears on the screen.

As an example, consider semiconductor silicon. Within the silicon crystal, each atom is bound tetrahedrally to its nearest neighbors. But when such a crystal is cleaved, bonds are broken along the cleavage surfaces. If no rearrangements occurred, the surface atoms would sport dangling bonds, valence electrons not involved in bonding. To reduce the

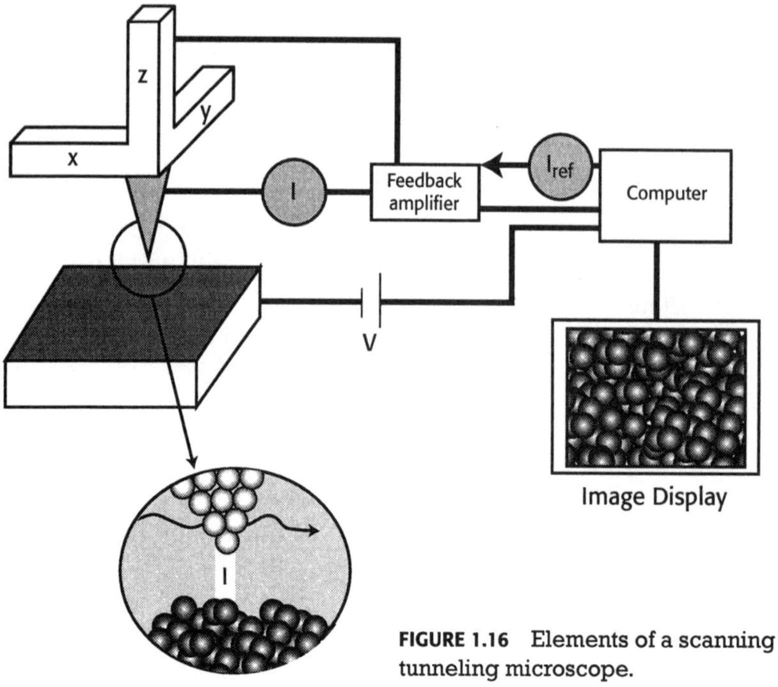

FIGURE 1.16 Elements of a scanning tunneling microscope.

number of dangling bonds, and thus to lower the surface energy, some atoms migrate and attach to three previously dangling bonds. These atoms are called *adatoms*. Surface atoms that do not bond to adatoms are called *rest atoms* (R). The rearrangement is referred to as a *reconstruction* of the surface.

The silicon (111) surface was first investigated by electron diffraction. But the results were not conclusive. Only STM studies revealed unambiguously how the surface atoms were rearranged. On the reconstructed (111) surface, one finds unit cells of 7 X 7 interior lattice units in size. See figure 1.17.

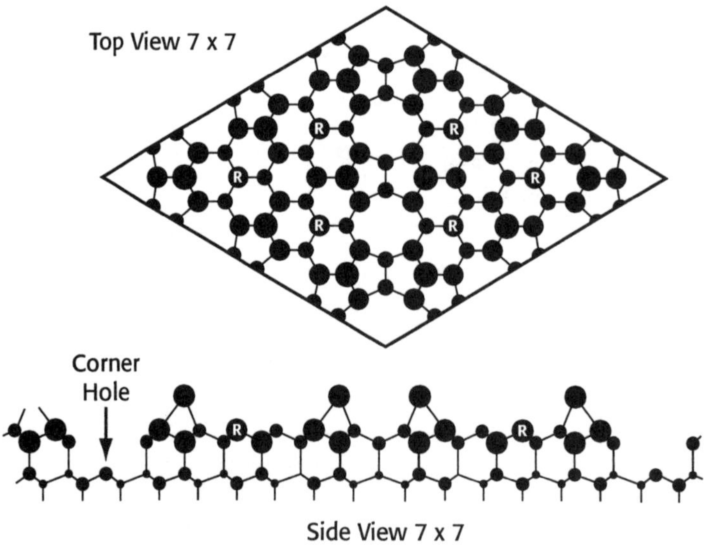

FIGURE 1.17 How a (111) surface of a pure silicon crystal is reconstructed.

1.14 The Quasicrystalline State

In an ideal crystal, a unit cell is repeated periodically in three independent directions. Such a system yields the Bragg diffraction results that we have described. It also yields Laue diffraction patterns of sharp spots. A completely disordered structure, on the other hand, such as we have in a glass or liquid, yields a Laue pattern of concentric circles.

A quasicrystal does not possess the periodicity in space of a crystal. Nevertheless, it produces a Laue diffraction pattern along certain axes of sharp spots. Furthermore, these indicate the presence of various "forbidden" symmetries.

For example, rapidly solidified alloys of Al-Mn, of Al-Li- Cu, and of Al-Fe-Cu exhibit 5-fold, 3-fold, and 2-fold symmetry axes. These are the symmetry axes of the icosahedron. And the Laue spots are as sharp as those for a crystal.

Some have suggested that quasicrystals are merely glasses of icosahedral clusters. Others that they are based on giant cell periodic structures. However, high quality single grains of the Al-Fe-Cu alloy have been grown with dodecahedral faceting. This can be described as a 3- dimensional slice of a 6-dimensional periodic structure. There are three possible 6-dimensional icosahedral periodic arrangements, a primitive, a face-cantered, and a body- centered lattice.

Questions

1.1 What limits exist on homogeneity?
1.2 Define the different states of matter.
1.3 Distinguish between a crystal, a quasicrystal, and a glass.
1.4 Distinguish between refraction and diffraction.
1.5 What are rays?
1.6 Define frequency, wavelength, phase velocity.
1.7 Distinguish a polychromatic wave from a monochromatic one.
1.8 Is a section of a sinusoidal wave strictly monochromatic? Explain.
1.9 Derive the law governing diffraction from a ruled grating.
1.10 How are X rays produced?
1.11 What does the reflection of X rays from a crystal indicate?
1.12 Derive the law governing Bragg diffraction.
1.13 Describe the possible symmetries of 3-dimensional crystal systems.
1.14 How are the Bravais lattices related to these?
1.15 Explain how Miller indices are defined.
1.16 What is the basis for a crystal?
1.17 What determines the structure of (a) a covalent crystal, (b) a molecular crystal, (c) an ionic crystal, (d) a metallic crystal?
1.18 How may Avogadro's number be determined from measurements on a crystal?
1.19 Why may the structure at the surface of a crystal differ from that in its interior?
1.20 How may this surface structure be determined?
1.21 What characterizes the quasicrystalline state?

Problems

1.1 When 1,540 Å X rays from a copper target fell on a given crystal, the maximum in a certain reflection was observed at $\theta = 10° 27'$. With a molybdenum target in the X-ray tube, the same maximum was found at $\theta = 4°48'$. What is the wavelength of the latter radiation?

1.2 How many orders of Bragg reflection can occur from the (100) planes in a CsC1 crystal if an edge of a unit cube in the crystal is 4.12 Å long and the wavelength of the incident radiation is (a) 1.540 Å, (b) O.712 Å?

1.3 Show that in a given crystal the density of atoms in a reflecting plane is proportional to the spacing between successive planes in its set.

1.4 If a (123) plane intersects the y axis at b, where does it intersect the z and x axes?

1.5 If equivalent solid spheres are packed to form the diamond lattice, what fraction of the total volume is occupied by the spheres?

1.6 X rays 1.540 Å in wavelength were reflected by powdered copper at the following θ's:

$$21.65°, 25.21°, 37.06°, 44.96°, 47.58°.$$

What kind of cubic lattice does the metal possess? What is the length of an edge of its unit cell?

1.7 At room temperature, sodium assumes a body-cantered cubic structure with unit cell edges 4.2906 Å in length. When cooled, it changes form. At -195° C, the unit cell edges are still mutually perpendicular, but each has become 5.350 Å long. The density has increased by 4.0%. Determine the lattice which the metal possesses at the low temperature.

1.8 Consider a sodium chloride lattice in which the cations touch neighboring anions but the anions do not touch each other(a stable situation). Suppose that the radius of each anion is increased while everything else is kept constant. Calculate the radius ratio at which the anions contact each other.

1.9 If a crystal exhibited periodic fivefold symmetry about parallel axes, the smallest nonzero translation **a** in one direction could be rotated by $\pm 2\pi/5$ radians to get other possible transitions. Show that such translations can add to give a nonzero resultant shorter than a. What does one conclude?

1.10 Show how a face-cantered cubic lattice can be described. as a rhombohedral lattice and calculate the angle α between the pertinent rhombohedral axes. Vector algebra may be employed.

– – – – –

1.11 A grating with 320 lines per centimeter was oriented so that the lines were perpendicular to an X-ray beam while the angle between the plane of the lines and the beam was 6' 0". If first— order diffraction appeared at a reflection angle of 12' 21", what was the wavelength of the X rays?

1.12 In figure 1.4, consider the distance between atoms along each dashed line to be c and the angle of reflection to be ϕ, rather than θ, the angle of incidence. Alter equations (1.4) and (1.1) to fit this situation. For what set of crystal planes are the diffracted rays the result of specular reflection satisfying the Bragg condition?

1.13 To the points in a simple cubic lattice is added a point at the middle of the base of each cube. What is the symmetry of the resulting lattice?

1.14 X rays of given wavelength reflect from NaF at a Bragg angle of 8° 47', while the corresponding reflection from KF is at 7° 40'. Calculate the ratio of the molar volume of KF to the molar volume of NaF.

1.15 If uniform solid spheres are packed to form the simple hexagonal lattice, what fraction of the total volume is occupied?

1.16 On entering a crystal, X rays are refracted slightly. The refractive index is given by

$$1-\delta = \frac{\cos\theta}{\cos\theta'} = \frac{\lambda}{\lambda'},$$

where θ is the angle of incidence and θ' the angle of refraction measured from the crystal surface, while λ is the external and λ' the internal wavelength. Correct the Bragg equation for this effect.

1.17 X rays 1.540 Å in wavelength were reflected most strongly by powdered molybdenum at the following angles θ:

$$20.25°, 29.30° , 36,82°, 43.81°, 50.69°.$$

What kind of cubic lattice does the metal possess? What is the length of an edge of its unit cell?

1.18 If α-iron has a body-cantered cubic lattice with unit cell width 2.8605 Å and density 7.865 g cm^{-3}, what is Avogadro's number?

1.19 Determine the radius ratio at which the cation is just small enough so the surrounding anions begin to touch each other in the zinc blende structure.

1.20 Consider the icosahedral arrangement of anions around a cation. Calculate the distance between the center of the cation and the center of the anion in terms of the distance between two neighboring anion centers. Then calculate the radius ratio at which these ions touch each other.

References

Books

Ashcroft, N. W., and Mermin, N. D.: 1976, *Solid State Physics*, Holt, Rinehart, and Winston, Philadelphia, pp. 63-129.

This text discusses the solid state in considerable detail, both descriptively and analytically. In the cited chapters, the standard crystal lattices are described and X-ray diffraction analyzed.

Burns, G.: 1985, *Solid State Physics*, Academic Press, Orlando, Fl, pp. 1-84.

This is an intermediate level text. Burns begins with a description of crystal group theory. He then considers simple crystal structures in some detail. In chapter 4, he discusses the use of X-ray diffraction. Bonding in crystals is considered later in the book.

Hahn, T. (editor): 1985, *International Tables for Crystallography*, Brief Teaching Edition of Volume A *Space-Group Symmetry*, Reidel, Dordrecht, pp. 1-119.

Because of the different bases that may be associated with the lattice points, the crystal lattices in tables 1.1 and 1.2 yield 230 different arrangements in space. In this text, 24 of the most commonly occurring examples are described in detail.

Kittel, C.: 1986, *Introduction to Solid State Physics*, 6th edn, John Wiley & Sons, Inc., New York, pp. 1-80.

This is a popular introductory text. In chapter 1, Kittel discusses the lattice structures of crystals. In chapter 2, there is a useful introduction to diffraction by crystals. In chapter 3, the bonding in crystals is treated.

Omar, M. A.: 1975, *Elementary Solid State Physics*, Addison-Wesley Publishing Co., Reading, MA, pp. 1-66.

This is another introductory text. In chapter 1, Omar describes both crystal structures and the bonding in crystals. Chapter 2 is a useful introduction to diffraction by crystals.

Rao, C. N. R., and Gopalakrishnan, J : 1997, New *Directions in Solid State Chemistry*, 2nd edn, Cambridge University Press, Cambridge, pp. 1-534.

The authors cover the field of solid state science in a comprehensive yet simple manner. Many references to the original literature are cited.

Articles

Breneman, G. L.: 1987, "Crystallographic Symmetry Point Group Notation Flow Chart," *J. Chem. Educ.* **64**, 216-217.

Dai, H., and Lieber, C. M.: 1993, "Scanning Tunneling Microscopy of Low-Dimensional Materials: Charge Density Wave Pinning and Melting in Two Dimensions," *Annu. Rev. Phys. Chem.* **44**, 237-263.

Francisco, E., Luana, V., Recio, J. M., and Pueyo, L.: 1988, "The Coulombic Lattice Potential of Ionic Compounds: The Cubic Perovskites," *J. Chem. Educ.* **65**, 6-9.

Goldman, A. I., and Kelton, R. F.: 1993, "Quasicrystals and Crystalline Approximants," *Rev. Mod. Phys.* **65**, 213-230.

Goldman, A. I., and Widom, M.: 1991, "Quasicrystal Structure and Properties," *Annul Rev. Phys. Chem.* **42**, 685-729.

Hardgrove, G. L.: 1997, "Teaching Space Group Symmetry through Problems," *J. Chem. Educ.* **74**, 797-799.

Janot, C., Dubois, J. -M., and de Boissieu, M.: 1989, "Quasiperiodic Structures: Another Type of Long-Range Order for Condensed Matter," *Am. J. Phys.* **57**, 972-987.

Kettle, S. F. A., and Norrby, L. J.: 1990, "The Brillouin Zone–An Interface between Spectroscopy and Crystallogrphy," *J. Chem. Educ.* **67**, 1022-1028.

Kettle, S. F. A., and Norrby, L. J.: 1993, "Really, Your Lattices are all Primitive, Mr. Bravais!" *J. Chem. Educ.* **70**, 959- 963.

Kettle, S. F. A., and Norrby, L. J.: 1994, "The Wigner-Seitz Unit Cell," *J. Chem. Educ.* **71**, 1003-1006.

Ladd, M. F. C.: 1997, "The Language of Lattices and Cells," *J. Chem. Educ.* **74**, 461-465.

Lessinger, L.: 1988, "Two Crystallographic Laboratory and Computational Exercises for Undergraduates," *J. Chem. Educ.* **65**, 480- 485.

Lieber, C. M.: 1994, "Scanning Tunneling Microscopy," *Chem. & Enq. News* **72** (16), 28-43.

Loehlin, J. H., and Norton, A. P.: 1988, "Crystallographic Determination of Molecular Parameters for K_2SiF_6," *J. Chem. Educ.* **65**, 486-490.

Nathan, L. C.: 1985, "Predictions of Crystal Structure Based on Radius Ratio," *J. Chem. Educ.* **62**, 215-218.

Pan, J., Jing, T. W., and Lindsay, S. M.: 1994, "Tunneliny Barriers in Electrochemical Scanning Tunneling Microscopy," *J. Phys. Chem.* **98**, 4205-4208.

Recio, J. M., Luana, V., and Pueyo, L.: 1989, "Ionic Crystals and Electrostatics," *J. Chem. Educ.* **66**, 307-310.

Schomaker, V., and Lingafelter, E. C.: 1985, "Crystal Systems," *J. Chem. Educ.* **62**, 219-220.

Stoneham, A. M., and Harding, J. H.: 1986, "Interatomic Potentials in Solid State Chemistry," *Annu. Rev. Phys. Chem.* **37**, 53- 80.

Vedamuthu, M., Singh, S., and Robinson, G. W.: 1994, "Properties of Liquid Water: Origin of the Density Anomalies," *J. Phys. Chem.* **98**, 2222-2230.

Welland, M.: 1994, "New Tunnels to the Surface," *Phys. World* **7** (3),32-36.

2

Structure in Molecules and Atoms

2.1 Introduction

TO EXPLAIN WHY PURE SUBSTANCES COMBINE in definite proportions, Dalton about 1807 suggested that each elementary substance is made up of identical atoms and that the atoms of different substances combine in definite ratios to form molecules. The different molecules of a given compound are equivalent. Presumably, each has the same structure, the same arrangement of atoms. Both chemical and physical properties are used to elucidate this structure.

Now, a given amount of a pure elementary substance is neutral. So if it contains only one kind of atom, each atom is neutral. However, rubbing the substance with a different material causes it to become charged. On heating the material, negative particles, electrons are driven out. Recall the action in an operating vacuum tube. A high speed particle on passing through a gas knocks highly mobile negative particles, electrons, out of molecules in its path. This action is employed in a counter tube.

In 1911, Rutherford reported that substantial foils of material were largely open to the free passage of alpha particles, high speed helium nuclei. But some would come near positive centers and be deflected through large angles. The results were explained by considering each atom to be composed of a small relatively massive positive nucleus surrounded by an electron cloud.

So a molecule consists of an array of nuclei, surrounded by core electron clouds, held together by the cloud of valence electrons. The nuclei of a two-atom molecule lie on a straight line. When the two atoms are the same, the centers of positive and of negative charge coincide and the molecule has no dipole moment. On the other hand, when they are different, these centers do not coincide and the molecule exhibits some dipole moment. In any case, we say there is a chemical bond between the two atoms. When this involves two valence electrons, it is called a single bond. When it involves four valence electrons, it is called a double bond. When it involves six valence electrons, it is called a triple bond. Each of these has an equilibrium length.

When the valence of an atom is not satisfied by its bonding to a second atom, it may bond to one or more additional atoms. Thus, many polyatomic structures exist. Some of these are linear; others nonlinear. In any case, an equilibrium length can be assigned to each bond. And, an equilibrium angle between any two bonds from a given atom. Furthermore, a definite dipole moment is associated with each bond, directed from the more negative to the more positive atom. The different dipole moments add vectorially to give the dipole moment of the molecule.

2.2 Symmetric Coordination Structures

The bond angles in small symmetric coordination structures are determined by the equivalence of the attached atoms.

Consider a molecule or ion of the type AB_n, where n B atoms are bonded to a single A atom. Let $\mathbf{r}_j$ be the vector drawn from the center of A to the center of the jth B.

Now, the dipole moment of the jth A- B bond lies along $\mathbf{r}_j$. Furthermore, we suppose that its magnitude varies with the bond length r_j. So when the ligands are arranged symmetrically around the central atom, the net dipole moment is zero and

$$\mathbf{r}_1 + \mathbf{r}_2 + \ldots + \mathbf{r}_n = 0. \quad [2.1]$$

The *dot product* of two vectors equals the projection of the magnitude of the first one on the second one times the magnitude of the second one. If θ_{jk} is the acute angle between $\mathbf{r}_j$ and $\mathbf{r}_k$, we have

$$\mathbf{r}_j \cdot \mathbf{r}_k = r_j r_k \cos\theta_{jk}. \quad [2.2]$$

When all the bonds are equivalent, they have the same length and

$$r_j = r. \quad [2.3]$$

When all the acute angles are the same, we also have

$$\theta_{jk} = \theta. \quad [2.4]$$

Let us now dot multiply both sides of equation (2.1) by vector $\mathbf{r}_j$ and introduce conditions (2.3) and (2.4) to construct

$$r^2 + (n-1)r^2\cos\theta = 0, \quad [2.5]$$

whence

$$\theta = \cos^{-1}\left[\frac{1}{(1-n)}\right]. \quad [2.6]$$

For $n = 2$, we have

$$\theta = \cos^{-1}(-1) = 180°. \quad [2.7]$$

The corresponding structure is linear. When $n = 3$, we find that

$$\theta = \cos^{-1}\left(-\frac{1}{2}\right) = 120°. \quad [2.8]$$

The ligands now lie at the corners of an equilateral triangle. When $n = 4$, we have

$$\theta = \cos^{-1}\left(-\frac{1}{3}\right) = 109.47°. \quad [2.9]$$

These ligands are arranged tetrahedrally about the central atom.

When the acute angle between $\mathbf{r}_1$ and $\mathbf{r}_n$ is 180°, dot multiplying both sides of equation (2.1) by $\mathbf{r}_1$ and setting the other acute angles involving $\mathbf{r}_1$ equal to θ_1 yields

$$r^2 + (n-2)r^2 \cos\theta_1 - r^2 = 0, \qquad [2.10]$$

whence

$$\theta_1 = \cos^{-1} 0 = 90°. \qquad [2.11]$$

A similar result follows from dot multiplying both sides of (2.1) by $\mathbf{r}_n$. But dot multiplying both sides of equation (2.1) by $\mathbf{r}_2$ yields

$$0 + r^2 + (n-3)r^2 \cos\theta_2 + 0 = 0, \qquad [2.12]$$

whence

$$\theta_2 = \cos^{-1}\left[\frac{1}{(3-n)}\right]. \qquad [2.13]$$

Here θ_2 is the other acute angle.

When $n = 4$, formula (2.13) yields

$$\theta_2 = \cos^{-1}(-1) = 180°. \qquad [2.14]$$

The configuration is now a square. When $n = 5$, we have

$$\theta_2 = \cos^{-1}\left(-\frac{1}{2}\right) = 120°. \qquad [2.15]$$

The configuration is a trigonal bipyramid.

For $n = 6$, the most symmetry prevails when the acute angles between $\mathbf{r}_1$ and $\mathbf{r}_6$, $\mathbf{r}_2$ and $\mathbf{r}_4$, $\mathbf{r}_3$ and $\mathbf{r}_5$ all equal 180°. Dot multiplying equation (2.1) by any $\mathbf{r}_j$. then yields equation (2.10). As a consequence, the other acute angle is 90° and the configuration is that of an octahedron.

Substituting other atoms or radicals for one or more of the B's attached to a given A would perturb the structure. But since the perturbations are generally small, the symmetric structure serves as a reference structure.

2.3 Momentum and Energy Relationships

To measure bond distances and angles in a molecule, one may employ diffraction methods. Approximately homogeneous beams of particles are generally employed. Electromagnetic radiation, as we have in X rays, is too penetrating.

Innumerable experiments show that a beam in which the momentum of each particle is $\mathbf{p}$ exhibits a wavevector $\mathbf{k}$ that is proportional:

$$\mathbf{p} = \frac{h}{2\pi}\mathbf{k} = \hbar\mathbf{k}. \qquad [2.16]$$

This relationship was introduced by Louis *de Broglie*. The wavevector has the magnitude

$$k = \frac{2\pi}{\lambda}, \qquad [2.17]$$

where λ is the wavelength. It has the direction of the beam.

Particles with a definite momentum may be obtained by passing a heterogeneous beam through a velocity selector. This may be a mechanical device consisting of a series of rotating discs containing openings that line up for the desired velocity. Or, when the particles are charged, the selector may employ a crossed electric and magnetic field.

Alternatively, charged particles may be accelerated from approximate rest by a measured voltage increment V. If $\mp e$ is the charge on the particle, the loss of potential energy, and the gain in kinetic energy, during the acceleration is $\pm eV$. Parameter e may be the magnitude of the charge on an electron or a small multiple of it. So a convenient unit of energy is the kinetic energy acquired by an electron on moving freely through a voltage rise of one volt, the *electron volt* (eV).

When m is the mass of the particle and p its momentum, the gain in kinetic energy is

$$\frac{p^2}{2m} = \pm eV, \qquad [2.18]$$

as long as Newtonian mechanics is applicable.

Example 2.1

Relate the wavelength of a homogeneous electron beam to the voltage accelerating the electrons.

From equation (2.18), we have

$$p = \left(2meV\right)^{1/2}.$$

Substituting this expression into the de Broglie equation

$$p = \frac{h}{\lambda}$$

obtained from (2.16) and (2.17), and solving for the wavelength yields

$$\lambda = \frac{h}{\left(2meV\right)^{1/2}}.$$

Now introduce accepted values of the fundamental constants to get

$$\lambda = \frac{6.6261 \times 10^{-34}\ \text{J s}}{\left[2\left(9.1094 \times 10^{-31}\ \text{kg}\right)\left(1.6022 \times 10^{-19}\ \text{C}\right)\right]^{1/2}} \frac{1}{V^{1/2}} = 12.264 \times 10^{-10}\ V^{-1/2}\ \text{m} = 12.264\ V^{-1/2}\ \text{Å}.$$

This can be rewritten in the form

$$\lambda = \left(\frac{150.41}{V}\right)^{1/2}\ \text{Å}.$$

2.4 Diffraction by Randomly Oriented Molecules

Bond distances and bond angles can be obtained from diffraction data. In a common setup, a homogeneous beam of particles passes through a low pressure jet of the material.

The beam has a wavelength determined by the momentum of a typical particle therein. This in turn is determined by the kinetic energy of the particle. The jet scatters the beam into a characteristic diffraction pattern.

Consider the scheme in figure 2.1. A potential between 30 and 70 kilovolts accelerates the electrons emitted by a hot filament. The electrons travel in a vacuum tube to a vapor stream containing the molecules to be studied. These diffract the beam. The resulting pattern is recorded on a photographic plate. Pertinent measurements are then made.

In analyzing these, we consider each atom in a target molecule to act as a scattering point. Since molecular distances are very small compared to the dimensions of the apparatus, we also consider the incident rays and the rays scattered at a certain angle by a molecule to be parallel. For simplicity, we also limit ourselves initially to diatomic molecules, those of type AB. A particular orientation of a scattering molecule then acts as shown in figure 2.2.

A reference Cartesian system is erected on atom A as shown. The corresponding spherical coordinates of atom B are (r, α, β). Line AC is drawn perpendicular to the incident rays; line BD perpendicular to the scattered rays. Then points A and C would be at the same phase. Also points B and D. But the ray scattered by A travels the extra length (which may be either positive or negative)

$$\delta = \mathrm{AD} - \mathrm{CB}. \qquad [2.19]$$

From the figure and trigonometric definitions, we find that

$$\mathrm{AD} = r \sin\alpha \cos\gamma = r \sin\alpha \sin(\beta + \theta), \qquad [2.20]$$

so

$$\mathrm{CB} = r \sin\alpha \sin\beta, \qquad [2.21]$$

$$\delta = r \sin\alpha\left[\sin(\beta + \theta) - \sin\beta\right] = 2r \sin\frac{1}{2}\theta \sin\alpha \cos\left(\beta + \frac{1}{2}\theta\right). \qquad [2.22]$$

In the coherent approximation, the electron wave has the amplitude

$$\Psi = Ne^{iks}e^{-i\omega t}, \qquad [2.23]$$

where N is a normalization constant, s the distance traveled, and t the time. Parameter k is the wavevector

$$k = \frac{2\pi}{\lambda}, \qquad [2.24]$$

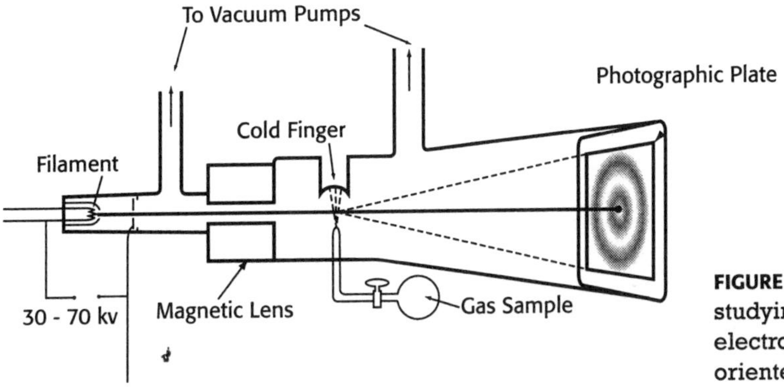

FIGURE 2.1 Setup for studying the diffraction of electrons by randomly oriented molecules.

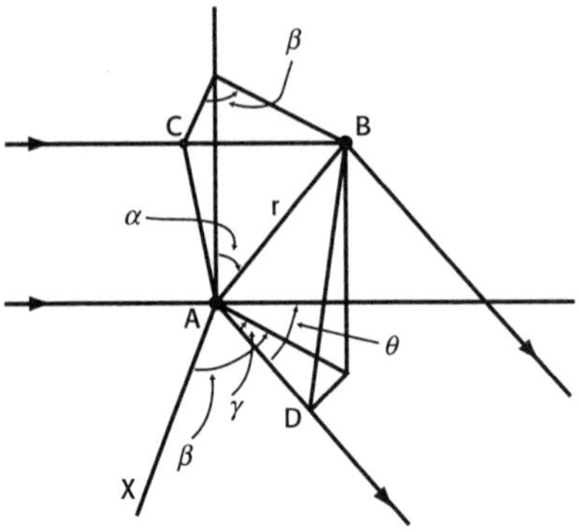

FIGURE 2.2 Rays scattered at angle θ by molecule AB oriented at angles α and β.

while the wavelength λ is given by the de Broglie equation in the form

$$\lambda = \left(\frac{150.41}{V}\right)^{1/2} \text{Å}, \tag{2.25}$$

where V is the accelerating voltage.

For simplicity, we consider A and B to be equivalent atoms. Then if a ray scattered by B arrives at an observation point as

$$\Psi_1 = Ae^{ika}e^{-i\omega t}, \tag{2.26}$$

the corresponding ray scattered by A arrives there as

$$\Psi_2 = Ae^{ik(a+\delta)}e^{-i\omega t}. \tag{2.27}$$

Superposing these yields the resultant

$$\Psi = Ae^{ika}e^{-i\omega t}\left(1+e^{ik\delta}\right). \tag{2.28}$$

The corresponding intensity is

$$I = \Psi^*\Psi = A^*A\left(1+e^{-ik\delta}\right)\left(1+e^{ik\delta}\right) = 2A^*A\left(1+\cos k\delta\right) = 4A_0^2\cos^2\frac{1}{2}k\delta. \tag{2.29}$$

In the diffracting jet, the molecules are oriented randomly. The resulting intensity is obtained on averaging expression (2.29) over the different orientations. Thus

$$\bar{I}(\theta) = \frac{4A_0^2}{4\pi}\int_0^{2\pi}\int_0^{\pi}\cos^2\left[kr\sin\frac{1}{2}\theta\sin\alpha\cos\left(\beta+\frac{1}{2}\theta\right)\right]\sin\alpha\, d\alpha\, d\beta, \tag{2.30}$$

whence

$$\bar{I}(\theta) = 2A_0^2\left(1+\frac{\sin sr}{sr}\right) \tag{2.31}$$

with

$$s = 2k\sin\frac{1}{2}\theta. \tag{2.32}$$

Here θ is the scattering angle, k the wavevector for the stream of electrons, r the interatomic distance, A_0 the scattering factor (the magnitude of the scattered wave function for angle θ), and I the intensity measured at angle θ. A plot of the scaled intensity against product sr appears in figure 2.3.

For a polyatomic molecule exhibiting scattering factors A_j, A_k from the jth and kth atoms a distance r_{jk} apart, the resulting intensity is

$$\bar{I}(\theta) = \sum_j \sum_k A_j A_k \frac{\sin sr_{jk}}{sr_{jk}} \quad [2.33]$$

with

$$s = 2k \sin \frac{1}{2}\theta. \quad [2.34]$$

Formula (2.33) is called the *Wierl equation.*

Example 2.2

Show that for a homonuclear diatomic molecule, equation (2.34) reduces to (2.31).

For a diatomic molecule, equation (2.34) becomes

$$\bar{I}(\theta) = A_1 A_1 \frac{\sin sr_{11}}{sr_{11}} + A_1 A_2 \frac{\sin sr_{12}}{sr_{12}} + A_2 A_1 \frac{\sin sr_{21}}{sr_{21}} + A_2 A_2 \frac{\sin sr_{22}}{sr_{22}}$$

But

$$r_{11} = r_{22} = 0$$

and

$$\lim_{\alpha \to 0} \frac{\sin \alpha}{\alpha} = \lim_{\alpha \to 0} \cos \alpha = 1.$$

Also

$$r_{12} = r_{21} = r.$$

So the equation for a diatomic molecule reduces to

$$\bar{I}(\theta) = A_1^2 + A_2^2 + 2A_1 A_2 \frac{\sin sr}{sr}.$$

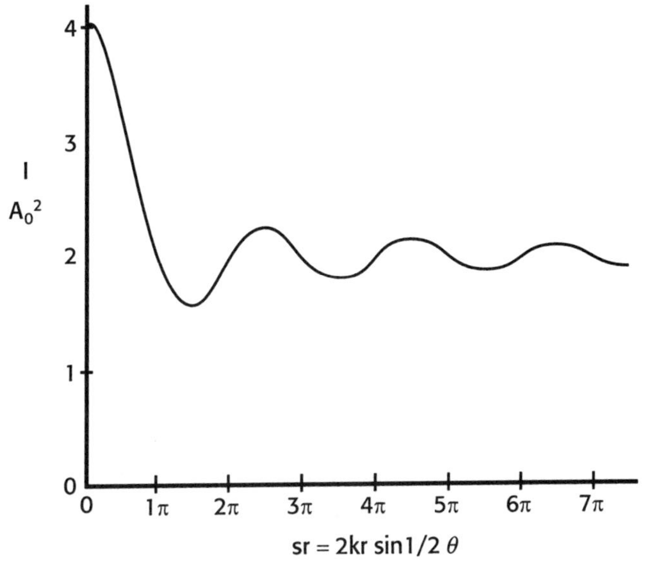

FIGURE 2.3 Variation of the scaled intensity with a function of the scattering angle.

When the two atoms are the same, the scattering factors are equal,

$$A_1 = A_2 = A_0,$$

and we obtain

$$\bar{I}(\theta) = 2A_0^2\left(1 + \frac{\sin sr}{sr}\right).$$

2.5 Electron Diffraction Molecular Parameters

When a monoenergetic beam of electrons is diffracted by randomly oriented molecules, a diffraction pattern in which the intensity varies only with deflection angle θ is produced. For a given screen position, maxima and minima in the pattern can be located and the corresponding θ's calculated. The wavevector k can be determined from the accelerating potential. Then the interatomic distances r_{jk} can be varied consistent with the molecule's geometry until the Wierl equation fits the data.

As an approximation, one may consider scattering factor A_j to be proportional to the number af electrons Z_j in the jth atom. Hydrogen atoms may be ignored because of their small scattering power.

Each independent bond distance is a separate parameter. The bond angles are related to these by geometric considerations. Good results are obtained only for the simpler molecules, those with only a few independent parameters.

Representative results appear in table 2.1.

2.6 Intensity in a Beam

Properties of molecules and atoms can be induced from absorption measurements, from how the intensity of a beam diminishes as it passes through the material under study. See figure 2.4. But what is intensity and how can it be measured?

Intensity I may be defined as (a) the number of particles with the chosen properties passing by a point per unit cross section per unit time, or (b) the kinetic energy of the pertinent kind of radiation passing by a point per unit cross section per unit time, or (c) a number proportional to either of these.

A traveling particle with a rest mass will interact with electrons, ions, or molecules along its path. A homogeneous beam of such particles will produce excitations proportional to the intensity and to the time of exposure.

With electromagnetic radiation, each frequency ν propagates independently. Furthermore, this component interacts with matter as if it were composed of particles (photons) with the energy

$$E = h\nu = \hbar\omega. \qquad [2.35]$$

TABLE 2.1 Equilibrium Electron Diffraction Parameters

Molecule	*Interatomic Distances, Å*		*Bond Angles*	
CO_2	C–O	1.16	O–C–O	180.00°
SO_2	S–O	1.43	O–S–O	119.3°
CCl_4	C–Cl	2.89	Cl–C–Cl	109.47°
$SiCl_4$	Si–Cl	2.00	Cl–Si–Cl	109.47°
C_6H_6	C–C	1.39	C–C–C	120.00°
P_4	P–P	2.21	P–P–P	60.0°

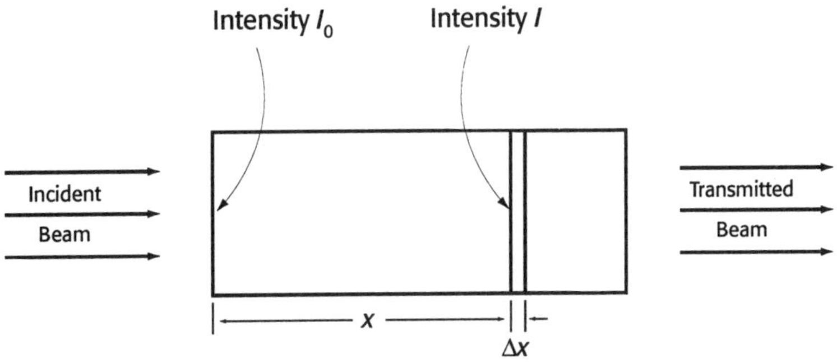

FIGURE 2.4 Passage of a beam of radiation through a medium with reduction of its intensity from I_0 to I in distance x.

Here h is Planck's constant, $\hbar$ equals $h/2\pi$, and ω is the angular frequency $2\pi\nu$. Thus, absorption occurs in steps. In each of these, the radiation loses a photon; the energy (2.35) is transferred to the absorbing electron, ion, or molecule, exciting it.

Both energetic particles and photons can be detected *photographically*. The active layer on photographic film consists of an emulsion of very small grains. These contain some photosensitive substance(s) such as silver bromide. An incident ray excites grains along its path. The resulting latent image is developed, fixed, and measured with a densitometer.

High energy particles and photons can be detected by the *ionization* they produce. The active agent may be a gas enclosed between two electrodes. Applying a potential difference then causes the electrons to move rapidly to the anode, the positive ions to move slowly to the cathode. The effect can be multiplied by increasing the potential applied until the mobile electrons produce further ionization before they reach the positive electrode. An external electric circuit serves to count the resulting pulses.

Alternatively, the active agent may be a momentarily *superheated* or *supersaturated* phase. The ionized molecules and electrons then serve as nuclei for evaporation or condensation. The resulting tracks can be illuminated and recorded with a camera.

In various materials, the excitation produced by a ray may be relieved by emission of one or more photons—a *scintillation*. This may be observed directly or multiplied in a photomultiplier tube.

The *photovoltaic* detector involves a p–n junction in silicon or some other semiconductor. The p material is produced by adding a small amount of impurity that combines with conduction electrons; the n material by adding impurity that supplies additional conduction electrons. Associated with the junction at equilibrium is an electric field. When an incident ray excites an electron, producing an electron-hole pair, these move to alter the field. A change in voltage across the junction results. This is measured.

2.7 The Cross Section Concept

The atoms in an absorbing material act to reduce the intensity of a beam by either deflecting or absorbing particles or photons. Insofar as the beam is concerned, it is as if each atom removes all incident particles falling on a fraction of the cross sectional area that it occupies, while it lets all others through. See figure 2.5.

The hypothetical shading area that an atom or molecule presents is called its *cross section* σ. This varies with the nature of both reactants, the absorbing atom or molecule, the projectile particle, and with the particle's kinetic energy.

When a beam traverses distance Δx of the sample, it sweeps over volume Δx per unit cross sectional area. If the number density of the absorbing atoms or molecules is n, the

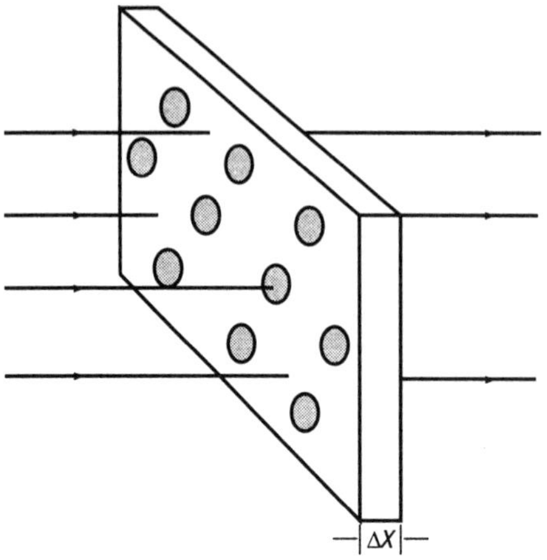

FIGURE 2.5 Apparent cross sections presented by the target atoms in thickness Δx of a material to a given beam.

number of atoms in this volume is $n\Delta x$. The corresponding shading area is $\sigma n\Delta x$. The fraction of incident particles that strike this obstruction, per unit area, must be $\sigma n\Delta x$. But intensity I is proportional to the number of projectile particles reaching this unit area in unit time. So we have

$$-\frac{\Delta I}{I} = \sigma n\Delta x. \qquad [2.36]$$

We suppose that the cross sections in one atomic layer do not cast effective shadows on those in successive atomic layers. We also suppose that the thickness of each of these layers is relatively small so that the increments can be replaced by differentials. In this *continuum approximation*, equation (2.36) becomes

$$-\frac{\mathrm{d}I}{I} = \sigma n\mathrm{d}x. \qquad [2.37]$$

Integrating (2.37) over the thickness x yields

$$\ln\frac{I}{I_0} = -\sigma nx. \qquad [2.38]$$

Here I_0 is the intensity of the beam at $x = 0$ while I is its intensity after traveling distance x through the absorbing medium.

A convenient unit for nuclear reactions is the *barn*, defined as 10^{-28} m^2 (10^{-24} cm^2).

Example 2.3

A beam of thermal neutrons is reduced to 0.0100 percent of its initial intensity by 0.083 cm of cadmium. Calculate the cross section presented by a cadmium atom to the beam.

Solve equation (2.38) for the cross section:

$$\sigma = \frac{-\ln\left(\frac{I}{I_0}\right)}{nx}.$$

To get n, divide the mass density of cadmium by the mass per mole and multiply by the number of atoms in a mole:

$$n = \frac{8.642 \text{ g cm}^{-3}}{112.41 \text{ g mol}^{-1}}\left(6.02 \times 10^{23} \text{ atoms mol}^{-1}\right) = 4.63 \times 10^{22} \text{ atoms cm}^{-3}.$$

Substitute this and the given I/I_0 and x into the formula for σ:

$$\sigma = \frac{-\ln 1.000 \times 10^{-4}}{\left(4.63 \times 10^{22} \text{ atoms cm}^{-3}\right)\left(0.083 \text{ cm}\right)} = 2.4 \times 10^{-21} \text{cm}^2$$

or

$$\sigma = 2400 \text{ barns}.$$

2.8 The Nuclear Atom

In the first decade of the twentieth century, the openness of materials to bombarding particles was discovered. In 1903, P. Lenard studied the penetration of matter by cathode rays (electrons). He found that they passed through foils and thin sheets with attenuation. In 1909, Ernest Rutherford's students, H. Geiger and E. Marsden, found that alpha rays readily penetrated thin sheets of material. However, a few of the helium nuclei were deflected through large angles.

In 1911, Rutherford introduced the nuclear atom to explain these results. According to this model, an atom consists of a small positive *nucleus* surrounded by a relatively large region through which the electrons move. Thus, most of the volume of an atom is open.

In determining how large nuclei are, one may advantageously employ neutrons. Electrostatic effects are thus avoided. And if the projectile energy is not too low, in the region where excessive absorption occurs, or too high, where the nuclei become transparent, the neutrons absorbed are those hitting the nucleus.

One does have to allow for the radius of the neutron. If both incident particle and target nucleus are spherical, every projectile whose center strikes within the area πR^2 of a nucleus, where

$$R = R_A + R_b \qquad [2.39]$$

does hit the nucleus. Here R_A is the radius of the nucleus while R_b is the radius of the projectile. See figure 2.6.

A beam of particles with definite momentum **p** is diffracted by a barrier as a ray with wavevector **k** given by formula (2.16). In the setup here, each nucleus diffracts projectiles to about the same extent that it absorbs them, So the total cross section is

$$\sigma = 2\pi R^2 = 2\pi\left(R_A + R_b\right)^2. \qquad [2.40]$$

Representative results, obtained with 14- and 25-MeV neutrons, appear in table 2.2.

Parameter R_0 is defined by the equation

$$R_A = R_0 A^{1/3} \qquad [2.41]$$

in which A is the *mass number*, the integer closest to the atomic mass of the isotope. The unit of distance is the *fermi* (fm), defined as 10^{-15}m. The data show that R_0 is nearly

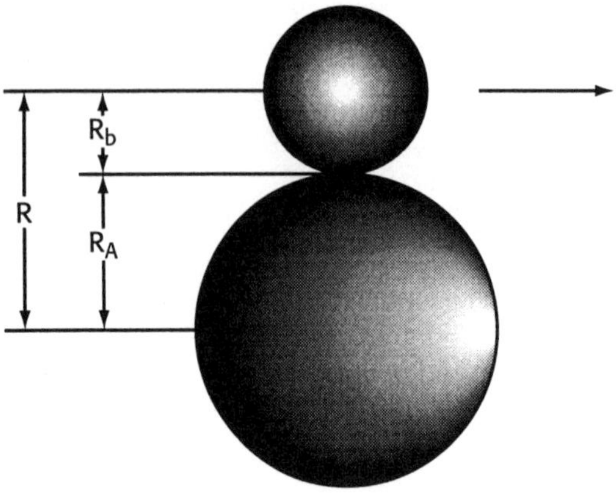

FIGURE 2.6 Beam particle just grazing a target nucleus.

TABLE 2.2 Nuclear Radii from Neutron Cross Sections*

Element	*Cross Section σ, barns*	*Neutron Energy, MeV*	*Nuclear Radius R_A, fermis*	*Parameter R_0, fermis*
Be	0.65	14	2.4	1.17
B	1.16	14	3.4	1.54
C	1.29	25	3.8	1.65
O	1.60	25	4.3	1.71
Mg	1.83	14	4.5	1.57
Al	1.92	14	4.6	1.52
Al	1.85	25	4.6	1.52
S	1.58	14	4.1	1.30
Cl	1.88	25	4.7	1.44
Fe	2.75	14	5.6	1.46
Cu	2.50	25	5.5	1.38
Zn	3.03	14	5.9	1.48
Se	3.03	14	6.3	1.46
Ag	3.82	14	6.8	1.44
Ag	3.70	25	6.9	1.46
Cd	4.25	14	7.2	1.48
Sn	4.52	14	7.4	1.52
Sb	4.35	14	7.3	1.46
Au	4.68	14	7.5	1.33
Hg	5.64	14	8.3	1.42
Hg	5.25	25	8.4	1.44
Pb	5.05	14	7.8	1.32
Bi	5.17	14	7.9	1.34

*H. Feshbach and V. F. Weisskopf, *Phys. Rev.* **76**, 1550-1560 (1949)

constant, with the value

$$R_0 = 1.4 \text{ fm}. \qquad [2.42]$$

Similar experiments have been performed with protons. After correcting for the scattering caused by electric interaction, one obtains similar results.

The radius R_A within which the charge seems to be distributed uniformly can be determined. One may use electrons having such a high kinetic energy (200 to 500 MeV) that each follows a nearly classical (nonquantal) trajectory. The nuclear radius R_b of the electrons is taken to be zero. Then one finds that R_0 ranges from 1.2 fm for heavy nuclei to 1.3 fm for light nuclei.

In the approximation that equation (2.41) holds with R_0 constant, the volume of a nucleus is proportional to its mass number

$$V_A = \frac{4}{3}\pi R_A{}^3 = \frac{4}{3}\pi R_0{}^3 A = (\text{const})A. \qquad [2.43]$$

This result suggests that a nucleus consists of A particles, each occupying the volume $(4/3)\pi R_0{}^3$. In modern theory, these are taken to be neutrons and protons. The number of protons is taken to equal the *atomic number Z*.

2.9 Composition of Atoms and Nuclei

Certain key experimental facts and their implications will now be summarized.

The charged particles that are easily driven out of materials by projectile particles and by heat are electrons. Presumably they come from the outer regions of atoms in the materials. The transparency of a substance to small projectiles shows how extensive these regions are. Under appropriate conditions, the radii of the absorbing regions, the nuclei, can be determined. These are small fractions of the atomic radii calculated from the observed interatomic spacings in molecules and in crystals.

The mass to charge ratio of particles can be determined from the observed deflections in a magnetic field. The charge on the highly mobile particle, the electron, can be determined from observations of the movements of small spheres in air in combined electric and gravitational fields.

One can observe the movements of an atom stripped of various numbers of its electrons in a mass spectrograph. It is found that the total number of electrons in an uncharged atom equals its number in the periodic table, the atomic number Z.

The unit for atomic mass is one-twelfth the mass of the atom ^{12}C. For other nuclides, the integer closest to the atomic mass is called the mass number A. The volume of a nucleus is approximately proportional to its mass number.

Questions

2.1 What chemical evidence do we have for the atomic theory of matter?
2.2 What evidence do we have that electrons are the agents that hold atoms together in molecules?
2.3 Define the term chemical bond.
2.4 Explain how the bond angles in small coordination structures may be determined.
2.5 How is the de Broglie equation justified?
2.6 How is a beam of particles with a definite wavevector obtained?
2.7 How can a jet of molecules produce a diffraction pattern?
2.8 How may the interatomic distances and angles be determined?
2.9 Define the intensity in a beam.
2.10 How is this intensity measured?
2.11 In what units does electromagnetic radiation interact with matter?
2.12 Define the cross section of an atom or molecule.
2.13 Show how the intensity of a beam decreases on passing through a target.
2.14 What is the continuum approximation?
2.15 Cite the evidence for the nuclear atom.
2.16 What is the conventional unit for atomic and molecular masses?

Problems

2.1 Molecule AB_3 has a trigonal pyramid structure. Show that for the dipole moment to be directed perpendicular to the base, the B's must lie at the corners of an equilateral triangle.

2.2 What accelerating voltage is needed to produce an electron beam with a wavelength of 0.100 Å?

2.3 Determine the angle *sr* at which the first minimum and the next maximum occur in formula (2.31) for $\bar{I}(\theta)$.

2.4 An electron beam with wavelength equal to 0.100 Å is diffracted by diatomic molecules in which the interatomic distance is 1.09 Å. At what deflection angle θ would the first maximum be observed?

2.5 Construct a Wierl equation for linear S—C—S, letting each scattering factor be proportional to the corresponding atomic number.

2.6 Calculate the thickness of a lead shield needed to reduce a beam of 14-MeV neutrons to 1.00 per cent of its initial intensity. The cross section for these neutrons is 5.05 barns.

2.7 A target of aluminum 9.00 cm thick absorbed 64.7 per cent of a beam of 14-MeV neutrons. What cross section is presented by an aluminum atom to the beam?

— — —

2.8 The electrons in a beam were accelerated from approximate rest by a potential rise of 30.0 kV. What is the resulting wavelength?

2.9 How is the wavelength of a beam of protons related to the accelerating potential drop *V*?

2.10 What is the wavelength of a beam of 150 V protons?

2.11 Construct a Wierl equation for benzene, neglecting the contribution of the hydrogen atoms.

2.12 What voltage applied to an X-ray tube will produce photons with wavelengths down to 0.311 Å?

2.13 How much is a beam of thermal neutrons reduced on passing through 2.50 cm zinc and 3.50 cm nickel? The pertinent cross section of zinc is 1.06 barns; that of nickel, 4.50 barns.

2.14 An iron shield 2.10 cm thick reduced the intensity of a beam of thermal neutrons by 35.0 per cent. What is the cross section of an iron atom to the beam?

References

Book

Eisberg, R., and Resmick, R.: 1974, *Quantum Physics of Atoms, Molecules, Solids, Nuclei, and Particles*, John Wiley & Sons, Inc., New York, pp. 29-134.

This book is essentially an introduction to modern physics. In the part cited, Eisberg and Resnick introduce the student to simple photon theory, the cross section concept, matter wave theory, and simple models of the atom.

Articles

Dantus, M., Kim, S. B., Williamson, J. C., and Zewail, A. H.: 1994, "Ultrafast Electron Diffraction. 5. Experimental Time Resolution and Applications," *J. Phys. Chem.* **98**, 2782-2796.

Hanson, R. FI., and Bergman, S. A.: 1994, "Data-Driven Chemistry: Building Models of Molecular Structure (Literally) from Electron Diffraction Data," *J. Chem. Educ.* **71**, 150-152.

Heilbronner, E.: 1989, "Why do Some Molecules Have Symmetry Different from that Expected?" *J. Chem. Educ.* **66**, 471-478.

Heyrovska, R.: 1992, "On Neutron Numbers and Atomic Masses," *J. Chem. Educ.* **69**, 742-743.

Kidd, R., Ardini, J., and Anton, A.: 1988, "Evolution of the Modern Photon," *Am. J. Phys.* **57**, 27-35.

King, R. B.: 1996, "The Shapes of Coordination Polyhedra," *J. Chem. Educ.* **73**, 993-997.

Nelson, P. G.: 1997, "Valency," *J. Chem. Educ.* **74**, 465- 470.

Pullen, B. P.: 1986, "Cerenkov Counting of ^{40}K in KC1 Using a Liquid Scintillation Spectrometer," *J. Chem. Educ.* **63**, 971.

Sacks, L. J.: 1986, "Coulombic Models in Chemical Bonding," *J. Chem. Educ.* **63**, 288-296.

Weinrach, J. B., Carter, K. L., Bennett, D. W., and McDowell, H. K.: 1990, "Point Charge Approximations to a Spherical Charge Distribution," *J. Chem. Educ.* **67**, 995-999.

Williamson, J. C., and Zewail, A. H.: 1994, "Ultrafast Electron Diffraction. 4. Molecular Structures and Coherent Dynamics," *J. Phys. Chem.* **98**, 2766-2781.

3

Gases and Collective Properties

3.1 Macroscopic Properties

FROM THE SIZE OF AVOGADRO'S NUMBER, we know that even small amounts of material contain an enormous number of formula units. As a consequence, a person cannot follow the detailed motions of the molecules, radicals, ions, or electrons therein. Nevertheless, one can get at the collective behavior of these.

The section of material chosen for study is called a *system*. The neighboring material that interacts with the system in various ways is called the *surroundings*. The collective behavior produces properties that can be observed. In a uniform material, some of these vary directly with the amount considered. Such a property is said to be *extensive*. As examples, we have volume, mass, energy of thermal agitation, moles of each constituent. In contrast, other cbservable properties are independent of the amount of material present. Such a property is said to be *intensive*. As examples, we have density, pressure, temperature, concentrations. The macroscopic *state* of a uniform system is defined when enough of these collective properties are given so that all other such properties are fixed.

Those that are picked to determine the state of the system form the *independent variables*. The properties dependent on the values of the chosen independent variables are called *dependent variables*.

The volume elements over which intensive properties vary continuously, if at all, form a *phase*. A system containing a single uniform phase is said to be *homogeneous*. A system containing two or more phases is said to be *heterogeneous*.

Phases are classified as solid, liquid, or gaseous following the criteria in Section 1.2, In any given phase, interactions among the molecules or ions tend to eliminate any local orderly motion or concentration of energy that might be introduced. Transients tend to disappear leaving a steady or equilibrium state.

3.2 Elements of Ideal-Gas Theory

A given gas may be considered at a low enough density so that its molecules do not interact appreciably with each other except during collisions. Such a phase, where the volume of the molecules and the attractions among them are negligible, is called an *ideal gas*.

Whether or not a given phase behaves as an ideal gas, collisions between molecules break down any orderly motion or concentration of energy that might be introduced. The collisions tend to create complete chaos in the system. A gas at rest is thus made isotropic, so that it exerts the same pressure P on each confining wall.

Consider an ideal gas containing N molecules, each of mass m, confined in a cubical box of edge l, at equilibrium under pressure P. Place rectangular axes along three edges of the cube as figure 3.1 shows. Then the coordinates of the ith molecule are represented as (x_i, y_i, z_i) at the given time t.

Let the velocity at which the ith molecule moves be u_i. Also represent the time derivatives of the coordinates as $(\dot{x}_i, \dot{y}_i, \dot{z}_i)$. Then with the Pythagorean theorem, we have

$$u_i^2 = \dot{x}_i^2 + \dot{y}_i^2 + \dot{z}_i^2. \qquad [3.1]$$

In an ideal gas the only role of collisions is to randomize the motions. So if we suppose that complete disorder reigns, we may neglect collisions in our derivation.

The intrinsic properties of a gas are independent of the nature of the walls confining it. Here for simplicity, we will consider that each wall is smooth and elastic. Then on striking the wall that coincides with the yz plane, molecule i is reflected, with $\dot{x}_i$ merely reversed in sign and with $\dot{y}_i$ and $\dot{z}_i$ unchanged. The only other wall where $\dot{x}_i$ is altered is the one opposite, where $\dot{x}_i$ is also reversed in sign. Neglecting collisions with other molecules, we find that the x component of the distance traveled by the molecule between strikes on the yz plane equals $2l$. If the molecule strikes the wall n_i times a second, it travels distance $2ln_i$ parallel to the x axis. Since speed is distance divided by time, we have

$$\left|\dot{x}_i\right| = \frac{2ln_i}{1}. \qquad [3.2]$$

Each time molecule i strikes the yz plane, its momentum p_i changes by $2m|\dot{x}_i|$. In unit time the change in momentum at this plane is

$$\frac{\Delta p_i}{\Delta t} = n_i 2m\left|\dot{x}_i\right| = \frac{m\dot{x}_i^2}{l}, \qquad [3.3]$$

where n_i has been eliminated using equation (3.2) and Δp_i is the momentum change in time Δt. Summing over all N molecules yields

$$\sum \frac{\Delta p_i}{\Delta t} = \frac{m}{l} \sum \dot{x}_i^2. \qquad [3.4]$$

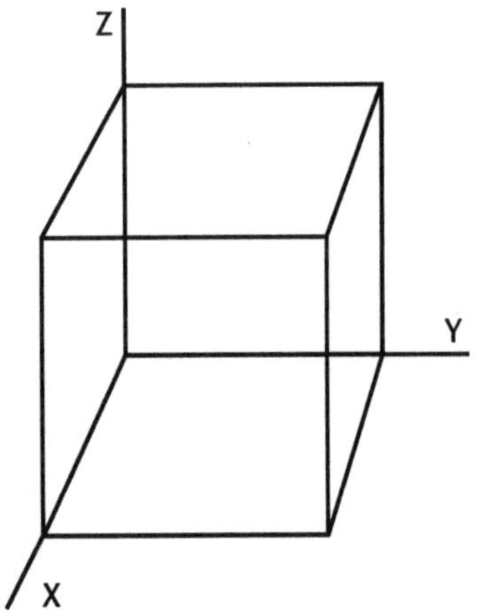

FIGURE 3.1 Cubical box confining the ideal gas.

Averaging $\dot{x}_i^2$,

$$\frac{\sum \dot{x}_i^2}{N} = \overline{\dot{x}^2}, \qquad [3.5]$$

and substituting the result into equation (3.4) gives us

$$\sum \frac{\Delta p_i}{\Delta t} = \frac{Nm\overline{\dot{x}^2}}{l}. \qquad [3.6]$$

From Newton's second law, the rate of change of momentum equals the force causing the change. So the sum on the left side here equals the net force exerted by the wall Pl^2:

$$Pl^2 = \frac{Nm\overline{\dot{x}^2}}{l}. \qquad [3.7]$$

Since l^3equals the volume of the gas V, equations (3.7) rearranges to

$$PV = Nm\overline{\dot{x}^2}. \qquad [3.8]$$

By symmetry,

$$PV = Nm\overline{\dot{y}^2}, \qquad [3.9]$$

$$PV = Nm\overline{\dot{z}^2}. \qquad [3.10]$$

Summing (3.1) over all N molecules and dividing by N yields

$$\frac{\sum u_i^2}{N} = \frac{\sum \dot{x}_i^2}{N} + \frac{\sum \dot{y}_i^2}{N} + \frac{\sum \dot{z}_i^2}{N} \qquad [3.11]$$

or

$$\overline{u^2} = \overline{\dot{x}^2} + \overline{\dot{y}^2} + \overline{\dot{z}^2}. \qquad [3.12]$$

Finally, combining (3.8), (3.9), (3.10) , and (3.12) gives us

$$PV = \frac{1}{3}Nm\overline{u^2}. \qquad [3.13]$$

Since the translational energy of the ith molecule is $1/2mu_i^2$, the total translational energy E_{tr} of the N molecules is

$$E_{\text{tr}} = \sum \frac{1}{2}mu_i^2 = \frac{1}{2}Nm\overline{u^2}. \qquad [3.14]$$

And (3.13) can be written in the meaningful form

$$PV = \frac{2}{3}E_{\text{tr}}. \qquad [3.15]$$

In the laboratory, atmospheric pressure is determined with a mercury barometer. The height of the mercury column balancing the air pressure is measured in millimeters. In accurate work, this is corrected to 0° C and to a standard gravity of 9.80665 m s^{-2}. Instead of reporting this in mm (of Hg), it is generally given in torrs, 1 *torr* being 1 mm (of Hg) under the standard conditions.

Example 3.1

Recall Boyle's law and interpret the E_{tr} in (3.15).

Boyle's law, an approximate law summarizing experimental results, states that the volume of a gas varies inversely with its pressure when the temperature is fixed:

$$PV = f(T).$$

Here T is the temperature on a suitable scale. On comparing this equation with (3.15), we see that the translational energy E_{tr} is a function of T alone in the ideal gas; thus

$$E_{tr} = \frac{3}{2} f(T).$$

Example 3.2

How many newtons per square meter are there in 1 atmosphere?

The standard 1 atmosphere pressure supports a mercury column 760.00 mm high at 0 °C, at a place where the acceleration due to gravity is 9.80665 m s^{-2}. The density of the mercury is 13595.1 kg m^{-3}.

Multiplying the mass per unit cross section of column by g, to convert the force, we have

$$P = (0.7600 \text{ m})(13595.1 \text{ kg m}^{-3})(9.80665 \text{ m s}^{-2}) = 101325 \text{ N m}^{-2}.$$

The newton per square meter is often called the *pascal* (Pa).

A most useful alternate standard is obtained on rounding off this value. The *bar* is defined as the pressure 10^5 Pa. Thus, it equals 750.06 torr.

Example 3.3

How many joules are there in 1 liter atmosphere?

In 1 liter, we have

$$V = (10^3 \text{ cm}^3)(10^{-2} \text{ m cm}^{-1})^3 = 10^{-3} \text{ m}^3.$$

Combining this with the result from example 3.2 gives us

$$PV = 101.325 \text{ J}.$$

Similarly, in 1 liter bar, we have 100 J; in 1 cm^3 bar, 0.1 J.

3.3 Temperature

Material systems consist of molecules, together with atoms, radicals, ions, free electrons. These are continually being agitated and churned. In a gas, the particle units travel more or less freely. In a liquid, the key ones oscillate about moving points; in a solid, about fixed points.

At an interface between systems, the particle units on one side interact with those on the other side. A molecule striking a wall with a high normal velocity and kinetic energy tends to lose some of the corresponding energy to a sluggishly moving low energy wall molecule. And a slowly moving attacking molecule tends to gain some kinetic energy from a fast moving high energy wall molecule. In the long run, the kinetic energy tends to become distributed over the different interacting degrees of freedom as randomly as possible.

The net energy transferred by such random processes across surfaces is called *heat*. The random motion of the particle units is called *thermal agitation*. When the energy of thermal agitation in contacting bodies has become distributed as randomly as possible, heat can no longer flow, on the average, and the bodies are said to have reached the same *temperature*. This induction is embodied in the *zeroth law of thermodynamics*: Whenever two or more bodies contact each other, heat flows between them until their temperatures become equal. One then says that thermal equilibrium has been established.

But what is temperature? From our standpoint, temperature is a statistical property to be defined in terms of the average behavior of molecules.

Consider a mixture of two gases A and B in which the molecules do not interact appreciably except at collisions. Such a mixture is said to be *ideal*.

For gas A in the mixture, equation (3.15) yields

$$\frac{P_A V}{N_A} = \frac{2(E_{tr})_A}{3N_A}, \tag{3.16}$$

while for gas B, we have

$$\frac{P_B V}{N_B} = \frac{2(E_{tr})_B}{3N_B}. \tag{3.17}$$

Here P_A and P_B are the partial pressures of A and B in the mixture, while N_A and N_B are the number of molecules of A and B and $(E_{tr})_A$ and $(E_{tr})_B$ are the translational kinetic energies of gas A and gas B in the mixture.

At equilibrium, the average pressure exerted by a molecule of A equals the average pressure exerted by a molecule of B:

$$\frac{P_A}{N_A} = \frac{P_B}{N_B} = \frac{P}{N}. \tag{3.18}$$

Here P and N are the total pressure and the total number of molecules. Solving for the partial pressures leads to

$$P_A = \frac{N_A}{N}P = X_A P, \qquad P_B = \frac{N_B}{N}P = X_B P, \tag{3.19}$$

where

$$\frac{N_A}{N} = X_A, \qquad \frac{N_B}{N} = X_B, \tag{3.20}$$

by definition. Quantities X_A and X_B give the mole fractions of A and B in the mixture.

Combining (3.18) with (3.16) and (3.17) yields the equality

$$\frac{(E_{tr})_A}{N_A} = \frac{(E_{tr})_B}{N_B}. \tag{3.21}$$

At equilibrium, the average translational energy of a molecule of A equals the average translational energy of a molecule of B.

Gases A and B are at the same temperature since they are intimately mixed. But there is no reason why the translational energy should be distributed differently were the gases separated by a heat conducting membrane. Consequently, two ideal gases are said to be at the same temperature when they exhibit the same average translational energy per molecule, E_{tr}/N. When they are at different temperatures, E_{tr}/N differs.

One can thus take E_{tr}/N as a measure of temperature T. By convention, a simple proportionality is assumed; so

$$\frac{E_{tr}}{N} = \frac{3}{2}kT. \tag{3.22}$$

Quantity k is called *Boltzmann's constant.* The gas constant R equals Avogadro's number times k, by definition; while N equals Avogadro's number times the number of moles n. So (3.22) may be rewritten in the form

$$\frac{E_{tr}}{n} = \frac{3}{2}RT. \quad [3.23]$$

Eliminating E_{tr} from (3.15) and (3.23) leads to the *ideal gas law*

$$PV = nRT. \quad [3.24]$$

A common laboratory thermometer consists of a glass bulb containing mercury attached to a uniform capillary tube. In operation, the surface of the mercury moves up or down the tube as the mercury expands or contracts more than the glass. The distance the column moves on heating the bulb from the freezing point of water to the boiling point at 1 atm pressure is determined. One hundredth this distance is taken as the *celsius* unit C and a linear scale is constructed. The freezing point of water is labeled 0° C; the boiling point at 1 atm, 100° C.

By experiment, it is found that the volume of a gas varies linearly with the mercury-in-glass celsius temperature t to a good approximation,

$$V = k\left(t + T_0\right), \quad [3\ 25]$$

when the pressure P of the gas is kept fixed at a low value. But equation (3.24) rearranges to

$$V = \frac{nR}{P}T. \quad [3.26]$$

Thus, the ideal-gas temperature T varies linearly with the t in (3.25). If we take the unit of temperature the same, we have

$$T = t + T_0. \quad [3.27]$$

The unit for this T is called the *kelvin* K.

3.4 Constants R and T_0

The gas constant R and the ice point T_0 can be determined from limiting measurements on gases.

The pressure-volume product for a real gas deviates from (3.24) because of interactions among its molecules. But whenever its pressure is lowered, the average distance between the molecules increases and this interaction decreases. As the pressure goes to zero, the behavior of the gas approaches that of an ideal gas. At a given temperature,

$$\lim_{P\to 0}\frac{PV}{n} = \left(\frac{PV}{n}\right)_{\text{ideal}}. \quad [3.28]$$

Combining this result with (3.24) yields

$$\lim_{P\to 0}\frac{PV}{n} = RT. \quad [3.29]$$

Here P is the observed pressure, V the measured volume, n the number of moles of the gas, and T the kelvin temperature.

In practice, this limit is determined at standard temperatures for such gases as nitrogen, hydrogen, helium, and neon. It is found that

$$T_0 = 273.15 \text{ K} \tag{3.30}$$

and that

$$R = 0.0820578 \text{ l atm K}^{-1} \text{ mol}^{-1} \tag{3.31}$$

or

$$R = 8.31451 \text{ J K}^{-1} \text{ mol}^{-1}. \tag{3.32}$$

Example 3.4

Change the energy unit in R (a) to the cubic centimeter bar, (b) to the calorie.
(a) Since the cubic centimeter bar equals 0.1 J, we have

$$R = \frac{8.31451}{0.1} \text{ cm}^3 \text{ bar K}^{-1} \text{ mol}^{-1} = 83.1451 \text{ cm}^3 \text{ bar K}^{-1} \text{ mol}^{-1}.$$

(b) A *calorie* is the energy needed to raise the temperature of 1 g water from 14.5° C to 15.5° C. In conversions, it is taken as 4.18400 J. So from (3.32), we get

$$R = \frac{8.31451 \text{ J K}^{-1} \text{ mol}^{-1}}{4.18400 \text{ J cal}^{-1}} = 1.98722 \text{ cal K}^{-1} \text{ mol}^{-1}.$$

Example 3.5

Calculate the root-mean-square velocity of a helium gas molecule at 20° C.
Solve equation (3.13) for the mean-square velocity:

$$\overline{u^2} = \frac{3PV}{Nm}.$$

Introduce formula (3.24) for PV and the molecular mass M for Nm/n:

$$\overline{u^2} = \frac{3RT}{M}.$$

Introduce the value of R in joules per degree per mole, of T in kelvins, and of M in kilograms per mole:

$$\overline{u^2} = \frac{3\left(8.31451 \text{ J K}^{-1} \text{ mol}^{-1}\right)\left(293.15 \text{ K}\right)}{0.0040026 \text{ kg mol}^{-1}} = 1.827 \times 10^6 \text{ m}^2 \text{ s}^{-2}.$$

Finally, take the square root

$$\left(\overline{u^2}\right)^{1/2} = 1.35 \times 10^3 \text{ m s}^{-1}.$$

3.5 Gaseous Solutions

Two or more gases can be mixed in any proportions to form a uniform solution. In the solution, the molecules generally interact as in a pure gas.

A solution in which this interaction is negligible, except insofar as it promotes randomness, is said to be *ideal*. For the *i*th gas in such a solution, we have

$$P_i V = n_i RT, \quad [3.33]$$

where P_i is the partial pressure exerted by the n_i moles of gas i in the volume V at temperature T.

Summing (3.33) over all gases in the solution yields

$$\sum P_i V = \sum n_i RT. \quad [3.34]$$

Since the sum of the partial pressures is the total pressure P and the sum of the partial moles the total moles n, equation (3.34) is equivalent to

$$PV = nRT. \quad [3.35]$$

But dividing (3.33) by (3.35) gives us

$$\frac{P_i}{P} = \frac{n_i}{n}, \quad [3.36]$$

whence

$$P_i = \frac{n_i}{n} P = X_i P. \quad [3.37]$$

In the last step, the *i*th mole fraction n_i/n is replaced by the symbol X_i.

This result checks (3.19), which was derived with the assumption that the average pressure caused by a molecule of A in volume V equals the average pressure caused by a molecule of B in that volume, under ideal conditions.

Example 3.6

Two moles N_2, ten moles O_2, and three moles CO_2 are mixed. Calculate the partial pressure of each component when the total pressure is 720 torr.

Apply equation (3.37) to each constituent. For the nitrogen,

$$P_{N_2} = \frac{2.0}{15.00} 720 \text{ torr} = 96 \text{ torr},$$

for the oxygen,

$$P_{O_2} = \frac{10.00}{15.00} 720 \text{ torr} = 480 \text{ torr},$$

and for the carbon dioxide,

$$P_{CO_2} = \frac{3.00}{15.00} 720 \text{ torr} = 144 \text{ torr}.$$

3.6 *Translational, Rotational, Vibrational Energies*

Each molecule in a gas executes various kinds of motion. These include translation of its center of mass, rotation about the center of mass, vibration in one or more modes. To describe translation in general requires three independent coordinates. To describe rotation of a linear molecule requires two additional coordinates; for a nonlinear molecule three independent additional coordinates are needed. The total number of independent coordinates for a molecule equals three times the number of atoms a. The number of vibrational modes equals $3a$ - 5 if the molecule is linear, $3a$ - 6 if it is not linear.

Each separate disjoint mode of motion corresponds to a *degree of freedom*. Molecular collisions distribute the available energy among the molecules and their degrees of freedom as randomly as possible. How much any one degree of freedom holds on the average is presumably determined by the mathematical expression for its energy.

The energy associated with translation of the ith molecule in the ±x direction is

$$\left(\varepsilon_x\right)_i = \frac{1}{2} m\dot{x}_i^2. \qquad [3.38]$$

For N molecules, we have the energy

$$E_x = \frac{1}{2} Nm\overline{\dot{x}^2} \qquad [3.39]$$

associated with this degree of freedom. Equations (3.8) and (3.24) transform this to

$$E_x = \frac{1}{2} nRT. \qquad [3.40]$$

Similarly, the energy associated with translation of the molecules in the $\pm y$ direction is

$$E_{\mathrm{y}} = \frac{1}{2} nRT, \qquad [3.41]$$

and that for translation in the $\pm z$ direction,

$$E_z = \frac{1}{2} nRT. \qquad [3.42]$$

The energy of rotation of the ith molecule at angular velocity $\dot{\theta}$ around an axis is

$$\left(\varepsilon_\theta\right)_{\mathrm{i}} = \frac{1}{2} I\dot{\theta}^2 \qquad [3.43]$$

where I is the moment of inertia about the axis. Since the expression for $(\varepsilon_\theta)_i$ is like that for (ε_x) we expect that the N identical molecules would have a rotational energy after randomnization like that for translation; thus

$$E_\theta = \frac{1}{2} nRT. \qquad [3.44]$$

For a linear molecule, we have two disjoint rotational modes. Then the rotational energy is

$$E_{\mathrm{rot}} = nRT. \qquad [3.45]$$

A nonlinear molecule has three disjoint rotational modes; then

$$E_{\mathrm{rot}} = \frac{3}{2} nRT. \qquad [3.46]$$

In classical mechanics, the energy associated with a simple harmonic vibration is

$$\left(\varepsilon_{\mathrm{vib}}\right)_i = \frac{1}{2}\mu\dot{q}_i^2 + \frac{1}{2}kq_i^2. \qquad [3.47]$$

Here μ is the reduced mass, k the force constant, and q_{i} the generalized coordinate for the mode. In the motion, the average potential energy over a cycle equals the average kinetic energy over the cycle. So we consider the energy associated with the last term to equal that associated with the preceding term. But this has the same form as (3.38), with which we associated 1/2 nRT of energy. We thus expect the total energy associated with one molecular vibrational degree of freedom to be

$$E_{\text{vib}} = nRT. \qquad [3.48]$$

Implicit in the derivations of this section is the assumption that a degree of freedom behaves as if it could absorb or emit any available amount of energy. So any quantization present must be small with respect to the average energy in the degree of freedom under consideration.

3.7 Quantum Restrictions

During an intermolecular collision, a degree of freedom can only be induced to shift from one quantized level to another. The energy involved equals the difference between the energy levels.

But from (3.22), the average kinetic energy in a classical degree of freedom at a given temperature T is

$$\varepsilon = \frac{1}{2}kT. \qquad [3.49]$$

For a degree of freedom to be excited classically, the separations between the pertinent energy levels must be small compared to this. Otherwise, in many of the encounters with more energetic molecules, the excess energy would not be large enough to cause excitation, and vice versa.

Translational eigenstates lie very close together. As a consequence, they are excited classically even at very low temperatures. The energy levels in rotational degrees of freedom are fairly close together; so generally they contribute classically. But at very low temperatures, the separation becomes significant for a molecule with a low moment of inertia, like H_2.

The separation between vibrational levels is generally larger, so that most molecular vibrations contribute less than nRT to E unless the temperature is high. The separation between electronic levels is usually so great that they do not contribute to the energy E at ordinary temperatures.

Example 3.7

Estimate the energy associated with thermal agitation in 1 mole water vapor at 25° C. The translational energy is given by equation (3.23):

$$E_{\text{tr}} = \frac{3}{2}nRT = \frac{3}{2}(1.000\ \text{mol})(8.3145\ \text{J K}^{-1}\ \text{mol}^{-1})(298.15\ \text{K}) = 3718\ \text{J}.$$

Since the H_2O molecule is nonlinear, the rotational energy is given by equation (3.46):

$$E_{\text{rot}} = \frac{3}{2}nRT = 3718\ \text{J}.$$

The three normal vibrations of H_2O have the wave numbers 1595, 3652, and 3756 cm^{-1}; so the least energy needed to excite a vibration is

$$E_{\min} = N_A h\nu = N_A hc\sigma = (0.119627\ \text{J m mol}^{-1})(1595 \times 10^2\ \text{m}^{-1}) = 19{,}080\ \text{J mol}^{-1}.$$

Since this is much larger than the average energy available for excitation, 1239 J, the vibrational degrees of freedom are essentially unexcited and

$$E_{\text{vib}} = 0 \text{ J}.$$

Similarly, not enough energy is available on the average to excite any electron.

Consequently, the energy associated with the thermal agitation is

$$E = E_{\text{tr}} + E_{\text{rot}} + E_{\text{vib}} + E_{\text{el}} = 3nRT = 7437 \text{ J}.$$

3.8 Allowing for Molecular Interactions

A gas situated so that either the volume occupied by the molecules is appreciable or the attraction between them must be considered is no longer ideal. Thus, equation (3.24) cannot account for the behavior at high pressures or at low temperatures. It cannot explain condensation. So let us introduce corrections for these omissions.

Consider 1 mole of gas, for which the ideal gas equation is

$$PV = RT. \qquad [3.50]$$

Let the volume which is not available for free movement be b. The V in (3.50) should now be replaced by $V - b$. When the gas is dilute enough, so that most encounters within are binary, parameter b is about four times the total volume of the molecules. See example 3.8.

Because of attraction between the molecules, each molecule about to strike a wall is pulled inward by a force proportional to d, the density of the gas doing the pulling. Furthermore, the number of molecules striking the wall per unit area per unit time is proportional to d. As a result, the pressure on the wall is lowered below the ideal value by an amount proportional to d^2. But d equals M/V, the molar mass divided by the volume of 1 mole. So the pressure lowering is set equal to a/V^2; and we substitute $P + a/V^2$ for the ideal P.

The ideal gas equation is thus transformed to

$$\left(P + \frac{a}{V^2}\right)\left(V - b\right) = RT, \qquad [3.51]$$

the *van der Waals equation*. Parameters a and b chosen to fit the behavior of various gases are listed in table 3.1. But in each case, compromises have been made. For each substance, a and b vary considerably, especially as the condensation region is approached.

In the next section, we will see that equation (3.51) is cubic in volume V. So it has three roots. Above a *critical temperature* T_c only one of these is real. But below T_c, all three roots are real. The highest root applies to the gas phase; the lowest one, to the liquid phase in equilibrium with it. Below the critical temperature, a phase of small molar volume V_1, a liquid phase, can exist in equilibrium with a phase of larger molar volume V_3, the gaseous phase. Above the critical temperature, no such equilibrium can exist; there we need postulate only one phase, the gaseous. The critical temperature is the highest temperature at which a distinct liquid phase can exist.

In studying a given substance, one might seal a sufficient amount of the substance in a glass tube so that both liquid and gas phases are present. Then one would heat it until the meniscus between the two phases disappears and their densities become equal. The temperature at which this happens is the experimental critical temperature T_c, the corresponding pressure is the critical pressure P_c, the corresponding molar volume is the critical volume V_c.

TABLE 3.1 Critical and van der Waals Constants

Gas	T_c, *K*	P_c, *atm*	V_c, *ml mol*$^{-1}$	*a*, l^2 *atm mol*$^{-1}$	*b*, *ml mol*$^{-1}$
He	5.3	2.26	57.8	0.0353	24.1
H_2	33.3	12.8	65.0	0.246	26.7
N_2	126.2	33.5	90.1	1.350	38.6
CO	133	34.5	93.1	1.456	39.5
O_2	154.8	50.1	78	1.359	31.7
CO_2	304.2	72.9	94.0	3.61	42.8
HCl	324.6	81.5	48	3.67	40.9
NH_3	405.5	111.3	72.5	4.20	37.4
Cl_2	417	76.1	124	6.49	56.2
CH_3OH	513.2	78.5	118	9.53	67.1
C_2H_5OH	516	63.0	167	12.00	84.0
CCl_4	556.4	45.0	276	19.54	126.8
C_6H_6	562	48.6	260	18.46	118.6
H_2O	647.4	218.3	56	5.45	30.4

Example 3.8

Calculate the radius of the helium atom from its van der Waals constant *b*.

The helium molecule contains a single atom. Consider each of these as a hard sphere of radius *r*. Then during a collision, the center of the second molecule gets to within distance $2r$ from the center of the first. Volume $(4/3)\pi(2r)^3$ is thus excluded about the center of the first molecule.

Dividing this excluded volume equally between the two molecules leads to

$$\frac{1}{2}\frac{4}{3}\pi\left(2r\right)^3 = 4\frac{4}{3}\pi r^3 = \frac{16}{3}\pi r^3$$

excluded volume per molecule. The total for Avogadro's number of molecules is N_A times as much, Setting this equal to *b*,

$$N_A\frac{16}{3}\pi r^3 = b,$$

solving for *r*,

$$r = \left(\frac{3b}{16\pi N_A}\right)^{1/3},$$

and introducing the pertinent numbers yields

$$r = \left[\frac{3\left(24.1\text{ cm}^3\text{ mol}^{-1}\right)}{16\left(3.1416\right)\left(6.022\times10^{23}\text{ mol}^{-1}\right)}\right]^{1/3} = 1.34\times10^{-8}\text{ cm} = 1.34\text{ Å}.$$

3.9 *Gas—Liquid Isotherms*

Multiplying van der Waals equation by V^2/P and rearranging leads to the form

$$V^3 - \left(b + \frac{RT}{P}\right)V^2 + \frac{a}{P}V - \frac{ab}{P} = 0. \qquad [3.52]$$

At a given temperature T and pressure P, this is a cubic equation in volume V. At high temperatures, only one root is real. But at a critical temperature and a critical pressure, the complex roots become real and equal. Going below this critical point on a pressure-volume plot, the three real roots gradually spread out. Equation (3.52) can be rewritten in the form

$$(V - V_1)(V - V_2)(V - V_3) = 0, \qquad [3.53]$$

in which V_1, V_2, V_3 are the three roots. At the critical point, where

$$V_1 = V_2 = V_3 = V_c, \qquad [3.54]$$

Equation (3.53) reduces to

$$(V - V_c)^3 = 0, \qquad [3.55]$$

whence

$$V^3 - 3V_cV^2 + 3V_c^2V - V_c^3 = 0. \qquad [3.56]$$

Comparing (3.52) with $T = T_c$ and $P = P_c$ to (3.56) yields

$$3V_c = b + \frac{RT_c}{P_c}, \qquad [3.57]$$

$$3V_c^2 = \frac{a}{P_c}, \qquad [3.58]$$

$$V_c^3 = \frac{ab}{P_c}. \qquad [3.59]$$

Solving for a, b and R gives us

$$a = 3P_cV_c^2, \qquad [3.60]$$

$$b = \frac{V_c}{3}, \qquad [3.61]$$

$$R = \frac{8P_cV_c}{3T_c}. \qquad [3.62]$$

With these formulas, one can calculate a and b from the critical constants P_C and V_C for a given substance. Alternatively, one can substitute experimental values of P, V, and T for a particular region into (3.52) and solve for a and b. Because they vary, these parameters should be determined for the temperature-pressure region to be represented.

Experimental data for a particular substance are plotted in figure 3.2. Each curve shows how the pressure varies with volume at a given temperature. Because the temperature is fixed along it, each curve is called an *isotherm*.

In Figure 3.3, van der Waals isotherms for the same material are plotted. Note the general similarity except in the liquefaction region. There the van der Waals equation yields S-shaped curves and three real roots. A dotted line has been drawn across each isotherm so that the highest root V_3 gives the molar volume of the gas and the lowest root V_1 the molar volume of the liquid in equilibrium with it. Note that by compressing the gas, one can go below point V_3 on the S-shaped curve. Also by expanding the liquid, one can go above point V_1 on this curve. However, the section where $(\partial P/\partial V)_T$ is positive is not accessible.

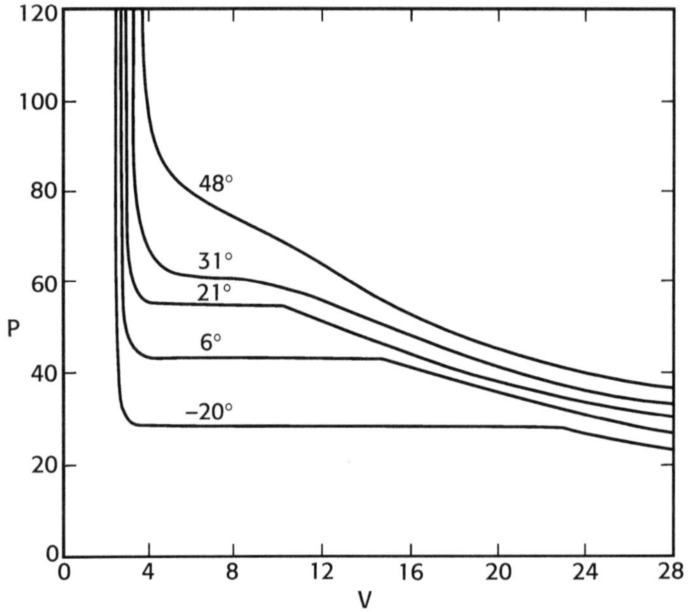

FIGURE 3.2 Experimental isotherms for a pure substance.

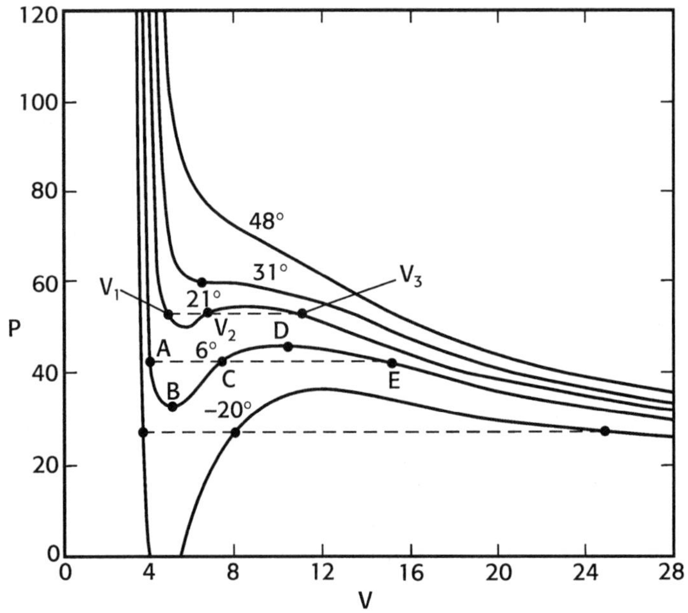

FIGURE 3.3 Isotherms for the same substance calculated using the van der Waals equation.

On each isotherm, the dotted line is drawn so that it cuts off equal areas above and below. Thus on the labeled curve, area ABC equals area CDE.

The work done on the system when it is taken reversibly from one point on an isotherm to another equals the negative of the area under the curve $-\int P\,dV$. The equal-area rule ensures that the area under the van der Waals curve between the liquid and the gaseous states equals the area under the dotted straight line joining those states, Thus, the work done would be the same. Along the van der Waals curve, the system would go homogeneously and reversibly; along the dotted line, it would go heterogeneously and reversibly.

Why the two works should be equal will be considered later.

3.10 Corresponding States

Describing mathematically the variation in a and b for various substances is beyond the scope of this book. Instead, we will merely investigate how van der Waals equation can be generalized to allow for some of the variation.

First, consider that a, b, and R are constant and given by (3.60), (3.61), (3.62). Substituting into (3.51) yields

$$\left(P+\frac{3P_{\mathrm{c}}V_{\mathrm{c}}{}^{2}}{V^{2}}\right)\left(V-\frac{V_{\mathrm{c}}}{3}\right)=\frac{8P_{\mathrm{c}}V_{\mathrm{c}}T}{3T_{\mathrm{c}}}, \qquad [3.63]$$

which can be reduced to the dimensionless equation

$$\left(P_{\mathrm{r}}+\frac{3}{V_{\mathrm{r}}^{2}}\right)\left(V_{\mathrm{r}}-\frac{1}{3}\right)=\frac{8}{3}T_{\mathrm{r}}. \qquad [3.64]$$

Here *reduced pressure* P_{r} is the ratio P/P_{c} , reduced volume V_{r} is the ratio V/V_{c}, and *reduced temperature* T_{r} is the ratio T/T_{c}.

In this form of van der Waals equation, all references to a specific substance have dropped out. The reduced volume is a universal function of the reduced pressure and reduced temperature. Also from van der Waals equation, we have the *critical ratio*

$$\frac{RT_{\mathrm{c}}}{P_{\mathrm{c}}T_{\mathrm{c}}}=\frac{8}{3}=2.67. \qquad [3.65]$$

As a generalization, let the reduced volume V_{r} be an empirical function of the reduced pressure P_{r} and the reduced temperature T_{r},

$$V_{\mathrm{r}}=V_{\mathrm{r}}\left(P_{\mathrm{r}},T_{\mathrm{r}}\right), \qquad [3.66]$$

and let the critical ratio be an empirical constant,

$$\frac{RT_{\mathrm{c}}}{P_{\mathrm{c}}V_{\mathrm{c}}}=\text{const.} \qquad [3.67]$$

Conditions (3.66) and (3.67) constitute a *law of corresponding states*. According to it, different substances at the same reduced pressure and reduced temperature correspond; they have the same reduced volume. We consider the law to apply to both gas and liquid phases. The constituent molecules must not be too abnormal in their shapes or interactions, however.

For normal substances, it is found that

$$\frac{RT_{\mathrm{c}}}{P_{\mathrm{c}}V_{\mathrm{c}}}\cong 3.75. \qquad [3.68]$$

Substances with low boiling points, such as hydrogen and helium, have low critical ratios, while substances that associate in the liquid state, such as ethyl alcohol and water, have high ratios.

3.11 Compressibility Factor for a Gas

Instead of plotting experimental V_r against P_r and T_r, to represent (3.66), one may conveniently plot PV/RT against these independent variables. This ratio is known as the *compressibility factor.*

First, we note that a general equation of state can be constructed by introducing a factor z on the right side of the ideal-gas equation:

$$PV = znRT. \qquad [3.69]$$

Solving for z with n equal to one mole leads to

$$z = \frac{PV}{RT} = \frac{P_c V_c}{RT_c} \frac{P_r V_r}{T_r}. \qquad [3.70]$$

But according to the law of corresponding states, the first fraction on the right is constant while the second fraction is a function of the reduced pressure and reduced temperature. Thus, z itself is a function of P_r and T_r.

In figure 3.4, experimental data on many gases have been summarized. Even a substance like water that has an abnormal critical ratio is described quite well by these curves. But for hydrogen and helium, poor results are obtained. For these substances, adding 8 atm to P_c and 8° to T_c before calculating P_r and T_r improves the fit.

3.12 Virial Equations of State

On the other hand, one may represent compressibility factor z as a power series (a) in pressure P, (b) in reciprocal volume $1/V$. The coefficients may be determined from

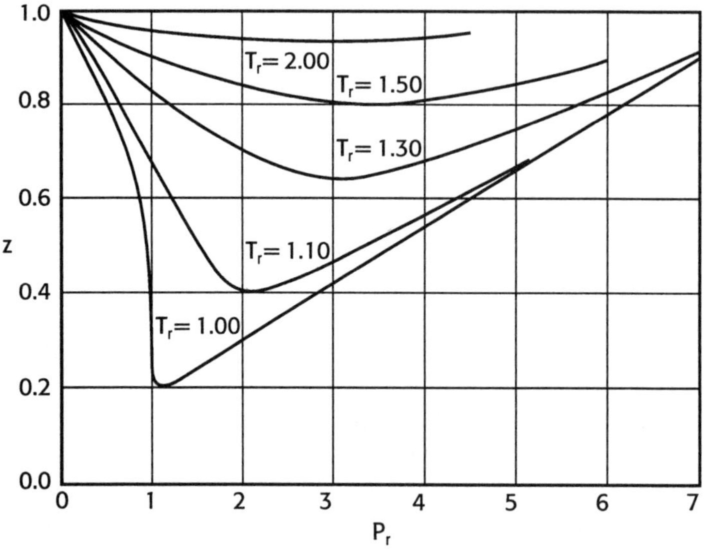

FIGURE 3.4 Empirical variation of the compressibility factor z with the reduced pressure P_r and the reduced temperature T_r.

pressure-volume-temperature measurements or from molecular parameters using statistical mechanics.

A suitable analytic form for the compressibility factor for a given gas is

$$z = \frac{PV}{RT} = 1 + A_2(T)P + A_3(T)P^2 + \ldots. \quad [3.71]$$

An alternate form for the gas is

$$z = \frac{PV}{RT} = 1 + \frac{B_2(T)}{V} + \frac{B_3(T)}{V^2} + \ldots. \quad [3.72]$$

Formulas (3.71) and (3.72) are called *virial equations of state*; expressions $A_2(T)$, $A_3(T)$, ... and $B_2(T)$, $B_3(T)$, ... are called *virial coefficients.*

When the series for different substances are compared at the same reduced temperature and reduced pressure, the coefficients are similar, following the law of corresponding states. The principal exceptions are hydrogen and helium at low temperatures, where quantum mechanical effects become significant.

Around 1900, D. Berthelot modified the van der Waals equation to mimic the behavior of gases at low pressures. Thus, he constructed the empirical equation

$$PV = RT\left[1 + \frac{9}{128}\frac{P_r}{T_r}\left(1 - \frac{6}{T_r^2}\right)\right]. \quad [3.73]$$

This yields the virial coefficients

$$A_2 = \frac{9}{128}\frac{1}{P_c T_r}\left(1 - \frac{6}{T_r^2}\right), \quad [3.74]$$

$$A_j = 0 \quad \text{for} \quad j = 3, 4, \ldots. \quad [3.75]$$

Formula (3.73) is known as the *Berthelot equation of state*; it is used for calculating deviations from ideality at low pressures.

3.13 Distribution of Equivalent Particles in a Potential Field

In the usual laboratory, all systems are in the gravitational field of the earth. At the edge of a plasma, a considerable electric field exists. About an ion, a fairly long range electric field may exist. The gravitational field acts on all particles; the electric field, on charged particles. In any case, the spatial distribution of the particles is affected.

Consider a gaseous mixture at temperature T subjected to a potential gradient. For simplicity, neglect the van der Waals interactions. The partial pressure P_i of the ith species at a given level of potential is then given by equation (3.33). Letting n_i/V be concentration c_i of the species, we have

$$P_i = c_i RT \quad [3.76]$$

and

$$\mathrm{d}P_i = RT\,\mathrm{d}c_i. \quad [3.77]$$

Let the potential of a molecule or ion of the ith species at a certain level be ε_i. The force acting in the z direction on the particle is then $-\partial\varepsilon_i/\partial z$. Multiplying this by the number of ith particles per unit area in thickness dz then gives the net pressure due to the change in potential. We have

$$dP_i = c_i N_A \left(-\frac{\partial \varepsilon_i}{\partial z}\right) dz = -c_i N_A \, d\varepsilon_i. \qquad [3.78]$$

Combining equations (3.77) and (3.78) leads to

$$RT \, dc_i = -c_i N_A \, d\varepsilon_i \qquad [3.79]$$

or

$$\frac{dc_i}{c_i} = -\frac{d\varepsilon_i}{kT} \qquad [3.80]$$

where k is the Boltzmann constant R/N_A. Integrating equation (3.80) from potential energy 0 to potential energy ε_i yields

$$c_i = c_{i0} e^{-\varepsilon_i/kT}. \qquad [3.81]$$

Since c_i and c_{i0} are proportional to the number of molecules of the ith species at levels ε_i and ε_{i0}, N_i and N_{i0} respectively, equation (3.81) implies that

$$N_i = N_{i0} e^{-\varepsilon_i/kT}. \qquad [3.82]$$

From the way it has been derived, we see that equation (3.82) gives us the *equilibrium* distribution of molecules in a potential field. In the next section, we will derive an equivalent form for the distribution over kinetic energy.

The form encompassing both potential and kinetic energies is known as the *Boltzmann distribution law*.

Example 3.9

A gold sol containing particles averaging 4.5×10^{-7} cm in radius reached equilibrium at 25° C. With an ultramicroscope, 89 particles were observed in 32 successive counts at a certain level. How many particles should be observed in 32 successive counts in a layer 0.200 cm higher?

For the volume of a typical particle, we have

$$V = \frac{4}{3}\pi r^3 = \frac{4}{3}\pi\left(4.5 \times 10^{-7} \text{ cm}\right)^3 = 3.82 \times 10^{-19} \text{ cm}^3.$$

Since the water buoys up the particle, its effective density equals its density in air minus the density of water:

$$\rho' = \rho_{\text{gold}} - \rho_{\text{water}} = 19.3 - 1.0 \text{ g cm}^{-3} = 18.3 \text{ g cm}^{-3}.$$

So the gravitational force acting on the particle is

$$F = \frac{18.3 \text{ g cm}^{-3}}{1000 \text{ g kg}^{-1}}\left(3.82 \times 10^{-19} \text{ cm}^3\right)\left(9.807 \text{ N kg}^{-1}\right) = 6.85 \times 10^{-20} \text{ N}.$$

On moving 0.200 cm up, its potential energy increases by

$$\varepsilon = \left(6.85 \times 10^{-20} \text{ N}\right)\left(0.200 \times 10^{-2} \text{ m}\right) = 1.37 \times 10^{-22} \text{ J}.$$

Substituting this energy and the given temperature into the Boltzmann distribution law tells us that

$$\ln\frac{N}{N_0} = -\frac{\varepsilon}{kT} = -\frac{1.37\times10^{-22}\ \text{J}}{\left(1.3807\times10^{-23}\ \text{J K}^{-1}\right)\left(298.15\ \text{K}\right)} = -3.33\times10^{-2}$$

whence

$$\frac{N}{N_0} = 0.967$$

and

$$N = \left(0.967\right)\left(89\right) = 86.$$

3.14 The Maxwell Distribution Law

Before Boltzmann, Maxwell employed a symmetry argument to construct a distribution law for the translational energy of molecules in an ideal gas.

The fundamental idea was that collisions between gas molecules establish a statistical distribution. Thus, there is no preferred direction in absence of an externally imposed field. Furthermore, each component of velocity of a molecule is independent of its orthogonal components.

Consider a pure ideal gas at equilibrium at temperature T and volume V. Let the probability that the rectangular components of the velocity of a molecule are between v_x and $v_x + dv_x$, between v_y and $v_y + dv_y$, and between v_z and $v_z + dv_z$ be $f(v_x)\ dv_x$, $f(v_y)\ dv_y$, $f(v_z)\ dv_z$, respectively.

Since these are independent, the joint probability $f(v_x, v_y, v_z)\ dv_x\ dv_y\ dv_z$ equals the product of the three probabilities:

$$f\left(v_x, v_y, v_z\right) dv_x\ dv_y\ dv_z = f\left(v_x\right) f\left(v_y\right) f\left(v_z\right) dv_x\ dv_y\ dv_z. \qquad [3.83]$$

Because of the assumed isotropy in space, the individual probabilities depend only on the pertinent magnitudes and so on the pertinent squares. Thus, one may rewrite equation (3.83) in the form

$$\phi\left(v_x^2 + v_y^2 + v_z^2\right) dv_x\ dv_y\ dv_z = \phi\left(v_x^2\right)\phi\left(v_y^2\right)\phi\left(v_z^2\right) dv_x\ dv_y\ dv_z. \qquad [3.84]$$

Equation (3.84) is satisfied when the independent variables in function ϕ are the powers of a common base. Without loss of generality, one may set

$$f\left(v_x\right) = \phi\left(v_x^2\right) = Ae^{-av_x^2}. \qquad [3.85]$$

Since each molecule must have an x-component of velocity, we see that

$$\int_{-\infty}^{\infty} f\left(v_x\right) dv_x = 1. \qquad [3.86]$$

Substituting form (3.85) into equation (3.86) and introducing the result from example 10.5 yields

$$A\int_{-\infty}^{\infty} e^{-av_x^2}\ dv_x = A\left(\frac{\pi}{a}\right)^{1/2} = 1, \qquad [3.87]$$

whence

$$A = \left(\frac{a}{\pi}\right)^{1/2}. \tag{3.88}$$

For the mean square velocity, we now get

$$\overline{v_x^2} = \int_{-\infty}^{\infty} v_x^2 \, f(v_x)\,\mathrm{d}v_x = \left(\frac{a}{\pi}\right)^{1/2} \int_{-\infty}^{\infty} v_x^2 \, e^{-av_x^2} \, \mathrm{d}v_x = \left(\frac{a}{\pi}\right)^{1/2} \frac{1}{2a}\left(\frac{\pi}{a}\right)^{1/2} = \frac{1}{2a}. \tag{3.89}$$

But from formulas (3.8) and (3.24), we have

$$\overline{v_x^2} = \frac{PV}{Nm} = \frac{nRT}{Nm} = \frac{kT}{m}. \tag{3.90}$$

Combining (3.89) and (3.90) leads to

$$a = \frac{1}{2\overline{v_x^2}} = \frac{m}{2kT}. \tag{3.91}$$

So distribution law (3.85) becomes

$$f(v_x) = \left(\frac{m}{2\pi kT}\right)^{1/2} e^{-mv_x^2/2kT}, \tag{3.92}$$

while equation (3.83) leads to

$$f(v_x, v_y, v_z) = \left(\frac{m}{2\pi kT}\right)^{3/2} e^{-mv^2/2kT}. \tag{3.93}$$

The probability $f(v)$ dv that a molecule has a speed between v and v + dv equals the sum of the probabilities that it is in any volume element in the spherical shell of radius v and thickness dv. The sum of these volume elements is $4\pi v^2$dv. So summing (3.93) over these volume elements yields

$$f(v) = 4\pi\left(\frac{m}{2\pi kT}\right)^{3/2} v^2 e^{-mv^2/2kT}. \tag{3.94}$$

Equations (3.92), (3.93), and (3.94) are forms of the *Maxwell distribution law*. Since $1/2mv_x^2$ and $1/2mv^2$ are the pertinent kinetic energies and the preexponential factors are normalized so that the fs are probabilities, these forms agree with equation (3.82). This equation holds regardless of the kind of energy ε_i.

Example 3.10

Evaluate the integral

$$\int_0^{\infty} x^2 e^{-ax^2} \, \mathrm{d}x.$$

Carry out an integration by parts:

$$\int_0^{\infty} x^2 e^{-ax^2} \, \mathrm{d}x = -\frac{1}{2a}\int_0^{\infty} x e^{-ax^2}\left(-2ax \, \mathrm{d}x\right) = -\frac{1}{2a} x e^{-ax^2}\Big|_0^{\infty} + \frac{1}{2a}\int_0^{\infty} e^{-ax^2} \, \mathrm{d}x = 0 + \frac{1}{4a}\left(\frac{\pi}{a}\right)^{1/2}$$

In the last step, the result from example 10.5 has been introduced. Since the integrand is even, we also have

$$\int_{-\infty}^{\infty} x^2 e^{-ax^2}\,dx = \frac{1}{2a}\left(\frac{\pi}{a}\right)^{1/2}$$

This formula was used in line (3.89).

Example 3.11

Evaluate the integral

$$\int_0^{\infty} x^3 e^{-ax^2}\,dx.$$

Again, integrate by parts:

$$\int_0^{\infty} x^3 e^{-ax^2}\,dx = -\frac{1}{2a}\int_0^{\infty} x^2 e^{-ax^2}\left(-2ax\,dx\right) = -\frac{1}{2a}x^2 e^{-ax^2}\Big|_0^{\infty} - \frac{1}{2a^2}\int_0^{\infty} e^{-ax^2}\left(-2ax\,dx\right)$$

$$) = 0 - \frac{1}{2a^2}e^{-ax^2}\Big|_0^{\infty} = \frac{1}{2a^2}.$$

Example 3.12

Calculate the most probable speed of a molecule in an ideal gas.

The most probable speed is the speed at which the probability density $f(v)$ is a maximum. But the derivative of expression (3.94),

$$\frac{df}{dv} = 4\pi\left(\frac{m}{2\pi kT}\right)^{3/2}\left(2v - \frac{m}{kT}v^3\right)e^{-mv^2/2kT},$$

vanishes when

$$v^2 = \frac{2kT}{m}$$

and

$$v = \left(\frac{2kT}{m}\right)^{1/2}.$$

3.15 The Collision Rate Density

For two molecules to react, they must get together with the proper orientations and with sufficient energy. From the theory developed so far, we can estimate the collision rate per unit volume in a gas. The other factors will be considered later.

In Newtonian mechanics, the relative movement of two particles, of mass m_A and mass m_B, respectively, is modeled by a single particle of mass

$$\mu = \frac{m_A m_B}{m_A + m_B} \qquad [3.95]$$

at the interparticle distance r from the origin. Parameter μ is called the *reduced mass.* The relative speed of the particles

$$v = \frac{dr}{dt} \qquad [3.96]$$

is represented by the radial speed of the model particle. The argument follows the treatment in examples 11.1 and 12.1.

But for the mean speed of a molecule, we have

$$\bar{v} = \int_0^\infty v f(v)\, dv. \qquad [3.97]$$

Introducing Maxwell distribution law (3.94) and integrating, following example 3.11, yields

$$\bar{v} = 4\pi\left(\frac{m}{2\pi kT}\right)^{3/2} \int_0^\infty v^3 e^{-mv^2/2kT}\, dv = 4\pi\left(\frac{m}{2\pi kT}\right)^{3/2} \frac{1}{2}\left(\frac{2kT}{m}\right)^2 = \left(\frac{8kT}{\pi m}\right)^{1/2}. \qquad [3.98]$$

We consider the model particle to behave as a molecule of an ideal gas. So for the mean relative speed of molecules at equilibrium, we have

$$\bar{v}_{rel} = \left(\frac{8kT}{\pi\mu}\right)^{1/2} \qquad \text{with} \qquad \mu = \frac{m_A m_B}{m_A + m_B} \qquad [3.99]$$

With respect to molecule B, molecule A is moving on the average at speed v_{rel}. But for A to strike B, its center must hit within the area

$$\sigma = \pi\left(r_A + r_B\right)^2 \qquad [3.100]$$

centered on molecule B. Here r_A is the effective radius of molecule A, r_B the effective radius of molecule B. Recall figure 2.6.

In time Δt, molecule A will sweep out volume $\sigma v_{rel}\Delta t$. The average number of B's per unit volume is N_2/V. So the average number of collisions per unit time effected by one A molecule is $\sigma v_{rel} N_2/V$. But the number of A molecules per unit volume on average is N_1/V. So the mean collision rate density between A's and B's is

$$Z_{AB} = \sigma\left(\frac{8kT}{\pi\mu}\right)^{1/2} \frac{N_1}{V}\frac{N_2}{V}. \qquad [3.101]$$

If every A reacted on colliding with B we would have

$$-\frac{d[A]}{dt} = \sigma\left(\frac{8kT}{\pi\mu}\right)^{1/2} N_A\left(10^3\ \mathrm{l\,m^{-3}}\right)[A][B] = k_{AB}[A][B]. \qquad [3.102]$$

Here N_A is Avogadro's number while [A] and [B] are the molar concentrations of A and B and k_{AB} is the specific reaction rate.

When molecules A and B are identical, the above expressions count each collision twice. Furthermore, the reduced mass is

$$\mu = \frac{mm}{m+m} = \frac{m}{2}. \qquad [3.103]$$

So the mean collision rate density between molecules of the same kind is

$$Z_{AA} = \frac{1}{2}\sigma\left(\frac{8kT}{\pi m/2}\right)^{1/2}\left(\frac{N}{V}\right)^2 = \sigma\left(\frac{4kT}{\pi m}\right)^{1/2}\left(\frac{N}{V}\right)^2 \qquad [3.104]$$

If every A reacted on colliding with another A we would have

$$-\frac{1}{2}\frac{d[A]}{dt} = \sigma\left(\frac{4kT}{\pi m}\right)^{1/2} N_A\left(10^3\ \mathrm{l\,m^{-3}}\right)[A]^2 = k_{AA}[A]^2. \qquad [3.105]$$

Considering r_A and r_B to be definite implies that molecules A and B act as hard spheres. Nevertheless, equations (3.102) and (3.105) yield fair upper limits to actual specific reaction rates. Only some monatomic molecules are spherical. Even these are relatively soft. Furthermore, an energy barrier must be surmounted with the reactants properly oriented with respect to each other. These effects can greatly reduce the reaction rates, as we will see later.

Example 3.13

Calculate the cross section, the mean relative speed, and the collision rate density for molecular oxygen at 25° C and 1.000 bar pressure, where the effective radius of O_2 is 1.805 Å.

Substitute the effective radius into formula (3.100):

$$\sigma = \pi\left(3.61\times10^{-10}\ \mathrm{m}\right)^2 = 4.094\times10^{-19}\ \mathrm{m}^2.$$

Insert the temperature and the reduced mass into (3.99):

$$\bar{v}_{rel} = \left(\frac{8kT}{\pi m/2}\right)^{1/2} = \left(\frac{16RT}{\pi M}\right)^{1/2} = \left(\frac{16\left(8.31451\ \mathrm{J\,K^{-1}\,mol^{-1}}\right)\left(298.15\ \mathrm{K}\right)}{\pi\left(31.999\times10^{-3}\ \mathrm{kg\,mol^{-1}}\right)}\right)^{1/2} = 628.13\ \mathrm{m\,s^{-1}}.$$

For the number density, we have

$$\frac{N}{V} = \frac{P}{kT} = \frac{1.000\times10^5\ \mathrm{N\,m^{-2}}}{\left(1.3807\times10^{-23}\ \mathrm{J\,K^{-1}}\right)\left(298.15\ \mathrm{K}\right)} = 2.429\times10^{25}\ \mathrm{m^{-3}}.$$

Multiplying these as in line (3.104) leads to

$$Z_{AA} = \frac{1}{2}\sigma\bar{v}_{rel}\left(\frac{N}{V}\right)^2 = 7.588\times10^{34}\ \mathrm{m^{-3}\,s^{-1}}.$$

Example 3.14

Calculate the hypothetical maximum specific reaction rate k_{AA} for the process in example 3.13.

From formulas (3.104) and (3.105), we see that

$$k_{AA} = \frac{1}{2}\sigma\bar{v}_{rel}N_A\left(10^3 \text{ l m}^{-3}\right) = \frac{1}{2}\left(4.094 \times 10^{-19} \text{ m}^2\right)\left(628.13 \text{ m s}^{-1}\right)\left(6.0221 \times 10^{23} \text{ mol}^{-1}\right)\left(10^3 \text{ l m}^{-3}\right)$$

$$= 7.74 \times 10^{10} \text{ mol}^{-1} \text{ l s}^{-1} = 7.74 \times 10^{13} \text{ cm}^3 \text{ mol}^{-1} \text{ s}^{-1}.$$

Questions

3.1 What do we mean by collective behavior?
3.2 Define macroscopic (thermodynamic) state, extensive property, intensive property.
3.3 Cite examples of extensive and intensive properties.
3.4 What is a phase?
3.5 Distinguish between steady and equilibrium states.
3.6 Define ideal gas.
3.7 What role do collisions play in a gas?
3.8 How can we neglect collisions in deriving the law $PV = (2/3)\ E_{tr}$?
3.9 Define heat, thermal agitation.
3.10 When is E_{tr}/N a measure of temperature T?
3 11 How did we introduce the Boltzmann constant?
3 12 What determines the unit of temperature?
3.13 How may (a) R, (b) T_0 be measured?
3.14 What is the ideal gas law? How does it apply to a gaseous solution?
3.15 What is the energy associated with (a) a translational mode, (b) a rotational mode, (c) a vibrational mode, classically excited?
3.16 How do quantum restrictions alter these results?
3.17 Give an approximate justification for the van der Waals equation.
3.18 How are parameters a and b for the van der Waals equation determined?
3.19 How is the law of corresponding states a generalization of the van der Waals equation?
3.20 Define the virial coefficients.
3.21 Show how equivalent particles are distributed in a potential field at equilibrium.
3.22 On what symmetries is the Maxwell distribution law based?
3.23 Derive the Maxwell distribution law.
3.24 What physical processes lead to the distributions described by the Boltzmann and Maxwell laws?
3.25 Show how the collision rate density is determined.

Problems

3.1 What is the translational energy of the molecules in an ideal gas that fills 3.25 l at 0.932 atm?
3.2 Calculate the root-mean-square speed of a benzene gas molecule at 100° C.
3.3 What is the density of molecules in a vacuum of 1.00×10^{-3} torr at 25° C?
3.4 At 730 torr pressure and 27° C a given amount of oxygen filled 5.00 1. Calculate (a) the number of moles, (b) the number of molecules, and (c) the weight in the system.
3.5 A 1.000 l vessel at 20° C contains 2.50 g N_2, 1.05 g O_2, and 0.56 g CO_2. Calculate the partial pressure of each gas and the total pressure, assuming ideal conditions.
3.6 The following densities were obtained for methyl chloride:

Pressure, atm	1.0000	0.5000	0.2500
Density at 0° C, g l^{-1}	2.3074	1.1401	0.5666

Calculate the ratio of density to pressure at each pressure. Consider this ratio to be a linear function of pressure and extrapolate to zero pressure. From the result, calculate the molecular mass of methyl chloride.

3.7 (a) What is the rotational energy of 1.000 g oxygen molecules at 25° C? (b) What is the rotational energy of 1,000 mol benzene molecules at 100° C?

3.8 What is the molecular diameter of a sulfur molecule, if its critical temperature is 1313 K and its critical pressure 116 atm?

3.9 Calculate the volume occupied by 1.000 mol water at 500° C and 200 atm pressure using (a) the ideal gas equation (b) the compressibility factor from figure 3.4.

3.10 For molecular nitrogen at 25° C, calculate (a) the cross section, (b) the mean relative speed, and (c) the collision rate density at 1.000 bar. Consider the effective radius of N_2 to be 1.90 Å under the given conditions.

3.11 Calculate the hypothetical. maximum specific reaction rate k_{AA} for the process in problem 3.10.

— — —

3.12 A strip of ideal rubber has the equation of state

$$F = T\phi(L)$$

where F is the force needed to stretch the strip to length L at absolute temperature T while $\phi(L)$ represents a function of L. If 100 g force stretches a rubber band 1.00 cm at 0° C, how much force is needed to cause the same elongation at 35° C?

3.13 At what temperature is the root-mean-square speed of an O_2 molecule the same as that of an H_2 molecule at -90° C? What is this speed?

3.14 Assume that the rate of flow of a gas through a small hole is nearly proportional to the mean speed of its molecules and that the mean speed is proportional to the root- mean-square speed. Calculate how long it will take for a millimole of He to flow through a pinhole under the same driving pressure and temperature used to force a millimole of H_2 through in 10.0 min.

3.15 What is the translational energy of 1.000 g oxygen molecules at 25° C?

3.16 Calculate the energy associated with thermal agitation in 1.000 mol hydrogen molecules at 100° c .

3.17 Calculate the volume occupied by 1.000 mol H_2O at 500° C and 200 atm pressure using van der Waals' equation. In this calculation, an approximate V may be employed in the a/V^2 term; the resulting linear equation is solved. An improved V may be estimated and used in a/V^2 . The new linear equation is then solved. Repeat until successive approximations check.

3.18 In a thermonuclear plasma, there are 1.00×10^{21} ions per cubic meter and the same concentration of electrons. The average energy of an ion and of an electron is 30 keV. Calculate the pressure P, remembering that 1 eV is 1.602×10^{-19} J.

3.19 Using the Berthelot equation, calculate the temperature at which 0.400 mol CO_2 fills 10.01 at 0. 900 atm pressure.

3.20 Suppose that the atmosphere is at equilibrium at 25° C and that the partial pressure of N_2 at ground level is 0.800 atm. Then what is its partial pressure at an altitude of 10.0 km?

3.21 Calculate the collision rate density for the reaction

$$\mathrm{H} + \mathrm{Br}_2 \rightarrow \mathrm{HBr} + \mathrm{Br}$$

at 25° C and 1.000 bar, when the mole fraction of Br_2 is 0.500 and that of H 0.000100. Take the effective radii of H and Br_2 to be 1.00 Å and 2.35 Å, respectively.

3.22 For the reaction in problem 3.21, calculate the hypothetical specific reaction rate k_{AB} at 25° C.

References

Books

Curtiss, C. F.: 1967, "Real Gases," in Eyring, H. (editor), *Physical Chemistry An Advanced Treatise*, vol. II, Academic Press, New York, pp. 285-338.

Curtiss considers how the virial coefficients for a given substance are related to various intermolecular potentials for the material. Statistical mechanical formulas are employed.

McGee, T. D.: 1988, *Principles and Methods of Temperature Measurement*, John Wiley & Sons, New York, pp. 1-534

McGee begins with a survey of the history of the temperature concept. Various fundamental scales are described. Details of the international practical scale are presented. Then the principal experimental methods appear in considerable detail. This is a very useful text.

Rowlinson, J. S.: 1958, "The Properties of Real Gases," in Flugge, S. (editor), *Encyclopedia of Physics*, vol. XII, Springer-Verlag, Berlin, pp. 1-72.

Rowlinson discusses both theory and experiment. Various equations of state are described. Formulas for the principal thermodynamic properties are constructed.

Articles

Bishop, M.: 1989, "A Kinetic Theory Derivation of the Second and Third Virial Coefficients of Rigid Rods, Disks, and Spheres," *Am. J. Phys.* **57**, 469-471.

Canayaratna, S. G.: 1992, "Intensive and Extensive: Underused Concepts," *J. Chem. Educ.* **69**, 957-963.

Castillo S., C.: 1991, "An Alternative Method for Calculating the Critical Compressibility Factor for the Redlich-Kwong Equation of State," *J. Chem. Educ.* **68**, 47.

Cuadros, F., Mulero, A., and Rubio, P.: 1994, "The Perturbative Theories of Fluids as a Modern Version of van der Waals Theory," *J. Chem. Educ.* **71**, 956-962.

Eberhart, J. G.: 1989a, "The Many Faces of van der Waals's Equation of State," *J. Chem. Educ.* **66**, 906-909.

Eberhart, J. G.: 1989b, "Applying the Critical Conditions to Equations of State," *J. Chem. Educ.* **66**, 990- 993.

Eberhart, J. G.: 1992, "A Least-Squares Technique for Determining the van der Waals Parameters from the Critical Constants," *J. Chem. Edu.* **69**, 220-221.

Eberhart, J. G.: 1994, "Solving Nonlinear Simultaneous Equations by the Method of Successive Substitution: Applications to Equations of State," *J. Chem. Educ.* **71**, 1038-1040.

Gorin, G.: 1994, "Mole and Chemical Amount: A Discussion of the Fundamental Measurements of Chemistry," *J. Chem. Educ.* **71**, 114-116.

Guerin, H.: 1992, "Influence of the Well Width on the Third Virial Coefficient of the Square-Well Intermolecular Potential," *J. Chem. Educ.* **69**, 203-205.

Hanson, M. P.: 1995, "The Virial Theorem, Perfect Gases, and the Second Virial Coefficient," *J. Chem. Educ.* **72**, 311-314.

Menon, V. J., and Agrawal, D. C.: 1986, "Maxwellian Distribution versus Rayleigh Distribution," *Am. J. Phys.* **54**, 1034- 1035.

Meyer, E. F., and Meyer, T. P.: 1985, "Supercritical Fluid: Liquid, Gas, Both, or Neither?" *J. Chem. Educ.* **63**, 463- 465.

Nash, L. K.: 1984, "Enduring Distributions that Deny Boltzmann," *J. Chem. Educ.* **61**, 22-25.

Ott, J. B., Goates, J. R., and Hall Jr., H. T.: 1971, "Comparisons of Equations of State in Effectively Describing PVT Relations," *J. Chem. Educ.* **48**, 515-517.

Peckham, G. D., and McNaught, T. J.: 1992, "Applications of Maxwell-Boltzmann Distribution Diagrams," *J. Chem. Educ.* **69**, 554-558.

Waite, B. A.: 1986, "Equilibrium Distribution Functions: Another Look," *J. Chem. Educ.* **63**, 117-120.

4

The First Law for Energy

4.1 Independent Thermodynamic Variables

TO CHARACTERIZE THE MACROSCOPIC STATE OF A SYSTEM, a person needs a complete set of independent variables. For a uniform *pure* substance at rest at a given height in the gravitational field, three are needed. These may be pressure P, temperature T, and number of moles n, set (P, T, n), or volume V, temperature T, and number of moles n, set (V, T, n).

But temperature T is a *thermal* variable proportional to the mean translational energy per molecule in an ideal gas thermometer at equilibrium with the system. As a consequence, it need not be uniquely determined by the macroscopic *mechanical* variables P and V. So P, V, and n, which make up set (P, V, n), are not generally suitable as independent variables. In the next chapter, a conjugate thermal variable S will be introduced. It may be substituted for T in the acceptable sets.

For impure materials and solutions at a given P and T, or at a given V and T, the amount of each chemical component is independent. For the ith component, this amount is measured by its number of moles n_i.

To relate different states, one employs processes. Of particular interest are those that are carried out reversibly. By definition, a process is *reversible* if its direction can be reversed at any stage by some infinitesimal shift in one or more of the independent variables.

The energy of a system may be broken down into (a) that due to its position and orientation in an externally imposed field, (b) that due to translation of its center of mass, (c) that due to the volume it displaces in surrounding material, and (d) that due to the relative positions and motions of the particles composing it. Part (d) is called the *internal energy* E of the system.

4.2 Work

In mechanics, the work done by a force $\mathbf{F}$ acting over a displacement $\mathrm{d}\mathbf{s}$ equals the component of the force in the direction of the displacement times the displacement, the scalar product

$$\mathrm{d}w = \mathbf{F}\cdot\mathrm{d}\mathbf{s} = F_x\,\mathrm{d}x + F_y\,\mathrm{d}y + F_z\mathrm{d}z. \qquad [4.1]$$

Here F_x, F_y, F_z are the components of the force for the rectangular coordinate changes $\mathrm{d}x$, $\mathrm{d}y$, $\mathrm{d}z$.

In thermodynamics, one *generalizes* this form. Any energy added to a system that can be represented as the scalar product of a generalized force and a generalized displacement is identified as *work*.

Always present when the volume changes is the work of compression or expansion. Consider a fluid enclosed by a cylinder and a movable piston, as figure 4.1 shows. Let the surroundings exert pressure on the piston, while the area of the piston is A. The corresponding force on the piston is $P_{ex}A$. When the piston moves distance -dz, the work done is

$$dw = (P_{ex}A)(-dz) = -P_{ex}\, dV. \tag{4.2}$$

Also, consider a rectangular solid subject to external pressure P_{ex}, as figure 4.2 shows. The work done on compressing the solid a small amount is

$$dw = P_{ex}A_x(-dx) + P_{ex}A_y(-dy) + P_{ex}A_z(-dz) = -P_{ex}\, dV. \tag{4.3}$$

For any other configuration, the *work of compression* is similarly

$$dw = -P_{ex}\, dV. \tag{4.4}$$

When the change is carried out reversibly, the internal pressure does not differ appreciably from the external pressure and

$$dw = -P\, dV. \tag{4.5}$$

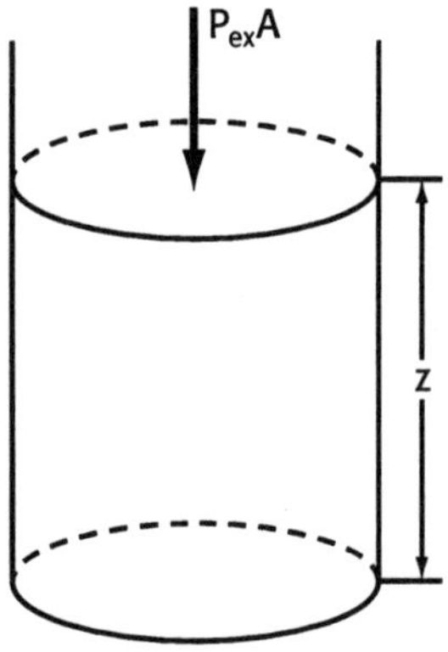

FIGURE 4.1 Cylinder fitted with a piston confining a fluid.

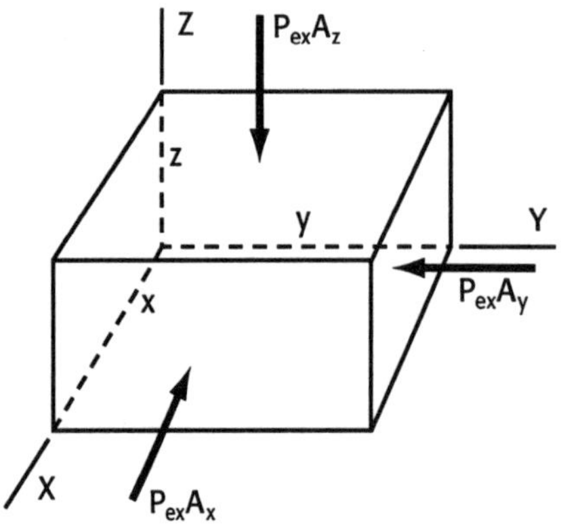

FIGURE 4.2 Rectangular solid that expands dx in the x direction, dy in the y direction, and dz in the z direction.

Wherever two phases meet, there is a surface, an interface, between them. If S is the area of this surface in a given system and γ is the energy associated with formation of unit area, the work needed to expand the area by dS is

$$dw = \gamma \, dS. \tag{4.6}$$

Property γ is the *surface tension* for the two phases. Note that if γ were negative, lower interfacial energy would result on making the surface more irregular. When both phases are fluid, the process would not stop until the surface wove back and forth, insofar as possible, through the whole system. In effect, only one phase would remain. Such a process does not occur spontaneously. So we expect γ in (4.6) to be positive.

The *electrical work* done by a voltage $\mathcal{E}$ driving charge d$\mathcal{Q}$ through a system is

$$dw = \mathcal{E} d\mathcal{Q}. \tag{4.7}$$

Similarly, the work done on increasing a magnetic field in volume V by dB is

$$dw = VH \, dB. \tag{4.8}$$

Here $B = \mu H$ and H is the magnetic intensity, B the magnetic induction. The work done on increasing an electric field in volume V by dD is

$$dw = VE \, dD. \tag{4.9}$$

Here $D = \varepsilon E$ and E is the electric intensity, D the electric displacement.

Qualitatively, work is energy that is being transported by some concerted movement.

4.3 Heat

The temperature is not necessarily uniform in a given system or between two systems in contact with each other. But the zeroth law tells us that energy in the form of heat then moves against the temperature gradient, acting to smooth out temperature differences.

Consider a small region in the given system through which the temperature T varies in the x direction. Across an interior surface dS, which is oriented perpendicular to the gradient $\partial T/\partial x$, the movement of heat in the positive x direction in time dt is expressed as

$$dq = -K \frac{\partial T}{\partial x} \, dS \, dt. \tag{4.10}$$

Factor K is a function of the conducting material, the temperature, and the pressure. It is called the *conduction coefficient.*

The process involves hot molecules bombarding colder neighboring molecules, passing energy on in a random fashion. Also, hot molecules may travel into colder regions and cold ones into hotter regions, adding to the transport of energy. Thirdly, hot molecules may radiate some of their energy as electromagnetic waves. This could then be absorbed in the colder regions, exciting the molecules in a random manner.

Heat flow can be controlled by altering the conducting material and its thickness, thus altering K and the gradient $\partial T/\partial x$. Or the speed of the process may be increased. A vacuum conducts only by radiation. Such transmission can be reduced to a negligible amount by silvering a dividing surface, as in a thermos or a Dewar flask. A process carried out without heat flow in or out is said to be *adiabatic.* Then at each stage, we have

$$dq = 0. \tag{4.11}$$

Qualitatively, heat is energy that is being transported by random chaotic movements.

4.4 The First Law

By no macroscopic process can energy be created or destroyed. Only transformations of energy occur. This limitation is embodied in the *first law of thermodynamics*: The amount of energy required to take a system from one state to another is independent of how the system goes between the two states.

The internal energy E of a given system depends only on the state of the system. This is altered by the heat transferred to the system and the work done on it. For an infinitesimal change in a system, we have

$$dE = dq + dw. \quad [4.12]$$

Integrating this from state 1 to state 2 yields

$$E_2 - E_1 = q + w \quad [4.13]$$

where q is the heat absorbed in the process and w the work done on the system in the process.

A given change in the system may be effected by various q's and w's. A person cannot say that a given system has a particular heat content q or a particular work content w.

However, there are constraints on the q's and w's needed to cause certain changes. By experiment, a pure uniform substance cannot have its temperature lowered at a given pressure without removing heat. But this involves the second law of thermodynamics, which will be considered later.

Since q and w are not functions of state, the differentials dq and dw are not exact. Their integrals depend on how the system is taken from the initial state to the final state.

4.5 Compression (P-V) Work

The work done on a given system depends not only on the initial and final states but on the path followed between the states. This is true whether or not a reversible path is followed.

Consider a given substance on which only work of compression is done. From (4.4), we may write

$$w = -\int_{V_1,T_1}^{V_2,T_2} P_{ex}\, dV. \quad [4.14]$$

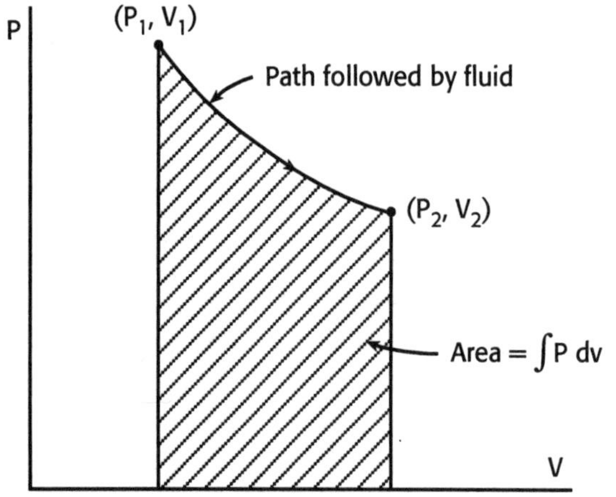

FIGURE 4.3 Representation of the work integral as an area.

Whenever P_{ex} differs only infinitesimally from P, (4.14) reduces to

$$w = -\int_{V_1,T_1}^{V_2,T_2} P\,dV. \tag{4.15}$$

When the pressure P is plotted against the volume V, this integral is represented by the area under the curve. See Figure 4.3.

For a process in which the pressure is kept constant (an *isobaric* process), (4.15) integrates without trouble:

$$w_P = -P\int_{V_1}^{V_2} dV = -P(V_2 - V_1) = -P\Delta V. \tag{4.16}$$

Processes in the laboratory are often carried out in this manner.

When the system is an ideal gas, equation (4.15) becomes

$$w = -\int_{V_1,T_1}^{V_2,T_2} nRT\frac{dV}{V}. \tag{4.17}$$

In general, the temperature is not determined by the volume alone. So this integral depends on the path, on how T does vary between the initial and the final states. When the change in volume occurs at a given temperature, isothermally, we have

$$w_T = -nRT\int_{V_1}^{V_2}\frac{dV}{V} = -nRT\ln\frac{V_2}{V_1} = nRT\ln\frac{P_2}{P_1}. \tag{4.18}$$

Results (4.16), (4.17), and (4.18) require that

$$P_{ex} \cong P. \tag{4.19}$$

Example 4.1

How much heat is absorbed by n moles of an ideal gas undergoing a reversible, isothermal process?

When an ideal gas absorbs heat, it may use the energy to do work against an external pressure and to increase the thermal agitation of its molecules. None is used to do work against intermolecular attraction as the volume increases, because this attraction is negligible.

As the temperature is kept constant, we have

$$\Delta E_T = 0$$

for the ideal gas, and

$$q_T = -w_T.$$

Substituting for w_T with (4.18) yields

$$q_T = nRT\ln\frac{V_2}{V_1} = -nRT\ln\frac{P_2}{P_1}.$$

4.6 Isochoric and Isobaric Heats

Similarly, the heat absorbed by a given system depends not only on the initial and final states but also on the path followed between the states.

Consider a uniform system for which all work is compressional. Then

$$dE = dq + dw = dq - P_{ex}\, dV. \quad [4.20]$$

If, in addition, the volume is kept constant (the *isochoric* condition), no *P-V* work is done and (4.20) reduces to

$$dE = dq_V. \quad [4.21]$$

Here as before, the subscript indicates the variable held constant. For a finite process, we have

$$\Delta E = q_V. \quad [4.22]$$

Heat absorbed at constant volume goes to increase the internal energy E of the system.

When the external pressure P_{ex} is kept constant, the internal pressure P tends to stay fixed and equal to it. Then (4.20) rearranges to

$$dq_P = dE + P\, dV. \quad [4.23]$$

With P constant, this equation yields the relation

$$dq_P = dE + d(PV) = d(E + PV). \quad [4.24]$$

Since E, P, and V are functions of state, the *enthalpy*

$$H = E + PV. \quad [4.25]$$

is also. With definition (4.25), equation (4.24) reduces to

$$dq_P = dH. \quad [4.26]$$

For a finite process, we have

$$\Delta H = q_P. \quad [4.27]$$

Heat absorbed at constant pressure goes to increase the enthalpy H of the system.

Whether or not a given system is kept at constant pressure, it possesses the property H defined by (4.25). Over an infinitesimal change,

$$dH = dE + d(PV), \quad [4.28]$$

and over a finite change,

$$\Delta H = \Delta E + \Delta(PV). \quad [4.29]$$

When the system consists of n moles ideal gas kept at a given temperature, the last term is zero and

$$\Delta H_T = \Delta E_T = 0. \quad [4.30]$$

Example 4.2

What is the significance of the PV term in the enthalpy?

The work needed to make a hole of volume V in a fluid with a constant pressure P is

$$w = \int_0^V P\, dV = P\int_0^V dV = PV.$$

So when a system of volume V is immersed in such a fluid (gas or liquid), it seemingly possesses this energy.

Indeed, expanding the system by volume ΔV requires the energy $\Delta(PV)$. On contracting to the original volume, this energy is reclaimed. Thus, PV is the energy a system *appears* to possess because it fills the volume V in a surrounding fluid at pressure P.

As a result, term PV may be called the *external energy* of the system.

Example 4.3

Calculate ΔE and ΔH for the vaporization of 1 mole water at 372.778 K and 1 bar pressure, where the heat of vaporization is 40,893 J mol^{-1}.

Since the vaporization occurs at constant pressure, the heat absorbed in the process is

$$q_P = (1.0000\text{ mol})(40{,}893\text{ J mol}^{-1}) = 40{,}893\text{ J}.$$

Also, the work done on the system is

$$w_P = -P\Delta V = -P(V_g - V_l)$$

where V_g is the volume of the final gas and V_l the volume of the initial liquid.

Neglect the volume of the liquid and approximate the volume of the gas with the ideal gas equation:

$$V \cong V_g \cong \frac{RT}{P}.$$

The work done reduces to

$$w_P = -P\frac{RT}{P} = -RT = -(8.3145\text{ J K}^{-1})(372.778\text{ K}) = -3099\text{ J}.$$

Substituting into equation (4.13) yields

$$\Delta E = q + w = 40{,}893 - 3099\text{ J} = 37{,}794\text{ J};$$

and into equation (4.27),

$$\Delta H = q_P = 40{,}893\text{ J}.$$

4.7 Energy Capacities

Any energy added to a system goes to increase the internal energy and the external energy. Varying with the former in a given phase is the average translational energy and the average kinetic energy in a degree of freedom. But the latter is proportional to the Kelvin temperature by section 3.3.

Let us consider a given uniform system at temperature T. Suppose that energy is added to the system in the form of heat q. Dividing this by the resulting temperature rise ΔT yields the *energy capacity* C:

$$\frac{q}{\Delta T} = C. \qquad [4.31]$$

Besides varying with the substance and its amount, C varies with the path followed and with the temperature. Because of the variation with T, one generally employs the ratio of differentials,

$$\frac{\mathrm{d}q}{\mathrm{d}T} = C, \qquad [4.32]$$

for each chosen temperature. Also by convention, determinations are made along paths with the volume constant or with the pressure constant. In any case, energy capacity C is an extensive property.

At constant volume, we have

$$\frac{dq_V}{dT} = C_V. \qquad [4.33]$$

Here dq_V is the infinitesimal heat needed to raise the temperature infinitesimal amount dT with the work w equal to zero.

In common parlance, C_V is called the *heat capacity at constant volume* of the system.

Combining (4.33) with (4.21) leads to

$$dE_V = \frac{dq_V}{dT} dT = C_V \, dT \qquad [4.34]$$

and

$$C_V = \left(\frac{\partial E}{\partial T}\right)_V. \qquad [4.35]$$

For a change in temperature from T_1 to T_2, we obtain

$$\Delta E_V = \int_{T_1}^{T_2} C_V \, dT. \qquad [4.36]$$

When the system is an ideal gas, a change in volume at a given temperature does not alter the internal energy; then

$$\left(\frac{\partial E}{\partial V}\right)_T = 0 \qquad [4.37]$$

and

$$dE = C_V \, dT \qquad [4.38]$$

whether the volume V is fixed or not.

At constant pressure, we have

$$\frac{dq_P}{dT} = C_P. \qquad [4.39]$$

Here dq_P is the infinitesimal heat needed to raise the temperature by dT when only work of expansion is done and the pressure P is kept constant. By convention, C_P is called the *heat capacity at constant pressure* of the system.

Combining (4.39) with (4.26) yields

$$dH_P = \frac{dq_P}{dT} dT = C_P \, dT \qquad [4.40]$$

and

$$C_P = \left(\frac{\partial H}{\partial T}\right)_P. \qquad [4.41]$$

Integrating (4.40) from T_1 to T_2 leads to

$$\Delta H_P = \int_{T_1}^{T_2} C_P \, dT. \qquad [4.42]$$

When the system is an ideal gas, condition (4.30) applies, enthalpy H does not vary with pressure P when the temperature T is fixed, and H varies only with T. Then we have

$$dH = C_P \, dT \qquad [4.43]$$

whether P is fixed or not.

Example 4.4

Determine how C_P is related to C_V when the system is an ideal gas.
The equation of state for n moles of ideal gas is

$$PV = nRT.$$

By definition, the enthalpy is given by

$$H = E + PV.$$

Combine these two equations,

$$H = E + nRT,$$

differentiate

$$dH = dE + nR\,dT,$$

and introduce relations (4.43) and (4.38):

$$C_P\,dT = C_V\,dT + nR\,dT = \left(C_V + nR\right)dT.$$

Thus, we find that

$$C_P = C_V + nR$$

for the ideal gas.

4.8 Reversible Adiabatic Change in an Ideal Gas

We now have the relationships needed to determine how an ideal gas behaves in a process involving negligible heat transfer.

The heat q may be kept small during a process by surrounding the system with enough insulation. Alternatively, the immediate surroundings may be heated or cooled so its temperature follows closely that of the system. Even without such precautions, there would be negligible heat flow if the process were rapid enough.

Consider a given system subject to an adiabatic change. In each infinitesimal step, we thus have

$$dq = 0. \qquad [4.44]$$

Let us also consider the process to be reversible with all work done work of compression. Then

$$dw = -P\,dV \qquad [4.45]$$

and

$$dE = dq + dw = -P\,dV. \qquad [4.46]$$

Furthermore, consider the system to be an ideal gas. The internal energy then depends only on the temperature following equation (4.38). Combining this with equation (4.46),

$$C_V\,dT = -P\,dV, \qquad [4.47]$$

and with the ideal gas equation yields

$$C_V\,dT = -\frac{nRT}{V}\,dV. \qquad [4.48]$$

Rearrange this equation,

$$C_V \frac{dT}{T} + nR\frac{dV}{V} = 0, \qquad [4.49]$$

and integrate to get

$$C_V \ln\frac{T_2}{T_1} + nR\ln\frac{V_2}{V_1} = 0 \qquad [4.50]$$

whence

$$T_2 V_2^{nR/C_V} = T_1 V_1^{nR/C_V}. \qquad [4.51]$$

Using the ideal gas equation to eliminate the temperature leads to

$$\frac{P_2 V_2}{nR} V_2^{nR/C_V} = \frac{P_1 V_1}{nR} V_1^{nR/C_V}, \qquad [4.52]$$

whence

$$P_2 V_2^{\gamma} = P_1 V_1^{\gamma} \qquad [4.53]$$

where

$$\gamma = \frac{C_P}{C_V} = \frac{C_V + nR}{C_V} = 1 + \frac{nR}{C_V}. \qquad [4.54]$$

In an isothermal compression of an ideal gas, product PV is constant. In a reversible adiabatic compression, product PV^{γ} is constant. A given volume decrease thus causes a much greater pressure rise when the process is adiabatic. Then energy that would otherwise escape as heat is trapped in the system to raise the temperature and cause the greater pressure increase.

Example 4.5

An evacuated 1-liter vessel is connected by a closed tube to a 1-liter vessel filled with argon at 1 atm and 25° C. The tube is opened and the gas is allowed to reach equilibrium at 1/2 atm and 25° C. What are w, ΔE, and q for the overall process?

In the approximation that the vessels do not change in size, no work is done on the system and

$$w = 0.$$

Since the argon behaves as an ideal gas, its internal energy depends only on T. But since the final temperature equals the initial temperature, we have

$$\Delta E = 0$$

and

$$q = \Delta E - w = 0 - 0 = 0.$$

The overall process is adiabatic. It is also highly irreversible. Thus it does not follow equation (4.53).

4.9 Conditions in a Planar Pressure Pulse

The energy-capacity ratio γ can be determined from the effects of adiabatic compression on the given gas. In a simple experiment, one measures the speed of sound in the gas and from this calculates γ. The relationship needed can be derived simply.

Consider a steady-state planar pressure pulse in the given fluid. Observe it from a point at rest with respect to the wave, as figure 4.4 illustrates. Locate plane 2 at a given phase of the wave and plane 1 a given distance in front of the wave.

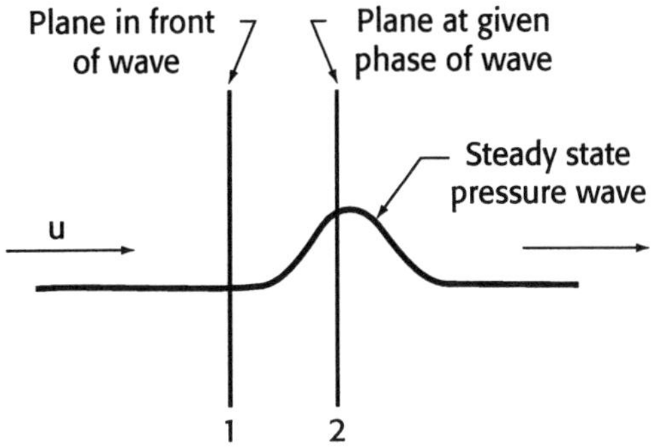

FIGURE 4.4 Variation of pressure with distance through an elementary sound wave.

Let ρ_1 be the density and u_1 the speed of the fluid at plane 1, while ρ_2 *is* the density and u_2 the speed at plane *2*. The volume of the fluid passing through unit area of plane j in unit time is then u_j, Multiplying this by the density as in $\rho_j u_j$ then gives the corresponding mass. In the steady state, there is no accumulation of mass between planes 1 and *2* and

$$\rho_1 u_1 = \rho_2 u_2. \tag{4.55}$$

The momentum of mass $\rho_1 u_1$ is $(\rho_1 u_1)u_1$; the momentum of mass $\rho_2 u_2$ is $(\rho_2 u_2)u_2$. The force acting to cause this change in momentum per unit area per unit time is $P_1 - P_2$. So from Newton's second law

$$\rho_2 u_2^2 - \rho_1 u_1^2 = P_1 - P_2 \tag{4.56}$$

or

$$\left(\rho_2 u_2\right)^2 \frac{1}{\rho_2} - \left(\rho_1 u_1\right)^2 \frac{1}{\rho_1} = P_1 - P_2. \tag{4.57}$$

Eliminating $\rho_2 u_2$ with equation (4.55) gives

$$\left(\rho_1 u_1\right)^2 \left(\frac{1}{\rho_2} - \frac{1}{\rho_1}\right) = P_1 - P_2. \tag{4.58}$$

Replacing $1/\rho_j$ with the specific volume V_j and solving for u_1^2 yields

$$u_1^2 = V_1^2 \frac{P_1 - P_2}{V_2 - V_1} = V_1^2 \left(-\frac{\Delta P}{\Delta V}\right). \tag{4.59}$$

In a typical wave, the process is so fast that negligible heat flows and

$$q = 0. \tag{4.60}$$

And when the amplitude is small, the ratio of increments may be replaced by the derivative to yield

$$u^2 = V^2 \left[-\left(\frac{dP}{dV}\right)_{q=0}\right] \tag{4.61}$$

or

$$u = V\left[-\left(\frac{dP}{dV}\right)_{q=0}\right]^{1/2}. \tag{4.62}$$

Here u is the speed of the incoming fluid with respect to the wave. In the laboratory it would equal the speed of the wave with respect to the fluid at rest. Thus, it can be identified as the *speed of sound* in the given fluid.

When the fluid can be approximated as an ideal gas, the adiabatic equation of state

$$PV^{\gamma} = \text{const} \tag{4.63}$$

applies. Differentiating this leads to the condition

$$\left(\frac{\mathrm{d}P}{\mathrm{d}V}\right)_{q=0} = -\gamma\frac{P}{V}. \tag{4.64}$$

which transforms formula (4.62) to

$$u = V\left(\gamma\frac{P}{V}\right)^{1/2} = \left(\gamma PV\right)^{1/2} = \left(\frac{\gamma RT}{M}\right)^{1/2}. \tag{4.65}$$

Here u is the speed of sound, M the molecular weight, T the absolute temperature, R the gas constant, and γ the ratio C_P/C_V.

4.10 Contributions to Energy Capacity C_V

Each degree of freedom of a molecule in a system absorbs energy as the temperature is raised. Thus each of these contributes to the parameter C_V.

By formula (4.35), the heat capacity at constant volume is given by the corresponding partial derivative of internal energy E:

$$C_V = \left(\frac{\partial E}{\partial T}\right)_V. \tag{4.66}$$

But the translational energy in an ideal gas is given by formula (3.23). Differentiating this yields

$$C_{V,\text{tr}} = \frac{3}{2}nR. \tag{4.67}$$

Each translational degree of freedom contributes one third of this.

For a diatomic rotator in an ideal gas, we have equation (3.45). Differentiating this yields

$$C_{V,\text{rot}} = nR. \tag{4.68}$$

For a nonlinear rotator in an ideal gas, equation (3.46) applies. Differentiating it gives us

$$C_{V,\text{rot}} = \frac{3}{2}nR. \tag{4.69}$$

Each rotational degree of freedom contributes $1/2nR$ to C_V.

For a classically excited vibrational degree of freedom, formula (3.48) applies. On differentiation, we obtain

$$C_{V,\text{vib}} = nR. \tag{4.70}$$

Some experimentally determined heat capacities appear in table 4.1.

TABLE 4.1 Constant-Volume Heat Capacities

Gas	C_V, $J\ K^{-1}\ mol^{-1}$ at 150 K	300 K	500 K	700 K	900 K
O_2	20.84	21.05	22.76	24.69	26.02
H_2	18.24	20.54	20.96	21.13	21.55
H_2O	—	25.27	26.90	29.16	31.67
N_2	20.79	20.79	21.25	22.43	23.77
NO	22.84	21.55	22.18	23.72	25.10
CO	20.79	20.84	21.46	22.84	24.27
CO_2	—	28.91	36.32	41.25	44.73

Example 4.6

Estimate the translational, rotational, and vibrational contributions to heat capacity C_V of H_2O at 500 K.

For the three translational degrees of freedom of the H_2O molecule, we have

$$C_{V,\text{tr}} = \frac{3}{2}R = \frac{3}{2}\left(8.3145\ \text{J K}^{-1}\ \text{mol}^{-1}\right) = 12.47\ \text{J K}^{-1}\ \text{mol}^{-1}.$$

Since H_2O is nonlinear, there are three rotational degrees of freedom. We expect these to be excited classically at 500 K; so

$$C_{V,\text{rot}} = \frac{3}{2}R = 12.47\ \text{J K}^{-1}\ \text{mol}^{-1}.$$

Subtracting the sum of these contributions from the value in table 4.1 yields

$$C_{V,\text{vib}} = 26.90 - 24.94 = 1.96\ \text{J K}^{-1}\ \text{mol}^{-1}.$$

Consistent with the data in example 3.7, the three vibrational degrees of freedom are only weakly excited.

4.11 Relating Differentials and Partial Derivatives

Because of its importance in thermodynamics, let us here establish the fundamental relationship linking differentials with the pertinent partial derivatives.

Consider a function *f* in which x and y vary independently:

$$f = f(x,y). \qquad [4.71]$$

For a change Δx in x and y in Δy, we find in *f* the change

$$\Delta f = f(x+\Delta x, y+\Delta y) - f(x,y) = f(x+\Delta x, y+\Delta y) - f(x, y+\Delta y) + f(x, y+\Delta y) - f(x,y)$$
$$= \frac{f(x+\Delta x, y+\Delta y) - f(x, y+\Delta y)}{\Delta x}\Delta x + \frac{f(x, y+\Delta y) - f(x,y)}{\Delta y}\Delta y. \qquad [4.72]$$

By definition, the limit of each ratio of changes is a derivative. Here we have

$$\lim_{\substack{\Delta x\to 0\\ \Delta y\to 0}} \frac{f(x+\Delta x, y+\Delta y) - f(x, y+\Delta y)}{\Delta x} = f_x'(x,y) = \frac{\partial f}{\partial x}, \qquad [4.73]$$

$$\lim_{\Delta y\to 0} \frac{f(x, y+\Delta y) - f(x,y)}{\Delta y} = f_y'(x,y) = \frac{\partial f}{\partial y}, \qquad [4.74]$$

so that as Δx, Δy, and Δf become infinitesimally small, equation (4.72) becomes

$$\mathrm{d}f = f_x'(x,y)\,\mathrm{d}x + f_y'(x,y)\,\mathrm{d}y = \left(\frac{\partial f}{\partial x}\right)_y \mathrm{d}x + \left(\frac{\partial f}{\partial y}\right)_x \mathrm{d}y. \qquad [4.75]$$

Note that x and y need only be functions of the independent variables. Also, an infinitesimal change in such a function is called an *exact differential*. Equation (4.75) expresses a fundamental relationship among such differentials.

When variables $x_1, x_2, \ldots, x_n$ enter independently into function f, we similarly find that

$$\mathrm{d}f = \sum \left(\frac{\partial f}{\partial x_j}\right)_{\text{other } x\text{'s}} \mathrm{d}x_j. \qquad [4.76]$$

Furthermore, $x_1, x_2, \ldots, x_n$ variables need only be functions of the independent variables for the overall problem.

If we consider the internal energy E as a function of T and V, we have

$$\mathrm{d}E = \left(\frac{\partial E}{\partial T}\right)_V \mathrm{d}T + \left(\frac{\partial E}{\partial V}\right)_T \mathrm{d}V. \qquad [4.77]$$

The subscript on the partial derivative indicates the other independent variable or variables; it lists the other variable or variables held constant in the differentiation. When energy E is considered a function of T and P, we similarly have

$$\mathrm{d}E = \left(\frac{\partial E}{\partial T}\right)_P \mathrm{d}T + \left(\frac{\partial E}{\partial P}\right)_T \mathrm{d}P. \qquad [4.78]$$

Example *4.7*

Relate

$$\left(\frac{\partial E}{\partial T}\right)_P \text{ to } \left(\frac{\partial E}{\partial T}\right)_V \text{ and } \left(\frac{\partial E}{\partial P}\right)_T \text{ to } \left(\frac{\partial E}{\partial V}\right)_T$$

for a uniform thermodynamic system.

In the system, volume V is a function of temperature T and pressure P; so

$$\mathrm{d}V = \left(\frac{\partial V}{\partial T}\right)_P \mathrm{d}T + \left(\frac{\partial V}{\partial P}\right)_T \mathrm{d}P.$$

Since we may consider the internal energy E to be a function of either T and V or T and P, both (4.77) and (4.78) are valid. Substituting the form for dV into (4.77) gives us

$$\mathrm{d}E = \left(\frac{\partial E}{\partial T}\right)_V \mathrm{d}T + \left(\frac{\partial E}{\partial V}\right)_T \left[\left(\frac{\partial V}{\partial T}\right)_P \mathrm{d}T + \left(\frac{\partial V}{\partial P}\right)_T \mathrm{d}P\right].$$

Comparing this result with (4.78) yields

$$\left(\frac{\partial E}{\partial T}\right)_P = \left(\frac{\partial E}{\partial T}\right)_V + \left(\frac{\partial E}{\partial V}\right)_T \left(\frac{\partial V}{\partial T}\right)_P$$

and

$$\left(\frac{\partial E}{\partial P}\right)_T = \left(\frac{\partial E}{\partial V}\right)_T \left(\frac{\partial V}{\partial P}\right)_T.$$

4.12 The Difference C_P - C_V

At constant volume, any energy supplied to a system goes to increase its internal energy E. But at constant pressure, some of the energy goes to do work against the external pressure P and some to do work against the molecular interaction. The latter exhibits itself as a cohesion in liquids and solids.

Consider a uniform thermodynamic system at temperature T, pressure P, and volume V. Let some energy be added as described in section 4.7. With formulas (4.41) and (4.35), we have

$$C_P - C_V = \left(\frac{\partial H}{\partial T}\right)_P - \left(\frac{\partial E}{\partial T}\right)_V. \qquad [4.79]$$

Introducing expression (4.25) for the enthalpy H gives

$$C_P - C_V = \left[\frac{\partial}{\partial T}(E + PV)\right]_P - \left(\frac{\partial E}{\partial T}\right)_V = \left(\frac{\partial E}{\partial T}\right)_P + P\left(\frac{\partial V}{\partial T}\right)_P - \left(\frac{\partial E}{\partial T}\right)_V. \qquad [4.80]$$

Combining this with the formula for $(\partial E/\partial T)_P$ found in example 4.7 leads to

$$C_P - C_V = \left(\frac{\partial E}{\partial V}\right)_T \left(\frac{\partial V}{\partial T}\right)_P + P\left(\frac{\partial V}{\partial T}\right)_P = \left[\left(\frac{\partial E}{\partial V}\right)_T + P\right]\left(\frac{\partial V}{\partial T}\right)_P. \qquad [4.81]$$

The internal property $(\partial E/\partial V)_T$ acts like pressure P in this expression. Consequently, it is often called the *internal pressure* in the system. On adding energy, work is done against this as well as against the external pressure P.

4.13 Coefficients of Thermal Expansion and Isothermal Compressibility

How the volume of a given system varies with temperature and pressure can be described by empirically measured coefficients.

Consider a system with the volume V at temperature T and pressure P. The fundamental form for an exact differential, (4.75), then becomes

$$\mathrm{d}V = \left(\frac{\partial V}{\partial T}\right)_P \mathrm{d}T + \left(\frac{\partial V}{\partial P}\right)_T \mathrm{d}P. \qquad [4.82]$$

The coefficients on the right vary directly with the volume. So we rewrite equation (4.82) as

$$dV = \alpha V\, dT - \beta V\, dP. \quad [4.83]$$

The relative rate of change of volume with temperature at constant pressure,

$$\frac{1}{V}\left(\frac{\partial V}{\partial T}\right)_P = \alpha, \quad [4.84]$$

is called the *coefficient of cubical expansion*. For a condensed phase, this is independent of pressure except at very high pressures. However, it does vary slowly with temperature.

The negative relative rate of change of volume with pressure at constant temperature,

$$-\frac{1}{V}\left(\frac{\partial V}{\partial P}\right)_T = \beta, \quad [4.85]$$

is called the *coefficient of compressibility*. For a condensed phase, this is nearly independent of temperature and pressure.

At constant volume, $dV = 0$. Then equation (4.83) yields

$$\left(\frac{\partial P}{\partial T}\right)_V = \frac{\alpha}{\beta}. \quad [4.86]$$

4.14 Exact Differentials

We have noted that dq and dw are not small changes in thermodynamic properties. Thus, they are not exact differentials. On the other hand, dE is an exact differential. Let us now consider how such differentials depend on the independent variables.

Suppose that u is a function of variables x and y such that

$$du = M(x,y)\, dx + N(x,y)\, dy. \quad [4.87]$$

But formula (4.75) tells us that

$$du = \frac{\partial u}{\partial x}\, dx + \frac{\partial u}{\partial y}\, dy; \quad [4.88]$$

so

$$M = \frac{\partial u}{\partial x}, \qquad N = \frac{\partial u}{\partial y}. \quad [4.89]$$

Differentiating M with respect to y and N with respect to x yields

$$\frac{\partial M}{\partial y} = \frac{\partial^2 u}{\partial y \partial x}, \quad \frac{\partial N}{\partial x} = \frac{\partial^2 u}{\partial x \partial y}. \quad [4.90]$$

Since the second derivative is independent of the order of differentiation, we have

$$\frac{\partial M}{\partial y} = \frac{\partial N}{\partial x}. \quad [4.91]$$

Next suppose that functions M and N are given satisfying condition(4.91). Construct the relation

$$M = \frac{\partial V}{\partial x}, \tag{4.92}$$

which differentiates to

$$\frac{\partial M}{\partial y} = \frac{\partial^2 V}{\partial y \partial x} = \frac{\partial}{\partial x}\left(\frac{\partial V}{\partial y}\right). \tag{4.93}$$

Use (4.91) to replace the derivative of M with that of N:

$$\frac{\partial N}{\partial x} = \frac{\partial}{\partial x}\left(\frac{\partial V}{\partial y}\right). \tag{4.94}$$

Integrate equation (4.94) with y constant:

$$N = \frac{\partial V}{\partial y} + A. \tag{4.95}$$

But A may vary with y. Write it in the form $\mathrm{d}\phi(y)/\mathrm{d}y$ to get

$$N = \frac{\partial V}{\partial y} + \frac{\mathrm{d}\phi}{\mathrm{d}y}. \tag{4.96}$$

Finally, construct $\mathrm{d}u$ as in (4.87) and reduce:

$$\mathrm{d}u = M\,\mathrm{d}x + N\,\mathrm{d}y = \frac{\partial V}{\partial x}\,\mathrm{d}x + \frac{\partial V}{\partial y}\,\mathrm{d}y + \frac{\mathrm{d}\phi}{\mathrm{d}y}\,\mathrm{d}y = \mathrm{d}V + \mathrm{d}\phi. \tag{4.97}$$

Since $\mathrm{d}V$ and $\mathrm{d}\phi$ are exact, differential $\mathrm{d}u$ is exact.

We see that equation (4.91) is a necessary and sufficient condition that $\mathrm{d}u$ in formula (4.87) be exact.

4.15 Partial Molar Properties

Any extensive thermodynamic property of a homogeneous material varies directly, linearly, with the amount of material considered. When the material is a solution, the property also varies linearly with the amount of each constituent.

As a simple example, consider the volume V of a homogeneous mixture of substances A and B. Let n_A be the number of moles of A, n_B the number of moles of B, and n the total number of moles in the solution.

At a given temperature T and pressure P, volume V is a function of n_A and n_B. Then application of formula (4.75) yields

$$\mathrm{d}V = \left(\frac{\partial V}{\partial n_A}\right)_{T,P,n_B} \mathrm{d}n_A + \left(\frac{\partial V}{\partial n_B}\right)_{T,P,n_A} \mathrm{d}n_B = \overline{V_A}\,\mathrm{d}n_A + \overline{V_B}\,\mathrm{d}n_B, \tag{4.98}$$

where

$$\left(\frac{\partial V}{\partial n_A}\right)_{T,P,n_B} = \overline{V_A}, \qquad \left(\frac{\partial V}{\partial n_B}\right)_{T,P,n_A} = \overline{V_B}. \tag{4.99}$$

Coefficient $\bar{V}_A$ is called the *partial molar volume* of A, coefficient $\bar{V}_B$ the partial molar volume of B, in the solution.

Next, divide the overall equation (4.98) by dn to get

$$\frac{dV}{dn} = \overline{V_A}\frac{dn_A}{dn} + \overline{V_B}\frac{dn_B}{dn}. \tag{4.100}$$

Then let the system be built up from nothing at constant temperature, pressure, and concentrations. We have

$$\frac{dn_A}{dn} = \frac{n_A}{n}, \qquad \frac{dn_B}{dn} = \frac{n_B}{n}, \qquad \frac{dV}{dn} = \frac{V}{n}, \tag{4.101}$$

and equation (4.100) becomes

$$\frac{V}{n} = \overline{V_A}\frac{n_A}{n} + \overline{V_B}\frac{n_B}{n} \tag{4.102}$$

whence

$$V = \overline{V_A}n_A + \overline{V_B}n_B. \tag{4.103}$$

The partial molar volume of a constituent is the contribution of that constituent per mole to the total volume. When the solution is homogeneous, this contribution depends on T, P, and the concentrations. It is independent of the total number of moles n. Thus, it is an intensive property.

In equation (4.102), n_A/n is the mole fraction of A, X_A, while n_B/n is the mole fraction of B, X_B. Thus, the equation can be rewritten as

$$\frac{V}{n} = \overline{V_A}X_A + \overline{V_B}X_B. \tag{4.104}$$

But since

$$X_A + X_B = 1, \tag{4.105}$$

we have

$$\frac{V}{n} = \overline{V_A} + \left(\overline{V_B} - \overline{V_A}\right)X_B. \tag{4.106}$$

If one plots the molar volume V/n against the mole fraction X_B, one generally obtains a curve. But the tangent line drawn at a given point on the curve represents formula

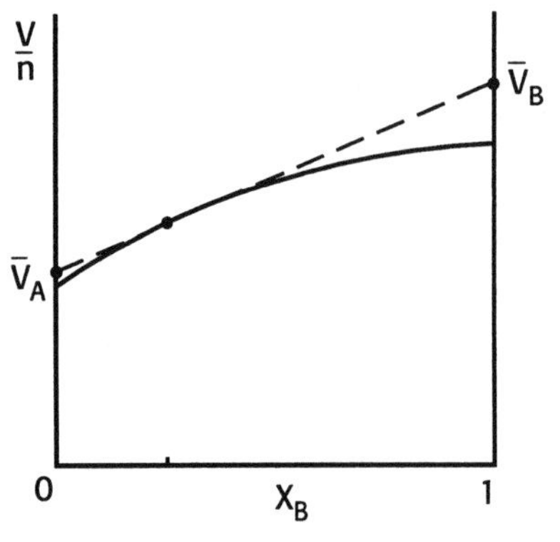

FIGURE 4.5 Plot of molar volume V/n against mole fraction X_B of B. At a particular X_B, a tangent line is drawn. This intersects the vertical axes at $\bar{V}_A$ and $\bar{V}_B$.

(4.106) with V_A and V_B fixed by their values at this point of tangency. The *intercepts* at $X_B = 0$ and at $X_B = 1$ then give numbers V_A and V_B. See figure 4.5.

In place of V, one can substitute any extensive thermodynamic property. If G is such a property, we have

$$G = \overline{G_A} n_A + \overline{G_B} n_B \qquad [4.107]$$

and

$$\frac{G}{n} = \overline{G_A} X_A + \overline{G_B} X_B. \qquad [4.108]$$

Furthermore, the method of intercepts may be used to find $\bar{G}_A$ and $\bar{G}_B$.

Questions

4.1 What independent variables may be used to characterize the state of a homogeneous thermodynamic System?
4.2 Show that for water, the set (P, V, n) is not sufficient to characterize uniquely each macroscopic state.
4.3 When is a process reversible?
4.4 Define internal energy. How is this energy varied?
4.5 What is work? Give examples.
4.6 How does heat flow?
4.7 Define the conduction coefficient.
4.8 Discuss the first law of thermodynamics.
4.9 Why doesn't a system in a particular state have a definite heat content or work content?
4.10 Define enthalpy.
4.11 What does the PV term in the enthalpy represent?
4.12 What does the vanishing of $(\partial E/\partial V)_T$ imply about the interactions between the molecules of a given system?
4.13 How are C_P and C_V related when the system is an ideal gas?
4.14 When is a process nearly adiabatic?
4.15 Derive the equation for the speed of sound in an ideal gas.
4.16 Discuss the contributions to the energy capacity C_V.
4.17 How does a partical derivative differ from an ordinary derivative?
4.18 Why is $(\partial E/\partial V)_T$ called the internal pressure in a system?
4.19 How is $(\partial P/\partial T)_V$ related to the coefficients of cubical expansion and compressibility?
4.20 How is an exact differential different from an inexact one? Give physical examples of each.
4.21 How is a partial molar property defined? Give an example.
4.22 Why are V_A and V_B constant at a given temperature, pressure, and concentration?

Problems

4.1 How much work is done on the surroundings when 0.86 mol Zn reacts with an acid to produce hydrogen at 25° C?
4.2 Calculate the work done on the surroundings when 1.35 mol water is decomposed by electrolysis at 1.00 atm and 25° C.
4.3 Calculate w when 100 g argon is heated from 25° to 75° C at 1.00 atm.
4.4 If 1.00 mole ideal gas travels down a tube at 2.00 atm pressure and 25° C, meets and passes through a porous plug, and emerges at 1.00 atm pressure and 25° C, what is w?
4.5 From a large volume of liquid benzene, 0.800 mol is vaporized at 80.1° C and 1.000 atm, the boiling point, where the heat of vaporization is 393.7 J g^{-1}. Calculate (a) q, (b) w, (c) ΔH, (d) ΔE.
4.6 The pressure on 1.00 mol helium is reduced reversibly at 25° C from 10.0 atm to 1.00 atm. Calculate (a) q, (b) w, (c) ΔH, (d) ΔE.

4.7 Calculate ΔE and ΔH of 100 g helium when it is heated from 0° to 100° C, assuming it is an ideal gas with a C_V of $3/2R$ per mole.

4.8 An ideal gas with C_V equal to $5/2R$ per mole initially filled 2.24 1 at 1.00 atm and 0° C, and finally filled 4.48 1 at 2.00 atm, Calculate q and w if the process was (a) a constant-volume increase of pressure to 2.00 atm followed by a constant-pressure increase of volume to 4.48 1, (b) a constant-pressure increase of volume to 4.48 1 followed by a constant-volume increase of pressure to 2.00 atm.

4.9 If 0.810 mol helium was compressed reversibly and adiabatically from 40.0 1 at 0° C to 30.0 1, what was the temperature increase? What was ΔE?

4.10 Calculate the speed of sound in hydrogen at 25° C.

4.11 How much do the (a) translational, (b) rotational, (c) vibrational degrees of freedom contribute to the heat capacity of carbon dioxide at 300 K where C_V is 28.91 J K^{-1} mol^{-1}, if CO_2 is a linear molecule?

4.12 For 1.000 mol of a van der Waals gas calculate

$$\text{(a)}\ \left(\frac{\partial P}{\partial T}\right)_V, \qquad \text{(b)}\ \frac{\partial^2 P}{\partial V \partial T}, \qquad \text{(c)}\ \left(\frac{\partial^2 P}{\partial T^2}\right)_V, \qquad \text{(d)}\ \left(\frac{\partial V}{\partial T}\right)_P.$$

4.13 Show that

$$C_P = \left(\frac{\partial H}{\partial T}\right)_V + \left(\frac{\partial H}{\partial V}\right)_T \left(\frac{\partial V}{\partial T}\right)_P.$$

4.14 Show that

$$\left(\frac{\partial E}{\partial P}\right)_V = \frac{-C_V\left(\partial V / \partial P\right)_T}{\left(\partial V / \partial T\right)_P}.$$

4.15 Derive

$$\left(\frac{\partial \alpha}{\partial P}\right)_T = -\left(\frac{\partial \beta}{\partial T}\right)_P.$$

— — —

4.16 A 1000 g weight falls 5.00 m to a concrete floor and cools to its original temperature. What are w and q?

4.17 If the surface tension of water is 72.75 ergs cm^{-2} at 20° C, how much surface work is done when a drop of water 0.042 ml in volume is split into 10 equal droplets at 20° C? If 80 ergs work was employed in this process, what is the corresponding q?

4.18 If one gram helium were constrained to expand at 25° C against an external 1.000 atm pressure from an initial 50.0 atm internal pressure, what would (a) q, (b) w, (c) ΔE, (d) ΔH be?

4.19 What are ΔE and ΔH when 1.000 mol water is heated from 20° to 100° C at 1.000 atm. The specific gravity of water is 0.9982 at 20° C and 0.9584 at 100° C while the specific heat is 1.000 in the temperature range. How large is the difference: ΔH - ΔE?

4.20 The volume V of 0.100 mol of an ideal gas varies linearly with its pressure P as it expands reversibly from 2.50 1 at 1.00 atm to 4.00 1 at 1.60 atm. Calculate w.

4.21 If C_V for the gas in problem 4.19 is $5/2R$ per mole, what is q for the process?

4.22 After 2.00 g helium was put into a cylinder at 25° C and 50 atm, the piston pressure was suddenly reduced to 1.00 atm. If the gas expanded adiabatically, what was its final temperature?

4.23 If the speed of sound in ethylene is 314 m s^{-1} at 0° C, what is it at 25° C?

4.24 If an ideal gas for which C_V is $3/2nR$ is heated so that its heat capacity C is 2 nR, what path is followed?

4.25 A rubber strip of uniform cross section is stretched to length L by a force F. Assuming the energy E of the strip is a function of its temperature T and length L, show that heat capacity C of the strip is given by the equation

$$C = \left(\frac{\partial E}{\partial T}\right)_L + \left[\left(\frac{\partial E}{\partial L}\right)_T - F\right]\frac{dL}{dT}.$$

4.26 Derive

$$\left(\frac{\partial P}{\partial V}\right)_T = -\frac{(\partial P / \partial T)_V}{(\partial V / \partial T)_P}.$$

4.27 Show that

$$\left(\frac{\partial T}{\partial P}\right)_H = -\frac{(\partial H / \partial P)_T}{(\partial H / \partial T)_P}.$$

4.28 Derive the relationship

$$\left(\frac{\partial T}{\partial P}\right)_H = -\frac{1}{C_P}\left[\left(\frac{\partial E}{\partial V}\right)_T\left(\frac{\partial V}{\partial P}\right)_T + \left(\frac{\partial (PV)}{\partial P}\right)_T\right]$$

4.29 Show that dq is not exact in

$$dq = C_V\, dT + P\, dV$$

when P is a function of T and V while C_V is a function only of T.

4.30 For water at 25° C the coefficient of cubical expansion α is 2.1×10^{-4} K^{-1} and the compressibility coefficient β 4.9×10^{-5} atm . If water is heated from 25 C but not allowed to expand, what temperature rise is needed to produce 100 atm pressure?

References

Books

De Heer, J.: 1986, *Phenomenological Thermodynamics*, Prentice-Hall, Englewood Cliffs, NJ, pp. 2-49, 78-100.

This text is a concise, readable, yet critical, presentation of chemical thermodynamics. The nature of thermodynamics, the use of models, the concepts of temperature, heat, work, reversibility, are all discussed. Both traditional and axiomatic approaches to the first law are surveyed. Important chemical applications are outlined. Key references are cited and annotated.

Haase, R.: 1971, "Survey of Fundamental Laws," in Jost, W. (editor), *Physical Chemistry An Advanced Treatise*, vol. I, Academic Press, New York, pp 1-37.

Here the basic concepts through the first law are covered in mathematical detail.

Lewis, G. N., Randall, M., Pitzer, K. S., and Brewer, L.: 1961, *Thermodynamics*, 2nd ea., McGraw-Hill Book Co., New York, pp. 1-52.

The first edition of this classic text, by Lewis and Randall alone, appeared in 1923. Because of its simple style and comprehensiveness, it served to introduce a whole generation of chemists to the use of thermodynamics. The second edition, appearing in 1961, was prepared by Pitzer and Brewer.

Truesdell, C.: 1980, *The Tragicomical History of Thermodynamics 1822-1854*, Springer-Verlag, New York, pp. 1-185.

Truesdell reviews the early development of thermodynamics rationally and critically.

5

Entropy and the Second Law

5.1 Accessibility of States

A MACROSCOPIC SYSTEM OF GIVEN COMPOSITION is made up of a very large number of molecules, atoms, radicals, and/or ions. The internal energy of the system is associated with random chaotic movements of these particles. Throughout a uniform region, the Kelvin temperature is proportional to the average kinetic energy in a classically excited degree of freedom.

Raising the temperature of a uniform section of a system requires adding energy to that section; lowering the temperature requires removing energy therefrom. These processes may be effected by work done on and/or heat transferred to the section.

A continuous manifold of states of a uniform system is generated from a given state by reversible adiabatic processes. In the space of the independent variables, this manifold forms a surface.

From a given point on such a surface, the given system may be moved at constant volume to lower temperatures or to higher temperatures. If only work of compression is allowed and the volume is kept constant, then to take the system to a lower temperature requires removal of heat. Similarly, to reach the point from the high temperature side at constant volume requires removal of heat. But through any given state of a uniform system, the adiabatic surface exists. Furthermore, the complete set of such surfaces exhausts the space of the independent variables. We are thus led to the *Principle of Caratheodory*:

Near any given state of a uniform system, there are states that cannot be reached by adiabatic processes. But if adiabatic transition from state 2 to state 1 is impossible, adiabatic transition from state 1 to state 2 can be carried out.

With the Caratheodory principle, one can order all states of a given uniform system along a line. Each point s on the line corresponds to a reversible adiabatic surface. A given s then separates the states that cannot be reached adiabatically starting on the corresponding surface from the states that cannot move to the surface adiabatically. See figure 5.1.

We have thus correlated all states of the system with points on the straight line. Let us assign the same entropy S to the states correlated with point s. These states are linked by adiabatic reversible processes. A larger S is assigned states to the right, those states for which removal of heat is needed to reach the surface identified with s. So if such a state is to be reached reversibly from this surface, heat would have to be added.

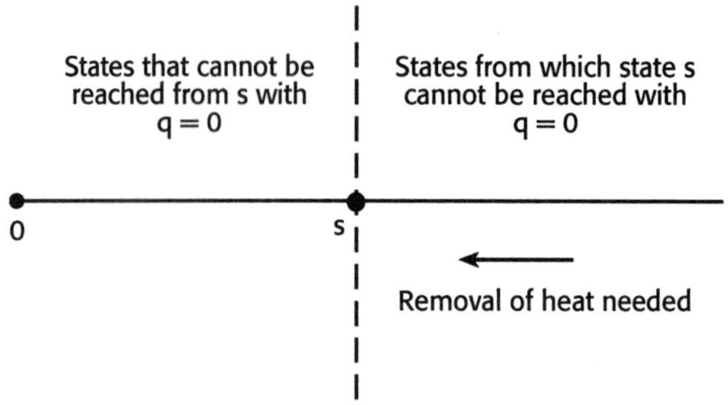

FIGURE 5.1 Correlation of states of a uniform system with points on a line.

We are thus led to write

$$dS = t^{-1}\, dq_{\text{rev}}. \qquad [5.1]$$

Since S is a function of state, expression t^{-1} is an integrating factor for dq_{rev}.

5.2 Integrability of the Reversible Heat

Let us now look at a general form for dq_{rev} and determine how it is integrable.
From section 4.2, the work done on a system in a reversible process has the form

$$dw_{\text{rev}} = \sum_{j=1}^{n-1} F_j\, dx_j \qquad [5.2]$$

in which F_j is the jth generalized force and x_j: is the jth generalized coordinate. Let the nth coordinate be the temperature.

The internal energy depends on all the coordinates; so

$$E = E(x_1, \ldots, x_n) = E(\mathbf{r}) \qquad [5.3]$$

and

$$dE = \sum_{j=1}^{n} \frac{\partial E}{\partial x_j}\, dx_j. \qquad [5.4]$$

With formula (4.12), we have

$$dq_{\text{rev}} = dE - dw_{\text{rev}} = \sum \frac{\partial E}{\partial x_j}\, dx_j - \sum F_j\, dx_j = \sum_{j=1}^{n} X_j\, dx_j \qquad [5.5]$$

where

$$X_j = \frac{\partial E}{\partial x_j} - F_j. \qquad [5.6]$$

The independent variables may be considered as Cartesian coordinates in an n-dimensional Euclidean space. At the origin these variables would all equal zero. A radius vector $\mathbf{r}$ drawn from the origin to the point $(x_1, x_2, \ldots, x_n)$ then defines this point. Coefficient X_j may be considered the jth component of vector function $\mathbf{R}$. See figure 5.2.

At a given entropy S, differential dq_{rev} equals zero and

$$\sum X_j\, dx_j = 0. \qquad [5.7]$$

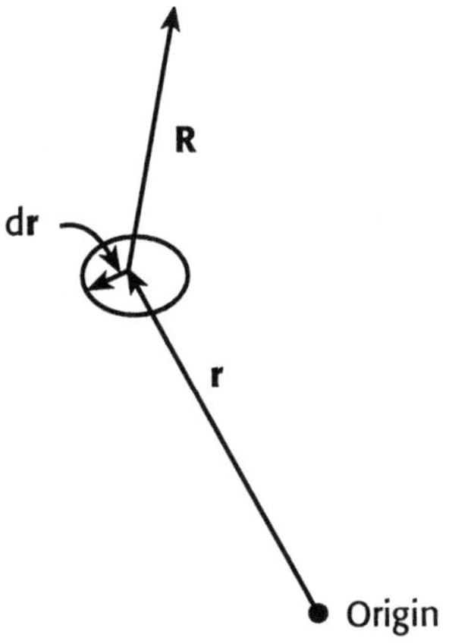

FIGURE 5.2 Plot of the n-dimensional **R**, **r**, and **dr**.

In the vector notation, equation (5.7) becomes

$$\mathbf{R} \cdot \mathbf{dr} = 0. \quad [5.8]$$

For the given system and conditions, vector **r** would determine the state while **R** would involve the behavior of the internal energy and the forces acting. Equation (5.8) implies that in an adiabatic reversible change, differential **dr** is perpendicular to **R**, as figure 5.2 shows.

Around the initial point, **dr** defines an element of surface. One can go to a point on the periphery, determine **R** there, construct the corresponding **dr**, and extend the surface. By iteration, one can in principle develop as much of the surface as necessary. Thus, equation (5.8) can be integrated to produce a locus of points on which S is constant:

$$S\left(x_1,\ldots,x_n\right) = c. \quad [5.9]$$

Any nonzero dq_{rev} involves going between points on two different surfaces, as from $\mathcal{A}$ to $\mathcal{C}$ in figure 5.3. For this change,

$$dS = dc. \quad [5.10]$$

However, differential dq_{rev} is not exact; it depends on the path. Two possible paths are sketched in figure 5.3. The crossings from $\mathcal{B}_1$ and $\mathcal{B}_2$ are set along the shortest paths to the varied surface. Because the sections of dq_{rev} along the constant S surfaces vanish, the nonzero contribution comes from the crossing. But this here depends on the choice of $\mathcal{B}$. So we have

$$dq_{rev} = t\left(\mathcal{B}\right) dS \quad [5.11]$$

or

$$dS = \frac{dq_{rev}}{t\left(\mathcal{B}\right)}, \quad [5.12]$$

where t($\mathcal{B}$) is a function to be determined.

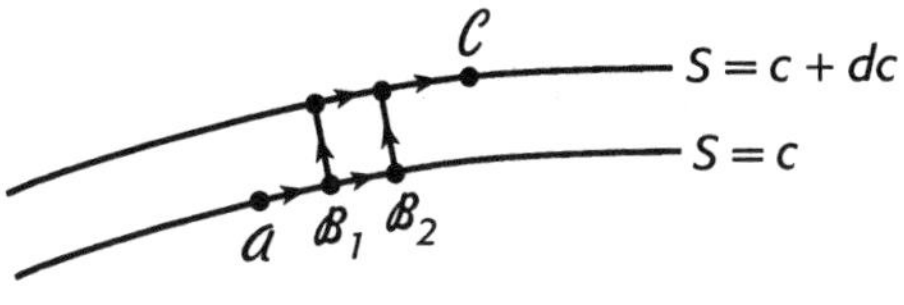

FIGURE 5.3 Two different paths from $\mathcal{A}$ to $\mathcal{C}$.

Consider a general uniform system A and a uniform ideal gas system B separated by a fixed heat conducting membrane. Thus, the systems are kept at the same temperature. The above argument can clearly be applied to changes in each system by itself. Furthermore, it can be applied to changes in the combined system.

Since the heat added to the combined system equals that added to A plus that added to B, we have

$$dq = dq_A + dq_B. \quad [5.13]$$

For a reversible step, equation (5.11) changes (5.13) to

$$t\, dS = t_A\, dS_A + t_B\, dS_B. \quad [5.14]$$

Because we consider entropy to be extensive and thus additive, we also have

$$dS = dS_A + dS_B. \quad [5.15]$$

Now, expression t_A is a function of the independent variables of system A, t_B a function of those of system B, and t those of the combined system. But for both (5.14) and (5.15) to hold, these must be the same. Furthermore, since the only common independent variable is the temperature T, this function must depend only on T.

5.3 Form for the Integrating Factor

The expression for the integrating factor can be induced from the behavior of an ideal gas.

Consider n moles of gas subject only to reversible changes with all work done work of compression. Then

$$dq_{rev} = dE - dw_{rev} = \left(\frac{\partial E}{\partial T}\right)_V dT + \left(\frac{\partial E}{\partial V}\right)_T dV + P\, dV. \quad [5.16]$$

Assume that the gas is ideal so that both (4.35) and (4.37) apply. Also introduce the ideal gas equation to construct

$$dq_{rev} = C_V\, dT + nRT\frac{dV}{V} \quad [5.17]$$

whence

$$\frac{dq_{rev}}{T} = C_V\frac{dT}{T} + nR\frac{dV}{V}. \quad [5.18]$$

Since C_V is a function of T alone, both terms on the right can be integrated without specifying the path. Thus, dq_{rev}/T is an exact differential. In equation (5.12), we can substitute T for t to get

$$dS = \frac{dq_{rev}}{T}. \quad [5.19]$$

From the argument at the end of section 5.2, we assume that relationship (5.19) holds for any uniform system or region.

5.4 Entropy Changes of Systems

Equation (5.19) can be applied to determine the entropy difference between any two states that can be linked by a reversible process.

The energy needed to raise the temperature of a uniform system by amount dT is $C\,dT$, where C is the energy capacity. If this were done reversibly, the energy would be in the form of heat:

$$dq_{rev} = C\,dT. \qquad [5.20]$$

Consequently, we have

$$dS = \frac{C\,dT}{T} \qquad [5.21]$$

and

$$\Delta S = \int_{T_1}^{T_2} \frac{C\,dT}{T}. \qquad [5.22]$$

Conventionally, C is called the heat capacity.

At a first-order phase change, coefficient C is infinite. But such a change occurs at a definite temperature. And if it were done reversibly, the energy required would be in the form of heat. Since this process occurs at constant pressure, we have

$$\Delta S = \frac{q_P}{T}, \qquad [5.23]$$

where q_P is the heat needed to effect the phase change at temperature T.

For a finite change in an ideal gas, equations (5.19) and (5.18) yield

$$\Delta S = \int_{T_1}^{T_2} C_V \frac{dT}{T} + \int_{V_1}^{V_2} nR\frac{dV}{V}. \qquad [5.24]$$

If we also consider C_V to be constant, we get

$$\Delta S = C_V \ln\frac{T_2}{T_1} + nR\ln\frac{V_2}{V_1}. \qquad [5.25]$$

At constant temperature, the first term on the right is zero and the entropy change is

$$\Delta S_T = nR\ln\frac{V_2}{V_1} = nR\ln\frac{P_1}{P_2}. \qquad [5.26]$$

The last equality arises because the pressure of an ideal gas is inversely proportional to its volume at a given temperature.

Example 5.1

Calculate the entropy change when NH_4NO_3 undergoes a reversible transition at 32.1° C, absorbing 1590 J mol $^{-1}$.

Substitute into (5.23) to obtain

$$\Delta S = \frac{1590 \text{ J mol}^{-1}}{305.25 \text{ K}} = 5.21 \text{ J mol}^{-1}\text{ K}^{-1}.$$

Example 5.2

Calculate the entropy change when 1.00 mol argon is heated from 50° to 100° C at 1.00 atm.

Since argon behaves as a monatomic ideal gas, we have

$$C_V = \left(\frac{\partial E}{\partial T}\right)_V = \frac{\partial}{\partial T}\left(\frac{3}{2}nRT\right) = \frac{3}{2}nR$$

and

$$C_P = C_V + nR = \frac{5}{2}nR.$$

Substituting into (5.22),

$$\Delta S = \int_{T_1}^{T_2} C_P \frac{\mathrm{d}T}{T} = C_P \ln\frac{T_2}{T_1} = \frac{5}{2}nR\ln\frac{T_2}{T_1},$$

with the given numbers, yields

$$\Delta S = \frac{5}{2}(1.00\ \text{mol})(8.3145\ \text{J K}^{-1}\ \text{mol}^{-1})\ln\frac{373.15}{323.15} = 2.99\ \text{J K}^{-1}.$$

Example 5.3

How does one calculate ΔS when 1.00 mol liquid water at -10° C and 1.00 atm freezes to ice at the same temperature and pressure?

Equation (5.19) and its integral can only be applied along a reversible path. Here one would determine integral (5.22) for heating the liquid water from -10° to 0° C. Then (5.23) would be applied to the reversible freezing at 0° C. Finally, integral (5.22) would be determined for the ice cooling from 0° to -10° C. These results would be added.

5.5 The Second Law

Any system behaving naturally moves toward equilibrium with its surroundings. In the process, energy that could appear as work is dissipated replacing heat that otherwise would be absorbed. In our discussion, we will consider the surroundings to behave reversibly, so that all of the dissipation occurs in the system itself.

Let E be the internal energy of the system. Then between two successive states in a natural process, the work done on the system would be equal to or greater than that for a reversible step, to allow for the dissipation:

$$\mathrm{d}w \geq \mathrm{d}w_{\text{rev}}. \qquad [5.27]$$

The change in internal energy is independent of whether the step is reversible or not:

$$\mathrm{d}q_{\text{rev}} + \mathrm{d}w_{\text{rev}} = \mathrm{d}E = \mathrm{d}q + \mathrm{d}w. \qquad [5.28]$$

Combining (5.27) and (5.28) leads to

$$\mathrm{d}q \leq \mathrm{d}q_{\text{rev}}. \qquad [5.29]$$

Substituting inequality (5.29) into formula (5.19) and adding the subscript 1 everywhere yields

$$\mathrm{d}S_1 \geq \frac{\mathrm{d}q_1}{T_1}. \qquad [5.30]$$

for the system.

Since the surrounding material behaves reversibly, for it we have

$$dS_2 = \frac{dq_2}{T_2} = -\frac{dq_1}{T_2}. \qquad [5.31]$$

Adding (5.30) and (5.31) gives us

$$dS \geq \left(\frac{1}{T_1} - \frac{1}{T_2}\right) dq_1 = \frac{T_2 - T_1}{T_1 T_2} dq_1 \geq 0 \qquad [5.32]$$

for the total entropy change. The last inequality follows from the zeroth law. .

The results we have obtained are embodied in the *second law of thermodynamics*: For a uniform system in contact with a reversibly acting surroundings, there is a function of state, entropy S, whose changes are governed by

$$dS \geq \frac{dq}{T}. \qquad [5.33]$$

The total entropy change of an isolated system is positive for a spontaneous change.

5.6 Entropy of Mixing

The essential features of a spontaneous process are exhibited in the mixing of two ideal gases to form an ideal solution.

Let us start with N_A molecules of ideal gas A at pressure P and temperature T in volume V_A on one side of a shutter and with N_B molecules of ideal gas B at the same P and T in volume V_B on the other side. Then open the shutter and let the gases mix at constant temperature as figure 5.4 indicates.

We suppose that the solution is ideal, with X_A the mole fraction of A, X_B the mole fraction of B, and n the total number of moles. By equation (3.37) the partial pressures are

$$P_A = X_A P, \qquad P_B = X_B P. \qquad [5.34]$$

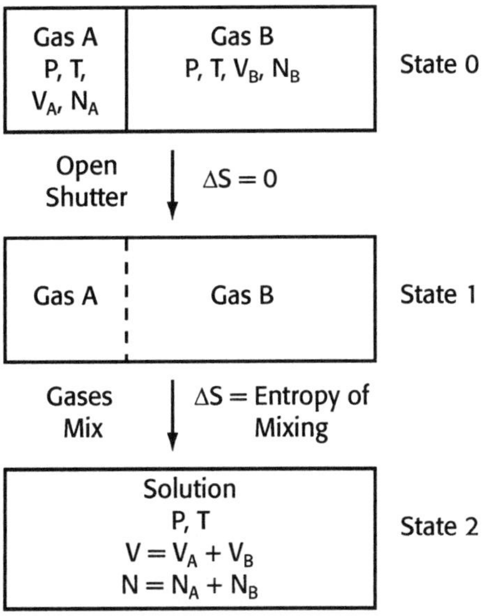

FIGURE 5.4 Stages in the mixing of two gases.

Since the mixture is ideal, gas A behaves as if gas B is not present, and vice versa. The entropy change for gas A in expanding from pressure P to pressure P_A is

$$\Delta S_A = nX_A R \ln \frac{P}{P_A} = -nX_A R \ln X_A, \quad [5.35]$$

according to equation (5.26). Similarly, the entropy change for gas B in expanding from pressure P to pressure P_B is

$$\Delta S_B = nX_B R \ln \frac{P}{P_B} = -nX_B R \ln X_B. \quad [5.36]$$

For the total entropy change on mixing, we obtain

$$\Delta S = -nR\left(X_A \ln X_A + X_B \ln X_B\right) = -\frac{nR}{N}\left(\ln X_A^{NX_A} + \ln X_B^{NX_B}\right) = k \ln \frac{1}{X_A^{NX_A} X_B^{NX_B}}. \quad [5.37]$$

Here k is Boltzmann's constant, X_A the mole fraction of A, X_B the mole fraction of B, NX_A the molecules of A, and NX_B the molecules of B in the mixture.

5.7 Probability Change on Mixing

Let us now go over the mixing process of figure 5.4, comparing a statistical weight for state 1 with that for state 2. The result will be used to relate the entropy to an appropriate statistical weight for a system.

In the final state, each molecule is somewhere in volume V. So its contribution to the statistical weight for the state can be taken as 1. But since statistical weights multiply when combined, the net weight of state 2 on the chosen scale is

$$w_2 = 1^N = 1. \quad [5.38]$$

In the state immediately after the shutter is opened, molecules of A occupy only volume V_A of V and molecules of B occupy only volume V_B of V. If each element of V is exactly like any other element of the same size, the probability of finding a given molecule in one is proportional to its volume. So the statistical weight for finding a molecule of A in V_A, rather than in V, is V_A/V. Since statistical weights multiply, the weight for NX_A molecules of A being in V_A is

$$w_A = \left(\frac{V_A}{V}\right)^{NX_A} = X_A^{NX_A}. \quad [5.39]$$

The second equality arises because in the ideal-gas solution

$$X_A = \frac{V_A}{V_A + V_B} = \frac{V_A}{V}. \quad [5.40]$$

Similarly, the statistical weight for finding NX_B molecules of B in volume V_B, rather than in the total volume V, is

$$w_B = \left(\frac{V_B}{V}\right)^{NX_B} = X_B^{NX_B}. \quad [5.41]$$

Consequently, the combined weight for state 1 is

$$w_1 = w_A w_B = X_A^{NX_A} X_B^{NX_B}. \quad [5.42]$$

Substituting (5.38) and (5.42) for the numerator and the denominator in the operand in formula (5.37) leads to

$$S_2 - S_1 = \Delta S = k \ln \frac{w_2}{w_1} = k \ln w_2 - k \ln w_1. \qquad [5.43]$$

For this equation to hold, we must have

$$S = k \ln w + \text{const}' \qquad [5.44]$$

where S is the entropy of the given state, w the statistical weight for the state, and k Boltzmann's constant.

An isolated system generally moves from a state of lower to a state of higher probability. In such a change, weight w increases and entropy S increases. Thus, the second law of thermodynamics is vindicated.

Example 5.4

In an isolated system, fluctuations from the equilibrium state occur. What entropy change is associated with going to a state whose statistical weight is 0.001 times that of the equilibrium state? It is given that

$$\frac{w_2}{w_1} = 0.001.$$

So formula (5.43) yields

$$\Delta S = k \ln \frac{w_2}{w_1} = \left(1.3807 \times 10^{-23}\ \text{J K}^{-1}\right) \ln 0.001 = -9.5 \times 10^{-23}\ \text{J K}^{-1}.$$

This change is not significant except in a very small system (one of colloidal size).

5.8 Effectively Contributing Microstates

Fixing the thermodynamic variables of a given system does not determine what individual molecules do. The number of variables is too small by many orders of magnitude. As a consequence, numerous disjoint (completely distinct) molecular states contribute to a given thermodynamic state. And the amount that a molecular state contributes may vary between 0 and 1 over time.

In the absence of evidence to the contrary, we presume that the average contribution of each of the disjoint molecular states is the same. In effect, each is equally probable. If we let W be the effective *number* of such *microstates* that contribute to the macrostate, the statistical weight w for the macrostate is proportional:

$$w = aW. \qquad [5.45]$$

This relationship transforms formula (5.44) to

$$S = k \ln aW + \text{const}' = k \ln W + \text{const}. \qquad [5.46]$$

Equation (5.46) has been established for ideal gases. Does it hold in general?

Consider a system C possessing entropy S_C and state number W_C in contact with an ideal-gas system D possessing entropy S_D and state number W_D. The entropy S of the combined system equals the sum of the entropies of the parts:

$$S = S_C + S_D. \qquad [5.47]$$

The number of states W equals the product of the state numbers for the parts:

$$W = W_C W_D. \tag{5.48}$$

We presume that each entropy depends only on the corresponding state number:

$$S = g(W). \tag{5.49}$$

Multiply both sides of equation (5.48) by $b_C b_D$ and then operate with k ln to get

$$k \ln b_C b_D W = k \ln b_C W_C + k \ln b_D W_D. \tag{5.50}$$

Subtract (5.50) from (5.47) to construct

$$(S - k \ln b_C b_D W) = (S_C - k \ln b_C W_C) + (S_D - k \ln b_D W_D). \tag{5.51}$$

This equation has the form

$$f(W) = f(W_C) + f(W_D). \tag{5.52}$$

System D consists of an ideal gas while system C consists of some other substance. Let us set k equal to Boltzmann's constant and adjust b_D so that $k \ln b_D$ equals the unprimed constant in equation (5.46). Then term $f(W_D)$ is zero identically and equation (5.52) reduces to

$$f(W) = f(W_C). \tag{5.53}$$

This equation implies a different relationship from (5.48) unless $f(W)$ and $f(W_C)$ are zero identically. But the condition $f(W) = 0$ makes

$$S = k \ln bW = k \ln W + k \ln b. \tag{5.54}$$

Here S is the entropy of the system, k Boltzmann's constant, W the state number, and b a constant.

In a uniform region at a given temperature, the energy tends to become distributed as randomly as possible among the allowed molecular states. Increasing the temperature increases the number of collective states available, increases W, and increases the randomness or disorder. Decreasing the temperature has the opposite effect. There is a lower limit on W. For a substance that forms a perfectly ordered crystal, it is 1. This would be approached as the temperature approached zero kelvin. Thus, there is a lower limit on the entropy S.

Now, there is no reason for taking the limit on S different from zero. This inference is embodied in the *third law of thermodynamics*: The entropy of a perfectly ordered crystal at 0 K is zero.

But for this to be true, constant b in (5.54) needs to equal 1 and the formula reduces to

$$S = k \ln W. \tag{5.55}$$

5.9 *Helmholtz Free Energy*

In the absence of dissipation, the work done on a system at constant temperature goes to increase an energy function labeled A. In the reverse process, this is recovered as work. But during a spontaneous process, dissipation prevails. Then at constant temperature and volume, the function A decreases.

Consider a given system. As long as it is behaving reversibly, we have

$$dw_{rev} = dE - dq_{rev} = dE - T\, dS. \tag{5.56}$$

When the temperature is kept constant, operator d commutes with T and

$$\mathrm{d}w_{\mathrm{rev}} = \mathrm{d}\left(E - TS\right)_T = \mathrm{d}A_T. \qquad [5.57]$$

Integration of this yields

$$w_{\mathrm{rev}} = \Delta\left(E - TS\right)_T = \Delta A_T. \qquad [5.58]$$

In the last steps, the label

$$A = E - TS \qquad [5.59]$$

has been introduced.

Function A is called the *Helmholtz free energy*. The energy is free in the sense that any increase in it may reappear as work when the pertinent process is reversed at the given temperature.

During a spontaneous process in a system, some dissipation occurs, some of the work replaces some of the heat, and

$$\mathrm{d}w > \mathrm{d}w_{\mathrm{rev}}. \qquad [5.60]$$

Combining inequality (5.60) with equality (5.57) gives us

$$\mathrm{d}w \geq \mathrm{d}A_T. \qquad [5.61]$$

When volume V of the system is also constant, all work is net work. But a spontaneous process does not require any net work to be done on the system. So we have

$$\mathrm{d}A_{T,V} \leq 0 \qquad [5.62]$$

in any such process at constant T and V. At constant temperature and volume, the Helmholtz free energy A of a system decreases spontaneously until it can decrease no more and equilibrium is reached.

5.10 Gibbs Free Energy

At constant temperature and pressure the external energy PV as well as the Helmholtz free energy A are sources for work energy. Thus, work is available for dissipation as long as $A + PV$ can decrease.

We again consider a specific system. The *net work* equals the total work minus the work of compression. For an infinitesimal change, we have

$$\mathrm{d}w_{\mathrm{net}} = \mathrm{d}w + P\,\mathrm{d}V. \qquad [5.63]$$

When this is reversible, the relation becomes

$$\mathrm{d}w_{\mathrm{net,rev}} = \mathrm{d}E - T\,\mathrm{d}S + P\,\mathrm{d}V. \qquad [5.64]$$

When both temperature and pressure are kept constant, operator d commutes with T and P and

$$\mathrm{d}w_{\mathrm{net,rev}} = \mathrm{d}\left(E - TS + PV\right)_{T,P} = \mathrm{d}\left(A + PV\right)_{T,P} = \mathrm{d}\left(H - TS\right)_{T,P} = \mathrm{d}G_{T,P}. \qquad [5.65]$$

Integration of this yields

$$w_{\mathrm{net,rev}} = \Delta\left(H - TS\right)_{T,P} = \Delta G_{T,P}. \qquad [5.66]$$

In the last steps, the label

$$G = H - TS \qquad [5.67]$$

has been introduced.

Function G is called the *Gibbs free energy*. Any increase in it during a process at constant T and P may reappear as net work when the process is reversed.

During an irreversible process, dissipation of work into heat occurs. At constant temperature and pressure, we have

$$dw_{\text{net}} > dw_{\text{net,rev}}, \tag{5.68}$$

Combining inequality (5.68) with equality (5.65) yields

$$dw_{\text{net}} \geq dG_{T,P}. \tag{5.69}$$

A spontaneous process does not require any net work to be done on the system. Then (5.69) reduces to

$$dG_{T,P} \leq 0. \tag{5.70}$$

At constant temperature and pressure, the Gibbs free energy of a system decreases spontaneously until it can decrease no more and equilibrium is reached. The condition for equilibrium is that

$$dG_{T,P} = 0 \tag{5.71}$$

for possible small changes.

Example 5.5

Calculate ΔA and ΔG for the vaporization of 1 mole water at 372.778 K and 1 bar pressure, where the heat of vaporization is 40,893 J mol^{-1}.

The increments of

$$A = E - TS \qquad \text{and} \qquad G = H - TS$$

at constant temperature are

$$\Delta A = \Delta E - T\Delta S \qquad \text{and} \qquad \Delta G = \Delta H - T\Delta S.$$

But with equation (5.23),

$$T\Delta S = q_P,$$

with equation (4.13),

$$\Delta E = q_P + w_P,$$

and with formula (4.27),

$$\Delta H = q_P,$$

we find that

$$\Delta A = q_P + w_P - q_P = w_P$$

and

$$\Delta G = q_P - q_P = 0 \text{ J}.$$

The work done was estimated in example 4.3 as

$$w_P = -3099 \text{ J}.$$

So

$$\Delta A = -3099 \text{ J}.$$

5.11 Key Thermodynamic Relations

Let us now summarize the main conditions relating the thermodynamic properties of a system when no material moves in or out and when no reaction occurs.

The energy added to such a system can be classified as heat absorbed q and as work done w. During an infinitesimal step, the increase in internal energy E is

$$dE = dq + dw. \qquad [5.72]$$

The heat absorbed in a reversible process goes to increase the entropy S by the formula

$$dS = \frac{dq_{rev}}{T}, \qquad [5.73]$$

where T is the absolute temperature. The first law tells us that dE is exact; the second law, that dS is exact.

A given system may be taken from one state to another along a reversible path with all work done work of compression. With P the pressure, V the volume, and S the entropy, we have

$$dw = dw_{rev} = -P\,dV \qquad [5.74]$$

and

$$dq = dq_{rev} = T\,dS. \qquad [5.75]$$

Substituting into equation (5.72) yields

$$dE = T\,dS - P\,dV. \qquad [5.76]$$

Relation (5.76) can be applied even though the actual process is irreversible and involves other kinds of work because an exact differential depends only on the initial and final points infinitesimally far apart.

By definition, the enthalpy H, Helmholtz free energy A, and Gibbs free energy G are given by

$$H = E + PV, \qquad [5.77]$$

$$A = E - TS, \qquad [5.78]$$

$$G = H - TS. \qquad [5.79]$$

Differentiating each of these equations and combining the result with condition (5.76) yields

$$dH = T\,dS + V\,dP, \qquad [5.80]$$

$$dA = -P\,dV - S\,dT, \qquad [5.81]$$

$$dG = V\,dP - S\,dT. \qquad [5.82]$$

Since differential dA is exact, equation (5.81) fits the form of (4.75) and

$$\left(\frac{\partial A}{\partial V}\right)_T = -P, \qquad \left(\frac{\partial A}{\partial T}\right)_V = -S. \qquad [5.83]$$

Differentiating the first of equations (5.83) with respect to P, the second with respect to V, and changing signs leads to

$$-\frac{\partial^2 A}{\partial V \partial T} = \left(\frac{\partial P}{\partial T}\right)_V = \left(\frac{\partial S}{\partial V}\right)_T. \qquad [5.84]$$

Since differential dG is exact, equation (5.82) similarly yields

$$\left(\frac{\partial G}{\partial P}\right)_T = V, \qquad \left(\frac{\partial G}{\partial T}\right)_P = -S. \qquad [5.85]$$

Differentiating the first of equations (5.85) with respect to T, the second with respect to P, leads to

$$\frac{\partial^2 G}{\partial P \partial T} = \left(\frac{\partial V}{\partial T}\right)_P = -\left(\frac{\partial S}{\partial P}\right)_T. \quad [5.86]$$

If we consider the entropy S to be a function of T and V, then

$$\mathrm{d}S = \left(\frac{\partial S}{\partial T}\right)_V \mathrm{d}T + \left(\frac{\partial S}{\partial V}\right)_T \mathrm{d}V. \quad [5.87]$$

Substituting this form into equation (5.76) yields

$$\mathrm{d}E = T\left[\left(\frac{\partial S}{\partial T}\right)_V \mathrm{d}T + \left(\frac{\partial S}{\partial V}\right)_T \mathrm{d}V\right] - P\,\mathrm{d}V. \quad [5.88]$$

But the coefficient of dV is

$$\left(\frac{\partial E}{\partial V}\right)_T = T\left(\frac{\partial S}{\partial V}\right)_T - P. \quad [5.89]$$

Eliminating $(\partial S/\partial V)_T$ with relation (5.84),

$$\left(\frac{\partial E}{\partial V}\right)_T = T\left(\frac{\partial P}{\partial T}\right)_V - P, \quad [5.90]$$

and $(\partial P/\partial T)_V$ with relation (4.86) yields

$$\left(\frac{\partial E}{\partial V}\right)_T = \frac{\alpha}{\beta}T - P. \quad [5.91]$$

Substituting this result and (4.84) into (4.81) gives us

$$C_P - C_V = \left[\frac{\alpha}{\beta}T - P + P\right]\alpha V = VT\frac{\alpha^2}{\beta}. \quad [5.92]$$

Example 5.6

Obtain $(\partial E/\partial V)_T$ for an ideal gas. An ideal gas obeys the equation

$$P = \frac{nRT}{V}.$$

Differentiate this with respect to T at constant V,

$$\left(\frac{\partial P}{\partial T}\right)_V = \frac{nR}{V},$$

and substitute into formula (5.90):

$$\left(\frac{\partial E}{\partial V}\right)_T = T\frac{nR}{V} - P = 0.$$

Example 5.7

Obtain C_P - C_V for an ideal gas. Rearrange the ideal gas equation to

$$V = \frac{nRT}{P}$$

and differentiate with respect to T at constant P:

$$\left(\frac{\partial V}{\partial T}\right)_P = \frac{nR}{P}.$$

Substitute this result and that from example 5.6 into formula (4.81)

$$C_P - C_V = \left(0 + P\right)\frac{nR}{P} = nR.$$

Note that this agrees with the result obtained in example 4.4,

5.12 *The Chemical Potential*

So far, we have considered systems in which the number of moles of each constituent is constant. But in general, this condition does not apply. Reactions go on and material may move into or out of the given system. Then one must introduce additional terms into formula (5.76). and the equations derived from it.

Material always carries energy. The reversible rate of change in internal energy caused by altering the number n_i of moles of the ith constituent, while keeping all other independent mole numbers, entropy S, and volume V constant, is called the *chemical potential* μ_i; thus

$$\mu_i = \left(\frac{\partial E}{\partial n_i}\right)_{S,V,\text{other } n\text{'s}} \qquad [5.93]$$

and

$$dE = T\,dS - P\,dV + \sum_i \mu_i\,dn_i. \qquad [5.94]$$

Combining formula (5.94) with the differentials of equations (5.77). (5.78), (5.79) leads to

$$dH = T\,dS + V\,dP + \sum_i \mu_i\,dn_i, \qquad [5.95]$$

$$dA = -P\,dV - S\,dT + \sum_i \mu_i\,dn_i, \qquad [5.96]$$

$$dG = V\,dP - S\,dT + \sum_i \mu_i\,dn_i. \qquad [5.97]$$

One may consider building up a system from nothing at constant pressure and temperature. Throughout such a process, equation (5.97) reduces to

$$dG = \sum_i \mu_i\,dn_i \qquad [5.98]$$

whence

$$\frac{dG}{dn} = \sum_i \mu_i \frac{dn_i}{dn}. \qquad [5.99]$$

where

$$n = \sum_i n_i, \qquad [5.100]$$

the total number of moles in the system.

If the concentrations are kept constant throughout the process, then dn_i/dn and dG/dn are constant and

$$\frac{dn_i}{dn} = \frac{n_i}{n}, \qquad \frac{dG}{dn} = \frac{G}{n}. \qquad [5.101]$$

Substituting into (5.99) and canceling n yields

$$G = \sum_i \mu_i n_i. \qquad [5.102]$$

Thus, μ_i equals the contribution per mole of the ith constituent to the Gibbs free energy, As a consequence, it is called the partial molar Gibbs free energy of constituent i in the solution.

5.13 Concentration Gradients

When a constituent diffuses through a viscous medium under the influence of a concentration gradient, dissipation of energy occurs. When the temperature and pressure are kept constant, this energy is Gibbs energy.

Consider a solution in which the concentration c_i of the ith constituent is a function of coordinate x alone. Suppose that this concentration gradient causes the i molecules to travel through a given cross section at an average speed $\dot{x}_i$. If the molecules continued to travel at this speed, they would travel distance $\dot{x}_i$ in unit time. So the molecules passing through unit cross section at position x in unit time equal

$$c_i \dot{x}_i = J_i \qquad [5.103]$$

in number.

Now, increasing the negative concentration gradient should increase J_i; decreasing it should decrease J_i. We set

$$J_i = -D_i \frac{\partial c_i}{\partial x}, \qquad [5.104]$$

where D_i is called the *diffusion coefficient* for the ith constituent. Combining equations (5.103) and (5.104) gives

$$\dot{x}_i = -\frac{D_i}{c_i}\frac{\partial c_i}{\partial x} = -D_i \frac{\partial \ln c_i}{\partial x}. \qquad [5.105]$$

We expect D_i to vary with the other constituents present and with temperature T and pressure P. When these influences are kept fixed, we expect it to vary only slowly with c_i. The empirical *Fick's law* states that D_i is independent of c_i.

In a first approximation, one may consider each i molecule to behave as a hard sphere moving through a viscous continuous fluid. But the force f needed to maintain a sphere of radius r at speed $\dot{x}$ is

$$f = 6\pi\eta r\dot{x}, \qquad [5.106]$$

where η is the viscosity of the ambient fluid. This is known as *Stokes' law.*

For one mole of the ith constituent, we have the force

$$F_i = N6\pi\eta r_i \dot{x}_i. \qquad [5.107]$$

The work done by this force is dissipated. But at constant P and T, the dissipated energy is Gibbs energy; we have

$$-\mathrm{d}\mu_i = -\left(\frac{\partial \mu_i}{\partial x}\right)_{T,P} \mathrm{d}x = F_i \, \mathrm{d}x. \qquad [5.108]$$

The force F_i derived from the chemical potential μ_i is known as the *thermodynamic force* acting on the ith constituent.

We will find that the chemical potential is related to the activity a_i of the constituent by the logarithmic relationship:

$$\mu_i = \mu_i^0 + RT \ln a_i. \qquad [5.109]$$

Furthermore, the activity is approximately represented by the concentration c_i:

$$\mu_i \cong \mu_i^0 + RT \ln c_i. \qquad [5.110]$$

Substituting the derivative of (5.110) into the expression for force F_i from (5.108) gives us

$$F_i = -\left(\frac{\partial \mu_i}{\partial x}\right)_{T,P} = -RT\left(\frac{\partial \ln c_i}{\partial x}\right)_{T,P}. \qquad [5.111]$$

But combining equations (5.107) and (5.105) yields

$$F_i = N6\pi\eta r_i \left(-D_i \frac{\partial \ln c_i}{\partial x}\right)_{T,P}. \qquad [5.112]$$

Eliminating F_i from (5.111) and (5.112), then solving for the diffusion coefficient leads to the result

$$D_i = \frac{RT}{N6\pi\eta r_i} = \frac{kT}{6\pi\eta r_i}, \qquad [5.113]$$

which is known as the *Einstein-Stokes relationship.*

Example *5.7*

Along the axis of a container, the concentration of a solution drops exponentially, decreasing by one-half in each 10.0 cm section. Calculate the thermodynamic force acting on the solute at 25° C.

The concentration distribution parallel to the axis of the container is given as

$$c = c_0 e^{-ax}$$

or

$$\ln c = \ln c_0 - ax,$$

whence

$$\frac{\partial \ln c}{\partial x} = -a$$

and

$$\ln \frac{c_0}{c} = ax.$$

But increasing x from 0 to 10.0 cm causes c to drop to $1/2c_0$. We have

$$a = \frac{1}{10.0\text{ cm}} \ln \frac{c_0}{\frac{1}{2}c_0} = \frac{\ln 2}{10.0\text{ cm}}.$$

So for the thermodynamic force, we find

$$F = -RT\frac{\partial \ln c}{\partial x} = aRT = \frac{(\ln 2)(8.31451\text{ J K}^{-1}\text{ mol}^{-1})(298.15\text{ K})}{10.0\times 10^{-2}\text{ m}} = 1.72\times 10^{4}\text{ N mol}^{-1}.$$

Example 5.8

Suppose that the axis of the container in example 5.7 is oriented vertically and determine the gravitational force acting on the solute if its molar mass is 100 g mol^{-1}.

We have

$$F = -Mg = -(0.100\text{ kg mol}^{-1})(9.807\text{ m s}^{-2}) = -0.98\text{ N mol}^{-1}.$$

Note that this is negligible with respect to the thermodynamic force due to any appreciable concentration gradient. In the laboratory, one can generally neglect gravitational effects on molecules in solution.

5.14 General Direction for a Process

According to the second law of thermodynamics, spontaneous changes in a system cause the total entropy of the system and its interacting surroundings to increase. In this way, the system moves from a less probable to a more probable state. When the total entropy can no longer increase, a point of equilibrium has been reached. Thus, we have

$$\Delta S_{\text{total}} \geq 0 \qquad [5.114]$$

on the average. Small deviations from (5.114) arise because of the fluctuations noted in example 5.4.

For a process occurring at constant temperature and volume, the argument in section 5.9 applies. For the system by itself, we have

$$\Delta A_{T,V} \leq 0 \qquad [5.115]$$

on the average. For a process occurring at constant temperature and pressure, the argument in section 5.10 applies. And for the system by itself, we have

$$\Delta G_{T,P} \leq 0 \qquad [5.116]$$

on the average,

In chapter 6, we will apply the equality in (5.116) to physical equilibria; in chapter 7, to chemical equilibria.

A qualitative interpretation of condition (5.116) is useful. From definition (5.67), we have

$$\Delta G_{T,P} = \Delta H - T\Delta S. \qquad [5.117]$$

At constant entropy, an enthalpy loss makes $\Delta G_{T,P}$ negative and so drives the given process. When there is entropy change, $T\Delta S$ has the same effect in driving the process as a - ΔH of the same size.

At the atomic level, an increase of entropy corresponds to an increase of state number W, in effect, an increase in the freedom for the system. This increase in freedom can counteract an increase in enthalpy energy H, following equation (5.117).

5.15 A Fourth Law

When the molecules of a system are distributed over the allowed states randomly, the entropy S is related to the state number W by formula (5.55). In an ideal gas, the only effect of the intermolecular interactions is to introduce this randomness. But when the interactions have a restrictive effect on the translational motions of the molecules, the result is similar to that produced by reducing the volume available. At a given temperature, this reduces the entropy, according to equation (5.26).

Consider a macroscopic system of molecules at temperature T in volume V. Suppose that all parameters of each species of molecule present are known. Then one can calculate the entropy S_i, assuming there is no interaction between the molecules, but that randomness prevails.

Any possible intermolecular interaction, attraction at large distances, repulsion at small intermolecular distances, has the effect of introducing a restriction on the movements of the molecules. As a result, it lowers the state number for formula (5.55). We conclude that the ideal S_i is an upper bound on the entropy.

This inference is embodied in the *fourth law of thermodynamics*: The entropy S of a real system at a given T and V is less than it would be if the system behaved as an ideal gas; we have

$$S \leq S_i. \qquad [5.118]$$

In section 7.7, we will calculate the difference S_i - S for a nonideal gas at low pressures. The ideal entropy S_i can be calculated from molecular parameters using the Boltzmann distribution law. Molecular parameters may be determined from spectroscopic measurements. Many may be calculated from first principles using quantum mechanics.

Questions

5.1 What empirical basis is there for the Caratheodory principle?
5.2 How does the Caratheodory principle allow one to correlate all macroscopic states of a uniform system to the points on a straight line?
5.3 How does this correlation imply that an integrating factor exists for dq_{rev} for the system?
5.4 What form does dw_{rev} for a uniform system assume?
5.5 What form does dq_{rev} for a uniform system assume?
5.6 How does this form allow one to construct a surface on which entropy S of the uniform system is constant?
5.7 How may dq_{rev} vary in going between neighboring points on two such neighboring surfaces?
5.8 How may the form for the integrating factor for dq_{rev} be found?
5.9 Construct formulas for the change in entropy in standard processes.
5.10 How is energy dissipated in a spontaneous process?
5.11 How does such dissipation affect the formula for the entropy change in (a) the system, (b) the system plus surroundings?
5.12 Show how the last inequality in (5.32) follows from the zeroth law.
5.13 Summarize the results obtained above in the second law of thermodynamics.
5.14 Derive the formula for the entropy of mixing.
5.15 Construct the corresponding expression for the probability change on mixing.

5.16 How do symmetry considerations let us relate the number of microstates in a macrostate (a) to the statistical weight of the macrostate and (b) to the entropy?
5.17 What is the third law of thermodynamics?
5.18 Define and interpret the Helmholtz free energy.
5.19 Define and interpret the Gibbs free energy.
5.20 Describe how the key thermodynamic properties of a closed uniform system are related.
5.21 Define the chemical potential of a constituent whose mass can vary independently in a homogeneous uniform region.
5.22 How do E, H, A, and G vary in an open uniform system?
5.23 Explain why the chemical potential is given by the partial molar Gibbs energy.
5.24 What does a concentration gradient cause?
5.25 How is the corresponding diffusion coefficient defined? What is Fick's law?
5.26 Describe the thermodynamic force that causes diffusion.
5.27 How is the chemical potential related to the activity of a constituent?
5.28 Construct the Einstein-Stokes expression for the diffusion coefficient.
5.29 Why do we generally have $\Delta S_{total} \geq 0$?
5.30 Under what circumstances is (a) $\Delta A \geq 0$, (b) $\Delta G \geq 0$?
5.31 What is the fourth law of thermodynamics?

Problems

5.1 Calculate the entropy change of 1.000 g helium when it is heated from 25° to 125° (a) at constant volume, (b) at constant pressure.
5.2 If 1.000 mol helium is compressed reversibly and adiabatically from 22.4 l at 0° C to 2.24 1, what is ΔS?
5.3 Calculate the entropy change when 10.00 g potassium is fused at its melting point, 336.35 K, where its heat of fusion is 2343 J mol^{-1}.
5.4 At their melting points, the entropy of fusion of chlorine is 37.20 J K^{-1} mol^{-1}, that of iodine 40.12 J K^{-1} mol^{-1}. Estimate the entropy of fusion and then the heat of fusion of bromine at its melting point 265.9 K.
5.5 A resistor is kept at 30° C by a stream of cooling water while 10,000 J of work is dissipated in it. (a) What is the change in entropy of the resistor? (b) What is the change in entropy of the water if it leaves the resistor at 25° C?
5.6 One kilogram water at 0° C contacts a large heat reservoir at 100° C and the water warms up to 100° C. Consider the specific heat of the water to be 1.000 and calculate the entropy change (a) of the water, (b) of the heat reservoir, and (c) of both.
5.7 Calculate the entropy of mixing 0.200 mol oxygen with 0.800 mol nitrogen at 25° C.
5.8 One mole benzene is vaporized at 80.1° C and 1.000 atm, its boiling point, where the heat of vaporization is 393.7 J g^{-1} Calculate (a) ΔS, (b) ΔG, and (c) ΔA.
5.9 Calculate w_{rev} for the electrolysis of 1.000 mol H_2O at 1.00 atm and 25° C, assuming that ΔG for the decomposition of H_2O is 237.14 kJ mol^{-1}.
5.10 Assuming that the liquid is incompressible, calculate ΔG when the pressure on 1.000 mol H_2O is increased from 1.00 atm to 20.0 atm at 25° C.
5.11 Calculate ΔG when the pressure on 10.0 mol ideal gas is reduced from 10.0 atm to 1.00 atm at 25° C.
5.12 Show that

$$\left(\frac{\partial A}{\partial P}\right)_T = -P\left(\frac{\partial V}{\partial P}\right)_T .$$

5.13 Show that

$$\left(\frac{\partial H}{\partial P}\right)_T = T\left(\frac{\partial S}{\partial P}\right)_T + V.$$

— — —

5.14 How much would the pressure on 1.000 mol argon initially at 0° C and 1.000 atm have to be raised to keep the entropy constant when the temperature was raised to 150° C?

5.15 Calculate the entropy of vaporization of acetic acid at its boiling point, 118.2° C, where its heat of vaporization is 24,390 J mol^{-1}.

5.16 An insulated 15.0 ohm resistor initially at 0° C carries a current of 10.0 amp for 1.00 s. If the mass of the resistor is 5.00 g and its specific heat 0.200, what is its entropy change?

5.17 After 10.0 g ice at 0° C is added to 65.0 g water at 45° C, the ice melts and equilibrium is set up without interchange of heat with the surroundings. If the heat of fusion of ice is 333.5 J g^{-1} and the specific heat of water is 1.000, what is the total entropy change in the process?

5.18 An insulated flask containing 1000 g liquid water at -3.8° C is disturbed so that freezing occurs until equilibrium is established. What is the total entropy change?

5.19 If 25.0 g water evaporates at 100° C, where its heat of vaporization is 2254 J g^{-1}, and mixes with 25.0 g nitrogen at 100° C and 1.00 atm, what is the entropy change?

5.20 For water at 25° C the coefficient of cubical expansion α is $2.1 \times 10^{-4}\ K^{-1}$ and the compressibility coefficient β $4.9 \times 10^{-5}\ atm^{-1}$. What is ΔS when the pressure on 1.00 mol H_2O is increased from 1.00 to 20.0 atm at 25° C?

5.21 Calculate $C_P - C_V$ for water at 25° C .

5.22 For a condensed phase, determine (a) $(\partial H/\partial P)_T$, (b) $(\partial H/\partial V)_T$, and (c) $(\partial V/\partial T)_S$

5.23 With a result from problem 5.22, calculate ΔH when 1.000 mol water is compressed at 25° C from 1.00 to 20.0 atm.

5.24 Show that

$$\left(\frac{\partial S}{\partial E}\right)_V = \frac{1}{T}.$$

What property must a system have for S to decrease as its energy E increases above a certain value? How would the temperature T vary with E in such a system?

5.25 Show that

$$\left(\frac{\partial H}{\partial P}\right)_T = 0$$

for an ideal gas.

5.26 A gas follows the equation

$$PV = RT + aP$$

where a is a function of T alone. Determine $(\partial E/\partial V)_T$. What does this imply about the function E?

References

Books

De Heer, J.: 1986, *Phenomenological Thermodynamics*, Prentice- Hall, Englewood Cliffs, NJ, pp. 50-77, 101-189.

In his presentation, De Heer incorporates the best features of various historical papers and textbooks. Both traditional and axiomatic approaches to the second law are surveyed. Key chemical applications are covered. The third law is introduced and its relation to the unattainability of absolute zero developed.

Haase, R.: 1971, "Survey of Fundamental Laws," in Jost, W. (editor), *Physical Chemistry An Advanced Treatise*, vol. I, Academic Press, New York, pp. 38-74, 86-97.

Haase omits any discussion of cyclic processes. The second law is stated mathematically and the connection to irreversibility pointed out. The conventional thermodynamic functions are introduced and the third law covered.

Lewis, G. N., Randall, M., Pitzer, K. S., and Brewer, L.: 1961, *Thermodynamics*, 2nd ed., McGraw-Hill Book Co., New York, pp. 53-157

Here we have conventional presentations of the second and third laws. But the choice of symbols and names for some of the thermodynamic functions are outdated.

Truesdell, C.: 1980, *The Tragicomical History of Thermodynamics 1822-1854*, Springer-Verlag, New York, pp 187- 339.

Truesdell criticizes how classical thermodynamics developed. But he does not consider how physical chemists have used the results.

Articles

Baierlein, R.: 1994, "Entropy and the Second Law: A Pedagogical Alternative," *Am. J. Phys.* **62**, 15-26.

Ben-Naim, A.: 1987, "Is Mixing a Thermodynamic Process?" *Am. J. Phys.* **55**, 725-733.

Bowen, L. H.: 1988, "Stokes' Theorem and the Geometric Basis for the Second Law of Thermodynamics," *J. Chem. Educ.* **65**, 50-52.

Buchdahl, H. A.: 1987, "A Variational Principle in Classical Thermodynamics," *Am. J. Phys.* **55**, 81-83.

Clugston, M. J.: 1990, "A Mathematical Verification of the Second Law of Thermodynamics from the Entropy of Mixing," *J. Chem. Educ.* **67**, 203-205.

Craig, N. C.: 1988, "Entropy Analyses of Four Familiar Processes," *J. Chem. Educ.* **65**, 750-764.

Cropper, W. H.: 1986, "Rudolf Clausius and the Road to Entropy," *Am. J. Phys.* **54**, 1068-1074.

Cropper, W. H.: 1987, "Walther Nernst and the Last Law," *J. Chem. Educ.* **64**, 3-8.

David, C. W.: 1988, "On the Legendre Transformation and the Sackur-Tetrode Equation," *J. Chem. Educ.* **65**, 876-877.

Djurdjevic, P., and Gutman, T.: 1988, "A Simple Method for Showing Entropy is a Function of State," *J. Chem. Educ.* **65**, 399.

Estevez, G. A., Yang, K., and Dasgupta, B. B.: 1989, "Thermodynamic Partial Derivatives and Experimentally Measurable Quantities," *J Chem. Educ.* **66**, 890-892.

Freeman, R. D.: 1997, "Collective Response by Robert D. Freeman to the Reply by Belandria to Criticism of Belandria's Paper on 'Entropy Coupling'" *J. Chem. Educ.* **74**, 290-292.

Frolich, H.: 1996, "On the Entropy Change of Surroundings of Finite and of Infinite Size: From Clausius to Gibbs," *J. Chem. Educ.* **73**, 716-717.

Granville, M. F.: 1985, "Student Misconceptions in Thermodynamics," *J. Chem. Educ.* **62**, 847-848.

Hong-Yi, L.: 1994, "A New Approach to Entropy and the Thermodynamic Temperature Scale," *J. Chem. Educ.* **71**, 853-854.

Kemp, H. R.: 1985, "Gibbs' Paradox for Entropy of Mixing," J. Chem. *Educ.* **62**, 47-49.

Leff, H. S.: 1987, "Thermal Efficiency at Maximum Work Output: New Results for Old Heat Engines," *Am. J. Phys.* **55**, 602-610.

Leff, H. S., and Rex, A. F.: 1994, "Entropy of Measurement and Erasure: Szilard's Membrane Model Revisited," *Am. J. Phys.* **62**, 994- 1000.

Li, J. C. M.: 1962, "Caratheodory's Principle and the Thermodynamics Potential in Irreversible Thermodynamics," *J. Phys. Chem.* **66**, 1414-1420.

Marcella, T. V.: 1992, "Entropy Production and the Second Law of Thermodynamics: An Introduction to Second Law Analysis," *Am. J. Phys.* **60**, 888-895.

Muschik, W.: 1990, "Second Law: Sears-Kestin Statement and Clausius Inequality," *Am. J. Phys.* **58**, 241-244.

Nelson, P. G.: 1988, "Derivation of the Second Law of Thermodynamics from Boltzmann's Distribution Law," *J. Chem. Educ.* **65**, 390-392.

Nelson, P. G.: 1994, "Statistical Mechanical Interpretation of Entropy, *J. Chem. Educ.* **71**, 103-104

Noyes, R. M.: 1992, "Thermodynamics of a Process in a Rigid Container," *J. Chem. Educ.* **69**, 470-472.

Noyes, R. M.: 1996, "Applications of the Gibbs Function to Chemical Systems and Subsystems," *J. Chem. Educ.* **73**, 404-408.

Rodewald, B.: 1990, "Entropy and Homogeneity," *Am. J. Phys.* **58**, 164-168.

Root, D. H., Haselton Jr., H. T., and Chou, I-M.: 1994, "An Alternative Method for the Reduction of Thermodynamic Derivatives," *J. Chem. Educ.* **71**, 303-306.

Rudakov, E. S.: 1991, "An Entropy Interaction Theorem as Addition to the Third Law of Thermodynamics," *Sov. Phys. Dokl.* **36**, 462-464.

Sanchez, K. S., and Vergenz, R. A.: 1994, "Understanding Entropy Using the Fundamental Stability Conditions," *J. Chem. Educ.* **71**, 562-566.

Schomaker, V., and Waser, J.: 1996, "Global Thermodynamics of Systems That Include Stressed Solids," *J. Chem. Educ.* **73**, 386-397.

Tansjo, L.: 1985, "Comments on the Closure of the Carnot Cycle," *J. Chem. Educ.* **62**, 585-591.

Turner, L. A.: 1960, 1962, "Simplification of Caratheodory's Treatment of Thermodynamics I, II," *Am. J. Phys.* **28**, 781-786; **30**, 506-508.

Tykodi, R. J.: 1995, "Spontaneity, Accessibility, Irreversibility, 'Useful Work'," *J. Chem. Educ.* **72**, 103-112.

Tykodi, R. J.: 1996, "The Gibbs Function, Spontaneity, and Walls," *J. Chem. Educ.* **73**, 398-403.

Weiner, J. H.: 1987, "Entropic versus Kinetic Viewpoints in Rubber Elasticity," *Am. J. Phys.* **55**, 746-749.

Wood, S. E., and Battino, R.: 1996, "The Gibbs Function Controversy," *J. Chem. Educ.* **73**, 408-411.

Xianmin, Z.: 1989, "The Free Energy Prediction of Reaction Direction and the Principle of Le Chatelier," *J. Chem. Educ.* **66**, 401- 402.

6

Relationships between Phases

6.1 Intensive Variables

IN A GIVEN SYSTEM, THE INTENSIVE THERMODYNAMIC properties may vary from point to point. But each part of the system through which these properties are constant or vary continuously constitutes a *phase*. A system may contain gaseous, liquid, and solid phases. A chemical constituent whose mass can vary independently in any small region of the system is called a *component*. Because of reactions, the number of chemical species present may be greater than the number of components.

As thermodynamic variables, a person may choose pressure P, temperature T, and concentrations of all but one component. The equation of state for each phase yields its specific volume V (and the reciprocal, the density ρ) from P and T. Since all the mole fractions add up to one in each phase, the mole fraction of the last component is obtainable from the other mole fractions.

6.2 The Gibbs Phase Rule

In studying systems, a person needs to know not only the various intensive variables but also how many are independent.

Consider a macroscopic system at equilibrium. Each phase is then uniform. Let p be the number of phases, c the number of components.

If we consider the pressure P, the temperature T, and c - 1 concentrations in each phase to be variables, the total is $2 + p(c - 1)$. But partition of the components between the phases would have proceeded until equilibrium was set up. There are p - 1 independent distribution constants for each component. For c components, we have $c(p - 1)$ independent conditions.

The total number of independent intensive variables is thus

$$f = 2 + p(c-1) - c(p-1) = c - p + 2. \qquad [6.1]$$

We call f the number of degrees of freedom for the system. Formula (6.1) is called the *Gibbs phase rule*.

Since a system must contain at least one phase, the maximum number of degrees of freedom is given by

$$f_{\max} = c - 1 + 2 = c + 1. \qquad [6.2]$$

In the laboratory, a person may determine the extreme conditions under which each phase of a given mixture is stable. A total of $c + 1$ coordinates would be involved. The boundaries between the phases may then be plotted. The result is called a phase diagram.

For a given system, there is only one gaseous phase. But the number of condensed phases is determined by how many mutual immiscibilities can be maintained. For liquids, this would equal the number of sufficiently different components present. With solids, the number of phases may exceed the number of components. A solid with a given composition may be able to assume several structures, depending on the pressure and temperature.

Example 6.1.

What does the Gibbs phase rule tell us about a one-component system?

When $c = 1$, equation (6.1) becomes

$$f = 3 - p.$$

When only one phase is present, this reduces to

$$f = 2.$$

On a pressure-temperature diagram, the region for such a phase appears as an area. When two phases are present, we have

$$f = 1.$$

On a pressure-temperature diagram, the points where the two phases coexist in equilibrium form a curved line. When three phases are present, we have

$$f = 0.$$

There is now no freedom; three phases coexist in equilibrium at a point.

6.3 Equilibrium States for Water

The conditions under which the different possible phases of a system are stable can be presented in a table and with a graph. Data which have been obtained for one of the most important chemical substances, water, will be summarized here in both ways.

In the laboratory, numerous samples of water have been studied at various pressures and temperatures. Notable particularly is the pioneering work of Percy W. Bridgman at high pressures.

The ranges over which the various phases are stable are presented by figure 6.1. In the area marked G, only the gas is stable. In the area marked L, only the liquid is stable. In the area marked I, only ice I is stable, and so on. The different areas are separated by curves, along which two phases may coexist. Where three curves meet, the corresponding three phases may coexist. These points of intersection are called *triple points*. Experimental triple points for water are listed in table 6.1.

The triple point vapor-liquid-ice I is at 0.0098° C rather than at 0.0000° C since zero on the Celsius scale is defined as the freezing point of water saturated with air at 1 atm. Under these conditions, the freezing point has been lowered 0.0024 by dissolved air and 0.0074 by the pressure.

At triple point B, pure water has no degree of freedom; no intensive variable can change without causing one of the phases to disappear. Liquid water in equilibrium with water vapor, on curve BA, has one degree of freedom. If the temperature is varied, the pressure also varies along the curve. Liquid water by itself has two degrees of freedom.

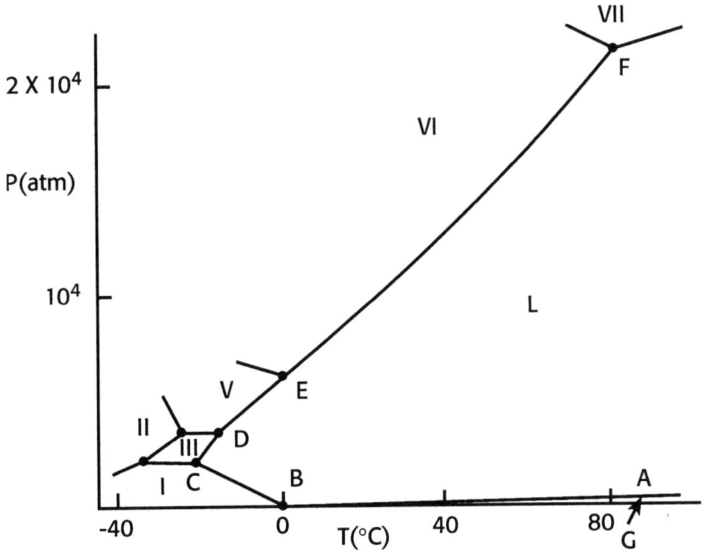

FIGURE 6.1 Phase diagram for water.

TABLE 6.1 Triple Points of Water

Phases in equilibrium	*Temperature, °C*	*Pressure, atm*
Vapor-liquid-ice I	0.0098	0.0060
Liquid-ice I-ice III	-22.0	2045
Liquid-ice III-ice V	-17.0	3420
Liquid-ice V-ice VI	0.16	6175
Liquid-ice VI-ice VII	81.6	21,680
Ice I-ice II-ice III	-34.7	2100
Ice II-ice III-ice V	-24.3	3400

Temperature and pressure can be varied independently within the area indicated without causing water vapor or any ice to appear.

6.4 Solutions

When substances A and B mix to form a thermodynamically uniform phase, they are said to be *miscible*. The resulting phase is a *solution*. In the usual gas phase, substances are miscible at all proportions. In the lower temperature liquid and solid phases, this property may or may not persist.

When substances A and B mix intimately at a given temperature T without changes in total volume V, internal energy E, and enthalpy H, the resulting phase is an *ideal solution*. When the ideal solution is uniform, the process is analogous to that occurring on mixing ideal gases and the entropy of mixing is given by formula (5.37):

$$\Delta S = -nR\left(X_A \ln X_A + X_B \ln X_B\right). \quad [6.3]$$

With the Gibbs free energy defined as

$$G = H - TS \tag{6.4}$$

and

$$\Delta H = 0, \qquad \Delta T = 0, \tag{6.5}$$

condition (6.3) yields

$$\Delta G = nRT\left(X_A \ln X_A + X_B \ln X_B\right). \tag{6.6}$$

Since X_A and X_B are fractions, this quantity is negative and the process is spontaneous.

For a general process at constant temperature, the definition of G yields

$$\Delta G = \Delta H - T\Delta S. \tag{6.7}$$

At constant pressure, we also have

$$\Delta H = q_P. \tag{6.8}$$

Now, ΔG for dissolving B in A may turn positive in a certain concentration range if in that range (a) the solution process is sufficiently endothermic and/or (b) the molecules form sufficiently organized combinations. Condition (a) makes ΔH sufficiently positive; condition (b) contributes negatively to ΔS. In such a concentration range, complete solution does not occur spontaneously; rather, immiscibility appears.

One may measure the nonideality of a solution by the deviations of the ΔS and ΔG from their ideal values. The excess entropy change on mixing substances A and B is

$$S^E = \Delta S_{mix} + nR\left(X_A \ln X_A + X_B \ln X_B\right), \tag{6.9}$$

where ΔS_{mix} is the actual entropy of mixing, while the excess Gibbs function change is

$$G^E = \Delta G_{mix} - nRT\left(X_A \ln X_A + X_B \ln X_B\right), \tag{6.10}$$

where ΔG_{mix} is the Gibbs function change in the solution process.

Generally, the average interaction among the A and B molecules in the solution is not the same as the average of the A - A and B - B interactions in the pure materials. Then the enthalpy of mixing is not zero,

$$\Delta H_{mix} \neq 0, \tag{6.11}$$

and we expect

$$G^E \neq 0. \tag{6.12}$$

But the effect on the arrangement of molecules is less. So as an approximation, one may consider the excess entropy of mixing to be small:

$$S^E \cong 0. \tag{6.13}$$

When approximation (6.13) holds, the solution is said to be *regular*.

Example 6.2

What endothermicity causes a regular solution of B in A to begin to separate into two phases at temperature 40° C and concentration X_B = 0.5000?

This critical point separates the temperatures over which B dissolves spontaneously in A to form the solution in which X_B = 0.5000 from a range in which the process is not spontaneous at a fixed T and P. So at this point, we have

$$\Delta G_{mix} = 0$$

and

$$q_P = \Delta H = T\Delta S = -nRT\left(X_A \ln X_A + X_B \ln X_B\right).$$

For dissolving 1 mole B in 1 mole A, we find

$$q_P = -(2.000 \text{ mol})(8.3145 \text{ J K}^{-1} \text{ mol}^{-1})(313.15 \text{ K})(\ln 0.5000) = 3609 \text{ J}.$$

6.5 *Aspects of Multicomponent Phase Diagrams*

From the Gibbs phase rule, $c + 1$ independent intensive variables are needed for each phase in a c component system. Thus, the diagram representing the conditions under which phases are stable or metastable is $c + 1$ dimensional. For convenience, however, one may employ 2-dimensional slices of the general diagram.

When c is 2, a person may look at a constant pressure slice, with perpendicular axes for temperature and composition. When c is 3, a person may look at a plane at constant pressure and temperature, with compositions referred to equilateral triangle axes.

In the laboratory, one would study the behavior of systems of representative compositions while being heated or cooled at constant pressure. High temperature phases generally possess more internal energy than lower temperature ones. Furthermore, transition of a given amount of material from a low temperature to a high temperature form requires energy.

The process may occur at a definite temperature. Or it may occur over a range of temperatures. In either case, the transition temperature is identified with the temperature at which the process may be completed reversibly. The absorption of the extra energy causes a slowing down or a halting of the temperature rise for a given rate of energy addition.

Generally, the transition temperature for a given phase change in pure B differs from that in pure A. When the mixtures are nearly ideal, the boundaries of the high temperature phase H and the low temperature phase L then appear as figure 6.2 shows. The vertical scale may be linear in temperature T; the horizontal scale, linear in mole fraction X_B. Alternatively, the horizontal scale may be linear in the weight per cent of B.

A point between the two curves represents a mixture of n_H moles phase H and n_L moles phase L. The total number n of moles is

$$n = n_H + n_L. \qquad [6.14]$$

Suppose at the given temperature the point is distance

$$l_H = X - X_H \qquad [6.15]$$

from the H boundary and distance

$$l_L = X_L - X \qquad [6.16]$$

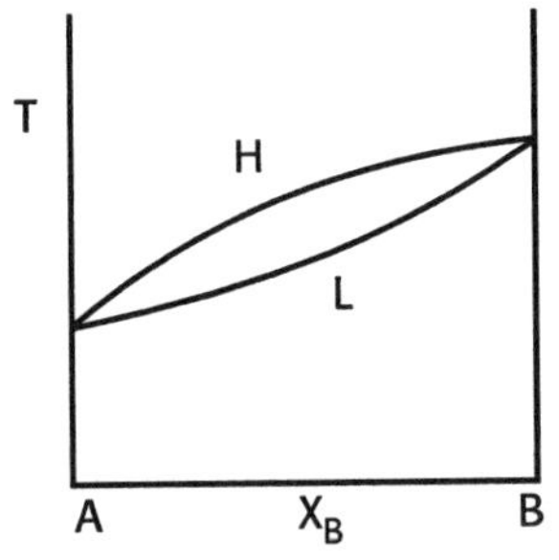

FIGURE 6.2 Representative constant pressure equilibrium diagram for A and B forming nearly ideal solutions in both phase H and phase L.

from the L boundary, as figure 6.3 shows. But the number of moles of substance B equals the sum in the two phases:

$$nX = n_H X_H + n_L X_L. \quad [6.17]$$

Substituting expression (6.14) into equation (6.17), rearranging,

$$n_H\left(X - X_H\right) = n_L\left(X_L - X\right), \quad [6.18]$$

and introducing expressions (6.15) and (6.16) gives us

$$n_H l_H = n_L l_L. \quad [6.19]$$

Formula (6.19) is known as the *lever rule*.

The mixtures of B in A may differ from ideality so that ΔH at constant T is considerably negative or positive. Then the curves joining the transition temperatures of pure A and pure B may exhibit a maximum or a minimum, even though B is still miscible in A at all concentrations and in both phases.

When the attraction and bonding between neighboring A and B molecules is considerably greater than the pertinent average A - A and B - B interactions, a maximum appears on the H and L curves, as figure 6.4 illustrates. Then heating a low temperature solution

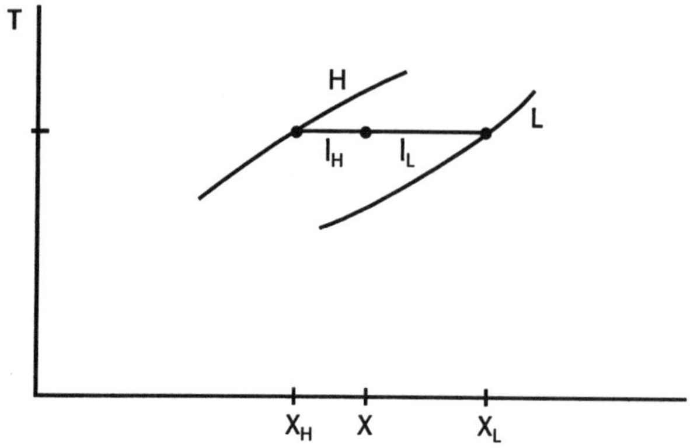

FIGURE 6.3 Tie line joining points on the H and L curves at a given temperature.

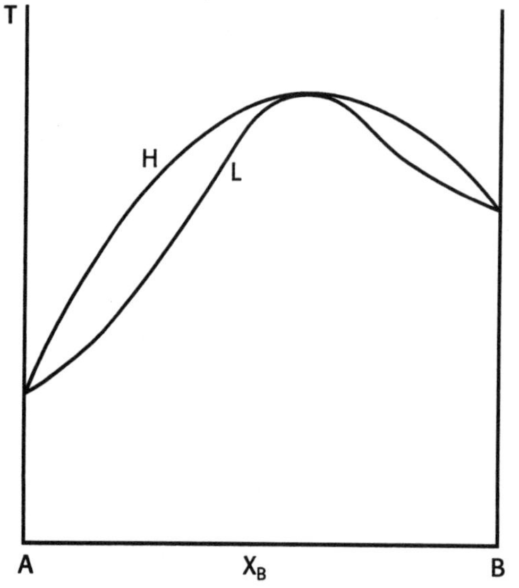

FIGURE 6.4 Representative constant pressure equilibrium diagram when the average A-B bonding interaction is much stronger than the pertinent average A-A and B-B interactions, with B miscible in A at all concentrations in both phases.

leads to fractionation when the L curve is reached. The low temperature phase then moves towards the maximum. When this point is reached, the resulting high temperature phase has the same composition as the low temperature one and transition continues as for a pure substance.

When the attraction and bonding between neighboring A and B molecules is considerably less than the pertinent average A - A and B - B interactions, a minimum appears on the H and L curves, as figure 6.5 illustrates. Then cooling a high temperature solution leads to fractionation when the H boundary is reached. The high temperature phase then moves along the bounding curve towards the minimum. When and if this point is reached, the forming low temperature phase has the same composition as the high temperature one and the transition continues as for a pure substance.

In many cases, the lessened interaction between A and B molecules leads to incomplete miscibility in the low temperature phase. The minimum in the H curve becomes a cusp and the L curve breaks into a horizontal segment and two separated curves giving the boundaries of phases L_1 and L_2. See figure 6.6. The minimum is called the *eutectic point* for the system. Removing heat from the H phase at this point produces an intimate mixture of phases L_1 and L_2.

On the other hand, the interaction between A and B molecules may be strong enough to lead to formation of the compound A_mB_n. And in the low temperature phases, the compound may be only slightly miscible with pure A and B. The resulting diagram might be described as one for substance A and compound A_mB_n joined to one for compound A_mB_n and substance B, as figure 6.7 shows.

6.6 Singularities Separating Phases

In principle, one may keep a given system homogeneous while raising its temperature at a certain pressure. As long as the system remains in a given phase, the process can take place continuously and reversibly. But at a boundary of the phase, singularities would appear. Various types of these occur. On the other hand, transitions in pure substances can be carried out reversibly at a given T and P. Then the Gibbs free energy G is continuous.

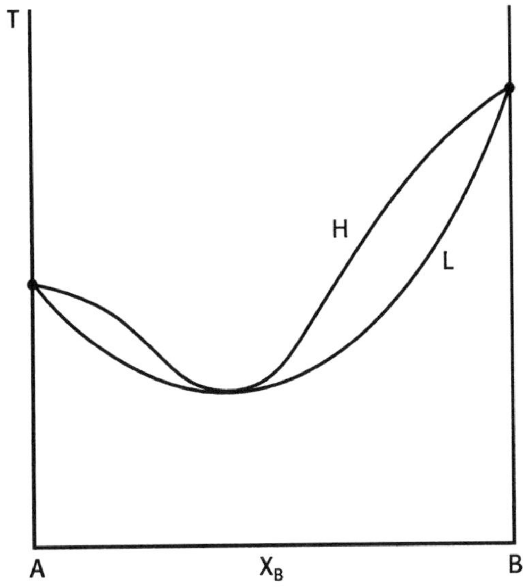

FIGURE 6.5 Representative constant pressure equilibrium diagram when the average A-B bonding interaction is much weaker than the pertinent average of the A - A and B - B interactions, with B miscible in A at all concentrations in both phases.

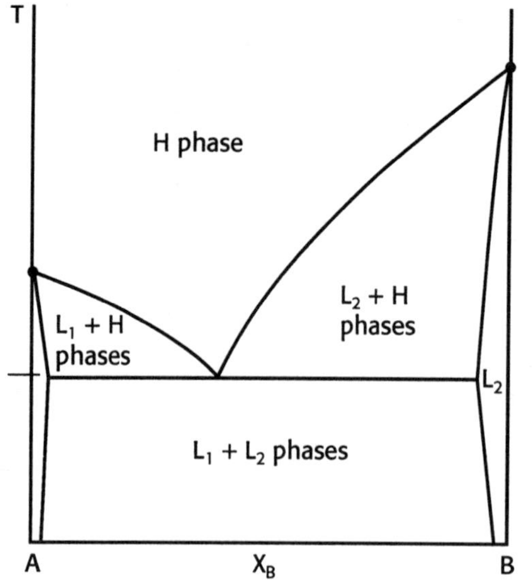

FIGURE 6.6 Representative constant pressure equilibrium diagram when the weak A - B bonding leads to extensive immiscibility in the low temperature phases.

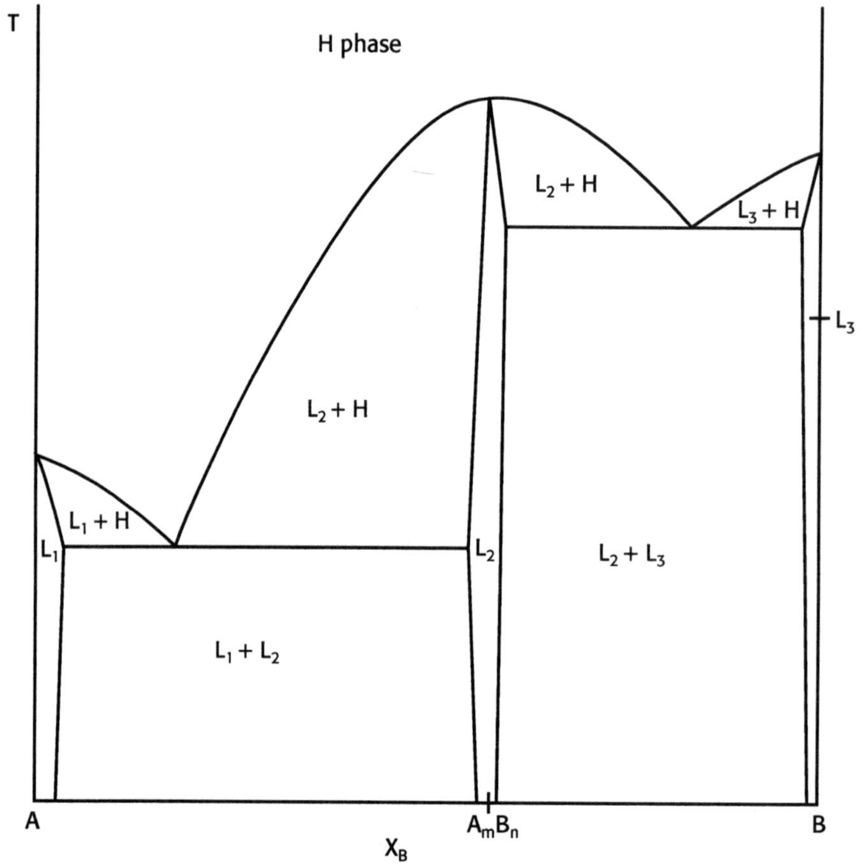

FIGURE 6.7 Representative constant pressure equilibrium diagram showing formation of the compound A_mB_n in the low temperature phase L_2.

However, there may be discontinuities in derivatives of G. When a discontinuity appears in a first derivative of G, as figure 6.8 illustrates, the transition is said to be *first order*. Then since

$$\left(\frac{\partial G}{\partial T}\right)_P = -S, \qquad [6.20]$$

from equation (5.82), entropy S jumps by ΔS at the phase change. See figure 6.9. Since the transition can be carried out reversibly, we have

$$\Delta S = \frac{q_P}{T} = \frac{\Delta H}{T} = \frac{L}{T}, \qquad [6.21]$$

where q_P is the heat absorbed to effect the change, ΔH is the jump in enthalpy, and L is the heat of transition.

From (5.82), we also have

$$\left(\frac{\partial G}{\partial P}\right)_T = V. \qquad [6.22]$$

At the first order transition, volume V jumps by ΔV. See figure 6.10. Some phase changes in the solid state and all fusions and vaporizations are first order.

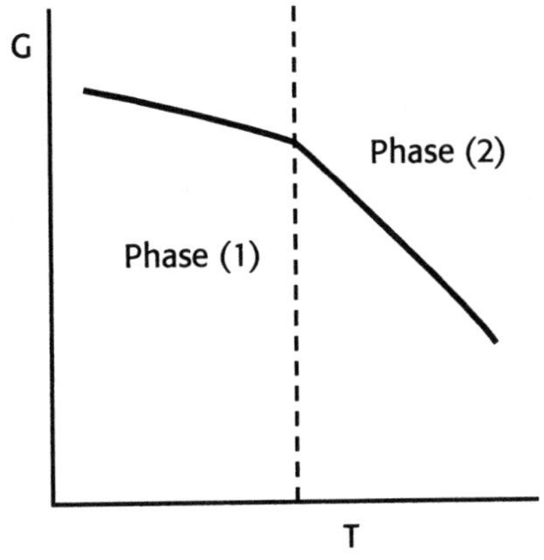

FIGURE 6.8 Abrupt change in slope of the free energy curve signaling a first order transition in a pure substance.

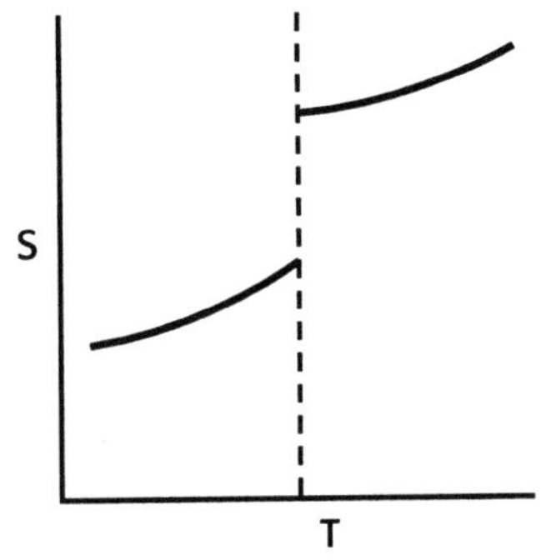

FIGURE 6.9 Jump in entropy indicating an abrupt increase in disorder at a first order transition.

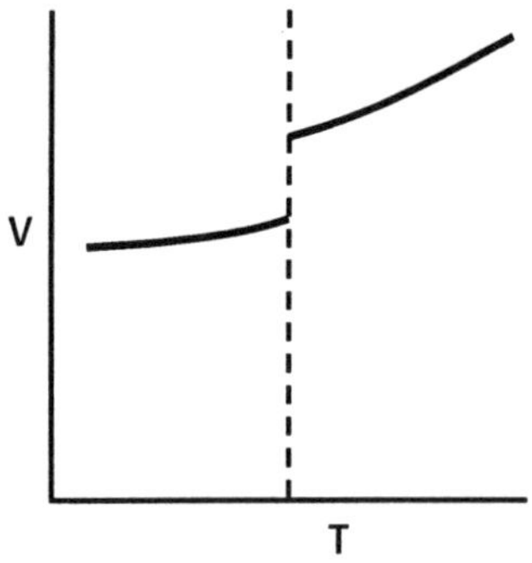

FIGURE 6.10 Variation in volume with temperature about a first order transition.

When all first derivatives of G are continuous but discontinuities appear in second derivatives, the transition is said to be *second order*. An example appears in figure 6.11. Differentiating equations (5.85) leads to the formulas

$$-\left(\frac{\partial^2 G}{\partial T^2}\right)_P = \left(\frac{\partial S}{\partial T}\right)_P = \frac{C_P}{T}, \tag{6.23}$$

$$\frac{\partial^2 G}{\partial T \partial P} = \left(\frac{\partial V}{\partial T}\right)_P = \alpha V, \tag{6.24}$$

$$-\left(\frac{\partial^2 G}{\partial P^2}\right)_T = -\left(\frac{\partial V}{\partial P}\right)_T = \beta V. \tag{6.25}$$

The last step in line (6.23) employs equation (5.21) at constant P; the last steps in (6.24) and (6.25), formulas (4.84) and (4.85). At a second order transition, volume V and entropy S are continuous, while energy capacity C_P, expansion coefficient α, and compressibility β jump by ΔC_P, $\Delta\alpha$, and $\Delta\beta$. See figures 6.12 and 6.13.

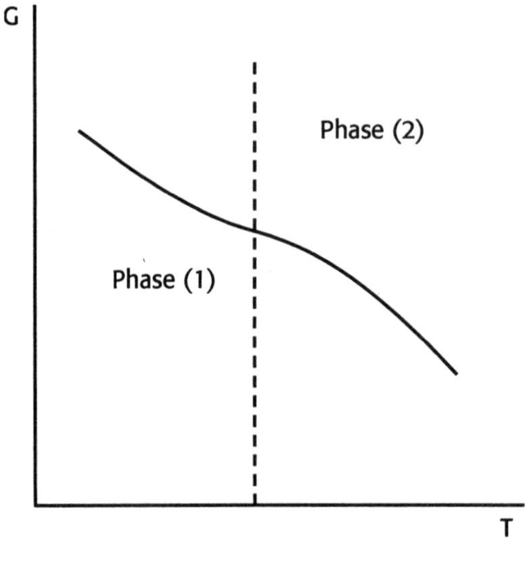

FIGURE 6.11 Change in curvature of the free energy curve at a second order transition.

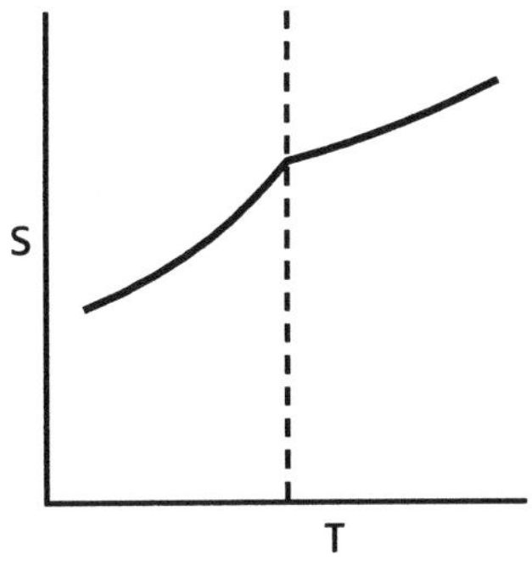

FIGURE 6.12 Change in slope of entropy curve locating a second order transition.

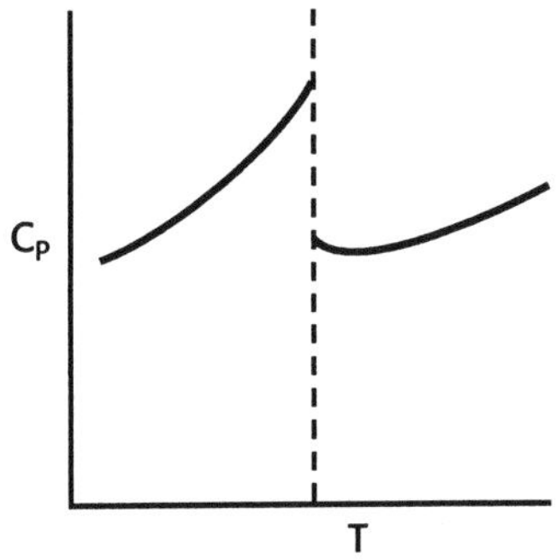

FIGURE 6.13 Jump in energy capacity indicating an abrupt drop in the temperature rate of entropy increase at a second order transition.

A higher order transition, one that is not first order, may not be cleanly second order because of macroscopic fluctuations and the resulting dispersion. As a consequence, the break in the curve in figure 6.12 may be rounded. Similarly, the peak of the curve in figure 6.13 may be rounded and the downward section precipitous but not missing.

The high energy capacity of the low temperature phase indicates that an extraordinary amount of energy is needed to raise the temperature a given amount in that region. Thus, an energy consuming conversion is taking place which is abruptly completed at the conventional transition temperature.

A common higher order transition involves introduction of disorder into a lattice. In alloys, this results from the random diffusion of atoms. In hydrogen bonded structures, it results from the random migration of protons. In magnetic materials, it involves changes in the orientation of spins.

Consider β brass, an alloy of copper and zinc with a body-centered cubic lattice. In the low-temperature ordered state, each zinc atom is surrounded by eight copper atoms and each copper atom by eight zinc atoms, insofar as the copper - zinc atom ratio allows. See figure 6.14. As the temperature is raised, the atoms begin to travel around and entropy of mixing is introduced. But this conversion occurs over a temperature interval. At the transition point, 742 K, the system reaches complete disorder as figure 6.15 shows. Disordering cannot contribute further to C_P and C_P drops abruptly.

In ammonium chloride, the nitrogen and chlorine atoms are also arranged on a body-cantered cubic lattice. Each nitrogen atom is surrounded by eight chlorine atoms and each chlorine atom by eight nitrogen atoms. Also around each nitrogen atom, four hydrogen atoms are arranged tetrahedrally as figures 6.16 and 6.17 illustrate.

In the lowest energy state, assumed at low temperatures, all unit cells in any given crystallite (a small homogeneous part of the crystal) exhibit the same orientation. But

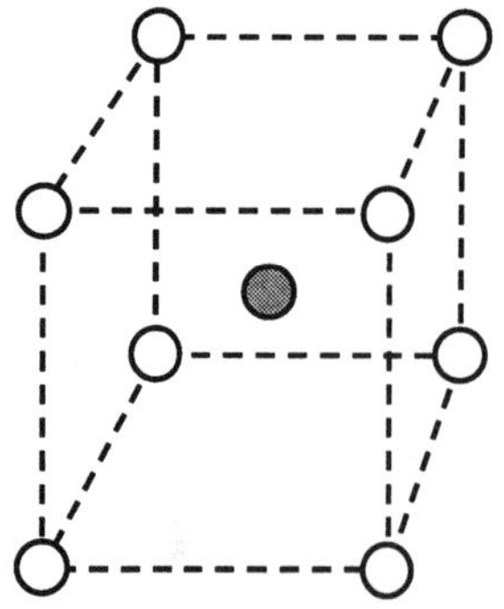

FIGURE 6.14 A unit cell of β brass, CuZn, in the completely ordered state.

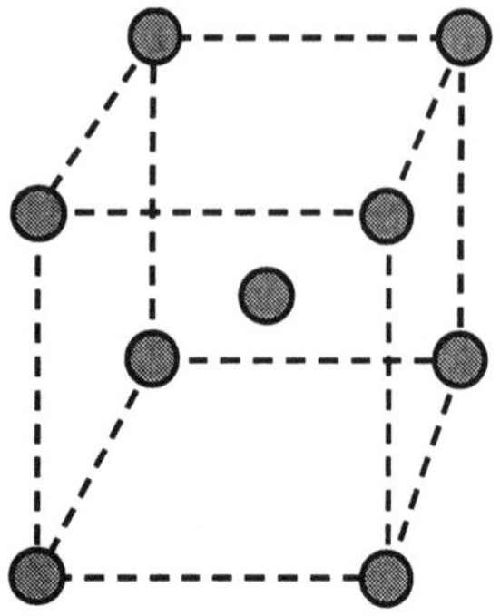

FIGURE 6.15 Same unit cell in the completely disordered state.

as energy is added, some of the hydrogen atoms move to the alternate positions, thus introducing disorder. At the transition temperature, 242.8 K, half of them have moved and the disordering is completed. It cannot contribute further to C_P and C_P drops abruptly.

A higher order transition is also observed in liquid helium. At very low temperatures and low pressures, helium exhibits peculiar flow properties that indicate the presence of a superfluid with no entropy or viscosity. At 0 K and low pressures, presumably only superfluid is present. As the temperature is raised, atoms are excited from the superfluid level and the fraction of normal fluid increases. As the so-called λ line is approached, this fraction increases rapidly to 1 and the energy capacity C_P increases dramatically. But on crossing this line, the excitation from the superfluid level is completed and C_P drops precipitously. At 38.65 torr, under pure vapor, the transition temperature is observed at 2.17 K.

6.7 The Clapeyron and Ehrenfest Equations

The slope dP/dT of the boundary between two phases of a substance depends on the discontinuities present.

Consider a given amount of a pure substance undergoing a transition from phase (1) to phase (2). Since the transition can be carried out reversibly at constant temperature and pressure, the change in the Gibbs function at the transition point is

$$\Delta G = G_{(2)} - G_{(1)} = 0. \qquad [6.26]$$

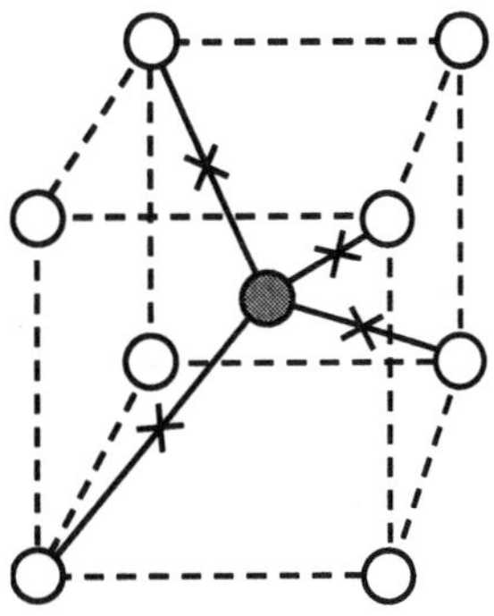

FIGURE 6.16 A unit cell of an ammonium chloride crystal exhibiting one orientation of the hydrogen atoms in the ammonium ion.

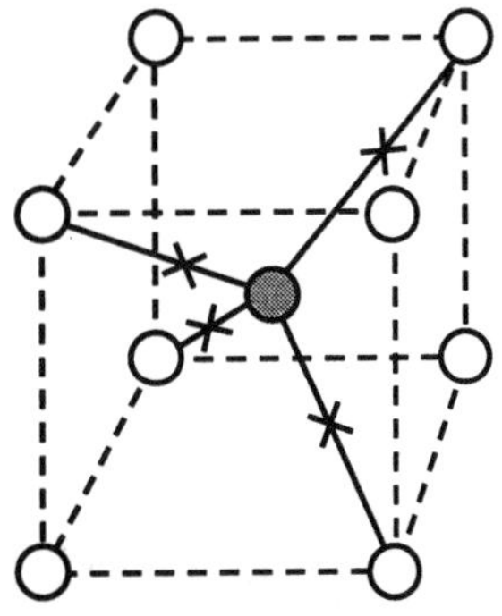

FIGURE 6.17 The other possible orientation of the hydrogen atoms in the unit cell. In a completely ordered state, all unit cells exhibit the same orientation, while in the completely disordered state, both occur with equal probability throughout the crystallite.

Here $G_{(2)}$ and $G_{(1)}$ are the Gibbs free energies of the given amount of substance in phases (2) and (1), respectively.

Condition (6.26) holds at each point on the boundary between the two phases. So for an infinitesimal change along the boundary, we have

$$\mathrm{d}G_{(1)} = \mathrm{d}G_{(2)}. \qquad [6.27]$$

Since formula (5.82) describes how G changes in each phase, equation (6.27) becomes

$$V_{(1)}\,\mathrm{d}P - S_{(1)}\,\mathrm{d}T = V_{(2)}\,\mathrm{d}P - S_{(2)}\,\mathrm{d}T. \qquad [6.28]$$

Here the subscripts identify the phase in which volume V and entropy S are determined. Since pressure P and temperature T are the same in both phases, they do not need subscripts.

Rearranging (6.28) yields

$$\frac{\mathrm{d}P}{\mathrm{d}T} = \frac{S_{(2)} - S_{(1)}}{V_{(2)} - V_{(1)}} = \frac{\Delta S}{\Delta V}, \qquad [6.29]$$

where ΔS is the change in entropy and ΔV the change in volume for the given amount of substance undergoing the transition.

At a *first order transition*, these increments differ from zero. Introducing relation (6.21) into (6.29) gives us the formula

$$\frac{\mathrm{d}P}{\mathrm{d}T} = \frac{L}{T\Delta V}, \tag{6.30}$$

which is known as the *Clapeyron equation*.

At a *second order transition*, volume V and entropy S are continuous and the right side of (6.29) is indeterminate. From the continuity of S, we have

$$\mathrm{d}S_{(1)} = \mathrm{d}S_{(2)} \tag{6.31}$$

for movements along the boundary between the phases. But since

$$\mathrm{d}S = \left(\frac{\partial S}{\partial T}\right)_P \mathrm{d}T + \left(\frac{\partial S}{\partial P}\right)_T \mathrm{d}P = \frac{C_P}{T}\,\mathrm{d}T - \alpha V\,\mathrm{d}P, \tag{6.32}$$

from formulas (6.23), (5.86), (6.24), equation (6.31) yields

$$\frac{C_{P(1)}}{T}\,\mathrm{d}T - \alpha_{(1)} V\,\mathrm{d}P = \frac{C_{P(2)}}{T}\,\mathrm{d}T - \alpha_{(2)} V\,\mathrm{d}P. \tag{6.33}$$

This rearranges to

$$\frac{\mathrm{d}P}{\mathrm{d}T} = \frac{C_{P(2)} - C_{P(1)}}{TV\left(\alpha_{(2)} - \alpha_{(1)}\right)} = \frac{\Delta C_P}{TV\Delta\alpha}, \tag{6.34}$$

where ΔC_P is the increment in energy capacity and $\Delta\alpha$ the increment in expansion coefficient associated with the transition.

From the continuity of V, we have

$$\mathrm{d}V_{(1)} = \mathrm{d}V_{(2)} \tag{6.35}$$

for movements along the boundary between the phases. But since

$$\mathrm{d}V = \left(\frac{\partial V}{\partial T}\right)_P \mathrm{d}T + \left(\frac{\partial V}{\partial P}\right)_T \mathrm{d}P = \alpha V\,\mathrm{d}T - \beta V\,\mathrm{d}P, \tag{6.36}$$

from formulas (6.24) and (6.25), equation (6.35) yields

$$\alpha_{(1)} V\,\mathrm{d}T - \beta_{(1)} V\,\mathrm{d}P = \alpha_{(2)} V\,\mathrm{d}T - \beta_{(2)} V\,\mathrm{d}P, \tag{6.37}$$

whence

$$\frac{\mathrm{d}P}{\mathrm{d}T} = \frac{\alpha_{(2)} - \alpha_{(1)}}{\beta_{(2)} - \beta_{(1)}} = \frac{\Delta\alpha}{\Delta\beta}. \tag{6.38}$$

Here $\Delta\alpha$ is the increment in expansion coefficient while $\Delta\beta$ is the increment in compressibility occurring during the transition.

Relationships (6.34) and (6.38) are known as the *Ehrenfest equations*.

Example 6.3

Calculate the change in melting point of ice produced by a decrease in pressure from 760 torr to 4.6 torr, the pressure at the triple point. The specific volumes of liquid water and of ice are 1.0001 and 1.0907 $\mathrm{cm^3\,g^{-1}}$, respectively, at 0° C. The heat of fusion is 333.5 $\mathrm{J\,g^{-1}}$.

From (6.30), we obtain the formula

$$dT = \frac{T\Delta V}{L}\,dP.$$

During the small change here, the coefficient of dP is approximately constant. So we have

$$\Delta T = \frac{T\Delta V}{L}\,\Delta P.$$

In applying this formula, the energy units in L and in $\Delta V\Delta P$ must be the same. From example 3.3, the number of joules in 1 cm^3 atm is 0.101325. But 760 torr is 1 atm. So here

$$\Delta P = \frac{-755.4\ \text{torr}}{760\ \text{torr atm}^{-1}}\,0.101325\ \text{J cm}^{-3}\ \text{atm}^{-1} = -0.100712\ \text{J cm}^{-3}.$$

Substituting into the formula now gives

$$\Delta T = \frac{(273.15\ \text{K})(-0.0906\ \text{cm}^3\ \text{g}^{-1})(-0.100712\ \text{J cm}^{-3})}{333.5\ \text{J g}^{-1}} = 0.0075\ \text{K}.$$

6.8 *Variation of Vapor Pressure with Temperature*

Below the pertinent critical point, transition from a condensed phase to vapor is first order. The Clapeyron equation then applies, with P the vapor pressure, T the temperature, L the heat of transition and V the volume for the given amount of pure substance. Here this amount is taken to be 1 mole.

To integrate equation (6.30), one needs the dependence of ΔV and L on T. As an approximation, we neglect the volume of 1 mole in the condensed phase:

$$\Delta V \cong V_{(2)}. \qquad [6.39]$$

Furthermore, we consider the vapor to be an ideal gas:

$$V_{(2)} \cong \frac{RT}{P}. \qquad [6.40]$$

These two approximations transform (6.30) to the form

$$\frac{dP}{dT} = \frac{LP}{RT^2}, \qquad [6.41]$$

which is called the *Clausius - Clapeyron equation*. Let us rearrange equation (6.41),

$$\frac{dP}{P} = \frac{L}{R}\frac{dT}{T^2}, \qquad [6.42]$$

and introduce a third approximation, that L be constant. Integration then leads to the formula

$$\ln P = -\frac{L}{R}\frac{1}{T} + C. \qquad [6.43]$$

Thus, a plot of the logarithm of the vapor pressure P against the reciprocal of the absolute temperature $1/T$ is approximately linear with the slope $-L/R$.

Example 6.4

Fitting equation (6.43) to vapor pressure data on benzene over the temperature range 40° to 100° C, with P in torrs, one finds that L/R is 3884 K and C is 17.6254. Calculate the vapor pressure of benzene at 60° C.

Here we have

$$\ln P = -\frac{3884 \text{ K}}{T} + 17.6254.$$

When

$$T = 333.15 \text{ K}$$

we obtain

$$\ln P = -\frac{3884 \text{ K}}{333.15 \text{ K}} + 17.6254 = 5.9670$$

and

$$T = 390 \text{ torr.}$$

6.9 *Variation of Vapor Pressure with Total Pressure*

Adding to the pressure on a pure condensed phase at a given temperature tends to squeeze molecules out of the phase and so acts to increase the vapor pressure of the given substance. The addition may be effected by mixing with the vapor a second gas that does not dissolve appreciably in the condensed phase.

Let the initial substance be A and the added one be B. Then for a given amount of A, equation (6.26) applies at the boundary between the two phases. For an infinitesimal increase in total pressure P, we have

$$\mathrm{d}G_{(1)} = \mathrm{d}G_{(2)}, \qquad [6.44]$$

as in equation (6.27).

Now, the pressure acting on the condensed form of A is P while the pressure of A in the gaseous phase is the partial pressure P_A. We consider the gas to be an ideal mixture; so A there behaves as if B were not present. We also consider the movement along the boundary to be at a fixed temperature. Condition (5.82) with $\mathrm{d}T$ equal to zero applies to each phase. So equation (6.44) becomes

$$V_{(1)}\,\mathrm{d}P = V_{(2)}\,\mathrm{d}P_A. \qquad [6.45]$$

Let $V_{(1)}$ be the molar volume of A in the condensed phase while $V_{(2)}$ is the volume of gas holding one mole of A. Also consider as an approximation that the gas is ideal. Equation (6.45) reduces to

$$V_{(1)}\,\mathrm{d}P = RT\frac{\mathrm{d}P_A}{P_A}. \qquad [6.46]$$

When the change in vapor pressure P_A and the change in molar volume $V_{(1)}$ are small, the coefficients of the differentials are approximately constant and (6.46) yields the formula

$$V_{(1)}\Delta P = RT\frac{\Delta P_A}{P_A}. \qquad [6.47]$$

Here ΔP is the excess pressure on a condensed phase with molar volume V(1) while ΔP_A is the resulting increase in vapor pressure P_A.

Example 6.5

Determine the partial pressure of water in a gas saturated with water vapor at 6.00 atm and 25° C. At this temperature the vapor pressure of pure water is 23.8 torr.

The change in total pressure is

$$\Delta P = 6.00 \text{ atm} - \frac{23.8 \text{ torr}}{760 \text{ torr atm}^{-1}} = 5.969 \text{ atm}.$$

Solve (6.47) for ΔP_A and insert the data together with the value of R from equation (3.31):

$$\Delta P_A = \frac{V_{(1)} P_A \Delta P}{RT} = \frac{(0.018 \text{ l mol}^{-1})(23.8 \text{ torr})(5.969 \text{ atm})}{(0.08206 \text{ l atm K}^{-1} \text{ mol}^{-1})(298.15 \text{ K})} = 0.10 \text{ torr}.$$

The final vapor pressure is

$$P_A = 23.8 \text{ torr} + 0.10 \text{ torr} = 23.9 \text{ torr}.$$

6.10 Variation of Vapor Pressure with Concentration

When two phases are in equilibrium, the various molecules leaving one phase per unit time balance those leaving the other phase and entering the first phase per unit time. For a solution of A and B in a condensed phase interacting with a gaseous phase, we have the equilibria

$$\text{A(c)} \leftrightarrows \text{A(g)} \qquad [6.48]$$

and

$$\text{B(c)} \leftrightarrows \text{B(g)} \qquad [6.49]$$

Here c identifies the condensed phase and g the gaseous phase.

Decreasing the rate of either process (6.48) or (6.49) to the right decreases the balancing rate to the left. But the latter decrease implies a decrease in the corresponding vapor pressure. In our discussion, let us keep the temperature T constant and neglect the effect of changing the total pressure on either vapor pressure.

If the average interactions between neighboring A molecules, neighboring B molecules, and neighboring A and B molecules are effectively the same and if A and B molecules occupy similar cross sections at the surface of the condensed phase, a molecule of A would exhibit the same tendency to leave as it would at the surface of pure condensed A. Furthermore, the number of A's per unit area of surface would be proportional to the mole fraction X_A in the condensed phase. So the resulting partial pressure P_A would be proportional to X_B. But since the vapor pressure of pure A is $P_A{}^0$, the constant of proportionality equals this number.

Raoult's law

$$P_A = X_A P_A^0 \qquad [6.50]$$

then holds. Similarly for constituent B, we have

$$P_B = X_B P_B^0, \qquad [6.51]$$

where X_B is the mole fraction of B in the condensed phase while P_B is the partial pressure of B in the gas phase and $P_B{}^0$ is the vapor pressure of pure B at the given temperature. When equations (6.50) and (6.51) hold at all concentrations, the solution is said to be *ideal*. A representative plot of pressures P, P_A, and P_B against the mole fraction X_B appears in figure 6.18.

From formula (3.37), the mole fractions in the gas phase are

$$X_A{}' = \frac{P_A}{P} \qquad [6.52]$$

and

$$X_B{}' = \frac{P_B}{P}. \qquad [6.53]$$

Here P is the total pressure.

When molecules A and B differ sufficiently, the conditions we have noted break down and deviations from (6.50) and (6.51) occur. Nevertheless, in dilute solutions of B in A, most A molecules are surrounded by A molecules and the tendency for an A to leave the surface is asymptotically the same as in pure A. So we expect Raoult's law

$$P_A = X_A P_A^0 \qquad [6.54]$$

to hold asymptotically as $X_A \to 1$. Similarly as $X_A \to 0$, the equation

$$P_B = X_B P_B^0 \qquad [6.55]$$

should apply asymptotically.

Also when the solution is dilute enough, each B molecule is surrounded by A molecules. Then further dilution does not affect its tendency to leave the surface and the resulting partial pressure is proportional to the mole fraction. Thus, the relationship

$$P_B = k_B X_B \qquad [6.56]$$

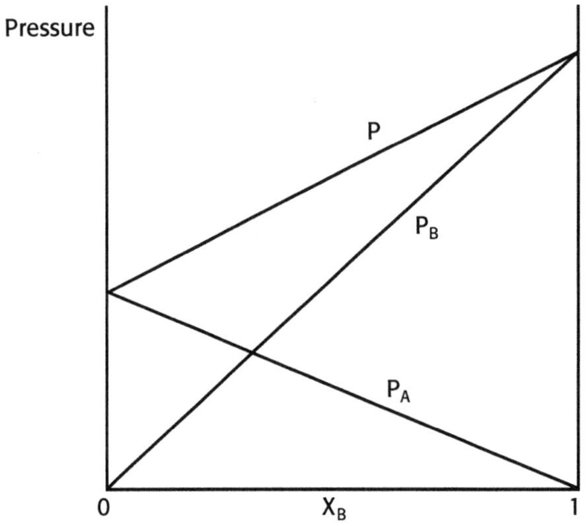

FIGURE 6.18 The linear variations of the partial pressures P_A, P_B and of the sum P with the mole fraction X_B when the solutions are ideal and the temperature is fixed.

holds asymptotically as $X_B \to 0$ ($X_A \to 1$), This is known as *Henry's law*. Similarly for constituent A, we have

$$P_A = k_A X_A \quad [6.57]$$

as

$$X_A \to 0 \left(X_B \to 1\right).$$

When the average interaction between neighboring A and B molecules is stronger than what it would be if the condensed solutions were ideal, the intermediate vapor pressures are below the ideal values. One then obtains results such as figure 6.19 depicts. When this average cross interaction is weaker than it would be if the condensed solutions were ideal, the intermediate vapor pressures are above the ideal values.

For a solute B that is practically nonvolatile, the Henry law constant vanishes:

$$k_B \cong 0. \quad [6.58]$$

Then at low concentrations, we have

$$P \cong P_A = X_A P^0 \quad [6.59]$$

and

$$P^0 - P = \left(1 - X_A\right)P^0 = X_B P^0. \quad [6.60]$$

Here P^0 - P is the vapor pressure lowering when the mole fraction of the nonvolatile solute is X_B.

Example 6.6

Methanol and ethanol are mixed to form a solution in which the mole fraction of methanol is 0.250. If a small amount of this solution is vaporized at 25° C, what is the initial concentration of methanol in the vapor? The vapor pressure of pure methanol is 96.0 torr; that of pure ethanol, 43.9 torr at 25° C.

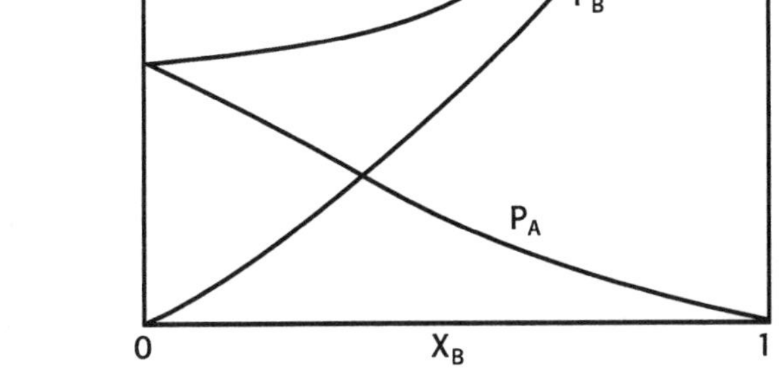

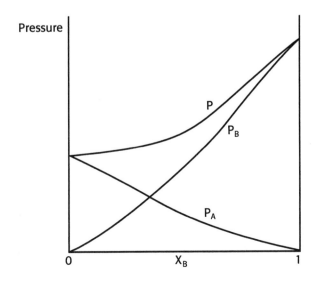

FIGURE 6.19 Representative variations of the partial pressures P_A, P_B, and of the sum P, with the mole fraction X_B for negative deviations from nonideality.

We consider the solution to be ideal so formulas (6.50) and (6.51) apply. Here

$$P_{CH_3OH} = (0.250)(96.0 \text{ torr}) = 24.0 \text{ torr},$$

$$P_{C_2H_5OH} = (0.750)(43.9 \text{ torr}) = 32.9 \text{ torr},$$

and

$$P = P_{CH_3OH} + P_{C_2H_5OH} = 56.9 \text{ torr}.$$

Equation (6.52) then yields

$$X'_{CH_3OH} = \frac{24.0 \text{ torr}}{56.9 \text{ torr}} = 0.422.$$

Note how the vapor is richer in the more volatile constituent than the liquid.

6.11 The Mole-Fraction Activity

One can quantify the tendency for a substance to escape from a phase in the following manner.

Consider a solution of A and B in a given phase at a given temperature. Imagine trying various *ideal-gas* mixtures in contact with the given phase until one is found that does not change with time. There is then a balance between the tendencies for A and B molecules to leave each phase. And the partial pressures P_A and P_B in the ideal gas mixture give one a measure of these tendencies.

Now, we define the *activity* a_i of a substance in a phase to be the effective concentration of the substance as measured by its tendency to escape. So for the given phase, we have the activities

$$a_A = (\text{const})P_A, \tag{6.61}$$

$$a_B = (\text{const})P_B. \tag{6.62}$$

In practice, the constants of proportionality are chosen so that the activities reduce to the concentrations in the infinitely dilute solution. With A the solvent and B the solute, equations (6.54) and (6.56) apply there. So for the activities of A and B, we obtain

$$a_A = \frac{P_A}{P_A^0} \tag{6.63}$$

$$a_B = \frac{P_B}{k_B}, \tag{6.64}$$

when the concentrations are expressed as mole fractions. Furthermore, equation (6.63) applies whenever the activity of the pure solvent is taken to be 1, regardless of how the concentration of the solute is expressed.

6.12 Concentration Units

The concentration of a constituent in a solution may be stated in various ways.

Commonly, it is expressed as a density. Thus, the molar concentration or *molarity* (c_i) of a constituent is the number of moles of the constituent in one liter of the phase. The unit mole per liter is abbreviated as M. The normality of a constituent equals the number of equivalents of the constituent in one liter of the phase. The unit equivalent per liter is abbreviated as N.

When a phase is kept at constant pressure, the densities of constituents vary with temperature. To circumvent this effect, a person bases concentrations on a mass or number of moles present. Thus, the molal concentration or *molality* (m_i) of a solute is the number of moles of the solute in 1000 g solvent. The unit mole per 1000 g solvent is abbreviated as m.

The *mole fraction* (X_i) for a constituent is the fraction of the total moles that is moles of the constituent, in the given region. This also equals the number fraction of the molecules that are the constituent molecules. The *mass fraction* equals the mass of the constituent divided by the total mass in the chosen region. This may be multiplied by 100 to give a *percentage*.

To construct X_i from m_i one calculates the total number of moles in the solution containing m_i moles of substance i, then divides m_i by this value. To construct X_i from c_i, one calculates the total number of moles in one lifer of solution, then divides c_i by this number. Thus, we get

$$X_i = \frac{m_i}{1000/M_s + m_i} = \frac{c_i}{\left(1000d - c_i M_i\right)/M_s + c_i}, \qquad [6.65]$$

where M_s is the molecular mass of the solvent, M_i the molecular mass of solute i, d the mass density of the solution, X_i the mole fraction of solute i, m_i the molality of i, and c_i the molarity of i.

6.13 Elevation of the Boiling Point

Adding a nonvolatile solute to a given solvent lowers the vapor pressure at each temperature. Consequently, the whole vapor pressure curve is lowered. Furthermore, the temperature at which the vapor reaches a given pressure is raised. See figure 6.20.

Consider a solvent whose vapor pressure is P^0. This varies with temperature following the Clausius - Clapeyron equation

$$\frac{dP^0}{P^0} = \frac{L_v}{RT^2}\, dT, \qquad [6.66]$$

in which L_v is the heat of vaporization while T is the temperature and R the gas constant.

Let a nonvolatile solute be added while the vapor pressure is kept constant. From formula (6.59). we have

$$X_A P^0 = P = \text{const.} \qquad [6.67]$$

Differentiating (6.67) yields

$$P^0\, dX_A + X_A\, dP^0 = 0 \qquad [6.68]$$

or

$$\frac{dX_A}{X_A} = -\frac{dP^0}{P^0}. \qquad [6.69]$$

Eliminating dP^0/P^0 from (6.66) and (6.69), then rearranging, leads to

$$dT = -\frac{RT^2}{L_v X_A}\, dX_A. \qquad [6.70]$$

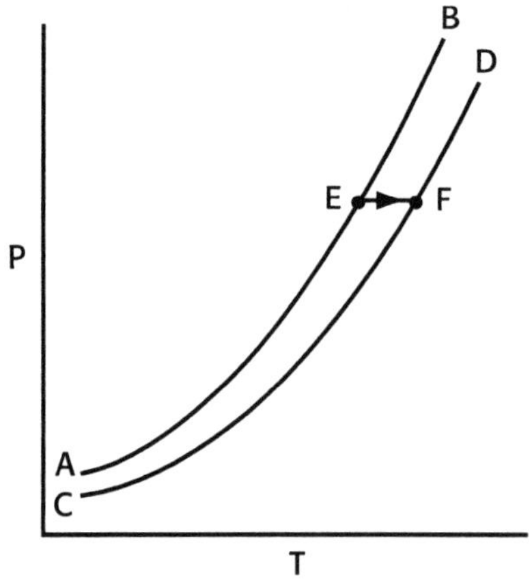

FIGURE 6.20 Vapor pressure of pure solvent, curve AB, vapor pressure of the given solution, curve CD, and the path of integration at 1 bar pressure, line EF.

As long as the final solution is dilute, T, X_A, and L_v do not change much from their initial values and we may integrate (6.70) with the coefficient of dX_A constant to get

$$\Delta T = -\frac{RT_0^2}{L_v}\left(X_A - 1\right) = \frac{RT_0^2}{L_v} X_B. \tag{6.71}$$

Also when the solution is dilute, we have the approximation

$$X_B = \frac{n_B}{n_A + n_B} \cong \frac{n_B}{n_A} = \frac{w_B / M_B}{w_A / M_A}, \tag{6.72}$$

where w_B is the mass (weight) of solute with molecular mass M_B and w_A is the mass (weight) of solvent with molecular mass M_A. Combining (6.71) and (6.72) yields

$$\Delta T = \frac{RT_0^2}{L_v} \frac{M_A}{w_A} \frac{w_B}{M_B}. \tag{6.73}$$

When the morality of the solute is m, one may set w_A equal to 1000 g and have w_B/M_B equal m. Thus

$$\Delta T = \frac{RT_0^2 M_A}{1000 L_v} m = K_b m, \tag{6.74}$$

where

$$K_b = \frac{RT_0^2 M_A}{1000 L_v}. \tag{6.75}$$

Coefficient K_b is called the *molal elevation constant* for the given solvent, R is the gas constant, T_0 the boiling point, M_A the molecular mass of the solvent, and L_v the heat of vaporization per mole of solvent. Experimental elevation constants for some common solvents are listed in table 6.2.

Example 6.7

Calculate the molal elevation constant for water at 1 bar pressure, under which water boils at 372.78 K with a heat of vaporization of 40,671 J mol^{-1}.

TABLE 6.2 Molal Elevation Constants

Solvent	K_b, K m^{-1}
Water	0.512
Methanol	0.83
Ethanol	1.22
Acetone	1.71
Benzene	2.53
Acetic Acid	3.07
Toluene	3.33
Chloroform	3.63
Carbon tetrachloride	5.03
Ethylene bromide	6.44
Iodobenzene	8.53
Stannic chloride	9.43

Since L_v is in joules per mole, the value of R in line (3.32) is employed. Thus, we have

$$K_b = \frac{\left(8.3145 \text{ J K}^{-1} \text{ mol}^{-1}\right)\left(372.78 \text{ K}\right)^2\left(18.015 \text{ g mol}^{-1}\right)}{\left(1000 \text{ g kg}^{-1}\right)\left(40{,}671 \text{ J mol}^{-1}\right)} = 0.512 \text{ K}\left(\text{mol kg}^{-1}\right)^{-1}.$$

6.14 Depression of the Freezing Point

The conditions for appreciable solubility in the crystalline phase of common solvents are very restrictive. So we will here consider the solid phase to be practically pure even when considerable solute is dissolved in the liquid phase from which it formed. And we can consider the vapor pressure curve of the solid phase to be unaffected. However, the vapor pressure curve of the liquid phase is shifted. When the solute is effectively non-volatile, the curves appear as figure 6.21 illustrates.

Again, let the vapor pressure of the pure liquid be P^0 while the vapor pressure of the solid is P_s. These vary with temperature following the Clausius - Clapeyron equations

$$\frac{dP^0}{P^0} = \frac{L_v}{RT^2}\, dT, \tag{6.76}$$

$$\frac{dP_s}{P_s} = \frac{L_s}{RT^2}\, dT, \tag{6.77}$$

in which L_v is the heat of vaporization of the pure liquid while L_s is the heat of sublimation of the solid.

For equilibrium between liquid and solid, the vapor pressure of the liquid equals that of the solid:

$$P = P_s. \tag{6.78}$$

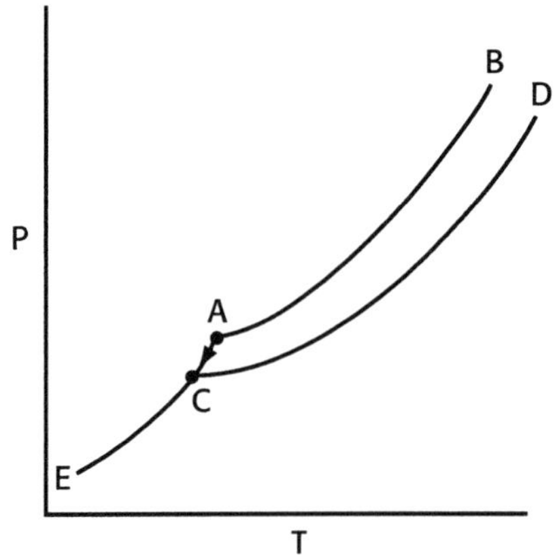

FIGURE 6.21 Vapor pressure of liquid solvent, curve AB, vapor pressure of the given solution, curve CD, vapor pressure of solid solvent, curve EA, and the path of integration, curve AC.

But at low concentrations, the vapor pressure of the liquid solution is given by formula (6.59):

$$P = X_A P^0. \qquad [6.79]$$

Now, eliminate P from equations (6.78) and (6.79),

$$P_s = X_A P^0, \qquad [6.80]$$

differentiate,

$$dP_s = X_A dP^0 + P^0 \, dX_A \qquad [6.81]$$

divide by $X_A P^0$, rearrange terms,

$$\frac{dP_s}{X_A P^0} - \frac{dP^0}{P^0} = \frac{dX_A}{X_A}, \qquad [6.82]$$

and introduce (6.76), (6.77) to get

$$\frac{L_s}{RT^2} \, dT - \frac{L_v}{RT^2} \, dT = \frac{dX_A}{X_A} \qquad [6.83]$$

or

$$\frac{1}{RT^2}\left(L_s - L_v\right) dT = \frac{dX_A}{X_A}. \qquad [6.84]$$

Solve for dT and identify $L_s - L_v$ as the heat of fusion L_f:

$$dT = \frac{RT^2}{L_f X_A} dX_A. \qquad [6.85]$$

As long as the final solution is dilute, T, X_A, and L_f do not change much from their initial values and we may integrate (6.85) with the coefficient of dX_A constant to get

$$\Delta T = \frac{RT_0^{\;2}}{L_f}\left(X_A - 1\right) = -\frac{RT_0^{\;2}}{L_f} X_B. \qquad [6.86]$$

Introducing the morality m as before and reducing leads to

$$\Delta T = -K_f m \qquad [6.87]$$

where

$$K_f = \frac{RT_0^{\,2} M_A}{1000 L_f}. \qquad [6.88]$$

Coefficient K_f is called the *molal depression constant* for the given solvent, R is the gas constant, T_0 the freezing point of the pure solvent, ΔT the change in freezing point caused by addition of the solute, M_A the molecular mass of the solvent, and L_f the heat of fusion of the solvent per mole. Experimental depression constants for some common solvents are listed in table 6.3.

Remember we assumed that the solid precipitating out was practically pure solvent. When the molecules of solute are nearly the same size and shape, and form the same kind of crystal, as the molecules of solvent, they enter the lattice to form a solid solution and alter ΔT from the value given by (6.87).

6.15 Permeability of Membranes

Two parts of a system may be separated by a sheet that allows some but not all chemical constituents to pass through. Such a sheet is called a *semipermeable membrane.* In principle, this could be either liquid, solid, or composite.

A solid film containing small holes may act merely as a sieve, letting small molecules and ions through but not larger ones. Such a porous barrier is called a *Donnan membrane.* Some collodion, cellophane, and inert animal membranes act approximately in this way.

But the film may attract and adsorb one or more of the constituents. The adsorbate(s) would then tend to line the walls of each hole. When the pores are small, other constituents may be blocked from passage. But the adsorbed layer may be mobile and move through. Furthermore, its molecular or ionic units may be larger than the units that are not allowed through.

The units making up a membrane may be mobile enough so that one or more constituents dissolve in them and so pass through. Rubber acts in this manner when it allows organic hydrocarbons to move through while it excludes water.

TABLE 6.3 Molal Depression Constants

Solvent	K_f, $K\ m^{-1}$
Water	1.86
Acetic acid	3.9
Benzene	5.12
Naphthalene	6.9
Phenol	7.27
Nitrobenzene	8.1
m-Dinitrobenzene	10.6
Bromoform	14.4
Cyclohexane	20.0
Stannic bromide	28.0
Camphor	37.7
Strontium chloride	107

The movement of a constituent through a membrane impermeable to other constituents is called *osmosis*. In each case, the flow proceeds until the tendency for the constituent to leave one side balances the tendency for it to leave the other side.

A difference in concentration of the mobile constituent can be maintained by applying an excess pressure on the low concentration side.

6.16 Osmotic Pressure

Molecules tend to move from a region where their concentration is high to a region where their concentration is lower. When pure A is separated from a solution of B in A by a barrier permeable only to A, a net movement of A molecules into the solution tends to occur. If the solution is confined, the pressure on it rises ΔP above that on the solvent. The excess pressure when equilibrium is reached is called the *osmotic pressure* of the solution.

The tendency for a substance to leave a phase is measured by its vapor pressure above the phase. But this vapor pressure is raised by the excess total pressure on the phase. In the equilibrium depicted in figure 6.22, the final vapor pressure of solvent above the solution equals that above the pure solvent.

Now, the effect on vapor pressure is governed by equation (6.46). If $\bar{V}_A$ is the partial molar volume of A in the solution, this becomes

$$\overline{V_A}\,dP = RT\frac{dP_A}{P_A}. \tag{6.89}$$

Let the vapor pressure of solvent above the solution be P_A when the total pressure is $P_A{}^0$, the conventional vapor pressure of the pure solvent. Then integrating equation (6.89) from P_A to $P_A{}^0$, the point of osmotic equilibrium, gives

$$\overline{V_A}\,\Delta P = RT\ln\frac{P_A^0}{P_A}, \tag{6.90}$$

if we consider $\bar{V}_A$ to be constant (incompressible, as an approximation).

As a further approximation, one may expand the logarithm into an infinite series, drop higher terms, and introduce Raoult's law:

$$\ln\frac{P_A^0}{P_A} = \left(\frac{P_A^0}{P_A} - 1\right) - \ldots \cong \frac{1}{X_A} - 1 = \frac{X_B}{X_A} = \frac{n_B}{n_A}. \tag{6.91}$$

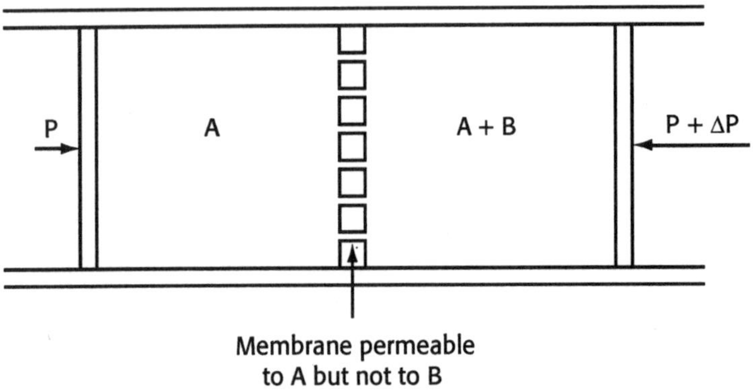

FIGURE 6.22 Equilibrium between solvent A and solution A + B maintained by applying the excess pressure ΔP to the solution.

For the final step, n_B is the number of moles of B, n_A the number of moles of A, in the solution. If we also set

$$n_A \overline{V_A} = V' \qquad \text{when} \qquad n_B = 1, \tag{6.92}$$

formula (6.90) reduces to

$$V' \Delta P = RT. \tag{6.93}$$

Interestingly, this has the form of the ideal gas law. Here V' is the volume of solvent containing one mole solute, ΔP is the osmotic pressure, R the gas constant, and T the temperature.

Example 6.8

Calculate the osmotic pressure for a 0.100 m sucrose solution at 30° C.
The weight of solvent containing one mole solute is

$$w = \frac{1000 \text{ g kg}^{-1}}{0.100 \text{ mol kg}^{-1}} = 10{,}000 \text{ g mol}^{-1},$$

whence the corresponding solvent volume is

$$V' = \frac{10{,}000 \text{ g mol}^{-1}}{1000 \text{ g l}^{-1}} = 10 \text{ l mol}^{-1}.$$

Solve (6.93) for ΔP. Then substitute in the temperature and the solvent volume:

$$\Delta P = \frac{RT}{V'} = \frac{(0.08206 \text{ l atm K}^{-1} \text{ mol}^{-1})(303.2 \text{ K})}{10 \text{ l mol}^{-1}} = 2.49 \text{ atm}$$

If the pressure on the pure solvent is 1.00 atm, then a pressure of 3.49 atm must be exerted on the 0.100 m sucrose solution to stop osmosis of water from a reservoir of pure water into the solution.

6.17 Equilibria among Phases

Let us now summarize the conditions governing the equilibria among differing homogeneous mixtures. We assume that the entire system is free from outside influences, particularly heat and work. So inequality (5.114) applies. In the immediate neighborhood of equilibrium, the system exhibits a given entropy, volume, and energy.

For simplicity, consider two uniform, composite phases separated by an interface. Starting from equilibrium, let us introduce a small fluctuation. During this, the internal energy of each phase changes following formula (5.94). But for bookkeeping purposes, we add the numerical superscripts (1) and (2) to the symbols to identify the pertinent phase.

For the energy change in an infinitesimal fluctuation, we now have

$$dE = dE^{(1)} + dE^{(2)} = T^{(1)}\,dS^{(1)} - P^{(1)}\,dV^{(1)} + \sum \mu_i^{(1)}\,dn_i^{(1)} + T^{(2)}\,dS^{(2)} - P^{(2)}\,dV^{(2)} + \sum \mu_i^{(2)}\,dn_i^{(2)}. \tag{6.94}$$

However, the total moles of constituent i are conserved:

$$dn_i^{(2)} = -dn_i^{(1)}. \tag{6.95}$$

Also, the total volume is constant,

$$dV^{(2)} = -dV^{(1)}, \tag{6.96}$$

the total entropy is constant,

$$dS^{(2)} = -dS^{(1)}, \tag{6.97}$$

and the total energy is constant,

$$dE = 0. \tag{6.98}$$

Using equations (6.95) - (6.98) to eliminate $dn_i^{(2)}$, $dV^{(2)}$, $dS^{(2)}$, dE from equation (6.94) leads to

$$0 = \left(T^{(1)} - T^{(2)}\right) dS^{(1)} - \left(P^{(1)} - P^{(2)}\right) dV^{(1)} + \sum \left(\mu_i^{(1)} - \mu_i^{(2)}\right) dn_i^{(1)}. \tag{6.99}$$

The fluctuations $dS^{(1)}$, $dV^{(1)}$, $dn_i^{(1)}$ are all independent (uncoupled). So to satisfy equation (6.99), we must have

$$T^{(1)} = T^{(2)}, \tag{6.100}$$

$$P^{(1)} = P^{(2)}, \tag{6.101}$$

$$\mu_i^{(1)} = \mu_i^{(2)}. \tag{6.102}$$

At equilibrium, the temperatures, the pressures, and the chemical potentials in the two phases must be equal.

A membrane placed between the two phases may prevent some of the constituents from moving between them. The corresponding chemical potentials would then not equilibrate. The membrane would also support a pressure difference ΔP as we saw in the discussion of osmotic pressure.

Questions

6.1 Define phase, component, thermodynamic degree of freedom.
6.2 Justify the Gibbs phase rule.
6.3 In a particular system, molecules of A_mB_n appear in two different electronic states. Can these be considered as separate components? Explain.
6.4 Why are there so many different solid phases for water?
6.5 How may one measure the nonideality of a solution?
6.6 What is a regular solution?
6.7 Derive and interpret the lever rule.
6.8 When and how does a eutectic point arise?
6.9 Distinguish between first order, second order, and higher order phase transitions.
6.10 Identify and describe examples of the different kinds of phase transitions.
6.11 Derive the Clapeyron and Ehrenfest equations.
6.12 Describe the approximations made in deriving the Clausius Clapeyron equation.
6.13 Explain why and how the vapor pressure of a component varies with the total pressure on the given system.
6.14 How may the vapor pressure of a component vary with its concentration in the condensed phase? Explain.
6.15 How is the activity of a component in a solution defined?
6.16 Explain the effects of a solute on the boiling point and on the freezing point of a system.
6.17 How may a semipermeable membrane act?
6.18 What conditions govern the equilibria between phases?

Problems

6.1 (a) The principal constituents in a system are H_2O, Na^+, K^+, Cl^-, and Br^-. How many components does it have? (b) Salts NaCl, NaBr, KCl, and KBr are mixed with water. How many components does the system have?

6.2 Hydrogen, bromine, and hydrogen bromide are mixed. How many components does the system possess (a) before and (b) after equilibrium is set up? (c) Salt $AlCl_3$ is added to water. How many components are there?

6.3 The freezing point of solutions of tin in lead was found to vary with the percentage of tin as shown in table 6.A.

TABLE 6.A

wt%	0	10	20	30	40	50	60	70	80	90	100
t, °C	326	295	276	262	240	220	190	185	200	216	232

Plot the data and graphically estimate the composition and temperature at the eutectic point.

6.4 Calculate the melting point of acetic acid at 10.0 atm if at 1.00 atm the melting point is 16.61° C, the heat of fusion, 180.8 J g^{-1}, and the specific volumes of liquid and solid phases are 0.9315 and 0.7720 $cm^3\ g^{-1}$, respectively.

6.5 The vapor pressure of benzene was found to vary with temperature as shown in table 6.B.

TABLE 6.B

P. torr	1.00	10.0	40.0	100	400	760
t, °C	-36.7	-11.6	7.6	26.1	60.6	80.1

Plot the data so that a straight line would be obtained if L were constant and from this curve get the vapor pressure at 40° C.

6.6 From the data in problem 6.5, determine the heat of vaporization of benzene at 40° C.

6.7 At 90° C, the vapor pressure of toluene is 400 torr while that of *o*-xylene is 150 torr. Estimate the composition of the liquid mixture that boils at 90° C with a pressure of 300 torr. What is the composition of the vapor coming off?

6.8 Adding 10.0 g $AgNO_3$ to 490 g H_2O yields a solution of density 1.0154 g cm^{-3}. What are the (a) mole fraction, (b) molality, and (c) molarity of silver nitrate in this solution?

6.9 Solutions of benzene (B) in acetic acid (A) yielded the partial pressures at 50° as shown in table 6.C.

TABLE 6.C

X_B	0.0069	0.1563	0.3396	0.4166	0.6304	0.7027
P_A, torr	54.7	50.7	40.2	36.3	28.7	24.8
P_B, torr	3.5	75.3	135.1	153.2	195.6	211.2
X_B	0.8286	0.8862	0.9165	0.9561	0.9840	
P_A, torr	18.4	14.2	11.51	7.25	3.63	
P_B, torr	231.8	244.8	249.6	257.2	262.9	

Plot P_i/X_i against X_i and extrapolate to $X_i = 1$ and to $X_i = 0$ to obtain the constants P_i^0 and k_i of equations (6.63) and (6.64) for both A and B. Then calculate the activity of benzene in the solutions on both the Raoult's law and the Henry's law basis.

6.10 What is the activity of CCl_4 in a solution made by adding 20.0 g CCl_4 to 30.0 g CBr_4?

6.11 When some ethyl benzoate was dissolved in benzene at 80° C, the vapor pressure was lowered from 753.6 to 745.9 torr. (a) What was the mole fraction of ethyl benzoate in the solution? (b) How much ethyl benzoate would have to be added to 100 g benzene to prepare such a solution?

6.12 How many grams of urea must be added to 25.0 g water to obtain a boiling point elevation of 0.200° C?

6.13 If the heat of fusion of acetic acid at its freezing point 16.6° C is 180.75 J g^{-1}, what is its molal depression constant K_f?

6.14 Solid iodine in equilibrium with its pure vapor at 20° C has a vapor pressure of 0.202 torr and a density of 4.93 g cm^{-3}. If the pressure on the iodine is raised to 100 atm by mixing an inert gas with its vapor, what is its new vapor pressure?

6.15 Calculate the osmotic pressure of a 0.130 m sucrose solution at 150° C.

— — —

6.16 Three solutions, 0.1 M NaCl, 0.1 M $NaNO_3$, and 0.1 M $AgNO_3$, were mixed. How many degrees of freedom does the resulting system exhibit?

6.17 When does sulfur fail to behave as a one-component system?

6.18 Show that during a first order transition at constant pressure, entropy S of the entire system varies linearly with the total volume V.

6.19 For the II - I transition in NH_4NO_3, volume change ΔV and temperature of transition t vary with pressure P as shown in table 6.D.

TABLE 6.D

P, atm	1	1000	2000	3000
t, °C	125.5	135.0	143.5	151.4
ΔV, cm^3 g^{-1}	0.01351	0.01166	0.01020	0.00906

Calculate the heat of transition at 1,500 atm.

6.20 If the vapor pressure of CCl_4 is 25.83 atm at 240° C and 33.59 atm at 260° C and if its critical temperature is 283.2° C, what is its critical pressure?

6.21 The vapor pressure P of solid ammonia follows the equation

$$\ln P = 23.03 - \frac{3754 \text{ K}}{T}$$

while that of liquid ammonia follows the equation

$$\ln P = 19.49 - \frac{3063 \text{ K}}{T}.$$

Calculate (a) the temperature of the triple point and (b) the heat of fusion of ammonia.

6.22 If a nonvolatile hydrocarbon added to benzene lowered its vapor pressure from 74.66 to 73.89 torr at 20° C, what was the activity of benzene in the solution?

6.23 What is the vapor pressure above a salt solution in which the mole-fraction activity of water is 0.9325, at 25° C, where the vapor pressure of pure water is 23.76 torr?

6.24 For a solution of 3.795 g sulfur in 100.0 g CS_2 the boiling point was 46.66° C. For pure CS_2 the boiling point is 46.30° C and the heat of vaporization 353.1 J g^{-1}. What is the molecular mass and the formula of sulfur in the CS_2?

6.25 If the osmotic pressure of an aqueous solution is 5.15 atm at 25° C, what is its freezing point?

6.26 Calculate the boiling point elevation when 1.000 g toluene is added to 100.0 g benzene. Benzene boils at 80.1° C with a heat of vaporization of 30,765 J mol^{-1}, while toluene has a vapor pressure of 290.6 torr at 80.1° C with a heat of vaporization of 33,472 J mol^{-1}.

6.27 Solutions of iodoethane (B) in ethyl acetate (A) yielded partial pressures at 50° C as shown in table 6.E. Obtain the Henry's law constants k_A and k_B by extrapolating P_A/X_A and P_B/X_B to $X_A = 0$ and to $X_B = 0$. Then calculate the activity of iodoethane (B) in the solutions on both the Raoult's law and the Henry's law basis.

6.28 At a given temperature, the vapor pressure of solute is related to that of solvent by the equation

$$\frac{X_A \, dP_A}{P_A} + \frac{X_B \, dP_B}{P_B} = 0.$$

TABLE 6.E

X_B	0.0000	0.0579	0.1095	0.1918	0.2353	0.3718
P_A, torr	280.4	266.1	252.3	231.4	220.8	187.9
P_B, torr	0.0	28.0	52.7	87.7	105.4	155.4
X_B	0.5478	0.6349	0.8253	0.9093	1.0000	
P_A, torr	144.2	122.9	66.6	38.2	0.0	
P_B, torr	213.3	239.1	296.9	322.5	353.4	

Show that in a concentration region where the solute obeys Henry's law, the solvent obeys Raoult's law.

6.29 If the third law applies to both liquid and solid helium and if P is the pressure at which the liquid is in equilibrium with the solid at temperature T, what should dP/dT approach as T approaches 0 K?

References

Books

Alper, A. M. (editor), *Phase Diagrams: Materials Science and Technology*, series, Academic Press, New York (abbreviated PD). The following sections are most pertinent:

MacChesney, J. B., and Rosenberg, P. E.: 1970, "The Methods of Phase Equilibria and Their Associated Problems," in PD vol. I, pp. 113-165.

The various experimental procedures employed in obtaining phase data are described. Numerous references are listed.

Newnham, R. E.: 1978, "Phase Diagrams and Crystal Chemistry," in PD vol. V, pp. 1-73.

Newnham considers how the observed phases for a given system can be related to the atomic and molecular structures involved. His results are approximate because of the simplifications introduced.

Nielsen, J. W., and Monchamp, R. R.: 1970, "The Use of Phase Diagrams in Crystal Growth," in PD vol. III, pp. 1-52.

The authors describe various crystal growth techniques. How the pertinent phase diagram is employed in choosing appropriate conditions is discussed in considerable detail.

Rao, Y. K.: 1970, "Thermodynamics of Phase Diagrams," in PD vol. I, pp. 1-43.

Rao presents the pertinent thermodynamics in easy-to-follow detail. Numerous diagrams illustrate the various possibilities. Representative experimental data is summarized.

Tiller, W. A.: 1970, "The Use of Phase Diagrams in Solidification," in PD vol. I, pp. 199-244.

Tiller describes the principles involved in crystallization work. Information on the chemical potentials of components is deduced from the pertinent phase diagrams. Normal freezing, zone melting, distribution coefficients, the forms of interfaces, are all discussed.

Yeh, H. C.: 1970, "Interpretation of Phase Diagrams," in PD vol. I, pp. 167-197.

Yeh picks out and interprets the different possible parts of one, two, and three component phase diagrams. Besides Gibbs phase rule, a law of adjoining phase regions is employed.

De Heer, J.: 1986, *Phenomenological Thermodynamics*, Prentice- Hall, Englewood Cliffs, NJ, pp. 192-219, 253-285, 310-346.

In the first section cited, de Heer constructs stability and equilibrium conditions for one-component, one- and two-phase, systems. In the second section, phase rules are constructed and applied. Then phase equilibria in binary systems are considered in some detail. In the third section, properties of ideal and simple solutions are developed. The various colligative properties of dilute solutions are derived.

Articles

Alberty, R. A.: 1995, "Components in Chemical Thermodynamics," *J. Chem. Educ.* **72**, 820.

Battino, R.: 1991, "The Critical Point and the Number of Degrees of Freedom," *J. Chem. Educ.* **68**, 276.

Binder, K.: 1992, "Phase Transitions in Reduced Geometry," *Annul Rev. Phys. Chem.* **43**, 33-59.

Bonicelli, M. G., di Giacomo, F., Cardinali, M. E., and Carelli, I.: 1984, "A New Derivation of the Approximate Phase Change Formula $\Delta T = K{\cdot}m$," *J. Chem. Educ.* **61**, 423-424.

Boyko, E. R., and Belliveau, J. F.: 1986, "Surface Tension and Avogadro's Number," *J. Chem. Educ.* **63**, 671-672.

Clerc, D. G., and Cleary, D. A.: 1995, "Spinodal Decomposition as an Interesting Example of Several Thermodynamic Principles," *J. Chem. Educ.* **72**, 112-115.

Craig, N. C.: 1996, "Entropy Diagrams," *J. Chem. Educ.* **73**, 710- 715.

Deumie, M., Henri-Rousseau, O., Boulil, K., and Boulil, B.: 1989, "Phase Equilibrium and *G* Minimum," *J. Chem. Educ.* **66**, 232- 237.

Franzen, H. F.: 1986, "The True Meaning of Component in the Gibbs Phase Rule," *J. Chem. Educ.* **63**, 948-949.

Franzen, H. F.: 1988, "The Freezing Point Depression Law in Physical Chemistry," *J. Chem. Educ.* **65**, 1077-1078.

Gilson, D. F. R.: 1992, "Order and Disorder and Entropies of Fusion," *J. Chem. Educ.* **69**, 23-25.

Gitterman, M.: 1988, "Are Dilute Solutions Always Dilute?" *Am. J. Phys.* **56**, 1000-1002.

Hawkes, S. J.: 1995, "Raoult's Law Is a Deception," *J. Chem. Educ.* **72**, 204- 205.

Haymet, A. D. J.: 1987, "Theory of the Equilibrium Liquid- Solid Transition," *Annul Rev. Phys. Chem.* **38**, 89-108.

Huque, E. M.: 1989, "The Hydrophobic Effect," *J. Chem. Educ.* **66**, 581-585.

Kildahl, N. K.: 1994, "Journey around a Phase Diagram," *J. Chem. Educ.* **71**, 1052-1055.

Kochansky, J.: 1991, "Liquid Systems with More than Two Immiscible Phases," *J. Chem. Educ.* **68**, 655-656.

Laughlin, R. G.: 1992, "The Use and Utility of Phase Science," *J. Chem. Educ.* **69**, 26-31.

Leyendekkers, J. V.: 1993, "The Scaled Particle Theory and the Tait Equation of State," *J. Phys. Chem.* **97**, 1220-1223.

Maaskant, W. J. A., and de Graff, R. A. G.: 1986, "Simulation of Two-Dimensional, Jahn-Teller Phase Transitions in Solids," *J. Chem. Educ.* **63**, 966-969.

MacCarthy, P.: 1983a, "A Novel Classification of Concentration Units," *J. Chem. Educ.* **60**, 187-189.

MacCarthy, P.: 1983b, "Ternary and Quaternary Composition Diagrams," *J. Chem. Educ.* **60**, 922-928.

MacCarthy, P.: 1986, "Application of Analytic Geometry to Ternary and Quaternary Diagrams," *J. Chem. Educ.* **63**, 40-42.

Marshall, D. B., McHale, J. L., Carswell, S., and Erne, D.: 1987, "Properties of Nonideal Binary Solutions," *J. Chem. Educ.* **64**, 369-370.

Mau, M., and Nckver Jr., J. W.: 1986, "Alternative Approach to Phase Transitions in the van der Waals Fluid," *J. Chem. Educ.* **63**, 880- 882.

Phelps, C. L., Smart, N. G., and Wai, C. M.: 1996, "Past, Present, and Possible Future Applications of Supercritical Fluid Extraction Technology," *J. Chem. Educ.* **73**, 1163-1168.

Queslel, J. P., and Mark, J. E.: 1987, "Advances in Rubber Elasticity and Characterization of Elastomeric Networks," *J. Chem. Educ.* **64**, 491- 494.

Santos, M. B.: 1986, "On the Distribution of the Nearest Neighbor," *Am. J. Phys.* **54**, 1139- 1141.

Scholsky, K. M.: 1989, "Supercritical Phase Transitions at Very High Pressure," *J. Chem. Educ.* **66**, 989- 990.

Secrest, D.: 1996, "Osmotic Pressure and the Effects of Gravity on Solutions," *J. Chem. Educ.* **73**, 998- 1000.

Sengers, J. V.: 1986, "Thermodynamic Behavior of Fluids Near the Critical Point," *Annul Rev. Phys. Chem.* **37**, 189-222.

Silverman, M. P.: 1985, "Theory of the Liquid-Vapor Transition of a Perfect Binary Solution," *J. Chem. Educ.* **62**, 112- 114.

Stead, R. J., and Stead, K.: 1990, "Phase Diagram for Ternary Liquid Systems," *J. Chem. Educ.* **67**, 385.

Treptow, R. S.: 1993, "Phase Diagrams for Aqueous Systems," *J. Chem. Educ.* **70**, 616-620.

Tykodi, R. J.: 1991, "Thermodynamics of an Incompressible Solid and a Thermodynamic Functional Determinant," *J. Chem. Educ.* **68**, 830- 832.

Udale, B. A., and Wells, J. D.: 1995, "A Ternary Phase Diagram for a Less Hazardous System," *J. Chem. Educ.* **72**, 1106.

Wloch, P., and Cherniak, E. A.: 1995, "Analysis of Cryoscopy Data," *J. Chem. Educ.* **72**, 59-61.

Zhao, M., Wang, Z., and Xiao, L.: 1992, "Determining the Number of Independent Components by Brinkley's Method," *J. Chem. Educ.* **69**, 539-542.

7

Relationships among Reactants

7.1 Introduction

A CHEMICAL REACTION GENERALLY INVOLVES rearrangement of the electron structures about atomic cores. Bonds between atoms may be broken and reformed. One or more elementary processes may be involved.

Traditionally, a reaction is studied in a macroscopic system. A state of a homogeneous region of the system is characterized by the intensive variables (a) pressure P, (b) temperature T, and (c) activity a_j of each j constituent. As the reaction proceeds, these vary. Furthermore, the region would possess the extensive properties volume V, an internal energy E, an enthalpy H, an entropy S, a Gibbs free energy G, a Helmholtz free energy A.

The pressure P, temperature T, volume V, and activities $a_1,..., a_n$ may be determined in the conventional ways. With energy capacity measurements from a calorimeter and the third law, the entropies of pure substances can be determined. The heat q for a reaction at a given temperature and pressure or volume can be determined with a calorimeter.

Thus, the change in G for reactants under certain initial conditions going to products under certain final conditions can be determined. And the equilibrium constant for the reaction can be related to calorimetric measurements.

Alternatively, molecular parameters can be found from spectroscopic measurements. From these, entropies can be calculated for the gas phase and the calorimetric entropies checked. Furthermore, equilibrium constants for the gas phase can be calculated. Since statistical mechanics is employed here, these calculations will not be developed at this stage of our studies.

7.2 Conventions

Thermodynamic data are customarily reported for specified standard states. For a liquid or solid substance, the *standard state* is the pure material at 1 bar pressure and the given temperature. For a gaseous substance, the standard state is the pure material in its hypothetical ideal gas condition at 1 bar pressure and the given temperature.

For a solvent, the standard state is the material at unit mole fraction under a pressure of 1 bar at the given temperature. For a solute, the standard state is the hypothetical ideal solution at unit molality (unit activity) under a pressure of 1 bar and the given temperature.

In writing the equation for a reaction, one needs to specify the initial and final states of each participant. Abbreviation g indicates the gaseous state, 1 indicates the liquid state, and s indicates the solid state. One may further distinguish a solid as cr for crystalline, am for amorphous, vit for vitreous. Where more than one crystalline form exists, the form employed may be indicated with a number or letter following s or cr. The temperature may appear as subscript on the thermodynamic variable employed. A superscript zero indicates that standard states were involved. Where necessary, the pressure may be added.

Example 7.1

Interpret the following:

$$6C\ (s) + 3H_2\ (g) \rightarrow C_6H_6\ (l), \qquad \Delta H^0_{298.15} = 49.04\ \text{kJ}.$$

The chemical equation represents 6 moles solid carbon and 3 moles gaseous hydrogen reacting to form 1 mole liquid benzene. When each of these is in its standard state at 298.15 K, the enthalpy increase is 49.04 kilojoules.

Example 7.2

The standard enthalpy of formation (a) of H_2O (g) is -241.83 kJ mol^{-1} , (b) of H_20_2 (g) is -136,11 kJ mol^{-1} , at 298.15 K. Determine $\Delta H^0{}_{298.15}$ for the reaction

$$2H_2O\ (g) \rightarrow H_2\ (g) + H_2O_2\ (g)$$

at 1 bar pressure.

It is given that

$$H_2\ (g) + \frac{1}{2}O_2\ (g) \rightarrow H_2O\ (g), \qquad \Delta H^0_{298.15} = -241.83\ \text{kJ}.$$

$$H_2\ (g) + O_2\ (g) \rightarrow H_2O_2\ (g), \qquad \Delta H^0_{298.15} = -136.11\ \text{kJ}.$$

Reverse the first equation of this pair and multiply by 2:

$$2H_2O\ (g) \rightarrow 2H_2\ (g) + O_2\ (g), \qquad \Delta H^0_{298.15} = 483.66\ \text{kJ}.$$

Then add to the second equation to get

$$2H_2O\ (g) \rightarrow H_2\ (g) + H_2O_2\ (g), \qquad \Delta H^0_{298.15} = 347.55\ \text{kJ}.$$

7.3 Measuring Reaction Heats

The heat evolved in a rapid exothermic reaction of a small amount of material can be readily determined. The heat $-q$ for other reactions may be obtained by combining these as in example 7.2.

For a determination, the chosen reaction is carried out in a small gas-tight container immersed in water in an *insulated* vessel. The initial temperature T_1 and the final temperature T_2 are measured. The overall process involves

$$\text{A}\ (T_1) + \text{calorimeter}\ (T_1) \rightarrow \text{B}\ (T_2) + \text{calorimeter}\ (T_2), q = 0. \qquad [7.1]$$

Here A symbolizes the reactants and B the products for the reaction.

One may break the process down into two hypothetical steps,

$$\text{A}\,(T_1) + \text{calorimeter}\,(T_1) \rightarrow \text{B}\,(T_1) + \text{calorimeter}\,(T_1),\ q_1, \qquad [7.2]$$

$$\text{B}\,(T_1) + \text{calorimeter}\,(T_1) \rightarrow \text{B}\,(T_2) + \text{calorimeter}\,(T_2),\ q_2, \qquad [7.3]$$

where $-q_1$ is the heat evolved during reaction at temperature T_1 while q_2 is the heat absorbed to raise the products and calorimeter from T_1 to T_2,

In the laboratory, one may cool the final system down to temperature T_1. Then it could be heated from T_1 to T_2 electrically. From the current, voltage, and time, one would calculate the electrical energy dissipated and thus obtain q_2. But since the reaction was run adiabatically, we have

$$q_1 = -q_2. \qquad [7.4]$$

When the reaction is carried out in an enclosed container (bomb), the reacting system is kept essentially at a fixed volume. It is thus kept from doing work, w is zero in formula (4.13), and we have

$$\left(q_1\right)_V = \Delta E_V \qquad [7.5]$$

at temperature T_1.

If the reaction were carried out at constant pressure and if all work done was compressional work, equation (4.27) would apply and

$$\left(q_1\right)_P = \Delta H_P \qquad [7.6]$$

at the given temperature T_1.

7.4 Relating the Isobaric and Isochoric Reaction Heats

Energy (7.6) is the isobaric heat of reaction while energy (7.5) is the isochoric heat of reaction for the given process. These are related through the definition of the enthalpy H for the given system.

Consider the process

$$\text{A}\,(P_1,T_1) \rightarrow \text{B}\,(P_2,T_1) \qquad [7.7]$$

where A represents the reactants and B the products of a reaction. Also, consider that all work done is work of compression.

When the final pressure is the same as the initial pressure, $P_2 = P_1$, the heat of reaction is given by equation (7.6). On the other hand, when P_2 is adjusted so that the final volume equals the initial volume, $V_2 = V_1$, the heat of reaction is given by equation (7.5).

According to its definition, the enthalpy H is

$$H = E + PV. \qquad [7.8]$$

Taking the increment with the pressure constant yields

$$\Delta H_P = \Delta E_P + P\Delta V. \qquad [7.9]$$

If the reaction were carried out at constant pressure and temperature and the products then compressed to the initial volume at constant temperature, the total change in E would be ΔE_V. If the products were ideal gases, there would be no change in E in the second step. Otherwise, there would be some change. But since this is generally small with respect to the heat of reaction, we may set

$$\Delta E_P \cong \Delta E_V \qquad [7.10]$$

and rewrite (7.9) as

$$\Delta H_P \cong \Delta E_V + P\Delta V. \quad [7.11]$$

In estimating $P\Delta V$, one may neglect the volumes of condensed phases and get the volumes of gases from the ideal gas law. At constant temperature, we would have

$$P\Delta V \cong (\Delta n)RT \quad [7.12]$$

where Δn is the change in number of moles of gas in the reaction as written. Then equation (7.11) becomes

$$\Delta H_P \cong \Delta E_V + (\Delta n)RT \quad [7.13]$$

Example 7.3

Relate the isobaric reaction heat to the isochoric one for the process

$$C_{10}H_8\ (s) + 12O_2\ (g) \rightarrow 10CO_2\ (g) + 4H_2O\ (l).$$

In this reaction there are 10 moles of gaseous product and 12 moles of gaseous reactant. Consequently,

$$\Delta n = 10 - 12 = -2$$

and

$$\Delta H_P = \Delta E_V - 2RT$$

whence

$$q_P = q_V - 2RT.$$

7.5 Variation in Heat of Reaction with Temperature

Whenever the heat capacity of the products in a reaction differs from the heat capacity of the reactants for the specified conditions, the heat q for the reaction varies with the temperature.

Let A represent the reactants and B the products of a reaction occurring at temperature T:

$$A\ (T) \rightarrow B\ (T). \quad [7.14]$$

When the volume of the products is the same as the volume of the reactants, the heat of reaction is

$$q_V = \Delta E = E_B - E_A. \quad [7.15]$$

Here E_B is the internal energy of the specified amount of products and E_A that of the corresponding amount of reactants.

Differentiating the second equality in (7.15) at constant volume yields

$$\left(\frac{\partial \Delta E}{\partial T}\right)_V = \left(\frac{\partial E_B}{\partial T}\right)_V - \left(\frac{\partial E_A}{\partial T}\right)_V = (C_V)_B - (C_V)_A = \Delta C_V. \quad [7.16]$$

In the second step, the energy capacities have been introduced with equation (4.35). Integrating over temperature then gives us the formula

$$\Delta E_2 = \Delta E_1 + \int_{T_1}^{T_2} \Delta C_V\ dT \quad [7.17]$$

relating the energy of reaction at temperature T_2 to that at temperature T_1, and by (7.15) the corresponding heats.

When the pressure on the products is the same as that on the reactants, the heat of reaction is

$$q_P = \Delta H = H_B - H_A. \quad [7.18]$$

Now H_B is the enthalpy of the specified amount of products and H_A that of the corresponding amount of reactants.

Differentiating the second equality in (7.18) at constant pressure yields

$$\left(\frac{\partial \Delta H}{\partial T}\right)_P = \left(\frac{\partial H_B}{\partial T}\right)_P - \left(\frac{\partial H_A}{\partial T}\right)_P = \left(C_P\right)_B - \left(C_P\right)_A = \Delta C_P. \quad [7.19]$$

In the second step, the energy capacities from formula (4.41) have been introduced. Integrating over temperature gives us the formula

$$\Delta H_2 = \Delta H_1 + \int_{T_1}^{T_2} \Delta C_P \, dT \quad [7.20]$$

Note that ΔH_2 equals the heat of reaction at T_2 and P while ΔH_1 equals the heat of reaction at T_1 and P following formula (7.18).

For the integrals in (7.17) and (7.20), one needs analytic expressions for the energy capacities. Polynomial coefficients for standard state energy capacities of common substances are listed in table 7.1.

Example 7.4

Construct an expression for the temperature dependence of the standard enthalpy change in the reaction

$$\frac{1}{2}H_2\ (g) + \frac{1}{2}Cl_2\ (g) \rightarrow HCL\ (g), \qquad \Delta H^0_{298.15} = -92{,}312\ J.$$

Table 7.1 Coefficients for the Empirical Function $C_P^0 = a + bT + cT^2\ (+dT^3)$ Valid from 300 K to 1500 K

Substance	*a*, *J K⁻¹ mol⁻¹*	*b x 10³*, *J K⁻² mol⁻¹*	*c x 10⁷*, *J K⁻³ mol⁻¹*	*d x 10⁹*, *J K⁻⁴ mol⁻¹*
H_2 (g)	29.066	-0.8364	20.117	
O_2 (g)	25.723	12.979	-38.62	
N_2 (g)	27.296	5.230	-0.04	
Cl_2 (g)	31.696	10.1437	-40.376	
Br_2 (g)	35.241	4.0748	-14.874	
H_2O (g)	30.359	9.615	11.84	
HCl (g)	28.166	1.8096	15.468	
CO_2 (g)	25.999	43.497	-148.32	
CO (g)	26.861	6.966	-8.20	
CH_4 (g)	14.146	75.496	-179.91	
C_2H_6 (g)	9.401	159.833	-462.29	
C_2H_4 (g)	11.841	119.667	-365.10	
C_6H_6 (g)	-1.711	324.766	-1105.79	
C (s)	-5.293	58.609	-432.25	11.510

With the empirical function

$$\Delta C_P^0 = \Delta a + \Delta bT + \Delta cT^2,$$

equation (7.20) becomes

$$\Delta H = \Delta H_0 + \int\left(\Delta a + \Delta bT + \Delta cT^2\right)\mathrm{d}T = \Delta H_0 + \Delta aT + \frac{\Delta b}{2}T^2 + \frac{\Delta c}{3}T^3.$$

Coefficients Δa, $\Delta b/2$, and $\Delta c/3$ calculated from the pertinent numbers in table 7.1 are

$$\Delta a = -2.215\ \mathrm{J\ K^{-1}}, \frac{\Delta b}{2} = -1.422\times10^{-3}\ \mathrm{J\ K^{-2}}, \frac{\Delta c}{3} = 8.533\times10^{-7}\ \mathrm{J\ K^{-3}}.$$

Then term ΔH_0 is

$$\Delta H_0 = \Delta H_{T_1} - \Delta aT_1 - \frac{\Delta b}{2}T_1^2 - \frac{\Delta c}{3}T_1^3$$
$$= -92{,}312 + 2.215(298.15) + 1.422\times10^{-3}(298.15)^2 - 8.533\times10^{-7}(298.15)^3\ \mathrm{J} = -91{,}548\ \mathrm{J}.$$

Inserting these parameters into the general form gives us

$$\Delta H_T^0 = -91{,}548 - 2.215T - 1.422\times10^{-3}T^2 + 8.533\times10^{-7}T^3\ \mathrm{J}.$$

7.6 Calorimetric Entropy

In determining the Gibbs free energy change ΔG for a reaction, a person needs not only ΔH but also ΔS for converting the reactants to products. The increment in entropy is obtained from the entropies of the pure participants. These may be found from energy capacity measurements made down to very low temperatures coupled with the third law of thermodynamics.

The entropy change occasioned by raising the temperature of a given phase at constant pressure from T_1 to T_2 is

$$\Delta S = \int_{T_1}^{T_2}\frac{C_P}{T}\,\mathrm{d}T, \qquad [7.21]$$

following formula (5.22). The entropy change occasioned by a first-order transition at temperature T_2 is

$$\Delta S = \frac{q_P}{T_2}, \qquad [7.22]$$

following formula (5.23).

Below about 15 K, energy capacity measurements become very difficult. But there a theoretical result of Paul Debye may be employed. By considering the possible elastic waves set up in a condensed phase, he determined that

$$C_P \cong aT^3 \qquad [7.23]$$

at the low temperatures. So we find that

$$\Delta S = \int_0^{T_1}\frac{C_P}{T}\,\mathrm{d}T = \int_0^{T_1} aT^2\,\mathrm{d}T = a\frac{T_1^3}{3} = \frac{\left(C_P\right)_{T_1}}{3}. \qquad [7.24]$$

Here temperature T_1 must be low enough so that (7.23) is a good approximation.

If we assume that the third law applies to the given pure material, then the entropy at 0 K is zero. The entropy at very low temperatures is then given by increment (7.24). At temperatures up to the lowest first-order transition, one adds the amount given by (7.21). At the transition, one adds amount (7.22). Then up to the second first-order transition, one adds the amount given by (7.21) for the pertinent temperatures. And so on. The result is known as the *calorimetric entropy* S_{calor} for the final temperature and amount of material.

When the final phase is gaseous, the entropy change on transforming it to the ideal gas state is generally added.

7.7 Correcting the Calorimetric Entropy to the Ideal Gas State

In an ideal gas, the molecules do not interact appreciably except during collisions. At a given temperature, this condition is approached as pressure P is reduced to zero. The change in S can be determined for taking the actual gas down to zero pressure and then taking the ideal gas form back up to the initial pressure.

Consider one mole of gas at a given temperature T. With its entropy S a function of P and T, we have

$$\mathrm{d}S = \left(\frac{\partial S}{\partial P}\right)_T \mathrm{d}P. \qquad [7.25]$$

But for the Gibbs free energy, equation (5.82) is

$$\mathrm{d}G = V\,\mathrm{d}P - S\,\mathrm{d}T. \qquad [7.26]$$

From this, we obtain

$$-\frac{\partial^2 G}{\partial P \partial T} = \left(\frac{\partial S}{\partial P}\right)_T = -\left(\frac{\partial V}{\partial T}\right)_P. \qquad [7.27]$$

In describing the actual gas, we employ Berthelot equation (3.73). From it, one gets

$$\left(\frac{\partial V}{\partial T}\right)_P = \frac{R}{P} + \frac{R}{P}\frac{27}{32}\frac{P}{P_c}\frac{T_c^3}{T^3}, \qquad [7.28]$$

while the ideal gas equation yields

$$\left(\frac{\partial V}{\partial T}\right)_P = \frac{R}{P}. \qquad [7.29]$$

Using (7.28) when lowering the pressure to P^* and (8.29) for raising it back to P gives us

$$\Delta S = -\int_P^{P^*}\left(\frac{R}{P} + \frac{R}{P}\frac{27}{32}\frac{P}{P_c}\frac{T_c^3}{T^3}\right)\mathrm{d}P - \int_{P^*}^{P}\frac{R}{P}\,\mathrm{d}P = \int_{P^*}^{P} R\frac{27}{32}\frac{1}{P_c}\frac{T_c^3}{T^3}\,\mathrm{d}P = \frac{27}{32}R\frac{P}{P_c}\frac{T_c^3}{T^3}\Bigg|_{P^*}^{P}. \qquad [7.30]$$

Letting P^* be zero, so the actual gas is ideal there, leads to

$$S_i - S = \frac{27}{32}R\frac{P}{P_c}\frac{T_c^3}{T^3}. \qquad [7.31]$$

Here S_i is the entropy of the hypothetical ideal gas at pressure P, S the entropy of the real gas at pressure P, T the absolute temperature, P the critical pressure, and T_c the critical temperature for the given gas. When P equals 1 bar, S_i equals S^0, the standard entropy.

Whenever the correction is small, the error introduced by using the Berthelot equation is very small.

Example 7.5

Calculate the entropy correction for gas imperfection of nitrogen at 77.32 K and 1.000 bar, if its critical temperature is 126 K and its critical pressure 33.9 bar.

Employ formula (7.31):

$$S^0 - S = \frac{27}{32}\left(8.3145 \text{ J K}^{-1} \text{ mol}^{-1}\right)\frac{1.000}{33.9}\left(\frac{126}{77.32}\right)^3 = 0.90 \text{ J K}^{-1} \text{ mol}^{-1}.$$

Thus, the entropy of 1.000 mol nitrogen in its standard state is 0.90 J K^{-1} more than its actual entropy at 1.000 bar and at 77.32 K. The gas imperfection imposes a slight amount of order on the system.

7.8 Correcting the Calorimetric Entropy for Frozen-in Disorder

The order in a crystal involves not only where the molecules or ions go but also how they are oriented. When two or more different orientations are almost equally stable, disorder is frozen in. This produces a residual entropy that must be added to the calorimetric entropy to get the absolute entropy for the given substance.

Consider a linear molecule that is nearly as stable when placed in the crystal lattice backward as forward. At the temperature of formation, the Boltzmann factor $e^{-\varepsilon/kT}$ is nearly the same for each configuration and the number put in each way is nearly the same.

Since each molecule may have two different configurations, N molecules in a crystal possess 2^N configurations. In the approximation that each of these is equally likely, state number W in formula (5.55) equals 2^N and the residual entropy at 0 K is

$$S_0 = k\ln 2^N = Nk\ln 2 = R\ln 2 = 5.76 \text{ J K}^{-1} \text{ mol}^{-1}. \qquad [7.32]$$

Using statistical mechanics, one can calculate state number W from molecular parameters. These may be determined from spectroscopic data. One can then compare the absolute entropies calculated from these data with the entropies that the third law provides. The discrepancies are attributed to the disorder frozen into the crystals. Some results appear in table 7.2.

The CO and N_2O molecules are linear; so number (7.32) gives the frozen-in entropy if there were complete disorder over the two orientations. The excess of this number over the corresponding number in table 7.2 indicates that there is a small amount of ordering.

In solid nitric oxide, the NO molecules dimerize to produce a diamagnetic crystal. As a consequence, the number in table 7.2 is for 1/2 mole of the solid. For 1.000 mol, the observed value is 6.28 J K^{-1}, somewhat higher than number (7.32).

In the interior of ice I, there are four close oxygen atoms to each given oxygen atom. These are at the corners of a tetrahedron centered on the given atom. Now, the four neighbors are linked to the central atom by hydrogen bonds. But an O — H–O bond is not symmetric; the H is 0.99 Å from one O and 1.77 Å from the other at equilibrium. Consequently, each H in the crystal has the choice of two positions. In a crystal of N molecules there are $2N$ hydrogen atoms, yielding a total of 2^{2N} possible configurations. However, many of these correspond to ionized structures that are unstable. If all configurations occurred in a given OH_4 unit, they would be 2^4 or 16 in number. But one of these is OH_4^{++}, four are OH_3^+, four are OH^-, and one is $O^=$. By difference, we see that six correspond to OH_2. So

TABLE 7.2 Spectroscopic Entropy Minus Calorimetric Entropy

Substances	*Entropy Excess, J K⁻¹ mol⁻¹*
CO	4.64
N_2O	4.77
NO	3.14
H_20	3.39

about each of the N oxygens, only (6/16) of the possible configurations are likely. For the crystal as a whole, this means that only $(6/16)^N$ of the 2^{2N} configurations are to be considered in calculating W:

$$W = 2^{2N}\left(6/16\right)^N = \left(3/2\right)^N. \qquad [7.33]$$

Putting this W into formula (5.55) yields the residual entropy

$$S_0 = k\ln\left(3/2\right)^N = Nk\ln\left(3/2\right) = R\ln\left(3/2\right) = 3.37\ \mathrm{J\ K^{-1}\ mol^{-1}}, \qquad [7.34]$$

which checks the observed S_0 in table 7.2.

7.9 Key Thermodynamic Properties

Data from many experiments on common substances are summarized in tables 7.3 and 7.4.

The entropies for the gases have been corrected to the ideal gas state. Also, allowances have been made for any frozen-in disorder. However, the contributions from mixing different isotopes of an element and the contributions from degeneracies arising from nonzero nuclear spins have been omitted since they do not contribute to S_{calor} and they cancel out in common chemical reactions.

The enthalpy of formation $\Delta H^0{}_f$ is the heat of reaction for the formation of 1 mole of the compound in its standard state from the elements in their standard states. From the tabulated entropies and enthalpies, standard Gibbs energy changes can be calculated with the formula

$$\Delta G^0 = \Delta H^0 - T\Delta S^0 \qquad [7.35]$$

which follows from definition (5.79). They can also be calculated from the listed Gibbs energies of formation $\Delta G^0{}_f$.

The entropy change in a reaction equals the entropies of the products minus the entropies of the reactants. Similarly, the enthalpy change equals the enthalpies of formation of the products minus the enthalpies of formation of the reactants. And the Gibbs energy change equals the Gibbs energy of formation of the products minus the Gibbs energy of formation of the reactants.

Example 7.6

Using tables 7.3 and 7.4, calculate ΔS^0, ΔH^0, and ΔG^0 for the reaction

$$\mathrm{NO\ (g)} + \frac{1}{2}\mathrm{O_2\ (g)} \rightarrow \mathrm{NO_2\ (g)}.$$

TABLE 7.3 Entropies and Energy Capacities at 298.15 K

Substance	S^0, $J\,K^{-1}\,mol^{-1}$	C_P^0, $J\,K^{-1}\,mol^{-1}$
H_2 (g)	130.68	28.84
O_2 (g)	205.15	29.38
N_2 (g)	191.61	29.12
Cl_2 (g)	223.12	33.94
Br_2 (l)	152.21	75.69
I_2 (s)	116.14	54.44
C (s)	5.740	8.512
H_2O (g)	188.96	33.59
HCl (g)	186.90	29.13
CH_4 (g)	186.21	35.64
C_2H_2 (g)	200.96	44.06
C_2H_4 (g)	219.33	42.89
C_2H_6 (g)	229.60	52.54
CO (g)	197.66	29.14
CO_2 (g)	213.77	37.13
SO_2 (g)	248.22	39.90
H_2S (g)	205.75	34.19
NH_3 (g)	192.78	35.65
N_2O (g)	219.98	38.84
NO (g)	210.76	29.84
N_2O_3 (g)	309.35	65.91
NO_2 (g)	240.02	36.66

The change in a thermodynamic property accompanying a given amount of reaction equals the value of the property for the given moles of products minus that for the given moles of reactants. From the standard entropies for the given participants, we obtain

$$\Delta S^0 = S^0_{NO_2} - S^0_{NO} - \frac{1}{2}S^0_{O_2} = 240.02 - 210.76 - \frac{1}{2}(205.15) = -73.32\ \mathrm{J\,K^{-1}}.$$

Similarly, from the standard enthalpies of formation of the participants, we find

$$\Delta H^0 = \Delta H^0_{f\,NO_2} - \Delta H^0_{f\,NO} - \frac{1}{2}\Delta H^0_{f\,O_2} = 33.10 - 90.29 - 0 = -57.19\ \mathrm{kJ}.$$

Also,

$$\Delta G^0 = \Delta G^0_{f\,NO_2} - \Delta G^0_{f\,NO} - \frac{1}{2}\Delta G^0_{f\,O_2} = 51.26 - 86.60 - 0 = -35.34\ \mathrm{kJ}.$$

TABLE 7.4 Enthalpies, Gibbs Energies, and Equilibrium Constants of Formation at 298.15 K

Substance	ΔH_f^0, *kJ mol⁻¹*	ΔG_f^0, *kJ mol⁻¹*	*log K_f*
H_2O (g)	-241.83	-228.62	40.053
HCl (g)	-92.31	-95.29	16.695
CH_4 (g)	-74.87	-50.76	8.892
C_2H_2 (g)	226.73	209.20	-36.651
C_2H_4 (g)	52.47	68.42	-11.987
C_2H_6 (g)	-84.68	-32.83	5.752
CO (g)	-110.54	-137.18	24.033
CO_2 (g)	-393.50	-394.36	69.091
SO_2 (g)	-296.81	-300.10	52.576
H_2S (g)	-20.50	-33.33	5.839
NH_3 (g)	-45.94	-16.41	2.875
N_2O (g)	82.05	104.17	-18.250
NO (g)	90.29	86.60	-15.172
N_2O_3 (g)	82.84	139.49	-24.437
NO_2 (g)	33.10	51.26	-8.981

7.10 Partial Molar Gibbs Energy

For a region that can be considered homogeneous, the contribution of each constituent to the Gibbs energy is proportional to the number of moles of the constituent there. Thus, equation (5.102),

$$G = \sum_i \mu_i n_i = \sum_i \overline{G_i} n_i, \qquad [7.36]$$

holds. Here n_i is the number of moles of the ith constituent in the region, while μ_i is the ith chemical potential or *partial molar Gibbs energy* $\overline{G}_i$.

If the given material were an ideal mixture of ideal gases, each constituent would behave as if the others were absent. Changing its partial pressure by dP_i. at constant temperature and moles of constituents would produce the change

$$dG_i = V_i \, dP_i. \qquad [7.37]$$

For one mole of the constituent in volume V_i, we would have

$$V_i = \frac{RT}{P_i} \qquad [7.38]$$

and

$$d\overline{G_i} = RT\frac{dP_i}{P_i}. \qquad [7.39]$$

Let us consider the *unit* of partial pressure P_i to be the standard-state pressure P_i^0. Then integrating (7.39) at constant temperature leads to

$$\mu_i = \overline{G}_i = \overline{G}_i^0 + RT\ln\left(P_i / P_i^0\right) = \mu_i^0 + RT\ln P_i. \qquad [7.40]$$

If the ideal gaseous solution were at equilibrium with another phase, the chemical potentials in the two phases would be equal by condition (6.102). Furthermore, the *activity* a_i of the *i*th constituent in the new phase is measured by the partial pressure in the gas phase. By definition, we take them to be proportional:

$$\frac{P_i}{P_i^0} = k_i \frac{a_i}{a_i^0}. \qquad [7.41]$$

Substituting the new μ_i and expression (7.41) into equation (7.40) gives us

$$\mu_i = \left(\mu_i^0\right)_{\text{gas}} + RT\ln k_i + RT\left(a_i / a_i^0\right) = \mu_i^0 + RT\ln a_i \qquad [7.42]$$

for the new phase. In the final form, quantity μ_i^0 has replaced the sum $(\mu_i^0)_{\text{gas}} + RT \ln k_i$ and the unit of activity has been taken to be the standard-state activity a_i^0. Expression μ_i^0 is now the *standard chemical potential* of the *i*th constituent in the given phase.

7.11 Gibbs Energy of Reaction

A common problem in chemistry is whether a particular reaction can proceed at a given temperature and pressure with a certain set of concentrations or activities. But from section 5.14, the net tendency for a reaction to go is measured by the pertinent negative Gibbs free energy change.

Let us consider a homogeneous system in which the reaction

$$a\text{A} + b\text{B} \rightarrow l\text{L} + m\text{M} \qquad [7.43]$$

moves forward by $d\lambda$ unit at a given temperature and pressure. The changes in moles of A, B, L, M are

$$dn_{\text{A}} = -a\ d\lambda,\quad dn_{\text{B}} = -b\ d\lambda,\quad dn_{\text{L}} = l\ d\lambda,\quad dn_{\text{M}} = m\ d\lambda. \qquad [7.44]$$

Applying formula (5.97) to the process gives us

$$dG = l\mu_{\text{L}}\ d\lambda + m\mu_{\text{M}}\ d\lambda - a\mu_{\text{A}}\ d\lambda - b\mu_{\text{B}}\ d\lambda \qquad [7.45]$$

whence

$$\left(\frac{\partial G}{\partial \lambda}\right)_{T,P} = l\mu_{\text{L}} + m\mu_{\text{M}} - a\mu_{\text{A}} - b\mu_{\text{B}}. \qquad [7.46]$$

Derivative $(\partial G/\partial \lambda)_{T,P}$ is the increase in Gibbs energy per unit of reaction when only an infinitesimal amount of reaction occurs. If the reacting system were of infinite extent, it would be the Gibbs energy change when *a* moles A reacted with *b* moles B to produce *l* moles L and *m* moles M. Thus it is called the *Gibbs energy of reaction* ΔG:

$$\Delta G = l\mu_{\text{L}} + m\mu_{\text{M}} - a\mu_{\text{A}} - b\mu_{\text{B}}. \qquad [7.47]$$

Consider the reacting mixture to be gaseous and introduce formula (7.40) for each chemical potential:

$$\Delta G = lG_{\text{L}}^0 + mG_{\text{M}}^0 - aG_{\text{A}}^0 - bG_{\text{B}}^0 + lRT\ln P_{\text{L}} + mRT\ln P_{\text{M}} - aRT\ln P_{\text{A}} - bRT\ln P_{\text{B}}. \qquad [7.48]$$

The first four terms on the right give the standard Gibbs energy change,

$$\Delta G^0 = lG_L^0 + mG_M^0 - aG_A^0 - bG_B^0 = l\Delta G_{fL}^0 + m\Delta G_{fM}^0 - a\Delta G_{fA}^0 - b\Delta G_{fB}^0, \qquad [7.49]$$

while the last four terms combine to give

$$RT\ln\frac{P_L^l P_M^m}{P_A^a P_B^b} = RT\ln Q_P. \qquad [7.50]$$

Thus, equation (7.48) has the form

$$\Delta G = \Delta G^0 + RT\ln Q_P. \qquad [7.51]$$

Expression Q_P is called the reaction quotient.

For a general homogeneous region, formula (7.42) replaces (7.40). Then equation (7.48) is replaced with

$$\Delta G = \Delta G^0 + RT\ln\frac{a_L^l a_M^m}{a_A^a a_B^b} = \Delta G^0 + RT\ln Q \qquad [7.52]$$

where

$$Q = \frac{a_L^l a_M^m}{a_A^a a_B^b}. \qquad [7.53]$$

Note that in expression Q the activities of products appear in the numerator while the activities of reactants appear in the denominator. Each activity is raised to a power equal to the coefficient of the constituent in the chemical equation.

Example 7.7

From 1.000 mol H_2 at 0.01000 bar, 25° C , and 0.500 mol O_2 at 0.1000 bar, 25° C, reaction produced 1.000 mol water vapor at 0.0500 bar and 25° C. What was the Gibbs energy change?

For the given reaction

$$H_2\,(g) + \frac{1}{2}O_2\,(g) \rightarrow H_2O\,(g)$$

we have

$$Q_P = \frac{P_{H_2O}}{P_{H_2}P_{O_2}^{1/2}} = \frac{0.0500}{0.01000(0.1000)^{1/2}} = 15.81.$$

From table 7.4

$$\Delta G^0 = -228.62 \text{ kJ}.$$

Substituting into equation (7.51) yields

$$\Delta G = -228.62 \text{ kJ} + (8.3145\times10^{-3} \text{ kJ K}^{-1})(298.15 \text{ K})\ln 15.81$$

$$= -228.62 \text{ kJ} + 6.84 \text{ kJ} = -221.78 \text{ kJ}.$$

The large negative value of ΔG indicates that there is a strong tendency for the reaction to proceed under the given conditions. However, the thermodynamics does not tell us anything about the rate of the process. This will be considered in the chapters on kinetics.

7.12 Energetic Conditions on Equilibria

A given reaction may proceed spontaneously at temperature T and pressure P as long as it releases net work. But net work is available for dissipation only as long as the Gibbs free energy G can decrease. When G can no longer decrease, the point of equilibrium has been reached.

At such a point, the derivative of G with respect to reaction coordinate λ must vanish. Since this equals the Gibbs energy of reaction, we have

$$\Delta G = \left(\frac{\partial G}{\partial \lambda}\right)_{T,P} = 0 \qquad [7.54]$$

and equation (7.52) reduces to

$$0 = \Delta G^0 + RT \ln K. \qquad [7.55]$$

Here K is the value quotient Q assumes at equilibrium.

Subtracting equation (7.55) from (7.52) yields

$$\Delta G = RT \ln \frac{Q}{K}. \qquad [7.56]$$

When $Q > K$, ΔG is positive and the reaction goes spontaneously to the left at the given T and P. Thus, Q is reduced until it equals K. When $Q < K$, ΔG is negative and the reaction proceeds spontaneously to the right at the given T and P. Now Q increases until it equals K.

At equilibrium, Q equals K and

$$\frac{a_L^l a_M^m}{a_A^a a_B^b} = K. \qquad [7.57]$$

From (7.55), we have

$$\ln K = -\frac{\Delta G^0}{RT} \qquad [7.58]$$

or

$$\log K = -\frac{\Delta G^0}{2.303RT}. \qquad [7.59]$$

Example 7.8

Determine the equilibrium constant at 25° C for the reaction

$$H_2\,(g) + \frac{1}{2}O_2\,(g) \rightarrow H_2O\,(g)$$

In table 7.4, log K_f is the logarithm for formation of 1 mole of the compound from its elements in their standard states. For the given reaction, we have

$$\log K = \left(\log K_f\right)_{H_2O} - \left(\log K_f\right)_{H_2} - \frac{1}{2}\left(\log K_f\right)_{O_2} = 40.053 - 0 - \frac{1}{2}0 = 40.053.$$

So

$$K = 1.130 \times 10^{40} \text{ bar}^{-1/2}.$$

7.13 Equilibria in Ideal Gaseous Phases

When a reacting gaseous solution is at a low enough pressure at the given temperature so that it is approximately ideal, the partial pressures represent the activities and the equilibrium

$$aA + bB \leftrightarrow lL + mM \tag{7.60}$$

is governed by the constant

$$\frac{P_L^l P_M^m}{P_A^a P_B^b} = K_P. \tag{7.61}$$

This is related to the standard Gibbs energy change by

$$\ln K_P = -\frac{\Delta G^0}{RT}. \tag{7.62}$$

In the ideal gaseous solution, the pressure exerted by the ith constituent is

$$P_i = \frac{n_i}{V} RT = c_i RT, \tag{7.63}$$

where c_i is the concentration of constituent i. Substituting (7.63) into (7.61) gives

$$K_P = \frac{c_L^l c_M^m}{c_A^a c_B^b} \left(RT\right)^{l+m-a-b} = K_c \left(RT\right)^{\Delta n}. \tag{7.64}$$

Here Δn is the change in moles in the reaction while the concentration equilibrium constant is

$$\frac{c_L^l c_M^m}{c_A^a c_B^b} = K_c \tag{7.65}$$

In a common problem, one starts with a given amount of each reactant and product in a given volume at a given temperature and lets the reaction proceed to equilibrium. From the final conditions, one can calculate the equilibrium constant. Or alternatively, from the equilibrium constant one can calculate the final conditions. The expressions needed depend on the form of the reaction.

As a first example, consider the water gas equilibrium

$$H_2O\ (g) + CO\ (g) \leftrightarrow CO_2\ (g) + H_2\ (g). \tag{7.66}$$

In a container of volume V, a moles H_2O, b moles CO, c moles CO_2, and d moles H_2 are mixed. On going to equilibrium, x moles H_2O react with x moles CO. One may organize the pertinent expressions as in table 7.A.

TABLE 7.A

Constituent	*Moles at Start*	*Equilibrium Moles*	*Equilibrium Concentration*
H_2O	a	$a - x$	$(a - x)/V$
CO	b	$b - x$	$(b - x)/V$
CO_2	c	$c + x$	$(c + x)/V$
H_2	d	$d + x$	$(d + x)/V$

Equation (7.65) now assumes the form

$$K_c = \frac{(c+x)(d+x)/V^2}{(a-x)(b-x)/V^2}, \quad [7.67]$$

which reduces to

$$K_c = \frac{(c+x)(d+x)}{(a-x)(b-x)}. \quad [7.68]$$

Note that here $K_c = K_p$ since $\Delta n = 0$.

Secondly, consider the dissociation of phosphorus pentachloride gas

$$PCL_5\ (g) \leftrightarrows PCL_3\ (g) + Cl_2\ (g). \quad [7.69]$$

In a container of volume V, a moles PCl_5, _ moles PCl_3, and c moles Cl_2 are mixed. On going to equilibrium, x moles PCl_5 dissociate. If n_0 is $a + b + c$, the number of moles in the beginning, the number at the end is

$$(a-x)+(b+x)+(c+x) = n_0 + x. \quad [7.70]$$

One may organize the pertinent expressions as in table 7.B.

TABLE 7.B

Constituent	*Moles at Start*	*Equilibrium Moles*	*Equilibrium Mole Fraction*
PCl_5	a	$a - x$	$\frac{a-x}{n_0+x}$
PCl_3	b	$b + x$	$\frac{b+x}{n_0+x}$
Cl_2	c	$c + x$	$\frac{c+x}{n_0+x}$

With these, equation (7.65) takes on the form

$$K_c = \frac{(b+x)(c+x)/V^2}{(a-x)/V} = \frac{(b+x)(c+x)}{(a-x)V}. \quad [7.71]$$

In formula (7.61), the partial pressure of each constituent equals its mole fraction times the total pressure P; so here

$$K_p = \frac{\dfrac{(b+x)(c+x)}{(n_0+x)^2}P^2}{\dfrac{a-x}{n_0+x}P} = \frac{(b+x)(c+x)P}{(a-x)(n_0+x)}. \quad [7.72]$$

Thirdly, consider the dissociation of nitrogen tetroxide gas

$$N_2O_4\ (g) \rightleftarrows 2NO_2\ (g). \quad [7.73]$$

Proceeding as before, we set up in table 7.C.

TABLE 7.C

Constituent	Moles at Start	Equilibrium Moles	Equilibrium Mole Fraction
N_2O_4	a	$a - x$	$\frac{a - x}{n_0 + x}$
NO_2	b	$b + 2x$	$\frac{b + 2x}{n_0 + x}$

and find that

$$K_c = \frac{(b+2x)^2 / V^2}{(a-x)/V} = \frac{(b+2x)^2}{(a-x)V}. \quad [7.74]$$

Also

$$K_p = \frac{\left(\frac{b+2x}{n_0+x}\right)^2 P^2}{\frac{a-x}{n_0+x}P} = \frac{(b+2x)^2 P}{(a-x)(n_0+x)}. \quad [7.75]$$

One may start with a pure substance and observe part of it dissociate in the gas phase. If the initial number of moles is n_0, the fraction of these dissociating α and the number of molecules produced from one molecule of reactant ν, then the total number of moles is

$$n = (1-\alpha)n_0 + \nu\alpha n_0 = [1+(\nu-1)\alpha]n_0. \quad [7.76]$$

The ideal gas equation then becomes

$$PV = [1+(\nu-1)\alpha]\frac{w}{M_1}RT, \quad [7.77]$$

where w is the mass of gas in volume V at pressure P and temperature T while M_1 is the molecular mass of the original undissociated material.

Example 7.9

What are K_P and K_c at 25° C for the reaction in example 7.8?

Because of our choice of standard states, the K calculated in example 7.8 is K_P. But from the chemical equation, we see that

$$\Delta n = -\frac{1}{2}.$$

So formula (7.64) yields

$$K_c = K_P(RT)^{-\Delta n} = (1.130 \times 10^{40}\ \text{bar}^{-1/2})\left[(0.083145\ 1\ \text{bar K}^{-1}\ \text{mol}^{-1})(298.15\ \text{K})\right]^{1/2} = 5.63 \times 10^{40}\ \text{M}^{-1/2}.$$

Example 7.10

How does the fraction of dissociation α vary with pressure P for (a) reaction (7.69), (b) reaction (7.73)?

(a) Begin with 1 mole PCl_5 and 0 mole PCl_3 and Cl_2. Then

$$n_0 = a = 1, \qquad b = c = 0, \qquad x = \alpha,$$

and equation (7.72) reduces to

$$K_P = \frac{\alpha^2 P}{(1-\alpha)(1+\alpha)} = \frac{\alpha^2 P}{1-\alpha^2}.$$

(b) Begin with 1 mole N_2O_4 and 0 mole NO_2. Then

$$n_0 = a = 1, \qquad b = 0, \qquad x = \alpha,$$

and equation (7.75) reduces to

$$K_P = \frac{4\alpha^2 P}{(1-\alpha)(1+\alpha)} = \frac{4\alpha^2 P}{1-\alpha^2}.$$

Example 7.11

For N_2O_4 (g) the log K_f is -17.132, while for NO_2 (g) it is -8.981. Calculate K_P for reaction (7.73).

Proceed as in example 7.8 to get

$$\log K_P = 2(\log K_f)_{NO_2} - (\log K_f)_{N_2O_4} = 2(-8.981) - (-17.132) = -0.830$$

whence

$$K_P = 0.148\ \text{bar}.$$

Example 7.12

What is the density of an equilibrium mixture of N_2O_4 and NO_2 at 25° C and 1.000 bar?

With the results in examples 7.10 (b) and 7.11, we have

$$\frac{4\alpha^2(1.000\ \text{bar})}{1-\alpha^2} = 0.148\ \text{bar}$$

whence

$$\alpha = 0.189.$$

With equation (7.77), we find that

$$\rho = \frac{w}{V} = \frac{PM_1}{RT[1+(\nu-1)\alpha]} = \frac{(1.000\ \text{bar})(92.011\ \text{g mol}^{-1})}{(0.083145\ 1\ \text{bar K}^{-1}\ \text{mol}^{-1})(298.15\ \text{K})(1.189)} = 3.12\ \text{g l}^{-1}.$$

7.14 Heterogeneous Gas Equilibria

In section 6.9, we found that a large increase in pressure on a pure condensed phase produces only a small increase in the tendency for the substance to leave the phase. Thus, common pressure changes produce only small changes in the substance's activity. But this activity is of unit value when the pressure equals 1 bar, since the material is then in its standard state. So when the pressure does not exceed a few bars, the substance is still approximately at unit activity. This insensitivity simplifies many calculations involving such a condensed phase.

As an example, consider the reaction

$$C\ (s) + CO_2\ (g) \rightleftarrows 2CO\ (g) \qquad [7.78]$$

for which equation (7.57) becomes

$$K = \frac{a_{CO}^2}{a_C a_{CO_2}}. \qquad [7.79]$$

Under moderate conditions, we have

$$a_C = 1, \qquad a_{CO_2} = P_{CO_2}, \qquad a_{CO} = P_{CO} \qquad [7.80]$$

and

$$K_P = \frac{P_{CO}^2}{P_{CO_2}}. \qquad [7.81]$$

Similarly, for the equilibrium

$$NH_4HS\ (s) \rightleftarrows NH_3\ (g) + H_2S\ (g), \qquad [7.82]$$

we find that

$$K_P = P_{NH_3} P_{H_2S}. \qquad [7.83]$$

Example 7.13

When solid ammonium hydrosulfide is vaporized into a vacuum at 20° C, the pressure rises to 0.474 bar. If it is vaporized into a container filled initially with H_2S at 0.500 bar at 20° C, what is the final pressure after equilibrium is set up?

For each molecule of NH_3 leaving the solid, one molecule of H_2S leaves. So in the initially empty container, the mole fraction of each constituent stays at 1/2 and its pressure rises to 1/2 (0.474 bar). Formula (7.83) becomes

$$K_P = (0.237\text{ bar})(0.237\text{ bar}) = 0.0562\text{ bar}^2.$$

In the second container, each molecule of NH_3 gas corresponds to an additional H_2S gas molecule. But in the approximation that the gaseous solution is ideal, each partial pressure is proportional to the number of moles of the constituent. Consequently, the increase in pressure of H_2S equals the partial pressure of NH_3. With

$$P_{NH_3} = x,$$

we have

$$K_P = P_{NH_3} P_{H_2S} = x(0.500 + x)\text{ bar}^2 = 0.0562\text{ bar}^2,$$

whence

$$x = 0.0945 \text{ bar}$$

and

$$P = 0.0945 + 0.5945 \text{ bar} = 0.689 \text{ bar}.$$

7.15 Variation of Equilibrium Constants with Temperature

The equilibrium constant K for a reaction is related to the temperature T by equation (7.58). So if we want to learn how $\ln K$ varies, we need to investigate the behavior of $\Delta G^0/T$.

Let us first consider properties of a single reactant or product. Its Gibbs energy G is defined by equation (5.79). Dividing this by temperature T,

$$\frac{G}{T} = \frac{H}{T} - S, \tag{7.84}$$

then differentiating, gives

$$\left(\frac{\partial}{\partial T}\frac{G}{T}\right)_P = -\frac{H}{T^2} + \frac{1}{T}\left(\frac{\partial H}{\partial T}\right)_P - \left(\frac{\partial S}{\partial T}\right)_P. \tag{7.85}$$

But at constant pressure, we have

$$\mathrm{d}S = \frac{\mathrm{d}q_P}{T} = \frac{\mathrm{d}H}{T} \tag{7.86}$$

and

$$\left(\frac{\partial S}{\partial T}\right)_P = \frac{1}{T}\left(\frac{\partial H}{\partial T}\right)_P. \tag{7.87}$$

So equation (7.85) reduces to

$$\left(\frac{\partial}{\partial T}\frac{G}{T}\right)_P = -\frac{H}{T^2}. \tag{7.88}$$

This is called the *Gibbs - Helmholtz equation.*

Since formula (7.88) applies to each reactant and product separately, one can construct the difference for the standard states:

$$\left(\frac{\partial}{\partial T}\frac{\Delta G^0}{T}\right)_P = -\frac{\Delta H^0}{T^2}. \tag{7.89}$$

Introducing formula (7.62) converts (7.89) to the *van't Hoff equation*

$$\left(\frac{\partial \ln K_P}{\partial T}\right)_P = \frac{\Delta H^0}{RT^2}. \tag{7.90}$$

Generally, the standard enthalpy change ΔH^0 varies slowly with temperature T. In the approximation that it is constant, integration of equation (7.90) yields

$$\ln K_P = -\frac{\Delta H^0}{R}\frac{1}{T} + C. \tag{7.91}$$

Thus, a plot of the logarithm of equilibrium constant K_P for a given reaction against the reciprocal of the absolute temperature $1/T$ is approximately linear with the slope $-\Delta H^0/R$. From the slope of the tangent line at a particular temperature, one obtains $-\Delta H^0/R$ for that temperature.

The concentration equilibrium constant K_c is related to the pressure equilibrium constant K_P by equation (7.64). Solving this equation for $\ln K_c$,

$$\ln K_c = \ln K_P - \Delta n \ln RT, \qquad [7.92]$$

and differentiating with respect to temperature at constant pressure yields

$$\left(\frac{\partial \ln K_c}{\partial T}\right)_P = \left(\frac{\partial \ln K_P}{\partial T}\right)_P - \frac{\Delta n}{RT} R = \frac{1}{RT^2}\left(\Delta H^0 - \Delta nRT\right)$$
$$= \frac{1}{RT^2}\Delta\left(H^0 - PV\right) = \frac{\Delta E^0}{RT^2}. \qquad [7.93]$$

7.16 Variation of Equilibrium Constants with Pressure

For the pressure equilibrium constant, the standard state for each reactant and product is the pure material in its hypothetical ideal gas condition at 1 bar pressure and the given temperature. The corresponding ΔG^0varies only with temperature, not with pressure. So we have

$$\left(\frac{\partial \ln K_P}{\partial P}\right)_T = -\left(\frac{\partial}{\partial P}\frac{\Delta G^0}{RT}\right)_T = 0. \qquad [7.94]$$

Thus expression K_P does not vary with pressure. With relation (7.64), K_c also does not vary with pressure.

But if one alters the definition of the standard states, the corresponding K may change. For instance, suppose that the standard state of each reactant and product is the constituent behaving in a specified manner ("ideally") at unit concentration at the given temperature. This state may change with pressure. Then since

$$\left(\frac{\partial}{\partial P}\frac{G}{T}\right)_T = \frac{1}{T}\left(\frac{\partial G}{\partial P}\right)_T = \frac{V}{T}, \qquad [7.95]$$

we now have

$$\left(\frac{\partial \ln K_c}{\partial P}\right)_T = -\left(\frac{\partial}{\partial P}\frac{\Delta G^0}{RT}\right)_T = -\frac{\Delta V^0}{RT}. \qquad [7.96]$$

Result (7.96) may be applied to reactions in condensed phases.

7.17 The Activity Coefficient Concept

We have seen that, at a given temperature in a gaseous solution, an equilibrium

$$a\mathrm{A} + b\mathrm{B} \rightleftarrows l\mathrm{L} + m\mathrm{M} \qquad [7.97]$$

is governed by a constant

$$K = \frac{a_\mathrm{L}^l a_\mathrm{M}^m}{a_\mathrm{A}^a a_\mathrm{B}^b}. \qquad [7.98]$$

This is related to the standard Gibbs energy change by the formula

$$\ln K = -\frac{\Delta G^0}{RT}. \tag{7.99}$$

At low pressures, where the solution is approximately ideal, the ith activity is measured by the corresponding partial pressure,

$$a_i = P_i. \tag{7.100}$$

Then the expression

$$K_P = \frac{P_L^l P_M^m}{P_A^a P_B^b} \tag{7.101}$$

equals the equilibrium constant K.

At higher pressures, deviations from the ideal gas equation (7.38) occur. However, formula (7.36) still applies. Furthermore, one can construct expression (7.42) for each constituent and so obtain (7.98) for the equilibrium condition.

But the activity of the ith constituent then differs from its partial pressure. We express this fact by the equation

$$a_i = \gamma_i P_i \tag{7.102}$$

in which γ_i is called the ith *activity coefficient*. Employing form (7.102) for each activity in (7.98) gives us

$$K = \frac{\gamma_L^l \gamma_M^m}{\gamma_A^a \gamma_B^b} \frac{P_L^l P_M^m}{P_A^a P_B^b} = Q_\gamma K_P \tag{7.103}$$

where

$$Q_\gamma = \frac{\gamma_L^l \gamma_M^m}{\gamma_A^a \gamma_B^b}. \tag{7.104}$$

Substituting form (7.102) into expression (7.42) leads to

$$\mu_i = \mu_i^0 + RT \ln \gamma_i + RT \ln P_i. \tag{7.105}$$

The deviation of μ_i from its ideal value is measured by the term RT in γ_i. If one could determine this deviation, one could then construct γ_i. This is a complicated development. Nevertheless, in the next chapter, we will thus estimate the effect of electric interaction in electrolytic solutions.

7.18 Helmholtz Energy of Reaction

At a given temperature and volume, the tendency for a reaction to go is measured by the pertinent Helmholtz free energy change.

Let us consider a homogeneous system in which the reaction

$$a\text{A} + b\text{B} \rightarrow l\text{L} + m\text{M} \tag{7.106}$$

moves forward by $d\lambda$ unit at a given temperature and volume. Equations (7.44) then apply and formula (5.96) yields

$$dA = l\mu_L\, d\lambda + m\mu_M\, d\lambda - a\mu_A\, d\lambda - b\mu_B\, d\lambda, \tag{7.107}$$

whence

$$\left(\frac{\partial A}{\partial \lambda}\right)_{T,V} = l\mu_{\mathrm{L}} + m\mu_{\mathrm{M}} - a\mu_{\mathrm{A}} - b\mu_{\mathrm{B}}. \quad [7.108]$$

Derivative $(\partial A/\partial\lambda)_{T,V}$ is the increase in Helmholtz energy per unit of reaction when only an infinitesimal amount of reaction occurs. If the reacting system were of infinite extent, it would be the Helmholtz energy change when a moles A reacted with b moles B to produce l moles L and m moles M. So it is called the Helmholtz energy of reaction ΔA:

$$\Delta A = l\mu_{\mathrm{L}} + m\mu_{\mathrm{M}} - a\mu_{\mathrm{A}} - b\mu_{\mathrm{B}}. \quad [7.109]$$

When each reactant and product is in its standard state, formula (7.109) becomes

$$\Delta A^0 = l\mu_{\mathrm{L}}{}^0 + m\mu_{\mathrm{M}}{}^0 - a\mu_{\mathrm{A}}{}^0 - b\mu_{\mathrm{B}}{}^0. \quad [7.110]$$

When the standard state for each substance is the substance at unit concentration and the corresponding activity coefficient is 1, we have

$$\mu_i = \mu_i^0 + RT\ln c_i. \quad [7.111]$$

Substituting this into formula (7.109) and reducing yields

$$\Delta A = \Delta A^0 + RT\ln Q_c \quad [7.112]$$

where

$$Q_c = \frac{c_{\mathrm{L}}^l c_{\mathrm{M}}^m}{c_{\mathrm{A}}^a c_{\mathrm{B}}^b}. \quad [7.113]$$

At equilibrium, this reaction quotient equals the equilibrium constant,

$$Q_c = K_c. \quad [7.114]$$

and the change in A vanishes,

$$\Delta A = 0. \quad [7.115]$$

Then equation (7.112) reduces to

$$\Delta A^0 = -RT\ln K_c. \quad [7.116]$$

When the activity coefficients differ from 1, equation (7.113) is replaced with

$$Q_{\gamma c} = \frac{a_{\mathrm{L}}^l a_{\mathrm{M}}^m}{a_{\mathrm{A}}^a a_{\mathrm{B}}^b}, \quad [7.117]$$

and equation (7.116) with

$$\Delta A^0 = -RT\ln K_{\gamma c}. \quad [7.118]$$

Questions

7.1 How are standard states defined?
7.2 How are reaction heats measured?
7.3 Explain how the heat of a reaction varies with temperature.
7.4 How is the calorimetric entropy determined?
7.5 How is the calorimetric entropy corrected to the ideal gas state?

7.6 What can cause the calorimetric entropy to be in error?
7.7 Estimate the entropy excess for (a) CO, (b) H_2O.
7. 8 How does one obtain (a) the standard Gibbs energy change, (b) the standard Helmholtz energy change, in a reaction from calorimetric data?
7.9 How is the chemical potential related to the activity of a constituent in a phase?
7.10 Define the Gibbs energy of reaction and the Helmholtz energy of reaction.
7.11 Distinguish between the pressure reaction quotient, the concentration reaction quotient, and the activity reaction quotient.
7.12 Distinguish between equilibrium constants K, K_P, and K_c.
7.13 How are these equilibrium constants related to changes in thermodynamic properties?
7.14 How does one employ an equilibrium constant in determining the equilibrium state for a gaseous system?
7.15 How are heterogeneous gas equilibria treated?
7.16 Explain how equilibrium constants vary with (a) temperature, (b) pressure.
7.17 Define and describe an activity coefficient (a) for partial pressure, (b) for concentration.

Problems

7.1 When 60.0 ml 0.500 M acetic acid was mixed with 60.0 ml 0.500 M sodium hydroxide, the temperature rose from 25.00° to 27.55° C. If the energy capacity of the 0.250 M sodium acetate formed was 4.029 J K^{-1} g^{-1}, its density 1.034 g ml^{-1}, and the effective energy capacity of the empty calorimeter 150.6 J K^{-1}, what is ΔH for the corresponding neutralization of 1 mole acetic acid?
7.2 When 1 mole liquid n-pentane, C_5H_{12}, is oxidized completely to gaseous CO_2 and liquid H_2O at 25° C, ΔH is -3509.5 kJ. Calculate ΔE for the reaction.
7.3 From the reaction heats

$$CO\ (g) + \frac{1}{2} O_2\ (g) \rightarrow CO_2\ (g), \qquad \Delta H^0_{298.15} = -282.96\ \text{kJ},$$

$$H_2\ (g) + \frac{1}{2} O_2\ (g) \rightarrow H_2O\ (g), \qquad \Delta H^0_{298.15} = -241.84\ \text{kJ},$$

calculate $\Delta H^0_{298.15}$ for

$$CO_2\ (g) + H_2\ (g) \rightarrow CO\ (g) + H_2O\ (g).$$

7.4 Calculate the entropy of solid NH_4OH at 15.0 K, where its energy capacity C_P is 1.753 J K^{-1} mol^{-1}.
7.5 Calculate the calorimetric entropy of titanium at 298.15 K from the data in table 7.D.

TABLE 7.D

T, K	C_P, J K^{-1} mol^{-1}	T, K	C_P, J K^{-1} mol^{-1}
0	0.000	150	19.598
15	0.167	175	21.100
25	0.657	200	22.263
50	4.753	225	23.175
75	10.050	250	23.903
100	14.368	275	24.535
125	17.385	298.15	25.004

7.6 Show that in a solid solution of AgCl in AgBr, the N_1 chloride ions and the N_2 bromide ions can be arranged in $N!/(N_1!N_2!)$ different ways. Then calculate the entropy at 0 K of a homogeneous phase containing 10.0 g AgCl and 50.0 g AgBr.

7.7 For a process between condensed substances obeying the third law, find the limiting slope, as T approaches zero, of a plot of ΔG against T at constant P.

7.8 Calculate a ΔS^0, ΔH^0, and ΔG^0 at 298.15 K for the reactions

(a) $$C_2H_4\ (g) + H_2\ (g) \rightarrow C_2H_6\ (g),$$

(b) $$2H_2O\ (g) + C\ (s) \rightarrow CO_2\ (g) + 2H_2\ (g),$$

from the numbers in tables 7.3 and 7.4.

7.9 From 1/2 mole nitrogen at 1.000 bar and 25° C and 3/2 mole hydrogen at 5.00 bar. and 25° C, 1 mole ammonia at 0.1000 bar. and 25° C was obtained. What was ΔG ?

7.10 Calculate the equilibrium constant at 25° C for

$$2NO\ (g) + O_2\ (g) \leftrightarrows 2NO_2\ (g)$$

from numbers in table 7.4.

7.11 Calculate K_P and K_c at 25° C for the reaction

$$\frac{3}{2}O_2\ (g) \leftrightarrows O_3\ (g), \qquad \Delta G^0_{298.15} = 163.602\ \text{kJ}.$$

7.12 If at 500 K the fraction of dissociation α of PCl_5 (g) is 0.644 at 1.000 bar, what is this α at 0.100 bar total pressure?

7.13 Determine K_P for the reaction

$$H_2O\ (g) \leftrightarrows H_2\ (g) + \frac{1}{2}O_2\ (g)$$

at 2500 K and 1.000 bar, where the percentage dissociation of the H_2O is 4.02.

7.14 Calculate the fraction of dissociation and the density of fluorine at 0.400 bar and 1000 K, where K_P for the reaction

$$F_2\ (g) \leftrightarrows 2F\ (g)$$

is 1.08×10^{-2} bar.

7.15 The equilibrium

$$LaCl_3\ (s) + H_2O\ (g) \leftrightarrows LaOCl\ (s) + 2HCL\ (g)$$

was established at 880 K with an HCl pressure of 72.6 torr and an H_2O pressure of 2.94 torr Determine K_P in bars.

7.16 For the reaction

$$NH_4HS\ (s) \leftrightarrows NH_3\ (g) + H_2S\ (g)$$

parameter K_P equals 5.62×10^{-2} bar^2 at 20° C. If solid ammonium hydrosulfide is vaporized at 20° C in a container filled initially with H_2S at 0.395 bar and NH_3 at 0.100 bar, what are the partial pressures when equilibrium is reached?

7.17 For the reaction

$$F_2\ (g) \leftrightarrows 2F\ (g)$$

log K_P is -5.612 at 700 K and -4.098 at 800 K. What is the average ΔH^0 over this temperature interval?

— — —

7.18 Construct an expression for the temperature dependence of the standard enthalpy change in the reaction

$$C\ (s) + \frac{1}{2}O_2\ (g) \rightarrow CO\ (g), \qquad \Delta H^0_{298.15} = -110{,}541\ \text{J}.$$

7.19 Show that for a condensed phase at absolute zero $(\partial S/\partial T)_V$ vanishes.

7.20 Calculate the entropy correction for gas imperfection of benzene at 400 K and 1.000 atm.

7.21 Molecules $^{35}Cl^{35}Cl$ and $^{37}Cl^{37}Cl$ need only be rotated by 180° to obtain a configuration indistinguishable from the original one, while $^{35}Cl^{37}Cl$ must be rotated by a full 360° . Thus, the latter has twice as many disjoint configurations as either of the former molecules. Determine the resulting state number ratio W_2/W_1 for the reaction

$$\frac{1}{2}{}^{35}Cl_2\ (g) + \frac{1}{2}{}^{37}Cl_2\ (g) \rightarrow {}^{35}Cl^{37}Cl\ (g).$$

Then calculate ΔS^0 and finally ΔG^0 at 25° C.

7.22 At 500 K the log K_f for H_2O (g) is 22.891 and for HCl (g) 10.150, while at 600 K these numbers are 18.637 and 8.530 respectively. Calculate K_P and K_c at these temperatures for the reaction

$$2HCL\ (g) + \frac{1}{2}O_2\ (g) \leftrightarrows H_2O\ (g) + Cl_2\ (g).$$

7.23 Show how unit pressure enters equation (7.40) to make it correct dimensionally. How does unit activity enter equation (7.42)?

7.24 At 50° C, K_P for

$$2NaHCO_3\ (s) \leftrightarrows Na_2CO_3\ (s) + H_2O\ (g) + CO_2\ (g)$$

is 4.000×10^{-4} bar^2 and K_P for

$$\frac{1}{2}CuSO_4 \cdot 5H_2O\ (s) \leftrightarrows \frac{1}{2}CuSO_4 \cdot 3H_2O\ (s) + H_2O\ (g)$$

is 6.049×10^{-2} bar. What is the pressure of CO_2 in the vapor at equilibrium with $NaHCO_3$ (s), Na_2CO_3 (s), $CuSO_4$ $5H_2O$ (s), and $CuSO_4$ $3H_2O$ (s) at 50° C?

7.25 If solid ammonium carbamate and solid ammonium hydrosulfide are both vaporized into a vacuum, what are the partial pressures when equilibrium is set up with the solids at 20° C, where K_P for

$$NH_4COONH_2\ (s) \leftrightarrows 2NH_3\ (g) + CO_2\ (g)$$

is 8.09×10^{-5} bar^3 and K_P for

$$NH_4HS\ (s) \leftrightarrows NH_3\ (g) + H_2S\ (g)$$

is 5.62×10^{-2} bar^2?

7.26 From the numbers in problem 7.22, calculate ΔH^0 at 550 K for the given reaction.

7.27 For the equilibrium

$$CO\ (g) + 2H_2\ (g) \leftrightarrows CH_3OH\ (g)$$

at 250° C and 170 atm, expression K_P equals 2.1×10^{-2} atm^{-2}. If under these conditions, $\gamma_{CO} = 1.09$, $\gamma_{H_2} = 1.06$, and $\gamma_{CH_3OH} = 0.45$, what is equilibrium constant K for the reaction?

References

Books

Barin, I.: 1989, *Thermochemical Data of Pure Substances*, Parts I and II, VCH, Weinheim, pp. I: 1-37, 1-1739

In the introduction, Barin surveys the basic thermodynamic principles and equations. The tables are then described. Representative examples of their use are developed.

The values of pertinent thermochemical functions for 2372 pure substances are tabulated at 298.15 K and at 100 K intervals from 300.00 K up. The substances include 91 elements and various compounds containing two, three, and four elements. Besides the inorganic compounds, nearly 100 organic compounds are described.

Chase Jr., M. W., Davies, C. A., Downey Jr., J. R., Frurip, D. J., McDonald, R. A., and Syverud, A. N.: 1986, *JANAF Thermochemical Tables*, 3rd ed., Parts I and II, Am. Inst. Phys., New York, pp. 4-1856.

The introduction describes techniques for evaluating data in the literature. Accompanying each table is a text outlining the sources, discrepancies, and how they were reconciled.

The substances that are important in studies of fuel combustion, jet and rocket combustion, and air pollution are covered. But many substances that metallurgical, ceramic, and chemical engineers study are omitted.

De Heer, J.: 1986, *Phenomenological Thermodynamics*, Prentice- Hall, Englewood Cliffs, NJ, pp. 220-252.

Here is a concise, easy-to-follow presentation of the principles governing chemical equilibria.

Articles

Alberty, R. A.: and Oppenheim, I.: 1993, "Thermodynamics of a Reaction System at a Specified Partial Pressure of a Reactant," *J. Chem. Educ.* **70**, 629-635.

Andersen, K.: 1994, "Practical Calculation of the Equilibrium Constant and the Enthalpy of Reaction at Different Temperatures," *J. Chem. Educ.* **71**, 474-479.

Boyko, E. R., and Belliveau, J. F.: 1990, "Simplification of Some Thermochemical Calculations," *J. Chem. Educ.* **67**, 743-744.

Canagaratna, S. G., and Witt, J.: 1988, "Calculation of Temperature Rise in Calorimetry," *J. Chem. Educ.* **65**, 126-129.

Combs, L. L.: 1992, "An .Alternative View of Fugacity," *J. Chem. Educ.* **69**, 218-219.

David, C. W.: 1988, "An Elementary Discussion of Chemical Equilibrium," *J. Chem. Educ.* **65**, 407-409.

D'Alessio, L.: 1993, "On the Fugacity of a van der Waals Gas: An Approximate Expression that Separates Attractive and Repulsive Forces," *J. Chem. Educ.* **70**, 96-98.

Deumie, M., Boulil, B., and Henri-Rousseau, 0.: 1987, "On the Minimum of the Gibbs Free Energy Involved in Chemical Equilibrium," *J. Chem. Educ.* **64**, 201-204.

Drago, R. S., and Wong, N. M.: 1996, "The Role of Electron- Density Transfer and Electronegativity in Understanding Chemical Reactivity and Bonding-," *J. Chem. Educ.* **73**, 123-129.

Dunitz, J. D.: 1994, "The Entropic Cost of Bound Water in Crystals and Biomolecules," *Science* **264**, 670.

Freeman, R. D.: 1985, "Conversion of Standard Thermodynamic Data to the New Standard-State Pressure," *J. Chem. Educ.* **62**, 681- 686.

Gerhartl, F. J.: 1994, "The A + B $\rightarrow$ C of Chemical Thermodynamics," *J. Chem. Educ.* **71**, 539-548.

Henri-Rousseau, O., Deumie, M., and Krallafa, A.: 1986, "Gibbs-Duhem Relation and Invariance of Mass Action Law toward Competitive Reactions," *J. Chem. Educ.* **63**, 682-684.

Jagannathan, S.: 1987, "On a Relation between Fugacity and Pressure," *J. Chem. Educ.* **64**, 677.

Kemp, H. R.: 1987, "The Effect of Temperature and Pressure on Equilibria: A Derivation of the van's Hoff Rules," *J. Chem. Educ.* **64**, 482-484.

MacDonald, J. J.: 1990a, "Equilibrium, Free Energy, and Entropy," *J. Chem. Educ.* **67**, 380-382.

MacDonald, J. J.: 1990b, "Equilibria and ΔG°," *J. Chem. Educ.* **67**, 745-746.

Molyneux, P.: 1991, "The Dimensions of Logarithmic Quantities," *J. Chem. Educ.* **68**, 467-469.

Ramshaw, J. D.: 1995, "Fugacity and Activity in a Nutshell," *J. Chem. Educ.* **72**, 601-603.

Rastogi, R. P., and Shabd, R.: 1983, "Thermodynamics of Stability of Nonequilibrium Steady States," *J. Chem. Educ.* **60**, 540- 545.

Smith, D. W.: 1986, "A Simple Empirical Analysis of the Enthalpies of Formation of Lanthanide Halides and Oxides," *J. Chem. Educ.* **63**, 228-231.

Spencer, J. N., Moog, R. S., and Gillespie, R. J.: 1996, "An Approach to Reaction Thermodynamics through Enthalpies, Entropies, and Free Energies of Atomization," *J. Chem. Educ.* **73**, 631-636.

Sutter, D. H.: 1996, "Free Enthalpy, Lagrange Multipliers, and Thermal Equilibrium," *J. Chem. Educ.* **73**, 718- 721.

Treptow, R. S.: 1996, "Free Energy versus Extent of Reaction: Understanding the Difference between ΔG and $\partial G/\partial \xi$," *J. Chem. Educ.* **73**, 51-54.

TyRodi, R. J.: 1986, "A Better Way of Dealing with Chemical Equilibrium," *J. Chem. Educ.* **63**, 582- 585.

Tykodi, R. J.: 1988, "Estimated Thermochemical Properties of Some Noble-Gas Monoxides and Difluorides," *J. Chem. Educ.* **65**, 981- 985.

Wai, C. M., and Hutchison, S. G.: 1989, "Free Energy Minimization Calculation of Complex Chemical Equilibria," *J. Chem. Educ.* **66**, 546- 549.

Yoder, C. H.: 1986, "Thermodynamic Analysis of Ionic Compounds: Synthetic Applications," *J. Chem. Educ.* **63**, 232- 235.

8

Equilibria in Condensed Phases

8.1 The Chemical Potential Revisited

BY EQUATION (5.102) THE CHEMICAL potential μ_i is the contribution per mole of the ith constituent to the Gibbs energy in a homogeneous region. For equilibrium to exist between this region and a different homogeneous region, the chemical potentials must be equal following condition (6.102). In an ideal gas phase, μ_i. is related to the pressure of the ith constituent by formula (7.40),

$$\mu_i = \mu_i^0 + RT \ln P_i. \tag{8.1}$$

In another phase at equilibrium with the ideal one, the chemical potential satisfies equation (7.42),

$$\mu_i = \mu_i^0 + RT \ln a_i, \tag{8.2}$$

with μ_i^0 the value of μ_i when a_i is 1, This is generally different from the μ_i^0 in equation (8.1), In this chapter, the pertinent phase is a condensed phase—liquid or possibly solid.

In the interval over which the solute in this phase obeys Henry's law, we have

$$P_i = k_i X_i \tag{8.3}$$

and the activity for (8.2) would be

$$a_i = X_i. \tag{8.4}$$

Here X_i is the ith mole fraction.

Alternatively, one may express Henry's law in the form

$$P_i = k_i' c_i, \tag{8.5}$$

where c_i is the molarity of i. Then in the very dilute solutions, we would have

$$a_i = c_i. \tag{8.6}$$

Or, one may express Henry's law in the form

$$P_i = k_i'' m_i, \tag{8.7}$$

where m_i is the molality of i. Then in the very dilute solutions, we would have

$$a_i = m_i. \tag{8.8}$$

Each of these procedures involves a different k_i in equation (7.42), a different $\mu^0{}_i$ in formula (8.2), and so a different standard state. But for any one of these choices, a person can construct equation (7.52) for ΔG, set it equal to zero as in equation (7.54), and obtain the conventional form for the equilibrium constant (7.57).

8.2 Useful Equilibrium Expressions

Equilibrium constants for the reaction

$$a\text{A} + b\text{B} \rightleftarrows l\text{L} + m\text{M} \tag{8.9}$$

have the form

$$\frac{a_\text{L}{}^l a_\text{M}{}^m}{a_\text{A}{}^a a_\text{B}{}^b} = K, \tag{8.10}$$

as we saw in equation (7.57).

When the concentrations are expressed in *mole fractions* and the corresponding standard states are employed, condition (8.3) on all reactants and products causes equality (8.4) to hold for them also. Then the K_X defined by the equation

$$\frac{X_\text{L}{}^l X_\text{M}{}^m}{X_\text{A}{}^a X_\text{B}{}^b} = K_X, \tag{8.11}$$

reduces to the equilibrium constant K. But when condition (8.3) breaks down for a constituent, we replace (8.4) by the relation

$$a_i = \gamma_i X_i \tag{8.12}$$

in which γ_i is the mole fraction activity coefficient. Then we have

$$Q_\gamma K_X = K_{\gamma X} = K \tag{8.13}$$

where

$$Q_\gamma = \frac{\gamma_\text{L}{}^l \gamma_\text{M}{}^m}{\gamma_\text{A}{}^a \gamma_\text{B}{}^b} \tag{8.14}$$

and K_X is given by formula (8.11).

When concentrations are expressed in *molarities*, we similarly define

$$\frac{c_\text{L}{}^l c_\text{M}{}^m}{c_\text{A}{}^a c_\text{B}{}^b} = K_c. \tag{8.15}$$

We also introduce the molarity activity coefficient γ_i by the relation

$$a_i = \gamma_i c_i, \tag{8.16}$$

so that

$$Q_\gamma K_c = K_{\gamma c} = K \tag{8.17}$$

with

$$Q_\gamma = \frac{\gamma_\text{L}{}^l \gamma_\text{M}{}^m}{\gamma_\text{A}{}^a \gamma_\text{B}{}^b}. \tag{8.18}$$

When concentrations are expressed in *molalities*, we define

$$\frac{m_L^{\,l}\, m_M^{\,m}}{m_A^{\,a}\, m_B^{\,b}} = K_m \qquad [8.19]$$

and employ

$$a_i = \gamma_i m_i. \qquad [8.20]$$

Then we have

$$Q_\gamma K_m = K_{\gamma m} = K \qquad [8.21]$$

with

$$Q_\gamma = \frac{\gamma_L^{\,l}\, \gamma_M^{\,m}}{\gamma_A^{\,a}\, \gamma_B^{\,b}}. \qquad [8.22]$$

The different equilibrium constants for a reaction are related in the following manner. For a very dilute solution, equations (6.65) reduce to

$$X_i = \frac{m_i}{1000 / M_s} = \frac{c_i}{1000d / M_s} \qquad [8.23]$$

or

$$\frac{1000}{M_s} X_i = m_i = \frac{c_i}{d}. \qquad [8.24]$$

So when A, B, L, and M are all solutes, we have

$$\left(\frac{1000}{M_s}\right)^{\Delta n} \frac{X_L^{\,l}\, X_M^{\,m}}{X_A^{\,a}\, X_B^{\,b}} = \frac{m_L^{\,l}\, m_M^{\,m}}{m_A^{\,a}\, m_B^{\,b}} = \left(\frac{1}{d}\right)^{\Delta n} \frac{c_L^{\,l}\, c_M^{\,m}}{c_A^{\,a}\, c_B^{\,b}} \qquad [8.25]$$

where

$$\Delta n = l + m - a - b. \qquad [8.26]$$

Introducing (8.11), (8.15), and (8.19) leads to

$$\left(\frac{1000}{M_s}\right)^{\Delta n} K_X = K_m = \left(\frac{1}{d}\right)^{\Delta n} K_c. \qquad [8.27]$$

From the last equation and the equalities

$$K_m = K_{\gamma m}, \qquad K_c = K_{\gamma c}, \qquad [8.28]$$

which hold in infinitely dilute solutions, we obtain

$$K_{\gamma m} = \left(\frac{1}{d_s}\right)^{\Delta n} K_{\gamma c}. \qquad [8.29]$$

Here d_s is the mass density of the solvent. The difference between these two constants is small in aqueous solutions, where d_s is nearly 1.

By convention, the activity of a solvent is generally expressed in the mole fraction system. So in dilute solutions, its activity is set equal to 1. For the so-called ionization of water

$$2H_2O\ (l) \rightleftharpoons H_3O^+\ (aq) + OH^-\ (aq), \qquad [8.30]$$

we have

$$K = \frac{a_{H_3O^+} a_{OH^-}}{a_{H_2O}^2} = a_{H_3O^+} a_{OH^-}. \qquad [8.31]$$

But note that writing (8.30) in the form

$$H_2O\ (l) \rightleftarrows H^+\ (aq) + OH^-\ (aq) \qquad [8.32]$$

yields

$$K = \frac{a_{H^+} a_{OH^-}}{a_{H_2O}} = a_{H^+} a_{OH^-}, \qquad [8.33]$$

essentially the same final form. For other reactions, a similar remark applies. So in equilibrium studies, we will symbolize the hydrogen ion in water merely as H^+ (aq).

When a crystal is effectively pure and under 1 bar pressure, it is in its standard state and its activity may be taken as 1. For the reaction

$$CaC_2O_4 \cdot H_2O\ (s) \rightleftarrows Ca^{++}\ (aq) + C_2O_4^{=}\ (aq) + H_2O\ (l) \qquad [8.34]$$

we have

$$K = \frac{a_{Ca^{++}} a_{C_2O_4^{=}} a_{H_2O}}{a_{CaC_2O_4 \cdot H_2O}} = a_{Ca^{++}} a_{C_2O_4^{=}}. \qquad [8.35]$$

Example 8.1

The mole fraction equilibrium constant for the reaction

$$\frac{1}{2} I_2 + \frac{1}{2} Cl_2 \rightleftarrows ICl$$

in carbon tetrachloride solution is 812 at 25° C. If 0.0200 mol ICl is added to 1.000 mol CCl_4, how much I_2 is present when equilibrium is reached at 25° C?

Because all molecules involved here are similar covalent ones, the solution is approximately ideal, K_X reduces to the equilibrium constant K, and

$$\frac{X_{ICl}}{X_{I_2}^{1/2} X_{Cl_2}^{1/2}} = 812.$$

But the chemical equation implies that reaction between two ICl molecules yields one I_2 and one Cl_2 molecule. For the given initial conditions, we have the results in table 8.A.

TABLE 8.A

Constituent	*Moles at Start*	*Equilibrium Moles*	*Equilibrium Mole Fraction*
I_2	0	x	x/n
Cl_2	0	x	x/n
ICl	0.0200	$0.0200 - 2x$	$(0.0200 - 2x)/n$

Putting these into the equilibrium expression yields

$$\frac{(0.0200\ \text{mol} - 2x)/n}{(x/n)^{1/2}(x/n)^{1/2}} = 812$$

or

$$812x + 2x = 0.0200\ \text{mol}$$

whence

$$x = \frac{0.0200\ \text{mol}}{814} = 2.46 \times 10^{-5}\ \text{mol}.$$

Example 8.2

For the ionization of acetic acid

$$CH_3COOH \rightleftharpoons CH_3COO^- + H^+$$

constant $K_{\gamma c}$ is 1.749 × 10^{-5} M at 25° C. Calculate the hydrogen ion concentration in a solution formed by adding 0.100 mol acetic acid and 0.0100 mol sodium acetate to enough water to give 1 liter of solution. Assume that Q_γ is 1 as an approximation. Taking Q_γ equal to 1 leads to the equation

$$K_c = \frac{c_{CH_3COO^-}c_{H^+}}{c_{CH_3COOH}} = 1.749 \times 10^{-5}\ \text{M}.$$

Let us neglect ionization of the acetic acid and hydrolysis of the acetate ion. Then

$$c_{H^+} = \frac{c_{CH_3COOH}}{c_{CH_3COO^-}}K_c = \frac{0.100}{0.0100}1.749 \times 10^{-5}\ \text{M} = 1.75 \times 10^{-4}\ \text{M}.$$

Example 8.3

Calculate the hydrogen ion concentration in 0.0100 M acetic acid at 25° C.

The acetic acid ionizes following the chemical equation in example 8.2. If we neglect the contribution to c_{H^+} from the ionization of water, we have

$$c_{CH_3COO^-} = c_{H^+} = x$$

and

$$c_{CH_3COOH} = 0.0100\ \text{M} - x.$$

Taking Q_γ, equal to 1, as in example 8.2, leads to

$$K_c = \frac{c_{CH_3COO^-}c_{H^+}}{c_{CH_3COOH}} = \frac{x^2}{0.0100\ \text{M} - x} = 1.749 \times 10^{-5}\ \text{M}$$

whence

$$x^2 = 1.749 \times 10^{-7} - 1.749 \times 10^{-5}\ x.$$

In the approximation that x is small with respect to 0.0100 M, we have

$$x = \left(1.749 \times 10^{-7}\ \mathrm{M}^2\right)^{1/2} = 4.18 \times 10^{-4}\ \mathrm{M}.$$

Then introducing this value for x on the right side of the preceding equation yields

$$x = \left(1.676 \times 10^{-7}\ \mathrm{M}^2\right)^{1/2} = 4.09 \times 10^{-4}\ \mathrm{M}.$$

Repeating the procedure gives us

$$x = 4.09 \times 10^{-4}\ \mathrm{M}.$$

Example 8.4

For the ionization of acetic acid, constant $K_{\gamma m}$ is 1.754 × 10^{-5} m, while for the ionization of water, constant $K_{\gamma m}$ is 1.008 × 10^{-14} at 25° C. Calculate the hydroxide ion concentration in 0.100 m sodium acetate.

The acetate ion hydrolyzes following the reaction

$$CH_3COO^- + H_2O \rightarrow CH_3COOH + OH^-.$$

At appreciable acetate ion concentrations, this swamps the contribution from the ionization of water and

$$m_{CH_3COOH} = m_{OH^-} = x.$$

As an approximation, we take the activity coefficients to be 1. So Q_γ for both equilibrium constants is set equal to 1. From the ionization of water, we now have

$$m_{H^+} = \frac{1.008 \times 10^{-14}\ \mathrm{m}^2}{x}.$$

The acetate ion concentration is reduced by the hydrolysis reaction so that

$$m_{CH_3COO^-} = 0.100\ \mathrm{m} - x.$$

For the acetic acid reaction, we obtain the relationship

$$\frac{m_{CH_3COO^-} m_{H^+}}{m_{CH_3COOH}} = \frac{\left(0.100\ \mathrm{m} - x\right)\left(1.008 \times 10^{-14}\ \mathrm{m}^2\right)/x}{x} = 1.754 \times 10^{-5}\ \mathrm{m}.$$

We expect x to be small with respect to 0.100 m, so we replace the last equality with

$$\frac{\left(0.100\ \mathrm{m}\right)\left(1.008 \times 10^{-14}\ \mathrm{m}^2\right)}{x^2} = 1.754 \times 10^{-5}\ \mathrm{m}$$

whence

$$x = 7.55 \times 10^{-6}\ \mathrm{m}.$$

8.3 Conditions Determining Equilibria

In a common problem, a person starts with a given amount of certain reactants and products in a given volume or mass of solvent kept at a certain temperature and pressure. One or more reactions proceed until equilibrium is established. The final concentrations are related to the initial ones through the coefficients in the chemical equations. The final concentrations are related to each other by the equilibrium expressions. Ionization constants for common weak electrolytes appear in tables 8.1 and 8.2.

As long as the amount of a substance does not exceed its solubility, the amount added is arbitrary. However, one cannot add cations independently of anions to an extensive region.

TABLE 8.1 Ion Product for Water in Molality Squared

Temperature, °C	$K_{\gamma m}$, m^2
0	0.1139×10^{-14}
10	0.2920×10^{-14}
20	0.6809×10^{-14}
25	1.008×10^{-14}
30	1.469×10^{-14}
40	2.919×10^{-14}
50	5.474×10^{-14}
60	9.614×10^{-14}

TABLE 8.2 Ionization Constants in Moles per 1000 g H_2O at 25°C

Weak Electrolyte	$K_{\gamma m}$, m
HBO_2	5.79×10^{-10}
HCOOH	1.772×10^{-4}
CH_3COOH	1.754×10^{-5}
CH_3CH_2COOH	1.336×10^{-5}
$CH_3CH_2CH_2COOH$	1.515×10^{-5}
$ClCH_2COOH$	1.379×10^{-3}
$HOCH_2COOH$ (glycolic acid)	1.475×10^{-4}
$CH_3CHOHCOOH$ (lactic acid)	1.374×10^{-4}
HCO_3^-	4.69×10^{-11}
$HC_2O_4^-$	5.18×10^{-5}
$COOHCH_2COO^-$	2.014×10^{-6}
H_3PO_4	7.516×10^{-3}
$H_2PO_4^-$	6.226×10^{-8}
HSO_4^-	1.01×10^{-2}
H_2SO_3	1.72×10^{-2}
HSO_3^-	6.24×10^{-8}
NH_4OH	1.75×10^{-5}
$NH_3OHCH_2COO^-$ (from glycine)	6.04×10^{-5}
$NH_3CH_2COOH^+$ (from glycine)	4.47×10^{-3}
$CH_3CHNH_3OHCOO^-$ (from alanine)	7.47×10^{-5}
$CH_3CHNH_3COOH^+$ (from alanine)	4.57×10^{-3}

Throughout such a region, electrical neutrality tends to exist. At a boundary, however, a charged sheath may form. Also, about individual ions, microscopic charged atmospheres form.

In working a problem, the expressions obtained from the chemical equations are substituted into the equilibrium constant formulas. For systems involving several ions, the electrical neutrality condition may also be invoked.

When the concentrations of all the ions are low, one may consider the activity coefficients to equal 1. A theory for ionic activity coefficients will be developed later and approximate results for common ions tabulated.

However the activity coefficients are chosen for a particular problem, a person faces a mixed set of nonlinear and linear equations. These may be combined to give a single equation

$$f(x) = 0, \qquad [8.36]$$

in which x represents the equilibrium extent of reaction measured in some manner while $f(x)$ is a polynomial. Equation (8.36) is then solved. But besides the desired positive root, this may have complex roots, negative roots, and positive roots that impose negative values on one or more concentrations.

Alternatively, a person may work with the simultaneous equations directly, rearranging them so that approximate concentrations on the right yield better values on the left. Letting the concentrations be w, x, y, and z, one constructs the set

$$w_i = w(w_{i-1}, x_{i-1}, y_{i-1}, z_{i-1}), \qquad [8.37]$$

$$x_i = x(w_i, x_{i-1}, y_{i-1}, z_{i-1}), \qquad [8.38]$$

$$y_i = y(w_i, x_i, y_{i-1}, z_{i-1}), \qquad [8.39]$$

$$z_i = z(w_i, x_i, y_i, z_{i-1}). \qquad [8.40]$$

The formulation is satisfactory if, as i increases, the concentrations approach reasonable limiting values. Each round in the calculations is called an *iteration*.

A person may improve the procedure by replacing the final equation, rewritten in the form

$$0 = f(x) \qquad [8.41]$$

with

$$y = f(x). \qquad [8.42]$$

The various assumed concentrations are implicitly in $f(x)$.

We then consider two values of x, say x_1 and x_2, near the root to be found and the corresponding values of y, y_1 and y_2. As long as the derivative of y does not change too fast near the root, we have

$$\left(\frac{dy}{dx}\right)_{y=0} \cong \frac{y_2 - y_1}{x_2 - x_1} \cong \frac{y_1 - 0}{x_1 - x_0} \qquad [8.43]$$

where x_0 is the root. Solving the second relation for x_0 yields

$$x_0 \cong x_1 - \frac{x_2 - x_1}{y_2 - y_1} y_1. \qquad [8.44]$$

This value may be corrected by using it together with the closest initial x in a repetition of the calculation.

Example 8.5

Calculate m_{H^+} in 0.010 m alanine using ionization constants from tables 8.2 and 8.1. Consider the Q_γ's to equal 1, The equation for ionization of the anion,

$$CH_3CHNH_3OHCOO^- \leftrightarrows CH_3CHNH_3COO + OH^-,$$

is abbreviated as

$$ZOH^- \leftrightarrows Z + OH^-,$$

while the equation for ionization of the cation,

$$CH_3CHNH_3COOH^+ \leftrightarrows CH_3CHNH_3COO + H^+,$$

is abbreviated as

$$ZH^+ \leftrightarrows Z + H^+.$$

Molecule Z is called a *zwitterion* since in it H^+ has migrated from the carboxyl to the amino group, leaving the first region anionic and the second region cationic.

In the bulk of the solution, electrical neutrality prevails; thus the concentration of positive charge equals the concentration of negative charge:

$$m_{ZH^+} + m_{H^+} = m_{ZOH^-} + m_{OH^-}.$$

But as long as the solution of alanine is concentrated enough, m_{H^+} is small with respect to m_{ZH^+} and m_{OH^-} is small with respect to m_{ZOH^-}. The preceding equation then reduces to

$$m_{ZH^+} = m_{ZOH^-}.$$

From table 8.2 with each Q_γ = 1, one obtains

$$\frac{m_Z m_{OH^-}}{m_{ZOH^-}} = 7.47 \times 10^{-5} \text{ m}$$

and

$$\frac{m_Z m_{H^+}}{m_{ZH^+}} = 4.57 \times 10^{-3} \text{ m}.$$

Dividing the second equation by the first and canceling the molalities that are equal yields

$$\frac{m_{H^+}}{m_{OH^-}} = \frac{4.57 \times 10^{-3}}{7.47 \times 10^{-5}} = 6.12 \times 10.$$

But in the aqueous solution at 25° C, table 8.1 tells us that

$$m_{H^+} m_{OH^-} = 1.008 \times 10^{-14} \text{ m}^2.$$

So we have

$$m_{H^+}{}^2 = \left(\frac{m_{H^+}}{m_{OH^-}}\right)\left(m_{H^+} m_{OH^-}\right) = \left(6.12 \times 10\right)\left(1.008 \times 10^{-14} \text{ m}^2\right) = 6.17 \times 10^{-13} \text{ m}^2$$

whence

$$m_{H^+} = 7.85 \times 10^{-7} \text{ m}.$$

Example 8.6

Calculate m_{H^+} in 0.0100 m ammonium formate.

Ammonium formate consists of ammonium ions and formate ions. These undergo the reactions

$$NH_4^+ + H_2O \rightarrow NH_4OH + H^+,$$

$$A^- + H_2O \rightarrow HA + OH^-,$$

$$H^+ + OH^- \rightarrow H_2O,$$

on dissolving in water. Here A^- represents the formate ion. Since the ammonium hydroxide and the formic acid formed do not react further, we have

$$0.0100 \text{ m} = m_{NH_4^+} + m_{NH_4OH} = m_{A^-} + m_{HA}.$$

Furthermore, electrical neutrality is maintained; so

$$m_{NH_4^+} + m_{H^+} = m_{A^-} + m_{OH^-}.$$

Because the excess hydrogen ions and hydroxide ions combine, a small difference in the first two reactions causes a large difference in m_{H^+} and m_{OH^-}:

$$m_{H^+} \neq m_{OH^-}.$$

But if enough ammonium formate is added to water, $m_{NH_4^+}$ is large with respect to m_{H^+} and m_{A^-} is large with respect to m_{OH^-}. Then the electrical neutrality condition reduces to

$$m_{NH_4^+} \cong m_{A^-}.$$

Substituting this into the equation for the overall molality yields

$$m_{NH_4OH} \cong m_{HA}.$$

For the ionization of ammonium hydroxide, we write

$$\frac{m_{NH_4^+} m_{OH^-}}{m_{NH_4OH}} = 1.75 \times 10^{-5} \text{ m},$$

and for the ionization of formic acid, we write

$$\frac{m_{A^-} m_{H^+}}{m_{HA}} = 1.772 \times 10^{-4} \text{ m},$$

assuming each $Q_\gamma = 1$. Dividing the last equation by the previous one and canceling the molalities that are nearly equal yields

$$\frac{m_{H^+}}{m_{OH^-}} = \frac{1.772 \times 10^{-4}}{1.75 \times 10^{-5}} = 1.013 \times 10.$$

But in the aqueous solution at 25° C, we have

$$m_{H^+}m_{OH^-} = 1.008 \times 10^{-14} \text{ m}^2.$$

Consequently, we find that

$$m_{H^+}{}^2 = \left(\frac{m_{H^+}}{m_{OH^-}}\right)\left(m_{H^+}m_{OH^-}\right) = \left(1.013 \times 10\right)\left(1.008 \times 10^{-14} \text{ m}^2\right) = 1.021 \times 10^{-13} \text{ m}^2,$$

whence

$$m_{H^+} = 3.19 \times 10^{-7} \text{ m}.$$

Example 8.7

Calculate m_{H^+} in 1.00×10^{-6} m acetic acid.
In this dilute solution we have to consider both equilibria

$$HA \rightleftarrows H^+ + A^-,$$

$$H_2O \rightleftarrows H^+ + OH^-,$$

simultaneously. Here HA represents the acetic acid molecule, A^- the acetate ion.

The 1.00×10^{-6} mol acetic acid per 1000 g H_2O appears as HA and A^-; there is no other contribution to m_{A^-}. So

$$1.00 \times 10^{-6} \text{ m} = m_{HA} + m_{A^-}.$$

From the electrical neutrality condition, we have

$$m_{H^+} = m_{A^-} + m_{OH^-}.$$

And governing the equilibria, we have

$$\frac{m_{H^+}m_{A^-}}{m_{HA}} = 1.754 \times 10^{-5} \text{ m},$$

$$m_{H^+}m_{OH^-} = 1.008 \times 10^{-14} \text{ m}^2.$$

Let us rearrange these simultaneous equations in a set of type (8.37) - (8.40):

$$m_{OH^-} = \frac{1.008 \times 10^{-14}}{m_{H^+}},$$

$$m_{A^-} = m_{H^+} - m_{OH^-},$$

$$m_{HA} = \frac{m_{H^+}m_{A^-}}{1.754 \times 10^{-5}},$$

$$m_{H^+} = 1.00 \times 10^{-6} - m_{HA} + m_{OH^-}.$$

If the acetic acid were completely ionized and the contribution from ionization of water were negligible, we would have

$$m_{H^+} = 1.00 \times 10^{-6} \text{ m}.$$

Put this into the first equation of the set and complete the round:

$$m_{OH^-} = \frac{1.008 \times 10^{-14}}{1.00 \times 10^{-6}} = 1.008 \times 10^{-8} \text{ m},$$

$$m_{A^-} = 1.00 \times 10^{-6} - 0.01 \times 10^{-6} = 0.99 \times 10^{-6} \text{ m},$$

$$m_{HA} = \frac{\left(1.00 \times 10^{-6}\right)\left(0.99 \times 10^{-6}\right)}{1.754 \times 10^{-5}} = 0.56 \times 10^{-7} \text{ m},$$

$$m_{H^+} = 1.00 \times 10^{-6} - 0.056 \times 10^{-6} + 0.010 \times 10^{-6} = 0.954 \times 10^{-6} \text{ m}.$$

Repeating the steps with this number gives us (in the final step)

$$m_{H^+} = 1.00 \times 10^{-6} - 0.051 \times 10^{-6} + 0.011 \times 10^{-6} = 0.960 \times 10^{-6} \text{ m}.$$

The average of these two values is

$$m_{H^+} = 0.957 \times 10^{-6} \text{ m}.$$

To illustrate use of equation (8.44), we replace the final equation in the set with

$$y = 1.00 \times 10^{-6} - m_{HA} - m_{A^-}.$$

Again, if the acetic acid were completely ionized but the contribution from ionization of water were negligible, we would have

$$x_1 = \left(m_{H^+}\right)_1 = 1.00 \times 10^{-6} \text{ m}.$$

If only the water ionized, we would have

$$x_2 = \left(m_{H^+}\right)_2 = 1.004 \times 10^{-7} \text{ m}.$$

With x_1 the first three equations are the same as in the first round before. So

$$m_{OH^-} = 1.008 \times 10^{-8} \text{ m},$$

$$m_{A^-} = 0.990 \times 10^{-6} \text{ m},$$

$$m_{HA} = 0.56 \times 10^{-7} \text{ m},$$

and

$$y_1 = 1.00 \times 10^{-6} - 0.056 \times 10^{-6} - 0.990 \times 10^{-6} = -0.046 \times 10^{-6} \text{ m}.$$

With x_2, we find that

$$m_{OH^-} = \frac{1.008 \times 10^{-14}}{1.004 \times 10^{-7}} = 1.004 \times 10^{-7} \text{ m},$$

$$m_{A^-} = 0,$$

$$m_{HA} = 0,$$

$$y_2 = 1.00 \times 10^{-6} \text{ m}.$$

Substituting these into formula (8.44) yields

$$m_{H^+} = 1.00 \times 10^{-6} - \frac{-0.90 \times 10^{-6}}{1.046 \times 10^{-6}}\left(-0.046 \times 10^{-6}\right) = 1.00 \times 10^{-6} - 0.040 \times 10^{-6} = 0.960 \times 10^{-6} \text{ m}.$$

Employing this as the new x_2, we obtain

$$y_2 = -0.002 \times 10^{-6} \text{ m}.$$

Using the same x_1 and y_1 as before, we now have

$$m_{H^+} = 1.00 \times 10^{-6} - \frac{-0.040 \times 10^{-6}}{1.044 \times 10^{-6}}\left(-0.046 \times 10^{-6}\right) = 1.00 \times 10^{-6} - 0.042 \times 10^{-6} = 0.958 \times 10^{-6} \text{ m}.$$

This checks the result found by the first method.

8.4 Key Thermodynamic Considerations

The standard state for a solid or liquid is the pure substance under 1 bar pressure at the chosen temperature. The standard state for a solute here is the hypothetical 1 molal solution obeying Henry's law at 1 bar total pressure and the chosen temperature. Thus, we are employing an activity system based on the infinitely dilute solution as the reference state, with concentrations in the molality system.

A difficulty arises with ions, however. For, a cation cannot be formed apart from the formation of an anion or an electron. Nor can an anion be formed apart from the formation of a cation or reaction with an electron. So a person can arbitrarily assign thermodynamic properties to a chosen ion or to the electron. By convention, the hydrogen ion is taken. One assigns to H^+ (aq) zero S^0, $\Delta H^0{}_f$, and $\Delta G^0{}_f$.

The results from many experiments with condensed phases are summarized in table 8.3.

Example 8.8

From table 8.3, determine ΔS^0, ΔH^0, log $K_{\gamma m}$, and $K_{\gamma m}$ for the reaction

$$\text{HCOOH (aq)} \rightleftharpoons \text{H}^+ \text{ (aq)} + \text{HCOO}^- \text{ (aq)}.$$

With the numbers in the table, we find that

$$\Delta S^0 = S^0_{H^+} + S^0_{HCOO^-} - S^0_{HCOOH} = 0.00 + 92 - 163 = -71 \text{ J K}^{-1}.$$

Also

$$\Delta H^0 = \Delta H^0_{f\,H^+} + \Delta H^0_{f\,HCOO^-} - \Delta H^0_{f\,HCOOH} = 0.00 - 425.55 + 425.43 = 0.12 \text{ J},$$

and

$$\log K_{\gamma m} = \left(\log K_f\right)_{H^+} + \left(\log K_f\right)_{HCOO^-} - \left(\log K_f\right)_{HCOOH} + 0.00 + 61.49 - 65.22 = -3.73,$$

whence

$$K_{\gamma m} = 1.86 \times 10^{-4} \text{ m}.$$

TABLE 8.3 Entropies, Formation Enthalpies, Formation Gibbs Energies, and Equilibrium Constants of Formations at 298.15 K

Substance	S^0, $J\,K^{-1}\,mol^{-1}$	ΔH_f^0, $kJ\,mol^{-1}$	ΔG_f^0, $kJ\,mol^{-1}$	$\log K_f$
Ag^+ (aq)	72.68	105.579	77.107	–13.5085
AgBr (cr)	107.1	–100.37	–96.90	16.976
AgCl (cr)	96.2	–127.068	–109.789	19.2341
AgI (cr)	115.5	–61.84	–66.19	11.596
Ba^{++} (aq)	9.6	–537.64	–560.77	98.242
BaF_2 (cr)	96.36	–1207.1	–1156.8	202.66
$BaSO_4$ (cr)	132.2	–1473.2	–1362.2	238.65
Br^- (aq)	82.4	–121.55	–103.96	18.213
Ca^{++} (aq)	–53.1	–542.83	–553.58	96.982
$CaC_2O_4{\bullet}H_2O$ (cr)	156.5	–1674.86	–1513.87	265.217
CaF_2 (cr)	68.87	–1219.6	–1167.3	204.501
$CaSO_4{\bullet}2H_2O$ (cr)	194.1	–2022.63	–1797.28	314.868
Cl^- (aq)	56.5	–167.159	–131.228	22.9900
CN^- (aq)	94.1	150.6	172.4	-30.20
CNO^- (aq)	106.7	–146.0	–97.4	17.06
$CO_3^=$ (aq)	–56.9	–677.14	–527.81	92.468
$C_2O_4^=$ (aq)	45.6	–825.1	–673.9	118.06
F^- (aq)	–13.8	–332.63	–278.79	48.842
H^+ (aq)	0.000	0.000	0.000	0.000
HCN (aq)	124.7	107.1	119.7	–20.97
HCNO (aq)	144.8	–154.39	–117.1	20.51
$HCOO^-$ (aq)	92	–425.55	–351.0	61.49
HCOOH (aq)	163	–425.43	–372.3	65.22
$HC_2O_4^-$ (aq)	149.4	–818.4	–698.34	122.343
H_2O (l)	69.91	–285.830	–237.129	41.5430
HS^- (aq)	62.8	–17.6	12.08	–2.116
H_2S (aq)	121	–39.7	–27.83	4.876
HSO_4^- (aq)	131.8	–887.34	–755.91	132.429
I^- (aq)	111.3	–55.19	–51.57	9.035
I_3^- (aq)	239.3	–51.5	–51.4	9.005
OH^- (aq)	–10.75	–229.994	–157.244	27.5478
Pb^{++} (aq)	10.5	–1.7	–24.43	4.280
PbF_2 (cr)	110.5	–664.0	–617.1	108.11
$PbSO_4$ (cr)	148.57	–919.94	–813.14	142.455
$S^=$ (aq)	–14.6	33.1	85.8	–15.031
$SO_4^=$ (aq)	20.1	–909.27	–744.53	130.435

8.5 Average Electric Atmosphere about an Ion

Each ion in an electrolytic solution attracts oppositely charged ions and repels like charged ions. This organizing influence is opposed by the random motion associated with the internal energy of the system. In the resulting compromise, each ion is surrounded by a neighborhood of considerable size containing, on the average, an excess of oppositely charged ions. A positive ion thus sports a *negative atmosphere*; a negative ion, a *positive atmosphere*. Since it is more difficult to remove an ion from its charged atmosphere than from an environment of neutral solvent molecules, the activity of the ion is reduced below what its concentration would indicate.

Consider a representative ion j about which the average electric potential is $\phi(r)$, where r is the distance from the center of the ion and the ion is taken to be spherically symmetric. The potential energy of an atmosphere ion carrying charge q_i is $q_i\phi$.

The average charge density at a given point in the atmosphere equals the charge q_i multiplied by the number density of ions $N_A c_i$ summed over the ionic species. With c_i the moles of i per 1000 cm^3, we multiply by 1000, the number of 1000 cm^3 in a cubic meter. In the potential field, average concentration c_i is given by the Boltzmann distribution law (3.81). Thus, we construct the formula

$$\rho = \sum 1000 N_A q_i c_i = \sum 1000 N_A q_i c_{i\infty} e^{-q_i\phi/kT}. \qquad [8.45]$$

As long as ϕ/T is small enough, the exponential may be expanded and the higher terms neglected:

$$\rho = \sum 1000 N_A q_i c_{i\infty}\left(1 - \frac{q_i\phi}{kT} + \ldots\right) = -\frac{1000 N_A \phi}{kT}\sum c_{i\infty} q_i^{\,2}. \qquad [8.46]$$

The first term vanishes because of the electrical neutrality condition.

If z_i is the number of units of charge on ion i while e is the charge on a proton, we have

$$q_i = z_i e. \qquad [8.47]$$

Also from (8.24), we have

$$c_i = m_i d \qquad [8.48]$$

where m_i is the molality of ion i in the solution and d is the density of the solution. Substituting these into the final form of (8.46) leads to the average charge density

$$\rho = -\frac{1000 N_A e^2 \phi d}{kT}\sum m_i z_i^{\,2} = -\frac{2000 N_A e^2 \phi d}{kT}\mu \qquad [8.49]$$

where

$$\mu = \frac{1}{2}\sum m_i z_i^{\,2}. \qquad [8.50]$$

Parameter μ is called the *ionic strength* of the solution.

Formula (8.49) describes how the electric potential ϕ entails a nonzero charge density ρ. Ions are distributed by thermal agitation up and down the potential slopes, following the Boltzmann equation (3.81).

8.6 An Integral Form of Coulomb's Law

The potential ϕ arises from the electric forces acting between the charges. These are described by Coulomb's law.

In principle, one could map out an electric field using a small test charge. The force per unit charge acting on the test charge, at rest at a certain point, is the *electric intensity* **E** at that point.

Coulomb's law, in its simplest form, states that the electric intensity produced by a point charge of magnitude q is

$$\mathbf{E} = \frac{q}{4\pi\varepsilon r^2}\mathbf{r}_1, \qquad [8.51]$$

where vector **r** is drawn from the source charge to the point at which the test charge is placed, $\mathbf{r}_1$ is the unit vector giving the direction of **r** ($\mathbf{r} = r\mathbf{r}_1$), and ε is the dielectric constant of the intervening medium.

One may surround the source with a surface S, as figure 8.1 shows. A differential element of the surface is represented by vector **dS** whose magnitude equals the size of the element and whose direction is perpendicular to the element in the outward direction. The solid angle $d\Omega$ subtended by the element equals the projection of **dS** on vector **r** divided by r^2. So we have

$$\mathbf{E}\cdot d\mathbf{S} = E\cos\theta\, dS = Er^2\, d\Omega, \qquad [8.52]$$

where θ is the angle between **E** and **dS**.

Combining equations (8.51) and (8.52), and integrating yields

$$\int_S \mathbf{E}\cdot d\mathbf{S} = \int_S Er^2\, d\Omega = \int_S \frac{q}{4\pi\varepsilon r^2} r^2\, d\Omega = \frac{q}{4\pi\varepsilon}\int_S d\Omega = \frac{q}{\varepsilon}. \qquad [8.53]$$

The integral of the solid angle over the closed surface equals 4π; since the area of a sphere of unit radius is 4π in magnitude.

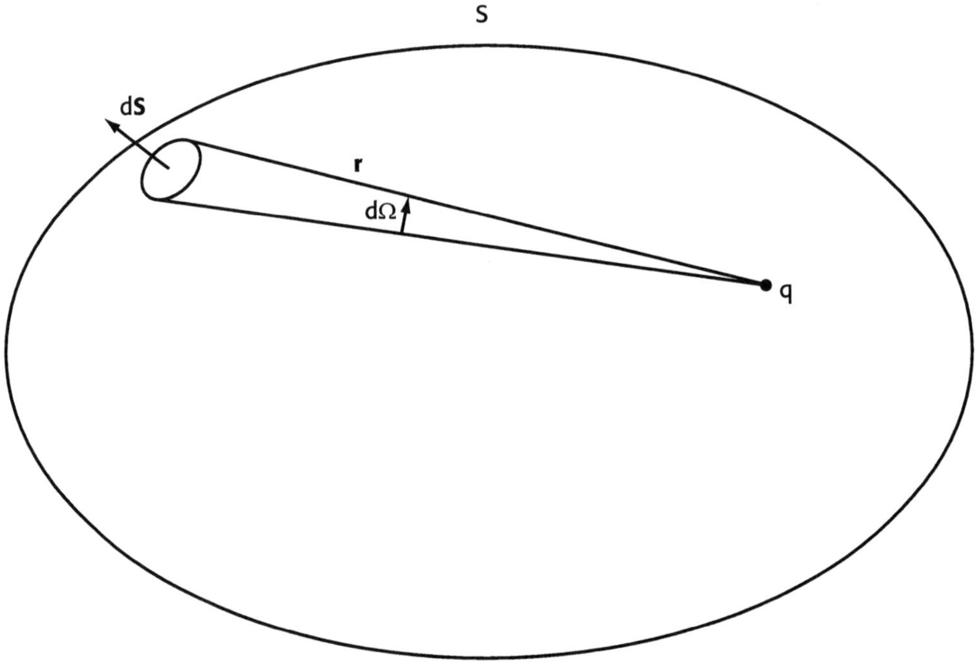

FIGURE 8.1 Surface element **dS** subtending the solid angle $d\Omega$ at point charge q.

From result (8.53), any charge q_i within a closed surface contributes q_i/ε to the integral $\int \mathbf{E}\cdot d\mathbf{S}$. Any charge q_j outside the closed surface does not contribute because $\int d\Omega$ for it is zero. For total charge Σq_i within the closed surface, we have

$$\int_S \mathbf{E}\cdot d\mathbf{S} = \frac{\sum q_i}{\varepsilon}. \qquad [8.54]$$

This form of Coulomb's law is called *Gauss's theorem.*

8.7 Application of Gauss's Theorem to the Ionic Atmosphere

We have considered the *j*th ion and its neighborhood to be spherically symmetric. Thus, the electric intensity $\mathbf{E}$ about the ion is considered to be directed radially. But any element $d\mathbf{S}$ of a sphere centered on the ion is also directed radially. So we have

$$\int_S \mathbf{E}\cdot d\mathbf{S} = \int_S E\cdot dS = E\int dS = E\left(4\pi r^2\right). \qquad [8.55]$$

According to Gauss's theorem, the left side equals the total charge q enclosed by the sphere of radius r divided by ε; thus

$$\frac{q}{\varepsilon} = 4\pi r^2 E. \qquad [8.56]$$

The change on going to a sphere infinitesimally larger is

$$\frac{dq}{\varepsilon} = d\left(4\pi r^2 E\right). \qquad [8.57]$$

But if ρ is the charge density at distance r from the center of the *j*th ion, we have

$$dq = \rho 4\pi r^2\, dr. \qquad [8.58]$$

Eliminating dq from (8.57) and (8.58) yields

$$d\left(4\pi r^2 E\right) = \frac{\rho 4\pi r^2\, dr}{\varepsilon}, \qquad [8.59]$$

whence

$$\frac{1}{r^2}\frac{d}{dr}\left(r^2 E\right) = \frac{\rho}{\varepsilon}. \qquad [8.60]$$

The electric intensity, the force per unit charge that would act on a test charge, is given by

$$E = -\frac{d\phi}{dr}. \qquad [8.61]$$

Inserting this into equation (8.60) leads to the form

$$\frac{1}{r^2}\frac{d}{dr}\left(r^2\frac{d\phi}{dr}\right) = -\frac{\rho}{\varepsilon} \qquad [8.62]$$

which can be rewritten as

$$\frac{d^2}{dr^2}\left(r\phi\right) = -\frac{\rho}{\varepsilon}r. \qquad [8.63]$$

8.8 Variation of Electric Potential in an Ionic Atmosphere

Equation (8.63) applies to any spherically symmetric distribution of charge. On the other hand, condition (8.49) describes the effect of thermal agitation on the distribution of ions in a potential field. Combining these equations gives us

$$\frac{d^2}{dr^2}(r\phi) = \frac{2000 N_A e^2 d}{\varepsilon kT}\mu\phi r = b^2(r\phi), \quad [8.64]$$

where we set

$$\frac{2000 N_A e^2 d}{\varepsilon kT}\mu = b^2. \quad [8.65]$$

Let us rearrange the overall equation (8.64),

$$\frac{d^2}{dr^2}(r\phi) - b^2(r\phi) = 0, \quad [8.66]$$

and factor the operator to get

$$(D - b)(D + b)(r\phi) = 0 \quad [8.67]$$

or

$$(D + b)(D - b)(r\phi) = 0, \quad [8.68]$$

where

$$D = \frac{d}{dr}. \quad [8.69]$$

Equation (8.67) is satisfied when

$$(D + b)(r\phi) = 0, \quad [8.70]$$

while equation (8.68) is satisfied when

$$(D - b)(r\phi) = 0. \quad [8.71]$$

The solution of (8.70) is

$$r\phi = Ae^{-br}, \quad [8.72]$$

while the solution of (8.71) is

$$r\phi = Be^{br}. \quad [8.73]$$

So equation (8.66) is satisfied by both (8.72) and (8.73); we have the general solution

$$r\phi = Ae^{-br} + Be^{br}, \quad [8.74]$$

or

$$\phi = A\frac{e^{-br}}{r} + B\frac{e^{br}}{r}. \quad [8.75]$$

If B were different from zero, the last term, and ϕ, would increase without limit as r increases. But this is not allowed. For the ionic atmosphere, we have

$$\phi = A\frac{e^{-br}}{r}. \quad [8.76]$$

Integration constant A is determined by the potential at the surface of the ion.

8.9 Free Energy of the Ionic Atmosphere

If only ion j were present, it would impose the potential

$$\phi_{\text{ion}} = \frac{z_j e}{4\pi\varepsilon r} \qquad [8.77]$$

on its surroundings. But when the ionic strength μ is zero, b is zero and formula (8.76) reduces to

$$\phi_{\text{ion}} = \frac{A}{r}. \qquad [8.78]$$

These expressions are the same when

$$A = \frac{z_j e}{4\pi\varepsilon}. \qquad [8.79]$$

So equation (8.76) becomes

$$\phi = \frac{z_j e}{4\pi\varepsilon r} e^{-br} = \frac{z_j e}{4\pi\varepsilon r}\left(1 - br + \ldots\right). \qquad [8.80]$$

For dilute solutions, parameter b is small. Higher terms in expansion (8.80) can then be neglected. Subtracting out ϕ_{ion} from the result yields

$$\phi_{\text{atm}} = -\frac{z_j eb}{4\pi\varepsilon}, \qquad [8.81]$$

the contribution to potential ϕ from the atmosphere of the ion.

To obtain the work done in setting up the atmosphere, one begins with the central ion discharged, Electric charge is then brought up from infinity continuously. At a given stage, the fraction of final charge on ion j is f. Then z_je in equation (8.81) is replaced with z_jef and the next charge brought up to the central ion is z_jedf.

The total work done against the potential due to the atmosphere is

$$w = \int_{f=0}^{1} \phi_{\text{atm}}\left(z_j e\, df\right) = -\int_0^1 \frac{z_j^2 e^2 b}{4\pi\varepsilon} f\, df = -\frac{z_j^2 e^2 b}{8\pi\varepsilon}. \qquad [8.82]$$

Since this is net work done reversibly, it is the Gibbs energy associated with setting up the ionic atmosphere:

$$\Delta G_{\text{atm}} = -\frac{z_j^2 e^2 b}{8\pi\varepsilon}. \qquad [8.83]$$

The approximation applied to expression (8.80) may be improved as follows.

In the argument, expression $e^{-br} - 1$ was replaced with $-br$. But a better average approximation to

$$e^{-br} - 1 = -\left[br - \frac{(br)^2}{2} + \frac{(br)^3}{6} - \frac{(br)^4}{24} + \ldots\right] \qquad [8.84]$$

is achieved by constructing

$$\frac{-br}{1 - ba} = -\left[br - (br)(ba) + (br)(ba)^2 - (br)(ba)^3 + \ldots\right]. \qquad [8.85]$$

Then a is chosen to yield the best weighted average fit to the preceding series. In place of (8.81), we would then get

$$\phi_{atm} = -\frac{1}{1+ba_j}\frac{z_j eb}{4\pi\varepsilon}. \tag{8.86}$$

And formula (8.83) would be replaced with

$$\Delta G_{atm} = -\frac{1}{1+ba_j}\frac{z_j^2 e^2 b}{8\pi\varepsilon}. \tag{8.87}$$

Parameter a is interpreted as the average distance of approach of the center of an oppositely charged ion to the center of the given ion. Parameter a_j may then be called the *effective diameter* of the *j*th ion.

8.10 Debye-Hückel Equations

From the energy needed to form an ionic atmosphere, a person can calculate the corresponding ionic activity coefficient.

The general form for the chemical potential of a constituent,

$$\mu_j = \mu_j^0 + RT\ln a_j, \tag{8.88}$$

leads to the usual expression,

$$K = \frac{a_L^{\,l} a_M^{\,m}}{a_A^{\,a} a_B^{\,b}}, \tag{8.89}$$

governing the equilibrium

$$a\text{A} + b\text{B} \rightleftarrows l\text{L} + m\text{M}, \tag{8.90}$$

as we saw in equation (7.57). When concentrations of the solutes are expressed in molalities, we have

$$a_j = \gamma_j m_j. \tag{8.91}$$

Then equation (8.88) becomes

$$\mu_j = \mu_j^0 + RT\ln m_j + RT\ln\gamma_j. \tag{8.92}$$

Dividing by Avogadro's number gives the Gibbs energy for a single ion:

$$\frac{\mu_j}{N_A} = \frac{\mu_j^0}{N_A} + kT\ln m_j + kT\ln\gamma_j. \tag{8.93}$$

In dilute solutions, the principal deviation from the ideal value is caused by the electric interaction. The effect on the Gibbs energy when the ionic strength is small is given by formula (8.83). So we obtain

$$kT\ln\gamma_j = \Delta G_{atm} = -\frac{z_j^2 e^2 b}{8\pi\varepsilon}, \tag{8.94}$$

whence

$$\ln\gamma_j = -z_j^2 \frac{e^3}{8\pi(\varepsilon kT)^{3/2}}(2000N_A d)^{1/2}\sqrt{\mu} \tag{8.95}$$

or

$$\log \gamma_j = -Az_j^2 \sqrt{\mu}. \qquad [8.96]$$

In equation (8.96), all the constants have been gathered together into expression A. This equation is known as the *Debye-Hückel limiting law.*

In actual systems, the ionic atmosphere about an ion can build up only in the region outside the ion. A large ion has a larger excluded volume than a smaller one and hence less of an ionic atmosphere, other things being equal. So it would have a larger activity coefficient in a solution of given ionic strength.

The difference is embodied in parameter a_j in formula (8.87). Employing (8.87) instead of (8.83) leads to the equation

$$\log \gamma_j = -\frac{Az_j^2 \sqrt{\mu}}{1 + Ba_j \sqrt{\mu}}. \qquad [8.97]$$

Here γ_j. is the activity coefficient of ion j, z_j the number of positive charges on j, as employed in equation (8.47), a_j the effective diameter of ion j in solution, μ the ionic strength, while A, B are constants depending on the temperature and solvent. This result is the single ion *Debye-Hückel equation.*

Parameters A and B for water solutions at various temperatures are listed in table 8.4. In table 8.5 empirical diameters for various ions appear. In table 8.6, activity coefficients calculated with formula (8.97) for solutions in water at 25° C are tabulated.

Example 8.9

Calculate the hydrogen ion concentration at 25° C in an aqueous solution 0.100 m in NaCl and 0.100 m in acetic acid.

From table 8.2, the equilibrium constant for

$$HA \rightleftarrows H^+ + A^-$$

TABLE 8.4 Debye-Hückel Parameters for an Ion in Water

Temperature, °C	*A*, $m^{-1/2}$	*B*, $Å^{-1}m^{-1/2}$
0	0.4883	0.3241
10	0.4960	0.3258
20	0.5042	0.3273
25	0.5085	0.3281
30	0.5130	0.3290
40	0.5221	0.3305
50	0.5319	0.3321
60	0.5425	0.3338
70	0.5537	0.3354
80	0.5658	0.3372
90	0.5788	0.3390
100	0.5929	0.3409

TABLE 8.5 Empirical Diameters for Ions Dissolved in Water at 25°C*

a_j, Å	*Inorganic Ions*
	Charge 1
9	H^+
6	Li^+
4-4.5	Na^+, $CdCl^+$, ClO_2^-, IO_3^-, HCO_3^-, $H_2PO_4^-$, HSO_3^-, $H_2AsO_4^-$, $Co(NH_3)_4(NO_2)_2^+$
3.5	OH^-, F^-, NCS^-, NCO^-, HS^-, ClO_3^-, ClO_4^-, BrO_3^-, IO_4^-, MnO_4^-
3	K^+, Cl^-, Br^-, I^-, CN^-, NO_2^-, NO_3^-
2.5	Rb^+, Cs^+, NH_4^+, Tl^+, Ag^+
	Charge 2
8	Mg^{++}, Be^{++}
6	Ca^{++}, Cu^{++}, Zn^{++}, Sn^{++}, Mn^{++}, Fe^{++}, Ni^{++}, Co^{++}
5	Sr^{++}, Ba^{++}, Ra^{++}, Cd^{++}, Hg^{++}, $S^{=}$, $S_2O_4^{=}$, $WO_4^{=}$
4.5	Pb^{++}, $CO_3^{=}$, $SO_3^{=}$, $MoO_4^{=}$, $Co(NH_3)_5Cl^{++}$, $Fe(CN)_5NO^{=}$
4	Hg_2^{++}, $SO_4^{=}$, $S_2O_3^{=}$, $S_2O_6^{=}$, $S_2O_8^{=}$, $SeO_4^{=}$, $CrO_4^{=}$, $HPO_4^{=}$
	Charge 3
9	Al^{3+}, Fe^{3+}, Cr^{3+}, Sc^{3+}, Y^{3+}, La^{3+}, In^{3+}, Ce^{3+}, Pr^{3+}, Nd^{3+}, Sm^{3+}
6	$Co(ethylenediamine)_3^{3+}$
4	PO_4^{3-}, $Fe(CN)_6^{3-}$, $Cr(NH_3)_6^{3+}$, $Co(NH_3)_6^{3+}$, $Co(NH_3)_5H_2O^{3+}$
	Charge 4
11	Th^{4+}, Zr^{4+}, Ce^{4+}, Sn^{4+}
6	$Co(S_2O_3)(CN)_5^{4-}$
5	$Fe(CN)_6^{4-}$
	Charge 5
9	$Co(SO_3)_2(CN)_4^{5-}$
	Organic Ions
	Charge 1
8	$(C_6H_5)_2CHOO^-$, $(C_3H_7)_4N^+$
7	$OC_6H_2(NO_2)_3^-$, $(C_3H_7)_3NH^+$, $CH_3OC_6H_4COO^-$,
6	$C_6H_5COO^-$, $C_6H_4OHCOO^-$, $C_6H_4ClCOO^-$, $C_6H_5CH_2COO^-$, $CH_2{=}CHCH_2COO^-$, $(CH_3)_2CHCH_2COO^-$, $(C_2H_5)_4N^+$, $(C_3H_7)_2NH_2^+$
5	$CHCl_2COO^-$, CCl_3COO^-, $(C_2H_5)_3NH^+$, $(C_3H_7)NH_3^+$
4.5	CH_3COO^-, CH_2ClCOO^-, $(CH_3)_4N^+$, $(C_2H_5)_2NH_2^+$, $NH_2CH_2COO^-$
4	$NH_3CH_2COOH^+$, $(CH_3)_3NH^+$, $C_2H_5NH_3^+$
3.5	$HCOO^-$, H_2 citrate$^-$, $CH_3NH_3^+$, $(CH_3)_2NH_2^+$
	Charge 2
7	$OOC(CH_2)_5COO^{=}$, $OOC(CH_2)_6COO^{=}$, congo red anion$^{=}$
6	$C_6H_4(COO)_2^{=}$, $H_2C(CH_2COO)_2^{=}$, $(CH_2CH_2COO)_2^{=}$
5	$H_2C(COO)_2^{=}$, $(CH_2COO)_2^{=}$, $(CHOHCOO)_2^{=}$
4.5	$(COO)_2^{=}$, H citrate$^{=}$
	Charge 3
5	Citrate^{3-}

*Kielland, J.: 1937. J. Am. Chem. Soc. **59**, 1675.

TABLE 8.6 Activity Coefficients of Ions in Water at 25°C for Concentrations in Moles per 1000 g Solvent

Ionic Diameter a_j, *Å*	*Ionic Strength*								
	0.00025	*0.0005*	*0.001*	*0.0025*	*0.005*	*0.01*	*0.025*	*0.05*	*0.1*
				Ion Charge 1					
9	0.982	0.976	0.967	0.950	0.934	0.914	0.881	0.855	0.825
8	0.982	0.976	0.966	0.950	0.933	0.911	0.877	0.85	0.815
7	0.982	0.975	0.966	0.949	0.931	0.909	0.872	0.84	0.805
6	0.982	0.975	0.966	0.948	0.930	0.907	0.868	0.835	0.795
5	0.982	0.975	0.965	0.947	0.928	0.904	0.863	0.825	0.785
4.5	0.982	0.975	0.965	0.947	0.928	0.903	0.861	0.82	0.775
4	0.982	0.975	0.965	0.947	0.927	0.902	0.858	0.815	0.77
3.5	0.982	0.975	0.965	0.946	0.926	0.900	0.855	0.81	0.76
3	0.982	0.975	0.965	0.946	0.926	0.899	0.852	0.805	0.755
2.5	0.982	0.975	0.965	0.945	0.925	0.897	0.849	0.80	0.745
				Ion Charge 2					
8	0.931	0.906	0.872	0.813	0.756	0.690	0.595	0.515	0.445
7	0.931	0.905	0.871	0.811	0.752	0.683	0.58	0.50	0.425
6	0.931	0.905	0.870	0.808	0.748	0.676	0.57	0.485	0.40
5	0.930	0.904	0.869	0.805	0.743	0.669	0.555	0.465	0.375
4.5	0.930	0.904	0.868	0.804	0.741	0.665	0.55	0.455	0.365
4	0.930	0.903	0.867	0.803	0.739	0.661	0.54	0.445	0.35
				Ion Charge 3					
9	0.853	0.802	0.737	0.632	0.540	0.445	0.32	0.24	0.18
6	0.851	0.798	0.731	0.619	0.520	0.415	0.28	0.195	0.13
5	0.850	0.797	0.728	0.614	0.513	0.405	0.265	0.18	0.11
4	0.849	0.795	0.726	0.610	0.506	0.395	0.25	0.16	0.095
				Ion Charge 4					
11	0.756	0.679	0.588	0.452	0.35	0.25	0.15	0.098	0.063
6	0.750	0.669	0.573	0.426	0.315	0.21	0.105	0.055	0.026
5	0.749	0.668	0.569	0.421	0.305	0.20	0.095	0.047	0.020
				Ion Charge 5					
9	0.643	0.541	0.429	0.28	0.18	0.105	0.043	0.019	0.008

is 1.754×10^{-5} m when HA is acetic acid. So we have

$$\frac{\gamma_{H^+}\gamma_{A^-}}{\gamma_{HA}}\frac{m_{H^+}m_{A^-}}{m_{HA}} = 1.754 \times 10^{-5} \text{ m}$$

whence

$$\frac{m_{H^+}m_{A^-}}{m_{HA}} = \frac{\gamma_{HA}}{\gamma_{H^+}\gamma_{A^-}} 1.754 \times 10^{-5} \text{ m.}$$

Since HA is a molecule at low concentration, we take

$$\gamma_{HA} = 1.00.$$

The ionic strength involved in reducing the ion activity coefficients is practically all due to the sodium chloride; we have

$$\mu = \frac{1}{2}\left(m_{Na^+}z_{Na^+}{}^2 + m_{Cl^-}z_{Cl^-}{}^2\right) = \frac{1}{2}\left(0.100 \times 1^2 + 0.100 \times 1^2\right) = 0.100 \text{ m.}$$

In table 8.5, one finds the ionic diameters

$$a_{H^+} = 9 \text{ Å}, \qquad a_{A^-} = 4.5 \text{ Å}.$$

Then from the 0.100 m column of table 8.6, one obtains

$$\gamma_{H^+} = 0.825, \qquad \gamma_{A^-} = 0.775.$$

Since the principal source of hydrogen ions and the only source of acetate ions is the ionization of the acetic acid, we have

$$m_{A^-} \cong m_{H^+} = x, \qquad m_{HA} \cong 0.100 - x.$$

Substituting into the equilibrium expression gives

$$\frac{x^2}{0.100 - x} = \frac{1.00}{(0.825)(0.775)} 1.754 \times 10^{-5} \text{ m} = 2.743 \times 10^{-5} \text{ m,}$$

whence

$$x^2 = 2.743 \times 10^{-6} - 2.743 \times 10^{-5} x.$$

Neglecting the last term yields

$$x = 1.656 \times 10^{-3} \text{ m.}$$

Introducing this approximate value into the last term and again solving for x leads to

$$x = 1.64 \times 10^{-3} \text{ m.}$$

Example 8.10

Calculate the hydrogen ion concentration in 0.100 m acetic acid at 25° C.

The ionic strength now is determined by the concentration of ionized acetic acid. But initially this is not known. So we follow an iteration procedure, first taking

$$\gamma_{H^+} = 1.000, \qquad \gamma_{A^-} = 1.000.$$

As in example 8.9, we also have

$$\gamma_{HA} = 1.00$$

and

$$m_{A^-} \cong m_{H^+} = x, \qquad m_{HA} \cong 0.100 - x.$$

Substituting into the equilibrium constant expression gives

$$\frac{x^2}{0.100 - x} = 1.754 \times 10^{-5}\ \text{m}$$

whence

$$x^2 = 1.754 \times 10^{-6} - 1.754 \times 10^{-5} x.$$

Neglecting the last term yields

$$x = 1.32 \times 10^{-3}\ \text{m}.$$

The corresponding ionic strength is

$$\mu = 1.32 \times 10^{-3}\ \text{m}.$$

Using the ionic diameters from table 8.5 and interpolating in table 8.6 leads to

$$\gamma_{H^+} = 0.963, \qquad \gamma_{A^-} = 0.961.$$

The equilibrium expression is now

$$\frac{x^2}{0.100 - x} = \frac{1.00}{(0.963)(0.961)} 1.754 \times 10^{-5}\ \text{m} = 1.895 \times 10^{-5}\ \text{m},$$

whence

$$x^2 = 1.895 \times 10^{-6} - 1.895 \times 10^{-5} x.$$

Neglecting the last term yields

$$x = 1.377 \times 10^{-3}\ \text{m}.$$

Then introducing this approximate value into the last term and solving for x gives us

$$x = 1.37 \times 10^{-3}\ \text{m}.$$

The activity coefficients calculated from the corresponding ionic strength do not differ in three significant figures from those used. So further iterations are not needed.

Example 8.11

Calculate the solubility of AgCl in 0.100 m KNO_3 at 25° C.
For the equilibrium

$$AgCl \rightleftarrows Ag^+ + Cl^-,$$

table 8.3 tells us that

$$\log K = -13.5085 + 22.9900 - 19.2341 = -9.7526$$

whence

$$K = 1.768 \times 10^{-10}\ \text{m}^2$$

and

$$m_{Ag^+}m_{Cl^-} = \frac{1}{\gamma_{Ag^+}\gamma_{Cl^-}} 1.768 \times 10^{-10} \text{ m}^2.$$

Since the principal contribution to the ionic strength is from the KNO_3, we have

$$\mu \cong 0.100 \text{ m}.$$

Taking the ionic diameters from table 8.5, table 8.6 now yields

$$\gamma_{Ag^+} = 0.745, \qquad \gamma_{Cl^-} = 0.755.$$

Since the only source of Ag^+ and Cl^- is the AgCl, we have

$$m_{Ag^+} = m_{Cl^-} = x.$$

So the equilibrium expression becomes

$$x^2 = \frac{1}{(0.745)(0.755)} 1.768 \times 10^{-10} \text{ m}^2 = 3.143 \times 10^{-10} \text{ m}^2,$$

whence

$$x = 1.77 \times 10^5 \text{ m}$$

and

$$w = \left(1.77 \times 10^{-5} \text{ m}\right)\left(143.321 \text{ g mol}^{-1}\right) = 2.54 \times 10^{-3} \text{ g per 1000 g } H_2O.$$

8.11 Equilibria Among Phases Revisited

In section 6.17, each constituent i was considered to move from phase to phase independently. But when the constituents are ions, such movement is subject to the electrical neutrality condition.

Suppose that the phases contain ions 1, 2, and 3 carrying z_1 positive charges, z_2 negative charges, and z_3 negative charges. Furthermore, we consider fluctuations in which only these ions travel from phase (2) to phase (1) or vice versa. Equation (6.99) then reduces to

$$\left(\mu_1^{(1)} - \mu_1^{(2)}\right) dn_1^{(1)} + \left(\mu_2^{(1)} - \mu_2^{(2)}\right) dn_2^{(1)} + \left(\mu_3^{(1)} - \mu_3^{(2)}\right) dn_3^{(1)} = 0. \qquad [8.98]$$

Electrical neutrality of the two phases is maintained if

$$z_1\, dn_1^{(1)} = z_2\, dn_2^{(1)} + z_3\, dn_3^{(1)}. \qquad [8.99]$$

Substituting this into equation (8.98) and rearranging yields

$$\left(\frac{\mu_1^{(1)}}{z_1} + \frac{\mu_2^{(1)}}{z_2} - \frac{\mu_1^{(2)}}{z_1} - \frac{\mu_2^{(2)}}{z_2}\right) z_2\, dn_2^{(1)} + \left(\frac{\mu_1^{(1)}}{z_1} + \frac{\mu_3^{(1)}}{z_3} - \frac{\mu_1^{(2)}}{z_1} - \frac{\mu_3^{(2)}}{z_3}\right) z_3\, dn_3^{(1)} = 0. \qquad [8.100]$$

Fluctuations $dn_2(1)$ and $dn_3(1)$ are independent. So the expressions in parentheses must vanish at equilibrium.

If the first two ions make up the strong electrolyte $A_{\nu_1}B_{\nu_2}$, then

$$\nu_1 z_1 = \nu_2 z_2 \qquad [8.101]$$

and

$$\mu_{A_{\nu1}B_{\nu2}} = \nu_1\mu_1 + \nu_2\mu_2. \quad [8.102]$$

The chemical potential per equivalent is

$$\frac{\mu_{A_{\nu1}B_{\nu2}}}{\nu_1 z_1} = \frac{\mu_1}{z_1} + \frac{\mu_2}{z_2}. \quad [8.103]$$

Thus, equation (8.100) tells us that the chemical potential of each strong electrolyte is the same in each phase at equilibrium; we have

$$\mu_{A_{\nu1}B_{\nu2}}^{(1)} = \mu_{A_{\nu1}B_{\nu2}}^{(2)}. \quad [8.104]$$

One may also define the activity of the electrolyte a by the equation

$$\mu_{A_{\nu1}B_{\nu2}} = \mu^{0}_{A_{\nu1}B_{\nu2}} + RT\ln a. \quad [8.105]$$

Then (8.104) can be rewritten as

$$\mu^{0(1)}_{A_{\nu1}B_{\nu2}} + RT\ln a^{(1)} = \mu^{0(2)}_{A_{\nu1}B_{\nu2}} + RT\ln a^{(2)}. \quad [8.106]$$

When the standard state is the same in both phases, the standard chemical potentials are the same and we have

$$a^{(1)} = a^{(2)}. \quad [8.107]$$

The activity of the electrolyte in one phase equals its activity in a phase in equilibrium with the first phase.

But for equilibrium between a pure crystal and a solution at 1 bar, we take the activity of the crystal to be 1 unit. Then for the solution, we have

$$\mu^{0(1)} + RT\ln a^{(1)} = \mu^{0(2)}, \quad [8.108]$$

whence

$$\ln a^{(1)} = -\frac{\mu^{0(1)} - \mu^{0(2)}}{RT} = -\frac{\Delta G^0}{RT} \quad [8.109]$$

or

$$a^{(1)} = K \quad [8.110]$$

where K is the equilibrium constant.

For the cation and for the anion in a given solution, one has

$$\mu_1 = \mu_1^{0} + RT\ln a_1, \quad [8.111]$$

$$\mu_2 = \mu_2^{0} + RT\ln a_2. \quad [8.112]$$

But substituting (8.105), (8.111), and (8.112) into (8.102) yields

$$\mu^{0}_{A_{\nu1}B_{\nu2}} + RT\ln a = \nu_1\mu_1^{0} + \nu_1 RT\ln a_1 + \nu_2\mu_2^{0} + \nu_2 RT\ln a_2. \quad [8.113]$$

Then subtracting the equation

$$\mu^{0}_{A_{\nu1}B_{\nu2}} = \nu_1\mu_1^{0} + \nu_2\mu_2^{0} \quad [8.114]$$

gives us

$$RT\ln a = RT\ln a_1^{\nu 1} + RT\ln a_2^{\nu 2}, \qquad [8.115]$$

whence

$$a = a_1^{\nu 1} a_2^{\nu 2}. \qquad [8.116]$$

The activity of a completely ionized electrolyte equals the activity of the cation raised to a power equal to the number of cations times the activity of the anion raised to a power equal to the number of anions in the formula of the electrolyte.

Example 8.12

A membrane is permeable to sodium ions and to chloride ions but not to an organic ion R^-. Calculate the equilibrium concentrations if initially 100 ml of a solution 0.100 M in NaCl and 0.0100 M in NaR were separated by the membrane from 1.00 1 of water. Neglect any movement of water and consider the molar activity coefficients to equal 1.

Let $[Na^+]_{(1)}$ and $[Cl^-]_{(1)}$ represent the sodium and chloride ion concentrations in phase (1), $[Na^+]_{(2)}$ and $[Cl^-]_{(2)}$ the sodium and chloride ion concentrations in phase (2). Initially we have

$$\left[Na^+\right]_{(1)} = 0.110 \text{ M}, \qquad \left[Cl^-\right]_{(1)} = 0.100 \text{ M},$$

$$\left[Na^+\right]_{(2)} = 0 \text{ M}, \qquad \left[Cl^-\right]_{(2)} = 0 \text{ M},$$

To preserve electrical neutrality in both phases, the amount of Na^+ migrating must equal the amount of Cl^- migrating. Let x equal this amount in moles when equilibrium has been established. If V_1 is the volume of phase (1) and V_2 the volume of phase (2), we then have

$$\left[Na^+\right]_{(1)} = 0.110 - \frac{x}{V_1} = 0.110 - \frac{x}{0.1} = 0.110 - 10x,$$

$$\left[Cl^-\right]_{(1)} = 0.100 - \frac{x}{V_1} = 0.100 - \frac{x}{0.1} = 0.100 - 10x,$$

$$\left[Na^+\right]_{(2)} = \frac{x}{V_2} = \frac{x}{1} = x, \qquad \left[Cl^-\right]_{(2)} = \frac{x}{V_2} = \frac{x}{1} = x.$$

For electrolyte NaCl, the cation and the anion are singly charged,

$$\nu_+ = \nu_- = 1.$$

So formulas (8.107) and (8.116) yield

$$a_{+(1)}a_{-(1)} = a_{+(2)}a_{-(2)}.$$

Setting the activities equal to the corresponding concentrations,

$$\left[Na^+\right]_{(1)}\left[Cl^-\right]_{(1)} = \left[Na^+\right]_{(2)}\left[Cl^-\right]_{(2)},$$

and introducing the equilibrium expressions already constructed leads to

$$(0.110 - 10x)(0.100 - 10x) = xx,$$

whence

$$99x^2 - 2.1x + 0.011 = 0.$$

Only the root

$$x = 0.00943$$

makes all concentrations positive. These include

$$\left[\mathrm{Na}^+\right]_{(1)} = 0.110 - 0.0943 = 0.0157\ \mathrm{M},$$

$$\left[\mathrm{Cl}^-\right]_{(1)} = 0.100 - 0.0943 = 0.0057\ \mathrm{M},$$

$$\left[\mathrm{Na}^+\right]_{(2)} = \left[\mathrm{Cl}^-\right]_{(2)} = 0.0094\ \mathrm{M}.$$

Since negative concentrations have no physical significance, the other root is discarded.

8.12 Determining Ionic Activities

In section 6.11, the activity of a constituent in a given phase was measured by the partial pressure of the constituent in an ideal gas phase at equilibrium with the given phase. One may imagine the two phases separated by a membrane permeable only to the given constituent.

For ions, the ideal gas phase would be a plasma. But the bulk of a plasma is electrically neutral. So in principle, the thermodynamic behavior of cations cannot be observed apart from anions or electrons. Activities of electrolytes can be measured but not those of individual ions.

For interpreting experimental results, one consequently defines a *mean ionic activity* $a_\pm$ by the formula

$$a_\pm = a^{1/\nu}, \qquad [8.117]$$

where a is the activity of the electrolyte and ν is the total number of ions in the formula for the electrolyte:

$$\nu = \nu_+ + \nu_-. \qquad [8.118]$$

One defines the *mean ionic molality* $m_\pm$ by the formula

$$m_\pm = \left(m_+^{\nu_+} m_-^{\nu_-}\right)^{1/\nu}, \qquad [8.119]$$

where m_+ is the molality of the cation and m_- the molality of the anion. Furthermore, one defines the *mean ionic activity coefficient* $\gamma_\pm$ by the equation

$$\gamma_\pm = \frac{a_\pm}{m_\pm}. \qquad [8.120]$$

Thus, $\gamma_\pm$ is the geometric mean of the individual ionic activity coefficients:

$$\gamma_\pm = \left(\gamma_+^{\nu_+} \gamma_-^{\nu_-}\right)^{1/\nu}. \qquad [8.121]$$

Combining the Debye-Huckel limiting law for the cation and for the anion of an electrolyte then yields

$$\log \gamma_\pm = -A z_+ z_- \sqrt{\mu}\,. \qquad [8.122]$$

Similarly, from (8.97) we construct

$$\log \gamma_{\pm} = -\frac{Az_+z_-\sqrt{\mu}}{1+Ba\sqrt{\mu}}, \qquad [8.123]$$

where a is the appropriate mean ionic diameter.

Example 8.13

How may one determine the equilibrium constant for a sparingly soluble electrolyte?

The procedure involves solubility measurements in the presence of various concentrations of an inert salt. By chemical analysis, the mean ionic molality of the given electrolyte and the ionic strength in each solution are determined.

Suppose the equilibrium is

$$A_{\nu_+}B_{\nu_-}\ (\text{cr}) \rightleftarrows \nu_+A^{z_+}\ (\text{aq}) + \nu_-A^{-z_-}\ (\text{aq}).$$

From formulas (8.110), (8.117), (8.120), the activity of the electrolyte is

$$a = a_{\pm}^{\nu} = \gamma_{\pm}^{\nu} m_{\pm}^{\nu} = K$$

But the logarithm of the last equality is

$$\nu \log \gamma_{\pm} + \nu \log m_{\pm} = \log K,$$

whence

$$\log m_{\pm} = -\log \gamma_{\pm} + \frac{1}{\nu}\log K.$$

Introducing the Debye-Huckel limiting law (8.122) then yields the formula

$$\log m_{\pm} = Az_+z_-\sqrt{\mu} + \frac{1}{\nu}\log K.$$

In practice, the experimental log $m_{\pm}$ is plotted against the experimental $\sqrt{\mu}$. A slightly curved line is obtained. This is extrapolated to zero $\sqrt{\mu}$ and the result identified as $(1/\nu)$ log K. From this, K itself is calculated.

8.13 Limitations of the Debye-Hückel Treatment

The Debye Hückel theory depicts the ionic atmosphere as a continuous unchanging structure. Furthermore, the structure is considered to be spherically symmetric. The question is, how well can this represent the *average* conditions about an ion? We suspect that the representation may be satisfactory in dilute solutions, those 0.1 m in ionic strength or less.

In evaluating the charge density ρ, we neglected the higher nonlinear terms in the expansion of exp $(-q_i\phi/kT)$. T. H. Gronwall and V. K. La Mer considered these terms in the integration of the Poisson equation and obtained a correction to the Debye-Hückel equation. This is significant in low dielectric constant solvents and for higher charged ions.

One may compensate for some of the error in the continuum approach by considering that there is association of ions. This becomes considerable at the higher concentrations and has the effect of raising γ_i. Authors who have considered this effect include N. Bjerrum, R. M. Fuoss and C. A. Kraus.

Solvation of the ions has several effects. First of all, it increases the effective diameter of the ion being considered. We see that in the large diameters ascribed to H^+ and Li^+ in water. Secondly, it decreases the effective concentration, the activity, of the solvent and so increases the activity of the solute ions. Thus, the mean ionic activity at 25° C in 3.0 m LiCl is reported to be 1.174, that in 3.0 m NaCl 0.714, that in 3.0 m KCl 0.571.

Questions

8.1 What does a chemical potential represent?
8.2 How do the different standard chemical potentials for a given constituent arise?
8.3 Describe the different equilibrium constants for a given reaction and show how they are related.
8.4 In a given equilibrium problem, what information is supplied by the chemical equations?
8.5 In a given equilibrium problem, what information is supplied by the equilibrium constants?
8.6 What information is supplied by the electrical neutrality condition?
8.7 In a typical equilibrium problem, what kind of equation does the equilibrium extent of reaction satisfy?
8.8 Does this equation generally have more than one root? How is the chemically significant root distinguished from the other roots?
8.9 What is the advantage in rearranging the initial equations so that they form an iteration sequence?
8.10 How may the final equation in the sequence be altered to improve the rate of convergence to the equilibrium state?
8.11 What approximations are involved in representing the potential about an ion as $\phi(r)$?
8.12 How does the Boltzmann distribution law determine the charge density at a given potential about a given ion?
8.13 How does the sum $\Sigma q_i c_{i\infty}$ appear in the discussion? Why does it vanish?
8.14 How does the charge density at potential ϕ vary with the ionic strength in the bulk of the solution?
8.15 State Coulomb's law in its discrete form.
8.16 Show how this may be recast in an integral form.
8.17 What approximations are involved in the derivation of the Yukawa type potential

$$\phi = A\frac{e^{-br}}{r} \;?$$

8.18 How is the free energy of the charged atmosphere about a given ion obtained?
8.19 How does one determine an ionic activity coefficient from the energy needed to form the pertinent ionic atmosphere?
8.20 What conditions govern how the ions in an electrolytic solution move from phase to phase?
8.21 What difficulty arises in measuring the activities of individual ions?
8.22 Discuss the limitations of the Debye-Hückel treatment of electrolytes.

Problems

In problems 8.1 to 8.6 and 8.19 to 8.25, let the activity coefficients equal 1, And when the temperature is not stated, take it to be 25° C.

8.1 At 100° C the equilibrium constant K_X for the reaction

$$CH_3COOH\ (l) + C_2H_5OH\ (l) \rightleftarrows CH_3COOC_2H_5\ (l) + H_2O\ (l)$$

is 4.00. Calculate the mole fraction of each substance present at equilibrium at 100° C if 1.08 mol acetic acid, 0.82 mol ethanol, and 0.08 mol water were present initially.

8.2 For the ionization of cyanoacetic acid

$$CNCH_2COOH \rightleftarrows CNCH_2COO^- + H^+,$$

K_{vc} is 3.360 × 10^{-3} M at 25° C. Calculate c_{H^+} in a buffer solution 0.050 M in cyanoacetic acid and 0.040 M in sodium cyanoacetate.

8.3 For the ionization of water

$$H_2O \rightleftarrows H^+ + OH^-,$$

K_{vc} , is 1.002 × 10^{-3} M at 25° C. Calculate c_{OH^-} in 0.0100 M HCl.

8.4 Calculate c_{OH^-} in 0.100 M potassium cynaoacetate.

8.5 Calculate m_{H^-} in 0.0100 m ammonium chloride.

8.6 Calculate m_{H^-} in 1.00 × 10-7 m HCl.

8.7 If the ionization constant of benzoic acid K_{vm} is 6.313 × 10^{-5} m, while the density of water is 0.9971 g ml at 25° C, what is K_{vc}, the ionization constant in moles per lifer at 25° C?

8.8 Calculate ΔH^0 for the ionization of water at 25° C from data in table 8.1.

8.9 From table 8,1, calculate ΔG^0 and ΔS^0 for the ionization of water at 25° C.

8.10 Combine the results from problems 8.8 and 8.9 with data from table 6.3 to calculate $\Delta H^0{}_f$, $\Delta G^0{}_f$, and $\Delta S^0{}_f$ for the OH^- ion.

8.11 (a) From entropies and enthalpies of formation in table 6.3 , calculate log K for the reaction

$$AgBr \rightleftarrows Ag^+ + Br^-.$$

(b) Calculate log K for this reaction from the Gibbs energies of formation.

(c) Calculate log K for this reaction from the tabulated log K_f.

8.12 For the reaction in problem 8.11, calculate (a) K_{vm},

(b) K_{vc}.

8.13 Calculate the solubility of AgBr in (a) pure water,

(b) 0.0333 m $CaBr_2$,

8.14 In thermodynamic studies of equilibria, the formulas NH_3 and NH_4OH usually describe the same constituent in water solutions. So instead of

$$NH_4^+ + OH^- \rightleftarrows NH_4OH$$

one may write

$$NH_4^+ \rightleftarrows NH_3 + H^+.$$

What is the equilibrium constant for the reaction written in this way?

8.15 Calculate m_{OH^-} in 0.200 m NH_4OH.

8.16 Calculate (a) the activity coefficients, (b) the mean ionic activity coefficient, (c) the activity of the electrolyte, and (d) the mean ionic activity in 0.0050 m $Fe_2(SO_4)_3$.

8.17 Calculate the solubility of CaF_2 in pure water.

8.18 If 100 g water containing 0.00100 mol NaCl and 0.0100 mol NaR was put on one side of a membrane permeable only to the ions Na^+ and Cl^- , and 200 g water was put on the other side, what is the equilibrium concentration of each ion in each phase?

— — —

8.19 If for the ionization of cyanoacetic acid, K_{vm} is 3.370 × 10^{-3} m, what is m_{H^+} in 0.0100 m cyanoacetic acid?

8.20 Calculate m_{H^+} in 0.100 m glycine.

8.21 Calculate m_{H^+} in 0.100 m ammonium cyanoacetate.

8.22 Calculate m_{H^+} in 2.00 × 10^{-7} m acetic acid.

8.23 Calculate m_{H^+} in 1.00 × 10^{-6} m HBO_2.

8.24 Calculate m_{H^+} in 1.00 × 10^{-7} m $NaBO_2$.

8.25 If a solution is 0.0100 m in formic acid and 0.0500 m in acetic acid, what is m_{H^+}?

8.26 Calculate the activity of the hydrogen ion in a solution 0.0100 m in HCl and 0.0150 m in NaCl.
8.27 Calculate m_{H^+} in a solution 0.0100 m in formic acid and 0.100 m in KCl.
8.28 Calculate m_{H^+} in a solution 0 100 m in KCl and 0.0100 m in glycine.
8.29 Calculate m_{H^+} in 0.0100 m ammonium formate.
8.30 If saturated aqueous NaCl is 6.12 m and has a solubility product of 38.42 m^2, what is (a) the activity of the electrolyte, (b) the mean ionic activity, and (c) the mean ionic activity coefficient in the saturated solution?
8.31 The solubility of $Cu(IO_3)_2$ in aqueous solutions of KCl was determined at 25° C with the results in table 8.B.

TABLE 8.B

KCl Concentration, m	*$Cu(IO_3)_2$ Solubility, m*
0.00000	0.003245
0.00501	0.003398
0.01002	0.003517
0.02005	0.003730
0.03511	0.003975
0.05017	0.004166
0.07529	0.004453
0.1005	0.004694

For each solution calculate μ and the $m_\pm$ of $Cu(IO_3)_2$. Then plot log $m_\pm$ against $\sqrt{\mu}$ and (a) extrapolate linearly to infinite dilution to find an approximate log K, (b) extrapolate the curve to infinite dilution to obtain a better log K.

8.32 (a) From the graph for problem 8.31, obtain log $m_\pm$ for a solution whose ionic strength is 0.100, Then calculate the corresponding mean ionic activity coefficient. (b) From tables 8.5 and 8.6, obtain the mean ionic activity coefficient for $Cu(IO_3)_2$ in a solution of ionic strength 0.100.

References

Books

Harned, H. S., and Owen, B. B.: 1958, *The Physical Chemistry of Electrolytic Solutions*, 3rd ed., Reinhold, New York, pp. 1-764.

This is a comprehensive text on properties of electrolytic solutions. The classic theories are presented in detail. Extensive experimental data is summarized in the tables.

Horvath, A. L.: 1985, *Handbook of Aqueous Electrolyte Solutlons*, John Wiley & Sons, New York, pp. 206-232.

Horvath summarizes the methods for estimating, predicting, and correlating physical properties of aqueous electrolyte solutions. Chapter 2.8 is devoted to ionic activity coefficients. Appropriate references are cited.

Meites, L.: 1981, *An Introduction to Chemical Equilibrium and Kinetics*, Pergamon, New York, pp. 8-32, 78-212.

Here we have a low level presentation of chemical thermodynamics and its application to equilibria. Solubilities of molecular and ionic compounds, acid-base reactions, and complex formation are all discussed in considerable detail.

Pitzer, K. S. (editor): 1991, *Activity Coefficients in Electrolyte Solutions*, CRC Press, Boca Raton, FL (abbreviated AC).

This book contains 8 different detailed contributions. Of most interest to us are the following:

Pitzer, K. S.: 1991, "Ion Interaction Approach: Theory and Data Correlation," in AC, pp. 75-153.

The charging process used by Debye and Huckel allows for the coulombic interactions but does not take into account the short- range interactions. Pitzer describes how these add terms to the formulas. Parameters in these terms are evaluated from experimental data.

Stokes, R. H.: 1991, "Thermodynamics of Solutions," in AC, pp 1-28.

Stokes surveys the pertinent thermodynamic theory and results. So this contribution is a fitting introduction to the volume.

Articles

Bodner, G. M.: 1986, "Assigning the pK_a's of Polyprotic Acids," *J. Chem. Educ.* **63**, 246-247.

Campbell, M. L., and Waite, B. A.: 1990, "The K_a Values of Water and the Hydronium Ion for Comparison with Other Acids," *J. Chem. Educ.* **67**, 386-388.

Cardinali, M. E., Giornini, C., and Marroau, G.: 1990, "The Hydrolysis of Salts Derived from a Weak Monoprotic Acid and a Weak Monoprotic Base," *J. Chem. Educ.* **67**, 221-223.

Cavaleiro, A. M. V. S. V.: 1996, "The Teaching of Precipitation Equilibrium," *J. Chem. Educ.* **73**, 423-425.

Chaston, S.: 1993a, "Calculating Complex Equilibrium Concentrations by a Next Guess Factor Method," *J. Chem. Educ.* **70**, 622-624.

Chaston, S.: 1993b, "Calculation of Titration Curves by the Next Guess Factor Method," *J. Chem. Educ.* **70**, 878-880.

Corao, E., Morales, D., and Araujo, 0.: 1986, "On the Existence and Uniqueness of a Chemical Root in Ionic Equilibrium Problems," *J. Chem. Educ.* **63**, 693-694.

Craig, N. C.: 1987, "The Chemists' Delta," *J. Chem. Educ.* **64**, 668-669.

Eberhart, J. G.: 1986, "Solving Equations by Successive Substitution-The Problem of Divergence and Slow Convergence," *J. Chem. Educ.* **63**, 576-578.

Gorders, A. A.: 1991a, "Chemical Equilibrium II. Deriving an Exact Equilibrium Equation," *J. Chem. Educ.* **68**, 215-217.

Gorders, A. A.: l991b, "Chemical Equilibrium III. A Few Math Tricks," *J. Chem. Educ.* **68**, 291-293.

Hawkes, S. J.: 1996, "Salts are Mostly Not Ionized," *J. Chem. Educ.* **73**, 421-423.

Hinckley, D. A., and Seybold, P. G.: 1987, "Thermodynamics of the Rhodamine B Lactone-Zwitterion Equilibrium," *J. Chem. Educ.* **64**, 362-364.

King, E. L.: 1986, "Sulfuric Acid Viewed as a 1-1 Electrolyte," *J. Chem. Educ.* **63**, 490-491.

Lainez, A., and Tardajos, G.: 1985, "Standard States of Real Solutions," *J. Chem. Educ.* **62**, 678-680.

Ludwig, 0. G.: 1992, "On the Chemically Impossible 'Other' Roots in Equilibrium Calculations, II," *J. Chem. Educ.* **69**, 884.

Michalowski, T., and Lesiak, A.: 1994, "Acid-Base Titration Curves in Disproportionating Redox Systems," *J. Chem. Educ.* **71**, 632-636.

Myers, R. J.: 1986, "The New Low Value for the Second Dissociation Constant for H_2S," *J. Chem. Educ.* **63**, 687-690.

Olivieri, A. C.: 1990, "Solution of Acid-Base Equilibria by Successive Approximations," *J. Chem. Educ.* **67**, 229-231.

Perimutter-Hayman, B.: 1984, "Equilibrium Constants of Chemical Reactions Involving Condensed Phases," *J. Chem. Educ.* **61**, 782-783.

Qureski, P. M., and Kamoonpuri, S. I. M.: 1991, "The Ionic Radii Problem," *J. Chem. Educ.* **68**, 109.

Saroff, H. A.: 1994, "Glycine: A Simple Zwitterion," *J. Chem. Educ.* **71**, 637-643.

Schmitz, G.: 1994, "The Uncertainty of pH," *J. Chem. Educ.* **71**, 117-118.

Schwartz, L. M.: 1995, "Ion-Pair Complexation in Moderately Strong Aqueous Acids," *J. Chem. Educ.* **72**, 823-826.

Starkey, R., Norman, J., and Hintze, M.: 1986, "Who Knows the K_a Values of Water and the Hydronium Ion?" *J. Chem. Educ.* **63**, 473-474.

Thompson, H. B.: 1987, "Good Numerical Techniques in Chemistry: The Quadratic Equation," *J. Chem. Educ.* **64**, 1005- 1010.

Tykodi, R. J.: 1986, "Thermodynamics and Reactions in the Dry Way," *J. Chem. Educ.* **63**, 107-111.

Weltin, E.: 1995, "Calculating Equilibrium Concentrations by Iteration: Recycle Your Approximations," *J. Chem. Educ.* **72**, 36-38.

Xijun, H., and Xiuping, Y.: 1991, "Influences of Temperature and Pressure on Chemical Equilibrium in Nonideal Systems," *J. Chem. Educ.* **68**, 295-296.

9

Electrochemistry

9.1 Definitions and Units

ELECTRIC CHARGE IS TRANSPORTED WITHIN A PHASE by the movement of charged particles. In a metal, these particles are the valence electrons. In a solution of an electrolyte, they include both cations and anions. In a plasma, the principal charged particles are electrons and cations.

In the body of a phase, electrical neutrality tends to prevail. At a boundary, however, a charged double layer may build up, negative on one side, positive on the other. A change in carrier imposed by a boundary is generally effected by chemical reaction.

Current may be introduced into a given phase through electrodes. We call the negative electrode the *cathode*, the positive electrode the *anode*. At each electrode, chemical reactions in which electrons are reactants or products occur. The electron conduction in one electrode is thus changed to ion conduction in the phase and back to electron conduction at the other electrode.

The *equivalent mass* of an electrolyte $A_{\nu_+}B_{\nu_-}$ that ionizes to give ν_+A^{z+} and ν_-B^{z-} is its formula mass divided by $\nu_+ z_+$ (or by $-\nu_- z_-$). When it is completely ionized, an equivalent of the electrolyte yields an equivalent of cations carrying one mole positive charge and an equivalent of anions carrying one mole negative charge. A mole of positive charge is called the *faraday F*.

If Q is the charge carried by n equivalents of an ion, we have

$$n = \frac{|Q|}{F}. \qquad [9.1]$$

So charge Q passing through an electrode causes n equivalents of chemical reaction there. By measuring such reaction, one finds that

$$F = 96{,}485.3 \text{ C mol}^{-1} \qquad [9.2]$$

Charge is related to force through Coulomb's law (8.51). So transporting charge between differing points generally requires work. The points are said to be at different electric potentials. The joules of work needed to take one coulomb from one point to another gives the *volts* of potential existing between the points. The unit of potential is the volt.

The coulombs Q of charge passing a given cross section of a conductor per second yield the electric current I there in *amperes*. We have

$$I = \frac{Q}{t}. \qquad [9.3]$$

When this ratio varies with the time t, it is replaced with the derivative

$$I = \frac{dQ}{dt}. \tag{9.4}$$

The ratio of the voltage $\mathcal{E}$ between two cross sections to the amperage I flowing between them is the resistance R between the cross sections in *ohms* (abbreviated Ω). We have

$$R = \frac{\mathcal{E}}{I}. \tag{9.5}$$

When this ratio varies with the current, it is replaced with the derivative

$$R = \frac{d\mathcal{E}}{dI}. \tag{9.6}$$

The reciprocal of resistance is known as the conductance L:

$$L = \frac{1}{R}. \tag{9.7}$$

The unit for conductance is the *siemens* (S), which is one reciprocal ohm (Ω^{-1}) or mho.

To measure conductance, a person may attach or insert two equivalent electrodes onto or into the phase to be studied, apply a potential, and determine the resulting current.

9.2 Electrode Reactions

A change in carrier of current generally occurs at the interface between an electrode and a solution. This is effected by a chemical reaction in which the electron is a product or a reactant.

In a conductance experiment, electrons are drawn away from the anode by an external source of potential. The necessary electrons may be produced in several different ways. These include the following possibilities:

1. An anion discharges. As an example

$$2I^- \rightarrow I_2 + 2e. \tag{9.8}$$

2. A cation is produced; thus

$$Ag \rightarrow Ag^+ + e, \tag{9.9}$$

$$Cu \rightarrow Cu^{++} + 2e. \tag{9.10}$$

3. A cation is oxidized; for example

$$Fe^{++} \rightarrow Fe^{3+} + e, \tag{9.11}$$

$$Sn^{++} \rightarrow Sn^{4+} + 2e. \tag{9.12}$$

4. The hydroxide ion from water is discharged when no other oxidation readily occurs:

$$4OH^- \rightarrow 2H_2O + O_2 + 4e. \tag{9.13}$$

In a conductance experiment, electrons are supplied to the cathode by an external source of potential. These may react in the following ways:

1. A cation discharges; for example

$$Ag^+ + e \rightarrow Ag, \qquad [9.14]$$

$$Cu^{++} + 2e \rightarrow Cu. \qquad [9.15]$$

2. A cation is reduced to another cation; thus

$$Fe^{3+} + e \rightarrow Fe^{++}, \qquad [9.16]$$

$$Sn^{4+} + 2e \rightarrow Sn^{++}. \qquad [9.17]$$

3. An anion is produced. Thus if one bubbles chlorine over a platinum electrode, the following reaction occurs:

$$Cl_2 + 2e \rightarrow 2Cl^-. \qquad [9.18]$$

4. The hydrogen ion from water is discharged. This reaction occurs when no other reduction can readily occur:

$$2H^+ + 2e \rightarrow H_2. \qquad [9.19]$$

Example 9.1

If 0.500 A current was passed for 55 min through a $CuSO_4$ solution using copper electrodes, how much copper plated out on the cathode and how much left the anode?

From equation (9.3), the number of coulombs passing any cross section in the circuit is

$$Q = It = (0.500\ \text{A})(55\ \text{min})(60\ \text{s min}^{-1}) = 1650\ \text{C}.$$

But following formula (9.1), the number of equivalents reduced at the cathode and oxidized at the anode is

$$n = \frac{1650\ \text{C}}{96,485.3\ \text{equiv}^{-1}} = 0.01710\ \text{equiv}.$$

Since the reaction at the cathode is

$$Cu^{++} + 2e \rightarrow Cu,$$

the equivalent mass of copper is its atomic mass divided by 2. So the weight of copper plating out is

$$w = (0.01710\ \text{equiv})(31.773\ \text{g equiv}^{-1}) = 0.543\ \text{g}.$$

At the anode, the reaction is

$$Cu \rightarrow Cu^{++} + 2e.$$

and the weight of copper leaving is

$$w = 0.543\ \text{g}.$$

9.3 Electrolyte Transference

When electric current is driven through a solution by an external source of potential, cations move toward the cathode and anions move toward the anode. Furthermore, reactions occur at the surface of each electrode, depleting some ions and augmenting others. As a result the electrolyte becomes more dilute near one electrode and more concentrated near the other. Associated with the flow of charge is a transport of the electrolyte.

A conductance cell may be designed so that convection and diffusion effects between a compartment around the anode, a middle compartment, and a compartment around the cathode are small. But electrolysis will cause concentration changes to occur. The divisions between compartments may be chosen so that there is no net change in the middle compartment, as figure 9.1 indicates.

Consider that the current carried by the cation in the body of the solution is I_+ while the current carried by the anion there is I_-. The *transference number* of an ion is then defined as the fraction of the total current I carried by the ion.

Thus the transference number of the cation in figure 9.1 is

$$t_+ = \frac{I_+}{I} = \frac{I_+ t / F}{It / F}, \qquad [9.20]$$

while the transference number of the anion is

$$t_- = \frac{I_-}{I} = \frac{I_- t / F}{It / F}. \qquad [9.21]$$

Here t is the time current I is allowed to flow while F is the faraday.

But multiplying current by time and dividing by the faraday yields the corresponding number of equivalents. We have

$$\begin{aligned} It / F = \Delta n_{\text{electrolysis}} &= \text{equivalents reduced at the cathode} \\ &= \text{equivalents oxidized at the anode}, \end{aligned} \qquad [9.22]$$

$$\begin{aligned} I_+ t / F = \Delta n_+ &= \text{equivalents cation migrating into cathode compartment} \\ &= \text{equivalents cation migrating out of anode compartment}, \end{aligned} \qquad [9.23]$$

$$\begin{aligned} I_- t / F = \Delta n_- &= \text{equivalents anion migrating into anode compartment} \\ &= \text{equivalents anion migrating out of cathode compartment}. \end{aligned} \qquad [9.24]$$

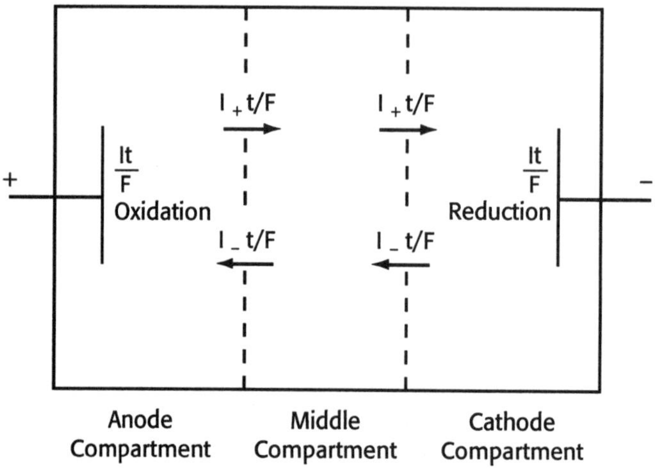

FIGURE 9.1 Equivalents of cation and anion transferring in time t due to a steady current I and the corresponding equivalents of reaction at the electrodes.

In practice, the compartments may be separated by porous membranes. Or better, the electrolysis tube may be shaped to keep the contents of the compartments separate.

For a run, electrolyte of known concentration is placed in the tube. Current is passed through for an interval of time and the number of faradays determined with a coulometer in series with the cell. About each electrode, all solution that differs from that in the middle compartment is removed and analyzed. The amount of electrolyte originally present in the water is calculated. The change in equivalents is then

$$\Delta n = n_{\text{final}} - n_{\text{initial}} \qquad [9.25]$$

for each ion.

For one ion and one compartment, the equivalents released at the electrode are $\Delta n_{\text{electrolysis}}$. The equivalents migrating are then

$$\Delta n_{\text{migration}} = \Delta n - \Delta n_{\text{electrolysis.}} \qquad [9.26]$$

The absolute value of $\Delta n_{\text{migration}}$ divided by $\Delta n_{\text{electrolysis}}$ equals the transference number of the ion.

This method was developed by J. W. Hittorf. For accurate work, one should correct for the water transported with the ions. Alternatively, one may employ the moving boundary method discussed in section 9.4. Or, the voltage of an electrolytic cell with transference can be compared to the same cell without transference.

Some empirical transference numbers are listed in table 9.1.

Example 9.2

A 0.2000 N copper sulfate solution was electrolyzed using copper electrodes. In the anode compartment, the solution contained 0.7532 g copper initially and 0.9972 g after the electrolysis, during which 0.4000 g copper plated out on the cathode. Calculate the transference numbers t_+ and t_-.

TABLE 9.1 Transference Numbers of Cations in Some Aqueous Solutions at 25° C

Concentration, N	*HCl*	*KCl*	*Solution* *NaCl*	*LiCl*	*NH_4Cl*
0.01	0.8251	0.4902	0.3918	0.3289	0.4907
0.02	0.8266	0.4901	0.3902	0.3261	0.4906
0.05	0.8292	0.4899	0.3876	0.3211	0.4905
0.10	0.8314	0.4898	0.3854	0.3168	0.4907
0.2	0.8337	0.4894	0.3821	0.3112	0.4911
Concentration, N	*KBr*	*KI*	*Solution* *$AgNO_3$*	*$NaC_2H_3O_2$*	*$CaCl_2$*
0.01	0.4833	0.4884	0.4648	0.5537	0.4264
0.02	0.4832	0.4883	0.4652	0.5550	0.4220
0.05	0.4831	0.4882	0.4664	0.5573	0.4140
0.10	0.4833	0.4883	0.4682	0.5594	0.4060
0.2	0.4841	0.4887	—	0.5610	0.3953

For the anion in the anode compartment, we have

$$\Delta n_{\text{electrolysis}} = 0$$

since anions are not oxidized at the electrode. But from the electrical neutrality condition, the change in equivalents of the cation equals the change in equivalents of the anion. For either ion,

$$\Delta n = \frac{0.9972}{31.773} - \frac{0.7532}{31.773} = \frac{0.2440}{31.773} \text{ equiv.}$$

Combining these results as in equation (9.26) gives us

$$\Delta n_{\text{migration}} = \frac{0.2440}{31.773} \text{ equiv}$$

for the anion. Dividing the equivalents migrating by the equivalents reduced at the cathode, as in (9.21), leads to

$$t_- = \frac{0.2440 / 31.773}{0.4000 / 31.773} = 0.6100.$$

Because the transference numbers of the ions add up to 1, we have

$$t_+ = 1 - t_- = 0.3900.$$

On the other hand, for the cation in the anode compartment, we have

$$\Delta n_{\text{electrolysis}} = \frac{0.4000}{31.773} \text{ equiv.}$$

As before, the change in equivalents in the compartment is

$$\Delta n = \frac{0.2440}{31.773} \text{ equiv.}$$

So for the equivalents of cation migrating into the anode compartment, equation (9.26) yields

$$\Delta n_{\text{migration}} = \frac{0.2440}{31.773} - \frac{0.4000}{31.773} = -\frac{0.1560}{31.773} \text{ equiv.}$$

To get the amount migrating out, one changes the sign. Dividing the result by the equivalents oxidized at the electrode then gives us

$$t_+ = \frac{0.1560 / 31.773}{0.4000 / 31.773} = 0.3900.$$

whence

$$t_- = 1 - t_+ = 0.6100.$$

9.4 Moving Ion Clouds and Boundaries

How a configuration of ions moves can be determined from the movement of a suitable boundary. This provides a second method for measuring transference numbers.

Through a region where an electrolyte is uniform, an imposed electric intensity $\boldsymbol{E}$ causes the cations to drift in the same direction at the average velocity $\boldsymbol{u}_+$ and the anions to drift in the opposite direction at the average velocity $\boldsymbol{u}_-$. In unit time, the cations travel distance $\boldsymbol{u}_+$ on average, while the anions travel distance $\boldsymbol{u}_-$ in the opposite direction on average. Passing through a given cross section would be the cations initially within distance u_+ and the anions initially within distance u_-, if the electrolyte were uniform throughout.

So the equivalents of cation passing a square centimeter of cross section in unit time equal $u_+c/1000$, where c is the concentration of the cation in equivalents per 1000 cm^3. If the cross sectional area of the electrolytic cell is A, then the current carried by the cation through that area is

$$I_+ = \frac{u_+c}{1000}FA. \quad [9.27]$$

Similarly, the current carried by the anion through the cross section is

$$I_- = \frac{u_-c}{1000}FA. \quad [9.28]$$

The corresponding transference numbers are

$$t_+ = \frac{I_+t}{It} = \frac{u_+tcFA/1000}{It} = \frac{x_+cFA}{1000Q} \quad [9.29]$$

and

$$t_- = \frac{x_-cFA}{1000Q}. \quad [9.30]$$

Here t is the time the steady current I flows, Q the total charge transported, x_+ the distance the average cation moves in time t, x_- the distance the average anion moves in time t.

One may determine either x_+ or x_- from movement of the boundary between the given electrolyte AB and an indicator electrolyte A'B or AB'. In practice, a vertical electrolysis tube kept at a given temperature is employed. A mechanical device may be used to put the less dense solution on top of the more dense solution. Or, the second electrolyte may be generated by electrolysis, as figure 9.2 shows.

The boundary can be seen because the two electrolytes generally differ in their refractive indices. Furthermore, a slight cloudiness might be found at the boundary, probably caused by hydrolysis. In some cases, the electrolytes differ in color. In any setup, however, the less dense solution must be on top to prevent convection.

In general, the boundary moves at the speed of the leading odd ion. For the setup in figure 9.2, the boundary travels upward at the speed of the cations in the upper solution. Thus, the distance the boundary moves up for passage of charge Q is x_+ for cation A^+. Equation (9.29) is then employed to calculate the empirical transference number for the cation in this solution.

Reversing the signs on the electrodes reverses the movement of the boundary. Cation A'^+ is now the leading odd ion; the boundary moves downward at its speed. Equation (9.29) is replaced with

$$t'_+ = \frac{x'_+c'FA}{1000Q}. \quad [9.31]$$

Here t'_+ is the transference number of cation A'^+ in its solution, x'_+ is the distance the boundary moves down when Q coulombs pass through the tube, c' is the concentration of A'B.

A person can improve the accuracy of the results by adjusting the concentration of the trailing odd ion so that it follows at the same speed as the leading ion. But an equation (9.29) rearranges to give

$$\frac{t_+}{c} = (\text{const})u_+ \quad [9.32]$$

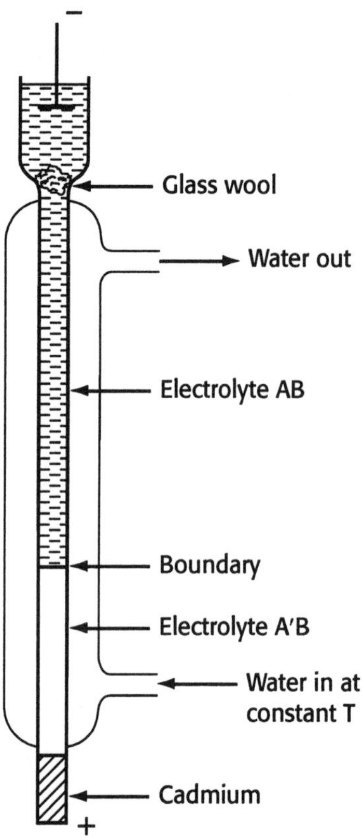

FIGURE 9.2 Simple moving boundary apparatus.

for electrolyte AB. Similarly, we have

$$\frac{t'_+}{c'} = (\text{const})u'_+ \qquad [9.33]$$

where u'_+ is the speed of A'^+ in A'B. Setting u'_+ equal to u_+ yields the condition

$$\frac{t_+}{c} = \frac{t'_+}{c'}. \qquad [9.34]$$

Here t_+ and c are the transference number and concentration of A^+, while t'_+ and c' are the transference number and concentration of A'^+.

The common ion B^- drifts in the direction opposite to that in which the boundary moves. On passing through this layer, each B^- speeds up or slows down so that its concentration is changed from c to c'.

9.5 *Ohm's Law*

A conducting medium is said to obey *Ohm's law* when the R defined by equation (9.5) is constant.

One may study the electrical conductivity of a medium by inserting two equivalent electrodes oriented parallel to each other a fixed distance apart and then applying a voltage $\mathcal{E}$. If the number of charge carriers per unit volume is constant, independent of

$\mathcal{E}$, and if the average velocity at which they move is proportional to $\mathcal{E}$, and to the pertinent potential gradient, then the total current I is also proportional to $\mathcal{E}$. We then have

$$I = L\mathcal{E} = \frac{\mathcal{E}}{R} \qquad [9.35]$$

with conductance L and resistance R constant. When I is in amperes and ε in volts, L is in mhos and R in ohms.

Consider a *uniform* conductor of length l and cross sectional area A. Since voltages in series add, resistances in series add and the resistance of the conductor is proportional to its length. Since charges add, parallel currents add. So conductances in parallel add and the conductance of the conductor is proportional to its cross sectional area. Se have

$$R = \rho \frac{l}{A}. \qquad [9.36]$$

The constant of proportionality ρ is called the *specific resistance*. The reciprocal of the specific resistance is the *specific conductance*

$$\kappa = \frac{1}{\rho} = \frac{l/A}{R}. \qquad [9.37]$$

To measure the conductance of an electrolytic solution, one fills a conductivity cell with the solution and inserts it in a Wheatstone bridge circuit. See figures 9.3 and 9.4. Direct current is not used because transference in the solution and reactions at the electrodes would then polarize the cell very rapidly, changing its apparent resistance. An

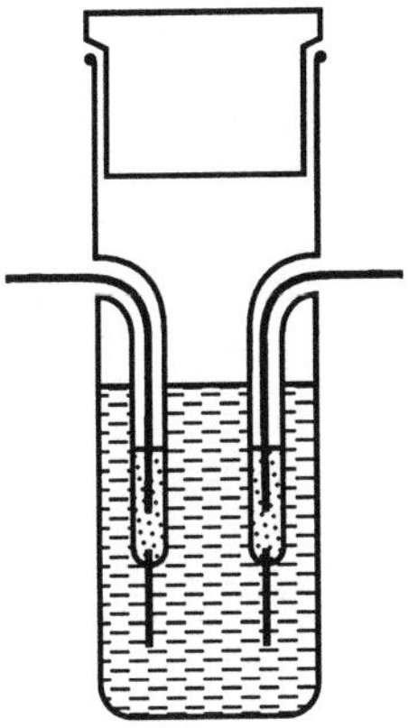

FIGURE 9.3 Conductivity cell.

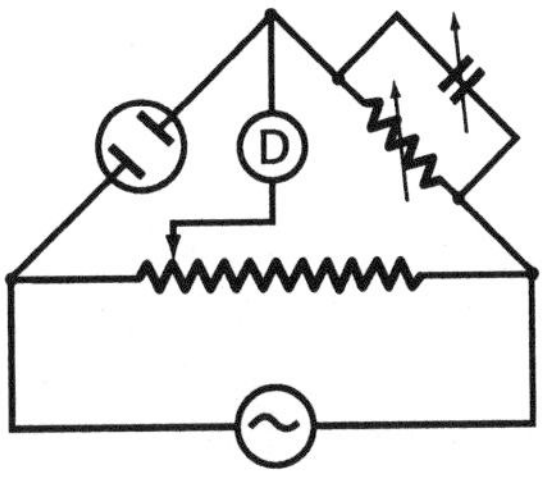

FIGURE 9.4 Wheatstone bridge circuit for measuring the resistance of electrolytes.

alternating current of 60 to 1000 hertz is suitable. But even then, some polarization would take place if the electrodes did not present sufficient active surface. Platinized platinum electrodes are suitable.

A person could calculate the specific conductance from the resistance of the cell and its linear dimensions, using equation (9.37). However, it is difficult to determine these dimensions accurately. Furthermore, some correction to the equation is generally needed because of edge effects.

For routine determinations, the equation is employed in the form

$$\kappa = \frac{k}{R} \tag{9.38}$$

where k is the *cell constant*, to be determined by measuring the resistance of the cell when it is filled with a solution of known specific conductance.

As a standard, one may use a solution of KCl, which is easily prepared. Careful absolute measurements show that the specific conductance of 0.02000 M KCl at 25° C is

$$\kappa = 0.002768\ \Omega^{-1}\ \text{cm}^{-1}. \tag{9.39}$$

In studying poorly conducting solutions, one must remove impurity ions, ammonia, and carbon dioxide from the distilled water and from the glassware that is used. Ordinary distilled water can be improved if it is redistilled in clean apparatus after a little $KMnO_4$ has been added. When in equilibrium with air at 25° C, such improved water may have a specific conductance as low as $7 \times 10^{-7}\ \Omega^{-1}\ \text{cm}^{-1}$. Pure water, on the other hand, has a κ of $6.2 \times 10^{-8}\ \Omega^{-1}\ \text{cm}^{-1}$ at 25° C.

9.6 Equivalent Conductance

The specific conductance κ of a strong electrolyte varies with concentration largely because a unit volume of a concentrated solution contains more ions than a unit volume of a dilute solution. But multiplying κ by the volume containing one equivalent yields a quantity that is free from this concentration effect. This new quantity can be broken down into the separate contributions of each ion.

By definition, the *equivalent conductance* Λ of an electrolyte equals its specific conductance times the volume containing one equivalent; thus

$$\Lambda = \frac{1000\kappa}{c} \tag{9.40}$$

where c equals the equivalents electrolyte in 1000 cm^3. To interpret Λ, consider the following ideal experiment.

A rectangular cell with electrodes covering two opposite vertical walls 1 cm apart is constructed. Into this is placed 1 equiv of the chosen electrolyte. Then water is added and the temperature is kept constant. After all is dissolved, the number of ions conducting the current does not vary unless the electrolyte is weak. Any change in the conductance of the cell is due to a changing average interionic interaction, as long as the electrolyte is completely ionized.

In the ideal cell, the cross sectional area conducting the current equals $1000/c$ while the distance between the electrodes is 1 cm. Consequently, the conductance of the cell equals the equivalent conductance of the electrolyte. Any change in it is caused by changing average interionic interaction and a varying degree of dissociation. In going from κ to Λ, we have gotten rid of the concentration effect and have obtained a quantity that is proportional to the conductivity of the ions from one molecule of electrolyte, on average.

Since charge is additive, the current carried by an electrolyte in solution equals the current carried by the cation plus the current carried by the anion:

$$I = I_+ + I_-. \quad [9.41]$$

Consequently, the conductance of an electrolyte equals the conductance of the cation plus the conductance of the anion. For equivalent conductance Λ, we have

$$\Lambda = \lambda_+ + \lambda_-, \quad [9.42]$$

where λ_+ is the equivalent conductance of the cation while λ_- is the equivalent conductance of the anion. An ion conductance depends on the nature of the ion and the interaction to which it is subjected. Thus, it varies with the concentration and the nature of the other ions present.

At extreme dilution, the ions are separated by such great distances that the average interionic interaction is negligible. The contribution of an ion to the equivalent conductance is then characteristic of the ion itself. We replace (9.42) with

$$\Lambda_0 = \lambda_{0,+} + \lambda_{0,-}, \quad [9.43]$$

where subscript 0 indicates that the concentrations of all ions are infinitesimal.

By Ohm's law, the current passing through a conductor is proportional to the conductance; here

$$\frac{I_+}{I} = \frac{\lambda_+}{\Lambda} \quad \text{and} \quad \frac{I_-}{I} = \frac{\lambda_-}{\Lambda}. \quad [9.44]$$

Solving for the ion conductances and replacing I_+/I with transference number t_+, I_-/I with transference number t_-, leads to

$$\lambda_+ = t_+\Lambda \quad \text{and} \quad \lambda_- = t_-\Lambda. \quad [9.45]$$

Through experiment, it is found that the equivalent conductance of a strong electrolyte is approximately a linear function of the square root of its concentration c:

$$\Lambda = \Lambda_0 - k\sqrt{c}. \quad [9.46]$$

So Λ_0 is found by extrapolating linearly a plot of Λ against $\sqrt{c}$ to zero concentration. With extrapolated transference numbers, one can then obtain ion conductances at infinite dilution. See tables 9.2 and 9.3 for representative results.

Example 9.3

When a conductivity cell contained 0.02000 N KCl, its resistance at 25° C was 312 Ω. When filled with 0.01000 N $NiSO_4$ at the same temperature, its resistance was 1043 Ω. Calculate the equivalent conductance of the nickel sulfate solution.

Solve equation (9.38) for the cell constant k and insert the resistance and specific conductance of the reference solution:

$$k = \kappa R = \left(0.002768\ \Omega^{-1}\ \text{cm}^{-1}\right)\left(312\ \Omega\right) = 0.8636\ \text{cm}^{-1}$$

Use this value together with the resistance of the nickel sulfate solution in formula (9.38):

$$\kappa = \frac{k}{R} = \frac{0.8636\ \text{cm}^{-1}}{1043\ \Omega} = 0.000828\ \Omega^{-1}\ \text{cm}^{-1}$$

TABLE 9.2 Equivalent Conductances of Aqueous Solutions at 25° C

c, N	*HCl,* $\Omega^{-1}cm^2$	*NaCl,* $\Omega^{-1}cm^2$	$AgNO_3$, $\Omega^{-1}cm^2$	$\frac{1}{2}BaCl_2$, $\Omega^{-1}cm^2$
0.000	426.2	126.4	133.4	140.0
0.001	421.4	123.7	130.5	134.3
0.005	415.8	120.6	127.2	128.0
0.01	412.0	118.5	124.8	123.9
0.02	407.2	115.8	121.4	119.1
0.05	399.1	111.1	115.2	111.5
0.1	391.3	106.7	109.1	105.2

TABLE 9.3 Ion Conductances at Infinite Dilution in Water at 25° C

Ion	λ_0, $\Omega^{-1}cm^2$	*Ion*	λ_0, $\Omega^{-1}cm^2$
H^+	349.8	OH^-	198
K^+	73.5	$\frac{1}{2}SO_4^=$	79.8
NH_4^+	73.4	Br^-	78.4
$\frac{1}{2}Ba^{++}$	63.6	I^-	76.8
Ag^+	61.9	Cl^-	76.3
$\frac{1}{2}Mg^{++}$	53.1	NO_3^-	71.4
Na^+	50.1	ClO_4^-	68.0
Li^+	38.7	HCO_3^-	44.5
		$C_2H_3O_2^-$	40.9

Then substitute this specific conductance and the given concentration into formula (9.40):

$$\Lambda = \frac{\kappa}{c/1000} = \frac{0.000828\ \Omega^{-1}\ \text{cm}^{-1}}{0.00001000\ \text{equiv cm}^{-3}} = 82.8\ \Omega^{-1}\ \text{cm}^2\ \text{equiv}^{-1}$$

In these calculations, we have neglected the small conductance of the water.

Example 9.4

From table 9.3, obtain Λ_0 for acetic acid.

We find that

$$\Lambda_{0,\ \text{HAc}} = \lambda_{0,\ \text{H}^+} + \lambda_{0,\ \text{Ac}^-} = 349.8 + 40.9 = 390.7\ \Omega^{-1}\ \text{cm}^2.$$

Since acetic acid is a weak electrolyte, its degree of dissociation changes rapidly with dilution in very dilute solutions. So its Λ_0 cannot be obtained on extrapolating a plot of Λ against *c*.

9.7 Determining Solubilities and Degrees of Ionization

Solving equation (9.40) for the equivalent concentration of an ion in an electrolytic solution yields

$$c_i = \frac{1000\kappa}{\Lambda_i}. \quad [9.47]$$

So if one knew the equivalent conductance of the ions together, Λ_i, one could obtain the equivalent concentration of the cation and the anion making up the solution from a measured specific conductance κ. But since Λ_i varies only slightly with c_i, an approximate value for c_i serves in determining it.

For a slightly soluble, or a weak, electrolyte, one may determine Λ_i from the conductances of related strong electrolytes. Suppose the electrolyte to be studied is AD. Also suppose that electrolytes AB, CD, and CB exist which are completely ionized and soluble enough so that equivalent conductances can be measured at the necessary concentrations. We then have

$$\Lambda_i = \lambda_{A^+} + \lambda_{D^-} = \left(\lambda_{A^+} + \lambda_{B^-}\right) + \left(\lambda_{C^+} + \lambda_{D^-}\right) - \left(\lambda_{C^+} + \lambda_{B^-}\right) = \Lambda_{AB} + \Lambda_{CD} - \Lambda_{CB} \quad [9.48]$$

In determining the solubility of a strong electrolyte AD, a person prepares a saturated solution in sufficiently pure water and measures its specific conductance κ. The equivalent conductance Λ_i is found using formula (9.48). Then the concentration $c = c_i$ is calculated with equation (9.47).

In determining the ionization of a weak electrolyte AD, a person prepares a solution of AD of known concentration c_{total} in sufficiently pure water. The specific conductance κ is then measured and substituted into equation (9.40). The conventional equivalent conductance Λ is calculated. This may be placed in the form

$$c_{total} = \frac{1000\kappa}{\Lambda}. \quad [9.49]$$

Then the equivalent conductance at the estimated ionic concentration is found employing formula (9.48). This is used in formula (9.47) in the form

$$c_{A^+} = c_{D^-} = \frac{1000\kappa}{\Lambda_i}. \quad [9.50]$$

For the reaction

$$AD \leftrightarrows A^+ + D^-, \quad [9.51]$$

we now have

$$\alpha = \frac{c_{A^+}}{c_{total}} = \frac{1000\kappa / \Lambda_i}{1000\kappa / \Lambda} = \frac{\Lambda}{\Lambda_i}. \quad [9.52]$$

As a first approximation to Λ_i one may employ Λ_0, the equivalent conductance at infinite dilution, as Arrhenius did in his seminal work.

Example 9.5

If a solution of $BaSO_4$ contains 1.00 mg l^{-1}, what is its specific conductance at 25° C?

From table 9.3, we obtain

$$\Lambda_{0,\,\mathrm{BaSO_4}} = \lambda_{0,\,\mathrm{Ba^{++}}} + \lambda_{0,\,\mathrm{SO_4^{=}}} = 63.6 + 79.8 = 143.4\ \Omega^{-1}\ \mathrm{cm^2\ equiv^{-1}}.$$

But the normality of the solution is

$$c = \frac{1.00 \times 10^{-3}\ \mathrm{g\ l^{-1}}}{116.70\ \mathrm{g\ equiv^{-1}}} = 8.57 \times 10^{-6}\ \mathrm{N}.$$

So formula (9.40) tells us that

$$\kappa = \frac{c}{1000} \cong \frac{8.57 \times 10^{-6} \times 143.4}{1000} = 1.23 \times 10^{-6}\,\Omega^{-1}\ \mathrm{cm^{-1}}.$$

To this number should be added the specific conductance of the water used to make up the solution.

9.8 Variation of Ion Conductance with Viscosity

On the average, an ion moving through a solvent under the influence of an electric field is like a solid sphere being forced to travel at its limiting velocity through a viscous fluid. As a result, its conductance tends to vary inversely with the viscosity of the solvent.

The steady continuous flow of a viscous fluid around a spherical solid can be treated mathematically. The resulting force **F** acting on the solid is

$$\mathbf{F} = 6\pi\eta r\boldsymbol{u}, \qquad [9.53]$$

where η is the viscosity coefficient of the fluid, r the radius of the solid, and $\boldsymbol{u}$ the velocity at which the fluid approaches the solid, or equivalently, the velocity of the solid with respect to the body of the fluid.

Equation (9.53) is known as *Stokes' law*. Measurements support its validity for macroscopic spheres moving in actual fluids.

By the analogy already cited, let us consider that formula (9.53) governs the drift velocity u of an ion subject to a potential gradient $d\mathcal{E}/dl$. The electric force acting on the ion is then

$$\mathbf{F} = -ze\frac{d\mathcal{E}}{dl}. \qquad [9.54]$$

But from (9.27) and (9.28), the current carried by an ion is proportional to its velocity $u_\pm$:

$$I_\pm = \frac{FAc}{1000}u_\pm. \qquad [9.55]$$

Eliminating $u_\pm$ and force **F** from equations (9.55), (9.53), (9.54) and rearranging yields

$$d\mathcal{E} = I_\pm\left(-\frac{6\pi r}{zeF}\right)\frac{1000\eta}{c}\frac{dl}{A} = I_\pm k_\pm \frac{1000\eta}{c}\frac{dl}{A}. \qquad [9.56]$$

Integrating and introducing $R_\pm$ as the constant of proportionality leads to

$$\mathcal{E} = I_\pm k_\pm \frac{1000\eta}{c}\frac{l}{A} = I_\pm R_\pm. \qquad [9.57]$$

But for resistance $R_{\pm}$, equations (9.37) and (9.40) yield the form

$$R_{\pm} = \frac{l / A}{\kappa} = \frac{1000 l}{A c \lambda_{\pm}}, \qquad [9.58]$$

which transforms the second equality in (9.57) to

$$I_{\pm} k_{\pm} \frac{1000\eta}{c} \frac{l}{A} = I_{\pm} \frac{1000 l}{A c \lambda_{\pm}}, \qquad [9.59]$$

whence

$$k_{\pm}\eta = \frac{1}{\lambda_{\pm}} \qquad [9.60]$$

or

$$\lambda_{\pm}\eta = \frac{1}{k_{\pm}} = \text{const.} \qquad [9.61]$$

Formula (9.60) with $k_{\pm}$ constant is called *Walden's rule*. Representative experimental data testing it appear in tables 9.4 and 9.5. We see that for a given ion $\lambda_{\pm}\eta$ varies considerably.

TABLE 9.4 Empirical Conductance-Viscosity Products at Infinite Dilution in Water

	$\lambda\eta$, Ω^{-1} cm^2 *poise*		
Ion	*0°C*	*25°C*	*100°C*
H^+	3.99	3.14	1.81
Na^+	0.466	0.459	0.043
$(C_2H_5)_4N^+$	0.287	0.295	0.293
Cl^-	0.741	0.682	0.59
$\frac{1}{2}SO_4^=$	0.73	0.724	
Picrate$^-$	0.269	0.274	0.27

TABLE 9.5 Empirical Conductance-Viscosity Products at Infinite Dilution in Various Solvents at 25°C, Except for NH_3, Where t = -33°C

	$\lambda\eta$, Ω^{-1} cm^2 *poise*					
Ion	H_2O	CH_3OH	C_2H_5OH	CH_3COCH_3	CH_3NO_2	NH_3 (at -33°C)
H^+	3.14	0.774	0.641	0.277	0.395	0.359
Na^+	0.460	0.250	0.204	0.253	0.364	0.333
K^+	0.670	0.293	0.235	0.259	0.383	0.430
Ag^+	0.563	0.274	0.195	—	0.326	0.297
$(C_2H_5)_4N^+$	0.295	0.338	0.310	0.284	0.310	—
I^-	0.685	0.334	0.290	0.366	0.403	0.437
ClO_4^-	0.606	0.387	0.340	0.366	—	—
Picrate$^-$	0.276	0.255	0.292	0.275	0.276	—

However, the constancy of $k_\pm$ depends on the constancy or r, the effective radius of the ion. We expect r to vary from solvent to solvent as the size of a solvent molecule and the number in the solvation layer vary, Furthermore, each solvated ion is only roughly spherical and the solvent is not a continuum as Stokes' law presumes.

Nevertheless, the behavior of the hydrogen ion is strange. From table 8.5, the effective radius of H^+ from activity measurements is much larger than that of any other univalent ion. So its $k_\pm$ is larger and its $\lambda_\pm\eta$ should be smaller than that of any other ion listed, contrary to what is found.

To explain the very high conductance of the hydrogen ion, one considers that it skips from solvent molecule to solvent molecule by a *Grotthuss mechanism*. In water, the charge flows from one H_3O to a neighboring H_2O by transfer of a proton:

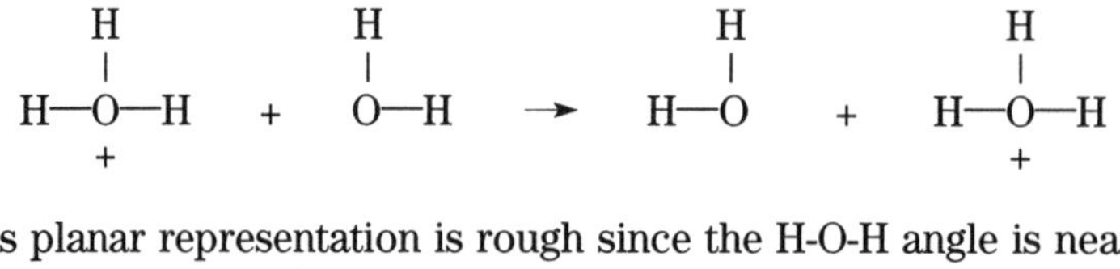

$$
\begin{array}{ccccccc}
\text{H} & & \text{H} & & \text{H} & & \text{H} \\
| & & | & & | & & | \\
\text{H--O--H} & + & \text{O--H} & \rightarrow & \text{H--O} & + & \text{H--O--H} \\
+ & & & & & & +
\end{array}
\qquad [9.62]
$$

This planar representation is rough since the H-O-H angle is near the tetrahedral value 109° 28'. Next, the product H_2O rotates to receive another proton. In the over-all process, a proton jumps about 0.86 Å while the positive charge moves about 3.1 Å.

In an alcohol, the transfer takes place as follows

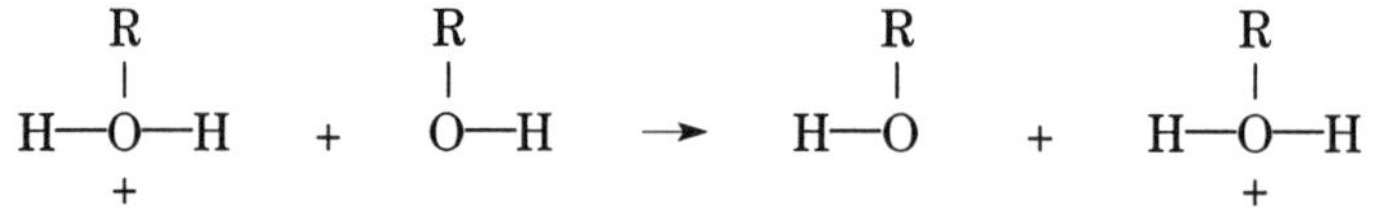

$$
\begin{array}{ccccccc}
\text{R} & & \text{R} & & \text{R} & & \text{R} \\
| & & | & & | & & | \\
\text{H--O--H} & + & \text{O--H} & \rightarrow & \text{H--O} & + & \text{H--O--H} \\
+ & & & & & & +
\end{array}
\qquad [9.63]
$$

This step is followed by rotation of the product alcohol molecule.

The high conductance of the hydroxyl ion in water is similarly explained. Thus, we have the reaction

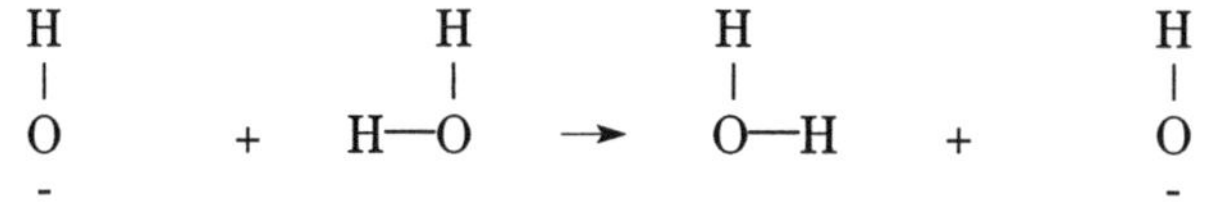

$$
\begin{array}{ccccccc}
\text{H} & & \text{H} & & \text{H} & & \text{H} \\
| & & | & & | & & | \\
\text{O} & + & \text{H--O} & \rightarrow & \text{O--H} & + & \text{O} \\
- & & & & & & -
\end{array}
\qquad [9.64]
$$

followed by rotation of the product water molecule.

In general, Grotthuss conduction is significant only if the reaction in which the charged particle is transferred is faster than the movement of the entire ion from cage to cage through the liquid.

9.9 *Cell Conventions*

An electrolytic cell is characterized by its resistance, governed by Ohm's law applied to each section, and by its voltage, governed by the phases present and how they react. When a cell is discharging, a spontaneous reaction producing electrons at the negative electrode and consuming electrons at the positive electrode proceeds.

A cell is defined by listing each phase in the order it appears in the circuit. For each aqueous phase, only the solute and its concentration may be stated. For a gas, the pertinent partial pressure needs to be given. Contact between successive phases is indicated by a vertical line I. But a hyphen - is used between phases forming a single electrode. A double vertical line II is employed when the junction potential between two phases is practically eliminated by the insertion of a suitable phase.

By convention, reactions are written with oxidation occurring at the left electrode, reduction at the right electrode. If these processes go spontaneously as written, the left electrode is negative, the right electrode positive, and a positive sign is assigned to the voltage of the cell $\mathcal{E}$. If they go spontaneously in the reverse direction, the left electrode is positive, the right electrode negative, and a negative sign is given $\mathcal{E}$.

Thus, the cell in figure 9.5 is defined by the scheme

$$\mathrm{Cu}|\mathrm{Zn}|\mathrm{ZnSO_4}\ (m_1)|\mathrm{KCl}\ (m_2)|\mathrm{CuSO_4}\ (m_3)|\mathrm{Cu}. \qquad [9.65]$$

The reaction supplying electrons to the left electrode and making it negative is

$$\mathrm{Zn} \rightarrow \mathrm{Zn}^{++} + 2e, \qquad [9.66]$$

while the reaction removing electrons from the right electrode and making it positive is

$$\mathrm{Cu}^{++} + 2e \rightarrow \mathrm{Cu} \qquad [9.67]$$

If discharging the cell makes these reactions go as written, voltage $\mathcal{E}$ is positive.

The tube containing KCl serves to reduce the junction potential between the two outer solutions to a negligible value. In such a *salt bridge*, concentrated KCl or concentrated NH_4NO_3 may serve.

The description of the cell may be shortened to the form

$$\mathrm{Zn}|\mathrm{ZNSO_4}\ (m_1)\|\mathrm{CuSO_4}\ (m_3)|\mathrm{Cu} \qquad [9.68]$$

or

$$\mathrm{Zn}|\mathrm{Zn}^{++}\ (m_1)\|\mathrm{Cu}^{++}\ (m_3)|\mathrm{Cu}. \qquad [9.69]$$

These shortened descriptions are misleading, however, since a considerable part of the observed potential may arise from the copper-zinc contact. Recall how a thermocouple acts. In general, cell potentials are measured between two contacts of the same metal at the same temperature.

An electrode with the solution surrounding it is called a *half cell*; the corresponding reaction, a half-cell reaction. The sum of the reactions occurring in a cell is called the *cell reaction*. For cell (9.65), the sum of (9.66) and (9.67) is

$$\mathrm{Zn} + \mathrm{Cu}^{++} \rightarrow \mathrm{Zn}^{++} + \mathrm{Cu}. \qquad [9.70]$$

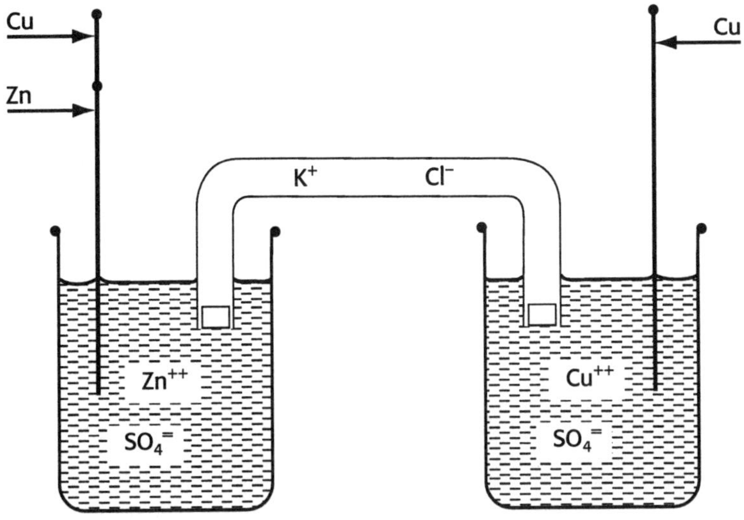

FIGURE 9.5 An electrolytic cell.

As written, equation (9.70) represents the transfer of 2 faradays of negative charge from 1 mole zinc to 1 mole copper. Thus, 2 faradays are involved per unit of reaction.

9.10 Measuring and Allocating Voltages

The voltage of a cell is a property of the system and is thermodynamically significant only when the cell is operating reversibly. Such operation may be found when the cell is balanced against an external potential, as the setup in figure 9.6 allows.

Here battery W drives the current I through resistance APB. When no current flows through galvanometer G, the potential between the electrodes of observation cell C is balanced against IR_{AP}, where R_{AP} is the resistance between points A and P. If reducing resistance R_{AP} infinitesimally causes the cell to discharge, while increasing R_{AP} infinitesimally charges the cell at an infinitesimal rate, the cell is reversible; for, an infinitesimal change then exists that reverses the direction of the process. If there is no single point of balance, the cell is not reversible.

Resistance APB may be a uniform wire with a movable contact at P. Resistance R_{AP} is then proportional to distance AP. With I fixed, the potential between A and P when G registers no current is proportional to the distance along the wire from A to P. The constant of proportionality is determined by balancing IR_{AP} against a cell of known potential $\mathcal{E}$ and reading the position of P.

The voltage between contacts of the same material attached to the electrodes of a cell, when the cell is operating reversibly, is called the *electromotive force* (emf), or $\mathcal{E}$, of the cell. Since voltages in series add, the emf of a cell is the sum of the emf's of its two parts. But the voltage of a single electrode is not measured,

So we assign a voltage of zero to the *standard hydrogen electrode* (SHE). In this, hydrogen gas at 1 atm pressure is passed over a platinized platinum electrode immersed in a solution containing hydrogen ions at unit activity and at 25° C. The potential of a conveniently reproducible reference electrode is then determined with respect to the hydrogen electrode. The potential of each unknown electrode is measured with respect to this reference.

The electrode potential when each reactant and product is at unit activity is called the *standard electrode potential* $\mathcal{E}^\circ$. Tables 9.6 and 9.7 present $\mathcal{E}^\circ$ and $(\partial\mathcal{E}^\circ/\partial T)_P$ for various half cells.

Example 9.6

Find $\mathcal{E}^\circ$ for the cell

$$\mathrm{Cu}|\mathrm{Zn}|\mathrm{ZnSO_4}||\mathrm{CuSO_4}|\mathrm{Cu}.$$

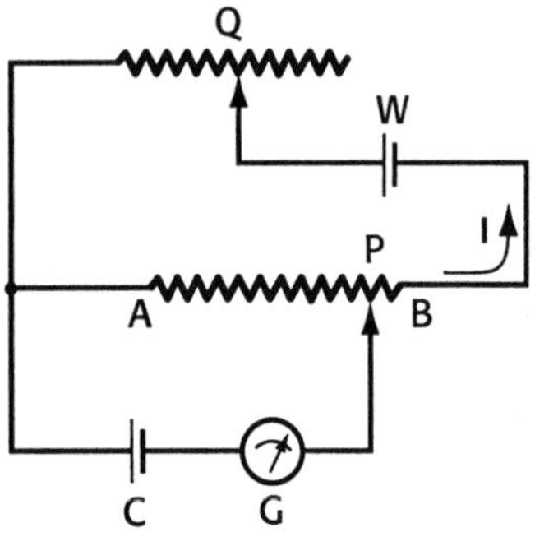

FIGURE 9.6 Potentiometer circuit for measuring the emf of cell C.

TABLE 9.6 Standard Potentials for Electrodes in Acidic Aqueous Solutions at 25°C

Half Cell Reaction	$\mathcal{E}^\circ$, V	$(\partial\mathcal{E}^\circ/\partial T)_P$, $mV\,K^{-1}$
$Li^+ + e \rightarrow Li$	–3.045	0.337
$K^+ + e \rightarrow K$	–2.925	–0.209
$Rb^+ + e \rightarrow Rb$	–2.925	–0.374
$Cs^+ + e \rightarrow Cs$	–2.923	–0.326
$Ra^{++} + 2e \rightarrow Ra$	–2.916	0.28
$Ba^{++} + 2e \rightarrow Ba$	–2.906	0.48
$Sr^{++} + 2e \rightarrow Sr$	–2.888	0.680
$Ca^{++} + 2e \rightarrow Ca$	–2.866	0.696
$Na^+ + e \rightarrow Na$	–2.714	0.099
$Ce^{3+} + 3e \rightarrow Ce$	–2.483	0.972
$Mg^{++} + 2e \rightarrow Mg$	–2.363	0.974
$Th^{4+} + 4e \rightarrow Th$	–1.899	1.15
$U^{3+} + 3e \rightarrow U$	–1.789	0.80
$Al^{3+} + 3e \rightarrow Al$	–1.662	1.375
$Mn^{++} + 2e \rightarrow Mn$	–1.180	0.79
$Zn^{++} + 2e \rightarrow Zn$	-0.7628	0.962
$Cr^{3+} + 3e \rightarrow Cr$	–0.744	1.339
$U^{4+} + e \rightarrow U^{3+}$	–0.607	2.27
$Fe^{++} + 2e \rightarrow Fe$	–0.440	0.923
$Cr^{3+} + e \rightarrow Cr^{++}$	–0.408	—
$Cd^{++} + 2e \rightarrow Cd$	–0.4029	0.778
$Co^{++} + 2e \rightarrow Co$	–0.277	0.93

Continued on next page.

By convention, the half cell reactions are written so oxidation is occurring at the left electrode, reduction at the right electrode. Also, reversing the direction of reaction from that shown in table 9.6 reverses the sign of $\mathcal{E}^\circ$. So we obtain

$$Zn \rightarrow Zn^{++} + 2e, \qquad 0.763\text{ V},$$

$$Cu^{++} + 2e \rightarrow Cu, \qquad 0.337\text{ V},$$

whence

$$\mathcal{E}^0 = 0.763 + 0.337 = 1.100\text{ V}$$

for the cell as a whole.

9.11 Cell Energetics

The electrical work that a cell does at constant temperature and pressure comes from its Helmholtz free energy A and its external energy PV. When the cell processes are carried out reversibly, so that none of the available work is dissipated therein, the electrical work

TABLE 9.6 *Continued from previous page.*

Half Cell Reaction	$\mathcal{E}°$, *V*	$(\partial\mathcal{E}°/\partial T)_P$, *mV K*$^{-1}$
$Ni^{++} + 2e \rightarrow Ni$	−0.250	0.93
$Mo^{3+} + 3e \rightarrow Mo$	−0.20	—
$Sn^{++} + 2e \rightarrow Sn$	−0.136	0.589
$Pb^{++} + 2e \rightarrow Pb$	−0.126	0.420
$2D^{+} + 2e \rightarrow D_2$	−0.0034	—
$2H^{+} + 2e \rightarrow H_2$	0.0000	0.871
$UO_2^{++} + e \rightarrow UO_2^{+}$	0.05	1.45
$Hg_2Br_2 + 2e \rightarrow 2Hg + 2Br^{-}$	0.1397	0.729
$S + 2H^{+} + 2e \rightarrow H_2S$	0.142	0.662
$Sn^{4+} + 2e \rightarrow Sn^{++}$	0.15	—
$Cu^{++} + e \rightarrow Cu^{+}$	0.153	0.944
$Hg_2Cl_2 + 2e \rightarrow 2Hg + 2Cl^{-}$	0.2676	0.554
$Cu^{++} + 2e \rightarrow Cu$	0.337	0.879
$I_2 + 2e \rightarrow 2I^{-}$	0.5355	0.723
$C_6H_4O_2 + 2H^{+} + 2e \rightarrow C_6H_4(OH)_2$	0.6994	0.140
$Fe^{3+} + e \rightarrow Fe^{++}$	0.771	2.059
$Hg_2^{++} + 2e \rightarrow 2Hg$	0.788	—
$Ag^{+} + e \rightarrow Ag$	0.7991	−0.0129
$2Hg^{++} + 2e \rightarrow Hg_2^{++}$	0.920	—
$Br_2 + 2e \rightarrow 2Br^{-}$	1.0652	0.242
$Cl_2 + 2e \rightarrow 2Cl^{-}$	1.3595	−0.389
$Ce^{4+} + e \rightarrow Ce^{3+}$	1.61	—
$F_2 + 2e \rightarrow 2F^{-}$	2.87	−0.959

equals the decrease in $A + PV$, that is, the decrease in the Gibbs function G. But when n equivalents of reaction cause nF coulombs to flow, the electrical work done is $nF\mathcal{E}$. So we have

$$nF\mathcal{E} = -\Delta G_{T,P} \quad [9.71]$$

and

$$nF\mathcal{E}^0 = -\Delta G^0. \quad [9.72]$$

The ith substance in the cell reaction

$$a\text{A} + b\text{B} \rightarrow l\text{L} + m\text{M} \quad [9.73]$$

has the chemical potential

$$\mu_i = \mu_i^{\ 0} + RT \ln a_i. \quad [9.74]$$

Putting these expressions into equation (7.47) gives

$$\Delta G = \Delta G^0 + RT \ln Q \quad [9.75]$$

where

TABLE 9.7 Standard Potentials for Electrodes in Basic Aqueous Solutions at 25°C

Half Cell Reaction	$\mathcal{E}°$, V	$(\partial\mathcal{E}°/\partial T)_P$, mV K^{-1}
$Ca(OH)_2 + 2e \rightarrow Ca + 2OH^-$	-3.02	-0.094
$Ba(OH)_2 + 2e \rightarrow Ba + 2OH^-$	-2.99	1.25
$Sr(OH)_2 + 2e \rightarrow Sr + 2OH^-$	-2.88	-0.09
$H_2BO_3^- + 3H_2O + 3e \rightarrow B + 4OH^-$	-1.79	-0.276
$SiO_3^= + 3H_20 + 4e \rightarrow Si + 6OH^-$	-1.697	—
$SO_4^= + H_20 + 2e \rightarrow SO_3^= + 2OH^-$	-0.93	-0.518
$Se + 2e \rightarrow Se^=$	-0.92	-0.02
$2H_2O + 2e \rightarrow H_2 + OH^-$	-0.82806	0.037
$S + 2e \rightarrow S^=$	-0.447	-0.06
$NO_3^- + H_20 + 2e \rightarrow NO_2^- + 2OH^-$	0.01	-0.388
$S_4O_6^= + 2e \rightarrow 2S_2O_3^=$	0.08	-0.24
$O_2 + 2H_2O + 4e \rightarrow 4OH^-$	0.401	-0.809
$HXeO_4^- + 3H_2O + 6e \rightarrow Xe + 7OH^-$	0.9	—

$$Q = \frac{a_L^{\,l}\, a_M^{\,m}}{a_A^{\,a}\, a_B^{\,b}}. \qquad [9.76]$$

Using equations (9.71) and (9.72) to eliminate ΔG and ΔG^0 in equation (9.75), then dividing by $-nF$, leads to

$$\mathcal{E} = \mathcal{E}^0 - \frac{RT}{nF}\ln Q, \qquad [9.77]$$

where n equals the number of electrons transferred in the cell reaction as written. At 25° C, we find that

$$\frac{(\ln 10)RT}{F} = \frac{(\ln 10)(8.31451\text{ J K}^{-1}\text{ mol}^{-1})(298.15\text{ K})}{96{,}485.3\text{ mol}^{-1}} = 0.059160\text{ V}, \qquad [9.78]$$

and equation (9.77) becomes

$$\mathcal{E} = \mathcal{E}^0 - \frac{0.059160\text{ V}}{n}\log Q. \qquad [9.79]$$

Example 9.7

Calculate the emf of the cell

$$\text{Cu}|\text{Zn}|\text{Zn}^{++}\ (a = 0.100\text{ m})||\text{Cu}^{++}\ (a = 0.100\text{ m})|\text{Cu}.$$

As in example 9.6, the half cell reactions are

$$\text{Zn} \rightarrow \text{Zn}^{++}\ (a = 0.100\text{ m}) + 2\text{e},$$

$$\text{Cu}^{++}\ (a = 0.100\text{ m}) + 2\text{e} \rightarrow \text{Cu}.$$

Adding these yields the cell reaction

$$\mathrm{Zn} + \mathrm{Cu}^{++}\ (a = 0.100\ \mathrm{m}) \to \mathrm{Cu} + \mathrm{Zn}^{++}\ (a = 0.100\ \mathrm{m}$$

in which two electrons are transferred. So

$$n = 2$$

and

$$Q = \frac{a_{\mathrm{Cu}} a_{\mathrm{Zn}^{++}}}{a_{\mathrm{Zn}} a_{\mathrm{Cu}^{++}}} = \frac{(1)(0.100)}{(1)(0.0100)} = 10.0\,.$$

From example 9.6, we have

$$\mathcal{E}^0 = 1.100\ \mathrm{V}.$$

Substituting into formula (9.79) now gives

$$\mathcal{E} = 1.100 - \frac{0.05910}{2}\log 10.0 = 1.100 - 0.030 = 1.070\ \mathrm{V}.$$

9.12 Cell with Transference

The junction potential between two solutions depends on the transference numbers of the pertinent electrolyte.

Consider the cell

$$\mathrm{A}\big|\mathrm{A}^+\ (a_{+(1)}),\ \mathrm{B}^-\ (a_{-(1)})\big|\mathrm{A}^+\ (a_{+(2)}),\ \mathrm{B}^-\ (a_{-(2)})\big|\mathrm{A} \qquad [9.80]$$

for which the electrode reactions are

$$\mathrm{A} \to \mathrm{A}^+\ (a_{+(1)}) + \mathrm{e}, \qquad [9.81]$$

$$\mathrm{A}^+\ (a_{+(2)}) + \mathrm{e} \to \mathrm{A}. \qquad [9.82]$$

In the transition layer between the two solutions, fraction t_+ of the current is carried by the cation, fraction t_- by the anion. The net reactions due to this division are

$$t_+\mathrm{A}^+\ (a_{+(1)}) \to t_+\mathrm{A}^+\ (a_{+(2)}), \qquad [9.83]$$

$$t_-\mathrm{B}^-\ (a_{-(2)}) \to t_-\mathrm{B}^-\ (a_{-(1)}). \qquad [9.84]$$

See figure 9.7. Furthermore, the transference numbers add to 1:

$$t_+ + t_- = 1. \qquad [9.85]$$

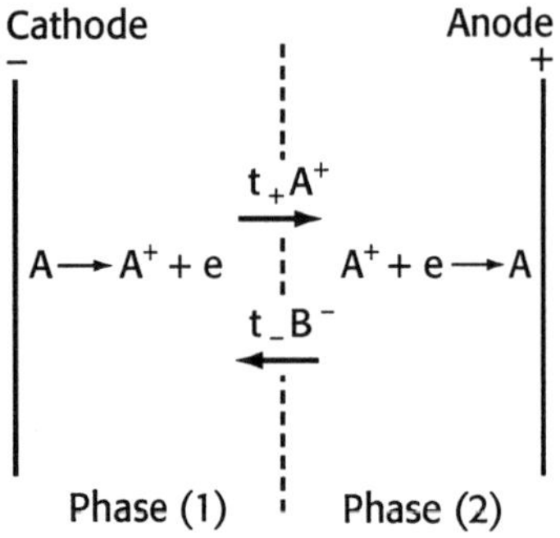

FIGURE 9.7 Representation of the processes occurring in the cell with transference.

Strictly, the transference numbers vary slightly with concentration through the transition layer. Furthermore, the effective thickness of the layer may vary with time. Here we will neglect these effects.

Summing equations (9.81) - (9.84) with condition (9.85) gives

$$t_- A^+ \left(a_{+(2)}\right) + t_- B^- \left(a_{-(2)}\right) \rightarrow t_- A^+ \left(a_{+(1)}\right) + t_- B^- \left(a_{-(1)}\right). \qquad [9.86]$$

For cell reaction (9.86), one electron moves through the external circuit from cathode to anode; consequently

$$n = 1. \qquad [9.87]$$

Furthermore, the anode reaction is the reverse of the cathode reaction; so

$$\mathcal{E}^0 = 0. \qquad [9.88]$$

Substituting into equation (9.79) yields the voltage

$$\mathcal{E} = -0.05916 \log \frac{a_{+(1)}^{t_-} a_{-(1)}^{t_-}}{a_{+(2)}^{t_-} a_{-(2)}^{t_-}} = 0.05916\, t_- \log \frac{a_{+(2)} a_{-(2)}}{a_{+(1)} a_{-(1)}}. \qquad [9.89]$$

One may introduce the mean ionic activity for each solution with the result

$$\mathcal{E} = 0.05916 \left(2t_-\right) \log \frac{a_{\pm(2)}}{a_{\pm(1)}} \qquad [9.90]$$

But the voltage of the cell without transference

$$A \left| A^+ \left(a_{+(1)}\right) \right\| A^+ \left(a_{+(2)}\right) \Big| A \qquad [9.91]$$

is

$$\mathcal{E} = 0.05916 \log \frac{a_{+(2)}}{a_{+(1)}}. \qquad [9.92]$$

When

$$\gamma_+ = \gamma_- = \gamma_\pm \quad \text{and} \quad t_- = \frac{1}{2}, \qquad [9.93]$$

these emf's are equal and the junction potential vanishes. In the general case, if condition (9.93) holds for the electrolyte that determines the potential, we consider that the junction potential vanishes. This result is employed in designing salt bridges. There the junction potential is determined by the electrolyte in the bridge since its concentration is relatively high.

9.13 Reference Cells and Electrodes

In calibrating a potentiometer, a person needs a standard emf source. Then in measuring the potential of a given electrode, a person needs a reference electrode whose potential is known or assumed.

For very accurate work, a solid reacting electrode is not suitable since its state depends not only on the impurities present but also on where they reside in the solid structure. Furthermore, the state depends on the temperature at which the substance solidified and on the subsequent mechanical treatment to which it has been subjected. The state of a liquid or gas, on the other hand, is not affected by such idiosyncrasies. In addition, a liquid may be readily purified by distillation.

A suitable standard is the saturated *Weston cell*

$$\text{Pt}\big|12.5\%\ \text{Cd in Hg}\big|3\text{CdSO}_4\cdot 8\text{H}_2\text{O (sat'd, in H}_2\text{O)}\big|\text{Hg}_2\text{SO}_4\ \text{(s), Hg}\big|\text{Pt.} \qquad [9.94]$$

This has an emf of 1.0186 V at 20° C. However, its $(\partial\mathcal{E}/\partial T)P$ is -0.0000406 V K^{-1}. In the *unsaturated* Weston cell, the electrolyte is saturated at 4° C and all excess cadmium sulfate is removed. Thus, the variation of $\mathcal{E}$ with T is made very small. In practice, such a cell is calibrated against a standard one and the measured emf recorded in a certificate.

As a reference half cell, the standard hydrogen electrode

$$\text{H}^+\ (a = 1.000)\big|\text{H}_2\ (1.000\ \text{atm}) - \text{Pt}, \quad \mathcal{E} = 0.0000\ \text{V} \qquad [9.95]$$

is suitable because its reactants are all in reproducible form. However, it does require a source of hydrogen gas of high purity which is passed over the platinized platinum electrode at a known pressure.

The calomel electrode is more convenient. It consists of mercury covered with a mixture of mercury and mercurous chloride (calomel) in contact with a potassium chloride solution, which may be 0.1000 or 1.0000 N. We have

$$\text{KCl (0.1000 N)}\big|\text{Hg}_2\text{Cl}_2\ \text{(s)} - \text{Hg}\big|\text{Pt}, \quad \mathcal{E} = 0.3354\ \text{V} \qquad [9.96]$$

and

$$\text{KCl (1.0000 N)}\big|\text{Hg}_2\text{Cl}_2\ \text{(s)} - \text{Hg}\big|\text{Pt}, \quad \mathcal{E} = 0.2825\ \text{V} \qquad [9.97]$$

9.14 Measuring the pH of Solutions

For a cell voltage to gauge the activity of an ion, quotient Q in formula (9.77) must vary only with this activity. If Q depends on other activities, these must either be constant or they must cancel. Here we will see how this happens in various pH determinations.

By definition, the pH of a solution is the negative of the power of ten needed to give the activity of H^+ ; thus,

$$\text{pH} = -\log a_{\text{H}^+}. \qquad [9.98]$$

In theory, the simplest choice is the *hydrogen electrode*. At it, the half cell reaction is

$$\frac{1}{2}\text{H}_2\left(P_{\text{H}_2}\right) \rightarrow \text{H}^+\left(a = x\right) + \text{e}. \qquad [9.99]$$

This process occurs reversibly on a clean platinized platinum surface. As a reference electrode, one may employ a hydrogen electrode with unit hydrogen ion activity. Its half cell reaction is

$$\text{H}^+\left(a = 1\right) + \text{e} \rightarrow \frac{1}{2}\text{H}_2\left(P_{\text{H}_2}\right). \qquad [9.100]$$

The complete cell has the structure

$$\text{Pt} - \text{H}_2\left(P_{\text{H}_2}\right)\big|\text{H}^+\left(a = x\right)\big\|\text{H}^+\left(a = 1\right)\big|\text{H}_2\left(P_{\text{H}_2}\right) - \text{Pt.} \qquad [9.101]$$

Summing (9.99) and (9.100)gives

$$H^+ \left(a = 1\right) \rightarrow H^+ \left(a = x\right), \qquad [9.102]$$

whence

$$Q = \frac{x}{1}. \qquad [9.103]$$

Since $\mathcal{E}^0 = 0$ and $n = 1$, formula (9.79) yields the result

$$\mathcal{E} = 0 - 0.05916 \ \log x = +0.05916 \text{ pH}. \qquad [9.104]$$

Employing hydrogen electrodes is troublesome because one needs to work with a gas. Furthermore, various substances are adsorbed strongly on the extensive platinum surface, poisoning it. When this is not significant, the adsorbed hydrogen can reduce oxidizing agents that may be present, thus altering the solution.

For routine work, the *glass electrode* is convenient. It consists of a glass bulb holding a solution of fixed pH around a small reference electrode. In use, it dips into the unknown solution together with another reference electrode. The complete cell has the structure

$$\begin{gathered}\text{Reference electrode}\,\|\,\text{solution of fixed pH}\,|\,\text{glass membrane}\,|\\ \text{solution of unknown pH}\,\|\,\text{reference electrode. (9.105)}\end{gathered} \qquad [9.105]$$

The glass is manufactured so that it acts as a carrier of hydrogen ions but not of the other ions that might be present. However, in highly alkaline solutions, this limitation breaks down and part of the current appears to be carried by other monovalent ions, such as Na^+.

One may make both reference electrodes the same and set the activity of H^+ in the fixed pH solution at 1, Then the reaction at the left electrode cancels the reaction at the right and the cell reaction is determined by the apparent transference through the membrane. As long as this involves only the hydrogen ions, the over-all reaction is

$$H^+ \left(a = 1\right) \rightarrow H^+ \left(a = x\right) \qquad [9.106]$$

and

$$Q = \frac{x}{1}. \qquad [9.107]$$

Formula (9.79) then gives us

$$\mathcal{E} = -0.05916 \ \log x = +0.05916 \text{ pH}. \qquad [9.108]$$

Altering the fixed pH solution merely adds a constant term to (9.108).

Because the glass membrane is a very poor conductor, one can draw very little current from cell (9.105) without introducing considerable error from the *IR* drop. Consequently, a vacuum tube voltmeter, or its equivalent, is employed in measuring the emf. The dial of the instrument is commonly calibrated in pH units rather than in volts so the pH can be read directly.

An alternative choice is the *quindydrone electrode*. This consists of a platinum wire dipping into the unknown solution to which quinhydrone has been added. Also dipping into the solution would be a reference electrode. When this is the normal calomel electrode, the cell has the structure

$$\text{Pt}\,|\,\text{unknown solution containing quinhydrone}\,\|\,\text{normal calomel electrode.} \qquad [9.109]$$

Each molecule of quinhydrone dissolves to give a molecule of quinone, $C_6H_4O_2$, and a molecule of hydroquinone, $C_6H_4(OH)_2$. These interact at the platinum surface with hydrogen ions and electrons:

$$C_6H_4(OH)_2 \rightarrow C_6H_4O_2 + 2H^+ + 2e, \qquad \mathcal{E}^0 = -0.6994\ \text{V}. \tag{9.110}$$

By formula (9.79), the potential of the electrode is

$$\mathcal{E} = -0.6994 - \frac{0.05916}{2}\log\frac{a_{H^+}^2\, a_{\text{quinone}}}{a_{\text{hydroquinone}}}. \tag{9.111}$$

For cell (9.109), we add the potential of the normal calomel electrode (9.97). Furthermore, we set

$$a_{\text{quinone}} = a_{\text{hydroquinone}} \tag{9.112}$$

since their concentrations are equal in the solution of quinhydrone and their activity coefficients should be nearly equal. Thus, we obtain the result

$$\mathcal{E} = 0.2825 - 0.6994 - 0.05916\log a_{H^+} = -0.4169 + 0.05916\ \text{pH}. \tag{9.113}$$

A quinhydrone cell is accurate only when the pH of the solution is less than 8. In highly alkaline solutions, hydroquinone is neutralized to some extent and it becomes oxidized by oxygen from the air.

Another possibility is the *antimony oxide electrode.* This is prepared by casting a stick of antimony in the presence of air. On its surface, both Sb and Sb_2O_3 are exposed. In the half cell, these react with hydrogen ions and electrons:

$$2Sb\ (s) + 3H_2O \rightarrow Sb_2O_3\ (s) + 6H^+ + 6e, \qquad \mathcal{E}^0 = -0.152\ \text{V}. \tag{9.114}$$

By formula (9.79), the potential of the electrode is

$$\mathcal{E} = -0.152 - \frac{0.05916}{2}\log\frac{a_{H^+}^6\, a_{Sb_2O_3}}{a_{Sb}^2\, a_{H_2O}^3} = -0.152 - 0.05916\log a_{H^+} = -0.152 + 0.05916\ \text{pH}, \tag{9.115}$$

where we have set

$$a_{Sb_2O_3} = 1, \qquad a_{Sb} = 1, \qquad a_{H_2O} = 1. \tag{9.116}$$

On combining with a normal calomel electrode, we obtain

$$\mathcal{E} = 0.130 + 0.05916\ \text{pH}. \tag{9.117}$$

The constant term in this equation is approximate since the states of the solid antimony and antimony trioxide vary with just how the electrode has been prepared.

9.15 Overpotential

A person often finds that a potential greater than the reversible potential needs to be applied to a given electrode to make the half cell reaction proceed at a given rate. This excess is called overvoltage or *overpotential.* It results from (1) the *IR* drop between the metal pole of the electrode and the point where the reference electrode is located, (2) concentration changes in the diffusion layer near the metal pole, (3) irreversibility of one or more stages in the half cell reaction. However, potential changes caused by (4) concentration changes in the body of the electrode compartment due to transference and (5) changes in the chemical nature of the electrode are ordinarily not included in the overpotential.

By definition, the *decomposition potential* is the smallest potential needed to cause observable electrolysis. The difference between the decomposition potential and the reversible electrode potential is called the *limiting overpotential*. Some results obtained for hydrogen and oxygen liberation appear in table 9.8.

For deposition of iron, cobalt, or nickel, limiting overpotentials of 0.2 to 0.3 V are found. Other common metals require little if any overpotentials. The deposition of gases is sensitive to the state of the electrode surface. Compare the results in table 9.8 for smooth platinum and for platinized platinum.

In the electrolytic evolution of a diatomic gas, the steps that may be slow and thus not reversible include (a) transfer of the ion to the electrode surface, (b) discharge of the adsorbed ion, (c) reaction on the surface to produce the diatomic product molecule, (d) desorption of this molecule to produce bubbles of the gas.

Example 9.8

If the reversible decomposition potential of an aqueous solution into hydrogen and oxygen is 1.23 V, what is the actual decomposition potential when the aqueous solution is electrolyzed using a silver cathode and a smooth platinum anode?

From table 9.8, the hydrogen overvoltage on silver is 0.15 V, while the oxygen overvoltage on smooth platinum is 0.45 V. So the decomposition potential is

$$\mathcal{E} = 1.23 + 0.15 + 0.45 = 1.83 \text{ V}.$$

9.16 Electrode Potentials for other Solvents

The junction potential between two solutions made up with different solvents cannot be eliminated with any practical salt bridge. So for each solvent, one defines a reference electrode. This may consist of hydrogen at 1 atm pressure passing over a platinized platinum surface immersed in the given solvent containing hydrogen ions at unit activity and 25° C.

The emf of the cell using the same solvent throughout,

$$\text{Cu}|\text{Pt - H}_2 \text{ (1 atm)}|\text{H}^+ \ (a = 1)\|\text{M}^+ \ (a = x)|\text{M}|\text{Cu}, \quad [9.118]$$

TABLE 9.8 Approximate Limiting Overpotentials

Electrode	*Hydrogen, V*	*Oxygen, V*
Platinized platinum	0.00	0.25
Iron	0.08	0.25
Smooth platinum	0.09	0.45
Silver	0.15	0.41
Nickel	0.21	0.06
Copper	0.23	
Cadmium	0.48	0.43
Lead	0.64	0.31
Zinc	0.70	
Mercury	0.78	

then gives the potential of the M^+ ($a = x$) | M electrode in that solvent. The cell reaction is

$$2M^+ \ (a = x) + H_2 \ (1 \text{ atm}) \to M + 2H^+ \ (a = 1). \qquad [9.119]$$

The standard potential appears when the activity of the M^+ is unity. In table 9.9, standard potentials for various electrodes in methyl alcohol, ethyl alcohol, and ammonia are listed.

The tendency for a reaction to go varies from solvent to solvent because the M^+ and H^+ are solvated. So reactant M^+ and product H^+ in the cell reaction change with the solvent. And the standard Gibbs energy increment, together with the emf, vary with the solvent.

9.17 Further Cell Thermodynamics

Since the emf $\mathcal{E}$ of a cell is the voltage between the terminals when the cell is operating reversibly, $\mathcal{E}$ equals the loss in the cell's Gibbs function per coulomb of reaction. But one unit of reaction corresponds to nF coulombs of flow. So for this unit, we have

$$\Delta G = -nF\mathcal{E} \qquad [9.120]$$

and

$$\Delta G^0 = -nF\mathcal{E}^0, \qquad [9.121]$$

as we noted in section 9.11.

When the cell reaction is written in the form

$$A \to B, \qquad [9.122]$$

the Gibbs energy change per unit of reaction is

$$\Delta G = G_B - G_A, \qquad [9.123]$$

where G_B is the Gibbs energy of the products and G_A that of the reactants.

TABLE 9.9 Standard Potentials for Electrodes in Some Nonaqueous Solvents

Electrode	$\mathcal{E}^0$ *in* CH_3OH *at 25°C, V*	$\mathcal{E}^0$ *in* C_2H_5OH *at 25°C, V*	$\mathcal{E}^0$ *in* NH_3 *at -50°C, V*
$Li^+ \mid Li$	-3.10	-3.04	—
$K^+ \mid K$	—	—	-1.98
$Na^+ \mid Na$	-2.73	-2.66	-1.84
$Zn^{++} \mid Zn$	—	—	-0.52
$Cd^{++} \mid Cd$	—	—	-0.18
$Tl^+ \mid Tl$	-0.38	-0.34	—
$Pb^{++} \mid Pb$	—	—	0.33
$H^+ \mid H_2$ - Pt	0.00	0.00	0.00
$Cu^{++} \mid Cu$	—	—	0.43
$I^- \mid I2$ - Pt	0.36	0.30	0.70
$Ag^+ \mid Ag$	0.76	0.75	0.83
$Br^- \mid Br_2$ - Pt	0.84	0.78	1.08
$Cl^- \mid Cl_2$ - Pt	1.12	1.05	1.28

But for a given amount of a substance, equation (5.82) holds:

$$dG = V\,dP - S\,dT. \qquad [9.124]$$

Consequently,

$$\left(\frac{\partial G_B}{\partial T}\right)_P = -S_B, \qquad \left(\frac{\partial G_A}{\partial T}\right)_P = -S_A. \qquad [9.125]$$

Differentiating equation (9.123) with respect to temperature at constant pressure yields

$$\left(\frac{\partial \Delta G}{\partial T}\right)_P = \left(\frac{\partial G_B}{\partial T}\right)_P - \left(\frac{\partial G_A}{\partial T}\right)_P. \qquad [9.126]$$

Introducing (9.120) on the left, (9.125) on the right, and changing signs, leads to

$$nF\left(\frac{\partial \mathcal{E}}{\partial T}\right)_P = S_B - S_A = \Delta S. \qquad [9.127]$$

This formula can be employed to calculate the entropy change in a cell reaction from the temperature dependence of its emf. By definition, we constructed

$$G = H - TS. \qquad [9.128]$$

So for the change in enthalpy at a fixed temperature, we have

$$\Delta H = \Delta G + T\Delta S. \qquad [9.129]$$

Now introducing formulas (9.120) and (9.127) gives us

$$\Delta H = -nF\mathcal{E} + nFT\left(\frac{\partial \mathcal{E}}{\partial T}\right)_P. \qquad [9.130]$$

One may use this formula to obtain the enthalpy of reaction from emf measurements. Note that here

$$\Delta H \neq q_P \qquad [9.131]$$

since the system does electrical work as well as work of expansion.

At equilibrium, Q in formula (9.76) equals the equilibrium constant K. Furthermore, the cell is run down and $\mathcal{E} = 0$. Then from equation (9.77), we obtain

$$\log K = \frac{nF\mathcal{E}^0}{(\ln 10)RT}. \qquad [9.132]$$

At 25° C, this reduces to

$$\log K = \frac{n\mathcal{E}^0}{0.059160}. \qquad [9.133]$$

Questions

9.1 How may charge be transported through conducting media?

9.2 Why may the carriers of current change (a) at an interfacial layer, (b) through a region?

9.3 In a conductance experiment, what kinds of reactions may occur (a) at the anode, (b) at the cathode?

9.4 What is the faraday?

9.5 Define transference number.

9.6 Describe how the transference number of an ion may be measured.
9.7 What is (a) resistance, (b) conductance?
9.8 State Ohm's law.
9.9 Explain how the resistance of a conductor varies with its (a) length, (b) cross sectional area.
9.10 Explain how the conductances of the different ions in a solution combine.
9.11 How may (a) solubilities, (b) degrees of ionization be determined from conductivity measurements?
9.12 Explain how an ion conductance varies with the nature of the solvent.
9.13 Why is the equivalent conductance of the hydrogen ion so high?
9.14 Describe the standard cell conventions.
9.15 How is the electrical work done by a cell interpreted thermodynamically?
9.16 Compare a cell with transference to one where transference effects have been made negligible.
9.17 Describe standard reference cells.
9.18 How is the pH of a solution measured electrically?
9.19 What is overpotential? What causes it?
9.20 Why do the standard electrode potentials vary with the solvent used?
9.21 How are equilibrium constants for cell reactions related to emf's?

Problems

When the temperature is not stated, take it to be 25° C.

9.1 If 0.250 A passed for 75 min through an $AgNO_3$ solution by way of platinum electrodes, how much silver deposited at the cathode? What is the volume of oxygen at 26° C and 720 torr set free at the anode?

9.2 A transference cell with Ag - AgCl electrodes was filled with solution, 100 g of which contained 3.6540 g KCl. After electrolysis, during which 1.9768 g silver plated out in the coulometer, the anode solution weighed 119.48 g and contained 3.1151 g KCl per 100 g solution. Calculate the transference number t_+ of the cation K^+.

9.3 A 0.1000 N KCl solution was studied in a moving boundary tube of 0.1142 cm^2 cross sectional area using 0.0650 N LiCl as indicator. A current of 0.005893 A caused the boundary to move 5.30 cm in 2016 s. Calculate the transference number of K^+ in the 0.1 N KCl.

9.4 In the experiment of problem 9.3, what should the concentration of the LiCl be so that the Li^+ would follow the K^+ at the same speed?

9.5 A conductance cell has parallel electrodes 0.95 cm^2 in area and 8.62 cm apart. Calculate (a) the cell constant and (b) the resistance of the cell when it is filled with 0.0200 N KCl having a specific conductance of 0.002768 Ω^{-1} cm^{-1}.

9.6 When a given conductance cell was filled with 0.02000 N KCl, its resistance was 155 Ω; but when filled with an unknown solution, its resistance was 895 Ω. Calculate the specific conductance of the solution.

9.7 From table 9.3 obtain Λ_0 for AgCl.

9.8 In 0.0500 N acetic acid, the equivalent conductance equals 7.358 Ω^{-1} cm^2 at 25° C. Calculate (a) the degree of ionization α and (b) the ionization constant K of the acetic acid, letting the activity coefficients be 1.

9.9 A saturated solution of AgCl has the specific conductance 1.80×10^{-6} Ω^{-1} cm^{-1} after subtracting out the conductance of the water. Calculate the solubility of the AgCl.

9.10 From table 9.5, obtain the ion conductance of the picrate ion in nitrobenzene, if the viscosity of the nitrobenzene is 0.0185 poise.

9.11 Construct the half cell reactions for each of the following:
(a) Cu I Zn I $ZnSO_4$ II $CdSO_4$ I Cd I Cu,
(b) Cu I Cd I Cd^{++} II Fe^{++}, Fe^{3+} I Pt I Cu,
(c) Cu I Pt - H_2 I H^+ II Cl^- I Cl_2 - Pt I Cu.
Then obtain the cell reaction and the standard potential $\mathcal{E}^0$ for each.

9.12 For each cell in problem 9.11, calculate the standard Gibbs energy change ΔG^0.

9.13 Calculate the enf $\mathcal{E}$ for each of the following cells, letting the activity coefficients be 1:
(a) Cu | Zn | $ZnSO_4$ (0.01000 m) || $CdSO_4$ (0.001000 m) | Cd | Cu,
(b) Pt - H_2 (1.000 atm) | HCl (0.02000 m) || $FeSO_4$ (0.01000 m), $Fe_2(SO_4)_3$ (0.001000 m) | Pt.

9.14 Calculate the emf of each of the following:
(a) Cu | Cu^{++} (a = 0.5000 m) || Cu^{++} (a = 0.00800 m) | Cu,
(b) Pt - H_2 (2.000 bar) | HCl (1.00 m) | H_2 (0.05000 bar) - Pt,
(c) Ag | Ag^+ ($a = 1.000 \times 10^{-3}$ m) || Ag^+ ($a = \sqrt{0.1000}$ m) | Ag.

9.15 If the mean transference number of the silver ion is 0.465, what is the emf of the cell with transference:
Ag | $AgNO_3$ ($a_+a_- = 1.000 \times 10^{-6}$ m^2) | $AgNO_3$ ($a_+a_- = 1.000 \times 10^{-1}$ m^2) | Ag?
Determine the junction potential between the silver nitrate solutions.

9.16 If the cell
Pt | quinhydrone solution || normal calomel electrode
has an emf of -0.203 V, what is the pH of the solution containing quindydrone?

9.17 If the cell
Sb - Sb_2O_3 | H^+ (a_{H^+}) || normal calomel electrode
has an emf of 0.368 V when the pH of the solution is 4.00, what emf would the cell have when the pH of the solution is 7.00?

9.18 If in the reaction

$$H_2O\ (l) \rightarrow H_2\ (g) + \frac{1}{2}O_2\ (g)$$

ΔG^0 is 237,142 J at 25° C, what is the reversible potential needed to electrolyze an aqueous solution to produce hydrogen gas at 1 bar and oxygen gas at 1 bar?

9.19 Using the result from problem 9.18 and numbers from table 9.8, calculate the potential needed to decompose water in the aqueous solution with a mercury cathode and a smooth platinum anode.

9.20 From the standard electrode potentials, calculate the equilibrium constant for the reaction

$$Zn + Fe^{++} \rightarrow Fe + Zn^{++}.$$

9.21 For the cell

$$Cu\,|\,Ag - AgBr\ (s)\,|\,KBr\ (aq)\,|\,Hg_2Br_2\ (s) - Hg\,|\,Cu$$

the emf is 0.06839 V at 25° C and 0.06630 V at 20° C. Calculate ΔS for the cell reaction at 22.5° C.

9.22 At 0° C a calorimetric determination of ΔH for the reaction

$$Zn + 2AgCl \rightarrow ZnCl_2 + 2Ag$$

yielded -217.78 kJ, while the emf of the corresponding cell was 1.015 V. What was $(\partial\mathcal{E}/\partial T)_P$ for the cell?

— — —

9.23 A 0.5000 N copper sulfate solution was electrolyzed at 18° C, where the transference number t_+ equals 0.327, using copper electrodes. If the solution in the anode compartment contained 1.3892 g copper before electrolysis and 1.5328 g after electrolysis, what was the mass of copper plated out on the cathode?

9.24 Current was passed through a cadmium iodide solution in a transference cell by way of cadmium electrodes as 0.5061 g silver deposited on the cathode of a coulometer in series with the cell. After transference, the anode solution weighed 301.700 g and contained 0.1868 per cent cadmium, while beforehand it had contained 0.1390 per cent cadmium. Calculate the transference numbers of cadmium and iodide ions in the cell.

9.25 If the equivalent conductance of 0.05000 N $AgNO_3$ is 115.2 Ω^{-1} cm^2, what is its specific conductance?

9.26 An electrode dipping into a solution acts like a capacitor when no chemical reaction takes place to transfer electrons. By definition, the capacity of a capacitor equals the ratio of charge Q on a plate to the potential $\mathcal{E}$ between the plates. Show that the capacity of an electrode equals $I(d\mathcal{E}/dt)$ when the electrode is being charged.

9.27 The capacitance of a parallel plate capacitor in coulombs per volt is given by the formula

$$C = 8.854 \times 10^{-14} \varepsilon_r \frac{A}{d}$$

in which A is the area of a plate in square centimeters, ε_r the relative dielectric constant, and d the distance between the plates in centimeters. If a clean inert electrode dipping into an aqueous solution has the capacitance 22×10^{-6} C V^{-1} cm^{-2} when it is a cathode, what is the apparent distance between the layers of positive and negative charge at the surface of the electrode? Let the relative dielectric constant of the material between the charges be 78.5 .

9.28 A hydrogen electrode was used in a dilute aqueous solution when the barometric pressure was 732 torr. Calculate the difference between its emf and the emf it would have if the partial pressure of hydrogen were 1.000 atm. At 25° C the vapor pressure of water is 23.8 torr

9.29 Combine two half cell reactions in table 9.6 to form the electrode reaction

$$Hg^{++} + 2e \rightarrow Hg.$$

Note that the corresponding Gibbs energies combine in the same way, while the emf's do not. Calculate the $\mathcal{E}^0$ for the overall reaction.

9.30 With activity coefficients from tables 8.5 and 8.6, recalculate the emf's for the cells in problem 9.13.

9.31 Calculate the emf of the cell
Cu | $CuSO_4$ (0.000250 m), KCl (0.100 m) || $CuSO_4$ (0.000250 m) | Cu.

9.32 Calculate the emf of the cell without transference
(a) Pt - H_2 (1 atm) | HCl (0.00100 m) || HCl (0.0100 m) | H_2 (1 atm) - Pt
and of the same cell with transference
(b) Pt - H_2 (1 atm) | HCl (0.00100 m) | HCl (0.0100 m) | H_2 (1 atm) - Pt.

9.33 On studying the cell
Pt - H_2 (1 atm) | HCl (m) | AgCl (s) - Ag | Pt,
researchers found that a plot of $\mathcal{E}$ + 0.1183 log m - 0.0602 m against m was linear. Extrapolation to zero molality yielded $\mathcal{E}^0$ = 0.2224 V. When m was 0.1238 m, $\mathcal{E}$ was 0.3420 V. For this HCl solution, calculate the mean ionic activity coefficient.

9.34 While being stirred, a 0.010 m $CuSO_4$ solution was electrolyzed between platinum electrodes. Calculate the copper ion concentration when H_2 begins to be evolved at 1 atm pressure.

References

Books

Bard, A. J., Parsons, R., and Jordan, J. (editors): 1985, *Standard Potentials in Aqueous Solution*, Marcel Dekker, New York (abbreviated SP).

After the first two general chapters, the following chapters present data for the elements grouped according to their positions in the periodic table. Each chapter was prepared by knowledgeable specialized experts. A summary of the results then appears in an appendix. Of general interest are the sections:

Parsons, R.: 1985, "Standard Electrode Potentials: Units, Conventions, and Methods of Determination," in SP, pp. 1-12.

This chapter identifies the symbols used, the units and constants employed, and the conventions followed. The relationship of electrode potentials to thermodynamic quantities is outlined.

Parsons, R.: 1985, "The Single Electrode Potential: Its Significance and Calculation," in SP, pp. 13-38.

The work done on moving a charged particle from a reference point where it is free to the body of a medium equals the sum of the chemical work and the electrical work. The former may be obtained from the work done when there is no net charge on the medium. When this chemical work is subtracted from the electrochemical potential, the electrical work is obtained. From this electrical work and the charge on the particle, one obtains the potential with respect to the reference point. Data for carrying out such calculations are tabulated.

Bard et al.: 1985, "Appendix: Synopsis of Standard Potentials," in SP, pp. 787-802.

Standard potentials for the most important electrodes are listed in order.

De Bethune, A. J., and Loud, N. A. S.: 1964, *Standard Aqueous Electrode Potentials and Temperature Coefficients at 25° C*, C. A. Hampel, Skokie, IL, pp, 1-19.

In the introduction, de Bethune discusses what calculations can be made and shows how to make them. The tables list 467 electrode potentials together with their temperature coefficients where known.

Horvath, A. L.: 1985, *Handbook of Aqueous Electrolyte Solutions*, John Wiley & Sons, New York, pp. 239-284.

Horvath lists key data on conductivities in water. Formulas for calculating conductances at various concentrations and temperatures are surveyed critically. Corresponding equations for representing transference numbers are given.

Articles

Arevalo, A., and Pastor, G.: 1985, "Verification of the Nernst Equation and Determination of a Standard Electrode Potential," *J. Chem. Educ.* **62**, 882-884.

Bard, A. J., et al.: 1993, "The Electrode/Electrolyte Interface-A Status Report," *J. Phys. Chem.* **97**, 7147-7173.

Bockris, J. O'M.: 1983, "Teaching the Double Layer," *J. Chem. Educ.* **60**, 265-268.

Donkersloot, M. C. A.: 1991, "Teaching Conductometry," *J. Chem. Educ.* **68**, 136-137.

Frant, M. S.: 1997, "Where Did Ion Selective Electrodes Come From?" *J. Chem. Educ.* **74**, 159-166.

Hertz, H. G., Braun, B. M., Muller, K. J., and Maurer, R.: 1987, "What is the Physical Significance of the Pictures Representing the Grotthus H^+ Conductance Mechanism?" *J. Chem. Educ.* **64**, 777-784.

Maloy, J. T.: 1983, "Factors Affecting the Shape of Current - Potential Curves," *J. Chem. Educ.* **60**, 285-289.

Matsen, F. A.: 1987, "Three Theories of Superconductivity," *J. Chem. Educ.* **64**, 842-846.

Michalowski, T.: 1994, "Calculation of pH and Potential E for Bromine Aqueous Solution," *J. Chem. Educ.* **71**, 560-562.

Millet, P.: 1996, "Electric Potential Distribution in an Electrochemical Cell," *J. Chem. Educ.* **73**, 956-958.

Moran, P. J., and Gileadi, E.: 1989, "Alleviating the Common Confusion Caused by Polarity in Electrochemistry," *J. Chem. Educ.* **66**, 912-916.

Power, G. P., and Ritchie, I. M.: 1983, "Mixed Potentials," *J. Chem. Educ.* **60**, 1022-1026.

Ramette, R. W.: 1987, "Outmoded Terminology: The Normal Hydrogen Electrode," *J. Chem. Educ.* **64**, 885.

Runo, J. R., and Peters, D. G.: 1993, "Climbing a Potential Ladder to Understanding Concepts in Electrochemistry," *J. Chem. Educ.* **70**, 708713.

Saslow, W. M.: 1994, "Consider a Spherical Battery ," *Am. J. Phys.* **62**, 495-501.

Sastre, M., and Santabella, J. A.: 1989, "A Note on the Meaning of the Electroneutrality Condition for Solutions," *J. Chem. Educ.* **66**, 403-404.

Shannon, C., Frank, D. G., and Hubbard, A. T.: 1991, "Electrode Reactions of Well-Characterized Adsorbed Molecules," *Annul Rev. Phys. Chem.* **42**, 393-431.

Smith, D. E.: 1983, "Thermodynamic and Kinetic Properties of the Electrochemical Cell," *J. Chem. Educ.* **60**, 299-301.

Solomon, T.: 1993, "The Polarity of Overpotential," *J. Chem. Educ.* **70**, 877-878.

Stock, J. T.: 1992, "A Century and a Half of Silver-Based Coulometry," *J. Chem. Educ.* **69**, 949-952.

Wetzel, T. L., Mills, T. E., and Safron, S. A.: 1986, "Chemical Potentials and Activities," *J. Chem. Educ.* **63**, 492- 495.

10

Basic Quantum Mechanics

10.1 The Statistical Nature of the Path of a Mass Element or Particle

ACCORDING TO CLASSICAL MECHANICS, each infinitesimal part of a given system is localized at a point at any given time. As time progresses, this point travels along a definite curve at a determinable smoothly varying rate.

However, no device can locate this point and determine its velocity or momentum with exactness at any time. Classical scientists considered that these uncertainties arose in the process of measurement. So no allowance for them was made in their theory.

Matter was found to be made up of atoms. Each atom was found to be made up of electrons and a relatively massive nucleus. The electrons had no apparent structure; they were presumed to be charged mass points. Each electron carried a unit of negative charge; the nucleus, Z units of positive charge.

The electrons were considered to move along definite orbits about the nucleus, being attracted by the positive nuclear charge. But such electrons are subjected to continuous acceleration toward the nucleus. The accelerating charges should radiate continuously according to classical theory. Associated with the loss in energy, the electrons should spiral in to the nucleus. But this collapse is not observed. A stable ground state exists. Furthermore, excited states with various half lives exist. Radiation occurs only during transitions between these discrete states.

Similarly, the valence electrons in a molecule were presumed to travel along definite orbits about the nuclei that were bonded together. These also should vary continuously, according to classical theory. Such changes are not observed. Instead, a molecule is found to exist in definite discrete states.

A nucleus also exists in definite states. Transitions between the states are effected by radiation or absorption of the requisite energy, charge, and spin.

Instead of associating all the uncertainties in position, velocity, and momentum with errors in measurement, considering an irreducible part of them to be an essential attribute of a particle leads to a viable theory. A person does violence to a system of particles when he or she considers each to follow a definite curve.

In our analysis, we presume that the space and time of the laboratory can be imposed on the submicroscopic system under consideration. Furthermore, when a particle acts at a certain point in space-time, it may be considered to have a particular kinetic energy T and a particular potential energy V given by the classical forms.

One can accelerate charged particles with a definite electric potential ϕ. Each particle then presumably acquires kinetic energy equal to $-q\Delta\phi$ where q is the charge on the particle. With a constant source and a constant accelerating potential, a directed beam can be produced. In the approximation that interaction among the particles is negligible, the beam may be considered uniform and homogeneous.

In addition, uncharged particles from an oven may be passed through a velocity selector to produce an approximately homogeneous beam. A velocity selector is constructed and operated so that successive slits in wheels are lined up only for particles of the desired velocity.

In the *ideal homogeneous beam*, all the particles are identical, with the same energy and momentum. However, the position of a particle cannot be specified at any time. But if A is a cross sectional area over which the beam is uniform and x is distance along the beam, the probability that a particle is in section $A\ dx$ can be specified. Dividing this probability by the volume of the section yields the probability density ρ for the particle in that section.

In an ideal gas at equilibrium, the direction of motion and the energy of individual molecules are distributed randomly over the possibilities. The allowed states may be determined from the behavior in homogeneous beams. Furthermore, the nature of the submicroscopic motions in atoms and molecules may be induced from the behavior in homogeneous beams.

10.2 Representing a Constituent State

A person can give the uncertainty in each particle's position, velocity, and momentum a fundamental role by describing the state of a system with appropriate functions.

Consider a system of identical particles traveling in an ideal homogeneous beam of given cross sectional area. To describe the state of the particles in a given section of the beam, one needs to specify the particle density and a parameter determining the energy, together with a specification of the direction.

For a definite momentum constituent of a nonhomogeneous beam, one similarly needs to specify the particle density function ρ, the direction, and the energy. The real function ρ by itself is not sufficient.

But a complex function has a magnitude and a phase at each independent variable point. The phase may determine the momentum and the energy. Thus, the complex function has the necessary flexibility.

This function describing the state of a system is designated Ψ. Furthermore, the density function ρ is considered to determine the square of the magnitude of Ψ by the equation

$$\rho = (\text{constant})\Psi^*\Psi. \qquad [10.1]$$

In most discussions, the constant of proportionality is taken to be 1. Expression Ψ is called the *state function* for the given system.

Since observable properties are determined by how the particles involved are distributed in space, on the average, and by how they propagate, such properties are derivable from the state function for the given system. The function Ψ represents the state of a given system to the extent that this state can be determined, as long as Galilean relativity is sufficient.

In constructing Ψ, we first of all apply the principle of *continuity*, the concept that small causes produce small effects. Thus, we consider that in the region of interest R, function Ψ and its derivatives vary smoothly. We say that function Ψ is analytic within the region R.

A motion in which each particle has a definite momentum at each point is said to be *pure* with respect to this momentum. We now presume that for such a motion, the fractional change in Ψ about a point is determined by symmetry considerations.

10.3 Introducing the Translational Symmetries

A particle moving freely at a definite speed does not distinguish between successive elements of the same length or duration along its path. So each element of given infinitesimal magnitude must be considered equivalent in its effects on the fractional change in the state function for the pure motion.

Let us examine the conditions in a homogeneous beam of effectively independent particles. Let the particle density in the beam be ρ, while the corresponding Ψ is related to ρ by equation (10.1). Let distance along the beam be measured by coordinate x and time be measured by variable t. Let the change in Ψ for infinitesimal changes $\mathrm{d}x$ and $\mathrm{d}t$ be $\mathrm{d}\Psi$.

By symmetry, successive elements of space and time, of given size, are equivalent. Consequently, the fractional change in Ψ produced by $\mathrm{d}x$ and $\mathrm{d}t$ is independent of x and t. And the effect varies linearly with $\mathrm{d}x$ and $\mathrm{d}t$. Thus

$$\frac{\mathrm{d}\Psi}{\Psi} = \kappa\,\mathrm{d}x + \gamma\,\mathrm{d}t \qquad [10.2]$$

with κ and γ parameters to be identified. Since these may be imaginary or complex, we may just as well write

$$\frac{\mathrm{d}\Psi}{\Psi} = ik\,\mathrm{d}x - i\omega\,\mathrm{d}t. \qquad [10.3]$$

The significance of constants k and ω are to be determined.

Equation (10.3) integrates to give

$$\ln\Psi = \ln A + ikx - i\omega t \qquad [10.4]$$

or

$$\Psi = Ae^{ikx}e^{-i\omega t} = Ae^{i(kx-\omega t)}, \qquad [10.5]$$

where A is called the *normalization constant* for Ψ.

Imposing condition (10.1) in the form

$$\rho = \Psi^*\Psi \qquad [10.6]$$

leads to

$$\rho = A^*(e^{ikx})^*(e^{-i\omega t})^* Ae^{ikx}e^{-i\omega t}. \qquad [10.7]$$

This reduces to an expression independent of x and t *only* if k and ω are real (κ and γ imaginary). Then (10.7) reduces to

$$\rho = A^*A. \qquad [10.8]$$

The magnitude of the normalization constant is determined by the particle density ρ.

The phase angle for A is not determined by the considerations so far. However, it would have a particular value for any one particle in the beam. But this could differ arbitrarily from that for the next particle. Then incoherence between the different particles would exist and the particles would be independent.

When the particles are not independent, as in a superfluid or in a superconducting stream, the phases of successive particles are related and *coherence* among the particles is said to prevail.

Example 10.1

In the state function for translation, find the distance between successive points at the same phase.

From equation (10.5), we have

$$\Psi = Ae^{i(kx-\omega t)} = Ae^{i\alpha}$$

with the phase angle

$$kx - \omega t = \alpha.$$

At a given time t, an increase in the phase angle appears as

$$k\,\Delta x = \Delta\alpha.$$

From the periodicity of the cosine and sine and from the identity

$$\exp(i\alpha) = e^{i\alpha} = \cos\alpha + i\sin\alpha,$$

we see that the exponential function goes through one cycle when α increases by 2π. Then

$$k\,\Delta x = 2\pi$$

and

$$\Delta x = \frac{2\pi}{k} = \lambda.$$

The distance λ that is required for a sinusoidal or an exponential function to go through one cycle is called its *wavelength*. The reciprocal of the wavelength gives the number of cycles per unit length, the *wave number*. For it, we write

$$\frac{1}{\lambda} = \frac{k}{2\pi} = \bar{k}$$

Example 10.2

At what speed does a given phase of the translational state function travel?

Solve the phase angle equation in example 10.1 for x:

$$x = \frac{\omega}{k}t + \frac{\alpha}{k}.$$

Then differentiate with respect to t at the given α:

$$\frac{\mathrm{d}x}{\mathrm{d}t} = \frac{\omega}{k} = w.$$

By convention, w is called the *phase velocity* or wave velocity, k the *wavevector*, and ω the *angular frequency* for Ψ. The entity is called a *de Broglie wave* after the man who introduced wave concepts to describe the motion of particles with a rest mass, in the early 1920's

Example 10.3

Interpret the product of the wavelength and the conventional frequency.

Wavelength λ is the distance between two successive points at the same phase along the periodic axis of a function. Furthermore, *frequency* ν is the number of cycles executed per unit time at a fixed point on that axis. Multiplying these yields the distance traveled in unit time, the phase velocity. Thus

$$w = \lambda \nu = \frac{2\pi}{k} \nu = \frac{\omega}{k}.$$

In the last step, we have introduced the relation between the conventional frequency and the angular frequency:

$$\nu = \frac{\omega}{2\pi}.$$

10.4 Translational Motion in a General Direction

In diffraction experiments, part of a beam is scattered in various directions by a barrier system. In an ideal gas, molecules travel in random directions about any given point within the gas. So we are led to rewrite equation (10.5) for such motions.

We still consider the particles to move freely along lines. So the symmetry argument applies independently for each of these lines. If s is the distance from a reference point on the line being considered, and if we replace A by N, then equation (10.5) becomes

$$\Psi = N e^{iks} e^{-i\omega t} = \psi e^{-i\omega t}. \quad [10.9]$$

Note that psi without serifs represents the spatial factor in the state function.

Let unit vector $\boldsymbol{e}$ define the direction of motion. Then expression

$$\mathbf{k} = k\mathbf{e} \quad [10.10]$$

is the actual *wavevector* for this motion. Also let the origin for s be the origin of a Cartesian coordinate system. Then if $\mathbf{r}$ is the radius vector locating the position s along the line, we have

$$\mathbf{e} \cdot \mathbf{r} = s, \quad [10.11]$$

$$ks = k\mathbf{e} \cdot \mathbf{r} = \mathbf{k} \cdot \mathbf{r} \quad [10.12]$$

and expression (10.9) may be written as

$$\Psi = N e^{i\mathbf{k}\cdot\mathbf{r}} e^{-i\omega t}. \quad [10.13]$$

With the Cartesian components of $\boldsymbol{k}$ as k_x, k_y, k_z and of $\boldsymbol{r}$ as x, y, z, we have

$$\mathbf{k} \cdot \mathbf{r} = k_x x + k_y y + k_z z. \quad [10.14]$$

Substituting this form into (10.13) yields

$$\begin{aligned} \Psi &= N e^{ik_x x} e^{ik_y y} e^{ik_z z} e^{-i\omega t} \\ &= X(x)Y(y)Z(z)T(t) \end{aligned} \quad [10.15]$$

and

$$\psi = X(x)Y(y)Z(z), \tag{10.16}$$

where

$$X(x) = Ae^{ik_x x}, \tag{10.17}$$

$$Y(y) = Be^{ik_y y}, \tag{10.18}$$

$$Z(z) = Ce^{ik_z z}, \tag{10.19}$$

$$T(t) = e^{-i\omega t}, \tag{10.20}$$

and

$$N = ABC. \tag{10.21}$$

Equations (10.17), (10.18), (10.19) represent motion in the x direction, the y direction, and the z direction, respectively. In free space, these motions are independent, Thus in product (10.15), the spatial factors for independent motions within a pure state *multiply*.

10.5 Diffraction Experiments Revisited

In sections 1.3, 1.5, and 1.6, diffraction theory was introduced and applied to the scattering of X rays by crystals. In sections 2.3 and 2.4, the scattering of a beam of electrons by randomly oriented molecules was analyzed. A beam of electrons is also diffracted by thin crystal films, yielding a pattern similar to that produced by electromagnetic radiation.

We have here *direct* evidence that the state function we have constructed is valid. One finds that the momentum p of a particle and its wavevector k are related by the de Broglie equation

$$\mathbf{p} = \hbar\mathbf{k}. \tag{10.22}$$

Here $\hbar$ is Planck's constant divided by 2π.

One also finds that where different rays meet, they combine *additively*. Only the complete effect is observed, not any constituent by itself. For n different alternative motions, described by waves Ψ_1, Ψ_2, ..., Ψ_n, we have

$$\Psi = \Psi_1 + \Psi_2 + \cdots + \Psi_{\underline{n}}. \tag{10.23}$$

Rule (10.23) is called the *superposition principle.*

10.6 Rectangularly Symmetric Free Motion

In an ideal gas, each molecule travels along a straight line until it comes near another molecule, in the interior or at the surface, and is deflected by it. In the free electron approximation for the conduction electrons in a metal, each valence electron travels along a particular straight line till it strikes another particle or the bounding surface. In each case, the net effect of the collisions between particles, or with the bounding surface, is to randomize the motions. So we will consider each particle to travel throughout along a straight line and take the collisions into account by assuming a random distribution over the different lines. But first, we will deduce the possible energies for a typical free particle.

Consider a system large enough so that adding more equivalent material at the same density and temperature has no effect on the intensive properties. For the purpose of discussion, consider this process to be continued until all space is filled uniformly.

Conditions in the added sections then merely duplicate those in the original section. For convenience, take the original section to be rectangular, as figure 10.1 shows. Then conditions in the first block are repeated periodically in the x direction, in the y direction, and in the z direction. *Periodic boundary conditions* are said to prevail.

In the composite system, a given particle experiences no change in potential along its straight line path. So it behaves as described in section 10.3 The pertinent wavevector k is related to its momentum by the de Broglie equation.

The assumed periodicity makes the de Broglie wave repeat itself on going from one unit cell to the next. An edge, $\mathcal{A}$, $\mathcal{B}$, or $\mathcal{C}$, is then an integral number of wavelengths. Letting λ_n be the wavelength for change in x, λ_y the wavelength for change in y, λ_z the wavelength for change in z, we obtain

$$\mathcal{A} = n_x \lambda_x = n_x \frac{2\pi}{k_x}, \qquad [10.24]$$

$$\mathcal{B} = n_y \lambda_y = n_y \frac{2\pi}{k_y}, \qquad n_y = \ldots, -2, -1, 0, 1, 2, \ldots, \qquad [10.25]$$

$$\mathcal{C} = n_z \lambda_z = n_z \frac{2\pi}{k_z}, \qquad n_z = \ldots, -2, -1, 0, 1, 2, \ldots, \qquad [10.26]$$

In the second equality on each line, we have made use of the result from example 10.1

Combining these equations with the de Broglie equation (10.22) yields the components of momentum

$$p_x = \frac{h}{2\pi} \frac{2\pi n_x}{\mathcal{A}} = h \frac{n_x}{\mathcal{A}}, \qquad [10.27]$$

$$p_y = \frac{h}{2\pi} \frac{2\pi n_y}{\mathcal{B}} = h \frac{n_y}{\mathcal{B}}, \qquad [10.28]$$

$$p_z = \frac{h}{2\pi} \frac{2\pi n_z}{\mathcal{C}} = h \frac{n_z}{\mathcal{C}}. \qquad [10.29]$$

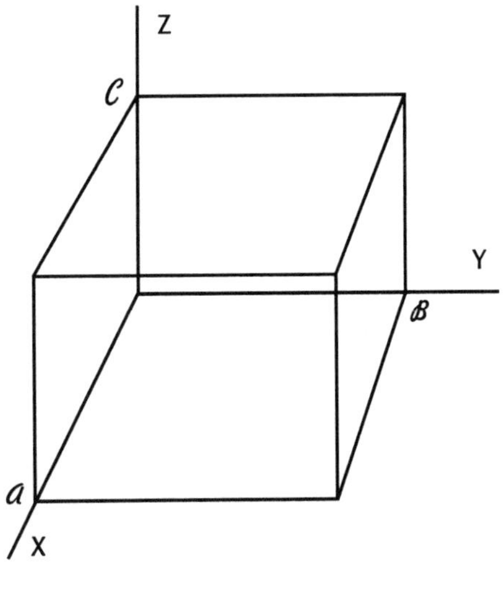

FIGURE 10.1 A rectangular unit of material with appropriate coordinate axes.

The classical form for the kinetic energy T of a particle with momentum p and mass m is

$$T = \frac{p^2}{2m}. \qquad [10.30]$$

We also take the potential energy V to be zero. So the total energy of the particle is

$$E = T + V = \frac{p^2}{2m} + 0 = \frac{1}{2m}(p_x^{\ 2} + p_y^{\ 2} + p_z^{\ 2}). \qquad [10.31]$$

Conditions (10.27) - (10.29) reduce (10.31) to

$$E = \frac{h^2}{2m}\left(\frac{n_x^{\ 2}}{\mathcal{A}^2} + \frac{n_y^{\ 2}}{\mathcal{B}^2} + \frac{n_z^{\ 2}}{\mathcal{C}^2}\right). \qquad [10.32]$$

Here E is the energy of the particle under consideration, h is Planck's constant, m is the mass of the particle, n_x, n_y, n_z are integers labeling the state, while $\mathcal{A}$, $\mathcal{B}$, $\mathcal{C}$ are the dimensions of the original rectangular unit. Any integer or half integer used to identify allowed states of a system is called a *quantum number*. Here n_x, n_y, n_z play this role.

When an energy level can be obtained with two or more different choices of quantum numbers, the level is said to be *degenerate*. The number of different choices that can contribute to a level is called the degeneracy of the level. Here degeneracy appears at some levels whenever two or three of the dimensions $\mathcal{A}$, $\mathcal{B}$, $\mathcal{C}$ are commensurable.

As long as $\mathcal{A}$, $\mathcal{B}$ and $\mathcal{C}$ are finite, the allowed states are discrete. The number of states up to any desired level can then be counted.

Example 10.4

What does coefficient $h^2/(2m)$ equal when the translating particle is an electron? Introduce accepted values of the fundamental constants

$$\frac{h^2}{2m} = \frac{(6.6261 \times 10^{-34}\,\text{J s})^2}{2(9.1094 \times 10^{-31}\,\text{kg})} = 2.40988 \times 10^{-37}\,\text{J m}^2$$

and change units to obtain

$$\frac{h^2}{2m} = \frac{2.40988 \times 10^{-37}\,\text{J m}^2}{1.60218 \times 10^{-19}\,\text{J eV}^{-1}} = 1.5041 \times 10^{-18}\,\text{eV m}^2$$

or

$$\frac{h^2}{2m} = 150.41\ \text{eV Å}^2.$$

Note how this quantity checks with the last number in example 2.1.

10.7 Translational Energy of the Molecules in an Ideal Gas

In a thermodynamic system, each disjoint particle state is occupied with a probability depending on its energy and on the temperature of the pertinent region. Knowing the probability distribution law, a person can calculate the system's thermodynamic properties. Here our concern is with the translational contribution to the internal energy.

Consider a pure ideal gas at equilibrium with its containing walls. Also consider the temperature to be high enough so that the probability for a molecule to occupy any one significant state is small. Then the Boltzmann distribution law, equation (3.82), applies. If N_j is the number of molecules in the jth disjoint state on the average, E_j the energy of this state, k the Boltzmann constant, and T the absolute temperature, we have

$$N_j = Ae^{-E_j/kT}. \quad [10.33]$$

Summing over all the disjoint states yields

$$N = \sum_j N_j = A\sum_j e^{-E_j/kT} = AZ. \quad [10.34]$$

In the third equality, the definition

$$Z = \sum_j e^{-E_j/kT} \quad [10.35]$$

has been introduced. Expression Z is called the *partition function* or state sum. In (10.34), we see that

$$A = \frac{N}{Z}. \quad [10.36]$$

So for the total energy of the states considered, we have

$$E = \sum_j N_j E_j = A\sum_j E_j e^{-E_j/kT} = (N/Z)\sum_j E_j e^{-E_j/kT}. \quad [10.37]$$

In the second equality, condition (10.33) has been employed; in the third equality, relation (10.36),

Differentiating equation (10.35) at constant volume V produces

$$\left(\frac{\partial Z}{\partial T}\right)_V = \frac{1}{kT^2}\sum_j E_j e^{-E_j/kT}. \quad [10.38]$$

Now solve for the sum and substitute it into the last expression in line (10.37) to get

$$E = \frac{N}{Z}kT^2\left(\frac{\partial Z}{\partial T}\right)_V = NkT^2\left(\frac{\partial \ln Z}{\partial T}\right)_V. \quad [10.39]$$

Consider the ideal gas subject to the rectangularly periodic boundary conditions. The disjoint translational energy states then satisfy equation (10.32), and the corresponding partition function is

$$Z_{\text{tr}} = \sum \exp\left[-\frac{h^2}{2mkT}\left(\frac{n_x^2}{\mathcal{A}^2} + \frac{n_y^2}{\mathcal{B}^2} + \frac{n_z^2}{\mathcal{C}^2}\right)\right]$$

$$= \sum \exp\left(-\frac{n_x^2 h^2}{2m\mathcal{A}^2 kT}\right)\sum \exp\left(-\frac{n_y^2 h^2}{2m\mathcal{B}^2 kT}\right)\sum \exp\left(-\frac{n_z^2 h^2}{2m\mathcal{C}^2 kT}\right)$$

$$= Z_x Z_y Z_z. \quad [10.40]$$

The first factor has the form

$$Z_x = \sum_{-\infty}^{\infty} e^{-\alpha n_x^2} \quad [10.41]$$

with

$$\alpha = \frac{h^2}{2m\mathcal{A}^2kT}. \qquad [10.42]$$

Except at low temperatures, α is relatively small and the sum can be approximated by the integral

$$Z_x = \int_{-\infty}^{\infty} e^{-\alpha n_x^2} \mathrm{d}n = \left(\frac{\pi}{\alpha}\right)^{1/2}. \qquad [10.43]$$

Inserting the expression for α gives us

$$Z_x = \left(\frac{\pi 2m\mathcal{A}^2kT}{h^2}\right)^{1/2} = \left(2\pi mkT\right)^{1/2}\frac{\mathcal{A}}{h}. \qquad [10.44]$$

Similarly,

$$Z_y = \left(2\pi mkT\right)^{1/2}\frac{\mathcal{B}}{h}, \qquad [10.45]$$

$$Z_z = \left(2\pi mkT\right)^{1/2}\frac{\mathcal{C}}{h}. \qquad [10.46]$$

Multiplying these three yields the expression

$$Z_{\mathrm{tr}} = \left(2\pi mkT\right)^{3/2}\frac{V}{h^3}, \qquad [10.47]$$

in which V is the volume $\mathcal{ABC}$.

Taking the logarithm of this partition function,

$$\ln Z_{\mathrm{tr}} = \frac{3}{2}\ln T + \ln V + \text{constant}, \qquad [10.48]$$

and substituting into formula (10.39) leads to

$$E_{\mathrm{tr}} = NkT^2\frac{3}{2T} = \frac{3}{2}NkT. \qquad [10.49]$$

When N is Avogadro's number, Nk is the gas constant R and (10.49) reduces to

$$E_{\mathrm{tr}} = \frac{3}{2}RT. \qquad [10.50]$$

In chapter 3, the same result was obtained using Newtonian mechanics. Recall equation (3.22).

Example 10.5

Evaluate the integral

$$I = \int_0^{\infty} \mathrm{e}^{-ax^2}\mathrm{d}x.$$

This integral may be taken along either the x or the y axis without affecting its value; so we also have

$$I = \int_0^{\infty} \mathrm{e}^{-ay^2}\mathrm{d}y.$$

Multiplying the two forms for I yields a double integral,

$$I^2 = \int_0^\infty e^{-ax^2} dx \int_0^\infty e^{-ay^2} dy = \int_0^\infty \int_0^\infty e^{-a(x^2+y^2)} dx\, dy,$$

over the first quadrant of the xy plane.

In polar coordinates, the element of area is

$$dA = r\, d\varphi\, dr,$$

while the square of the radius vector locating the area is

$$r^2 = x^2 + y^2.$$

So the double integral can be rewritten as

$$I^2 = \int_0^\infty \int_0^{\pi/2} e^{-ar^2} r\, d\varphi\, dr,$$

whence

$$I^2 = \int_0^{\pi/2} d\varphi \int_0^\infty e^{-ar^2} r\, dr = \varphi \left(-\frac{e^{-ar^2}}{2a} \right) \Bigg|_0^\infty = \frac{\pi}{4a}$$

and

$$I = \frac{1}{2}\left(\frac{\pi}{a}\right)^{1/2}.$$

Because the given integrand is even, the integral from - ∞ to 1 equals that from 0 to ∞? and

$$\int_{-\infty}^\infty e^{-ax^2} dx = \left(\frac{\pi}{a}\right)^{1/2}.$$

This value is used in line (10.43).

When the nonzero limit is finite, one employs the definition of the *error function* (erf):

$$\int_0^u e^{-w^2} dw = \frac{(\pi)^{1/2}}{2} \text{erf}\, u.$$

For $w = a^{1/2}x$ and $u = a^{1/2}b$, we have

$$\int_0^b e^{-ax^2} dx = \frac{1}{2}\left(\frac{\pi}{a}\right)^{1/2} \text{erf}\, a^{1/2}b$$

and

$$\int_{-b}^b e^{-ax^2} dx = \left(\frac{\pi}{a}\right)^{1/2} \text{erf}\, a^{1/2}b.$$

Standard tables of the error function are available.

10.8 Enumerating and Filling Translational States

To a first approximation, the potential energy of a valence electron within a metal, and of a nucleon within a nucleus, is constant. So our discussion of free motion may be applied. Let us first determine the number of states below a certain energy. Then we will consider how the states are filled.

Suppose a large number of equivalent independent particles of a given kind are moving freely within a rectangular volume of dimensions $\mathcal{A} \times \mathcal{B} \times \mathcal{C}$. Let us impose periodic boundary conditions as in section 10.6.

From equations (10.24) - (10.26), the components of the wave number for a particular state are

$$\bar{k}_x = \frac{k_x}{2\pi} = \frac{n_x}{\mathcal{A}}, \quad n_x = ..., -2, -1, 0, 1, 2, ..., \qquad [10.51]$$

$$\bar{k}_y = \frac{k_y}{2\pi} = \frac{n_y}{\mathcal{B}}, \quad n_y = ..., -2, -1, 0, 1, 2, ..., \qquad [10.52]$$

$$\bar{k}_z = \frac{k_z}{2\pi} = \frac{n_z}{\mathcal{C}}, \quad n_z = ..., -2, -1, 0, 1, 2, ..., \qquad [10.53]$$

Plotting the allowed wave numbers in a three dimensional Euclidean space produces a rectangular lattice of points with the cell size

$$\frac{1}{\mathcal{ABC}}. \qquad [10.54]$$

Each cell has eight corners and each corner is shared by eight cells. So the number of cells equals the number of lattice points, the number of translational states.

For unit volume of the physical system, we have

$$\mathcal{ABC} = 1, \qquad [10.55]$$

Each cell in the wave number plot is then of unit size. Furthermore, each translational state is associated with a separate unit volume

A fundamental particle, such as an electron, a proton, or a neutron, may assume two different orientations in a magnetic field. These are associated with a spin of $+1/2\hbar$ and a spin of $-1/2\hbar$ with respect to the positive direction of the field. The allowed components of spin have the form $m_s\hbar$, where the *spin magnetic quantum number* m_s equals either $+1/2$ or $-1/2$.

In a multiparticle system, it is found that no two fundamental particles of a given kind can have the same set of quantum numbers. This law, which we will here consider as empirical, is called the *Pauli exclusion principle.* As a consequence, no more than two such particles can occupy a translational state for which n_x, n_y, and n_z are given. The corresponding gas is called a Fermi gas, and the particles are called *fermions.*

Let us consider that unit volume of a given system contains N fermions of a given kind. Let us also suppose that the temperature of the system is low enough so that thermal agitation cannot excite any appreciable number of these particles out of the lowest $N/2$ translational states. Since the kinetic energy of a particle is

$$E = \frac{p^2}{2m} = \frac{h\bar{k}^2}{2m}, \qquad [10.56]$$

these states have the smallest $\bar{k}$'s. The 1/2 N th smallest wave number, the highest utilized $\bar{k}$ at 0 K, is labeled $\bar{k}_{max}$.

All utilized $\bar{k}$'s now lie within a sphere of radius $\bar{k}_{max}$ centered on the origin of the plot. When the particle density N is large, the number of these is large and the volume of the rectangular cells for $\bar{k} \leq \bar{k}_{max}$ is very close to the volume of the sphere. We have

$$\int_0^{\bar{k}_{max}} 4\pi\bar{k}^2 d\bar{k} = \frac{4}{3}\pi\bar{k}^3_{max} = \frac{N}{2}. \quad [10.57]$$

Solving for the highest employed wave number at 0 K gives us

$$\bar{k}_{max} = \left(\frac{3N}{8\pi}\right)^{1/3} = \frac{1}{2}\left(\frac{3N}{\pi}\right)^{1/3}. \quad [10.58]$$

The kinetic energy of a particle at this level is

$$E_F = \frac{h\bar{k}^2_{max}}{2m} = \frac{h^2}{8m}\left(\frac{3N}{\pi}\right)^{1/3}. \quad [10.59]$$

We call E_F the Fermi energy for the particle.

The potential energy of a valence electron outside its metal and the potential energy of a nucleon removed from its nucleus are taken to be 0. But when the particle is well within the metal or nucleus, its potential energy is approximately constant at V_0. See figure 10.2.

The minimum energy needed to remove a particle from the low potential region is the work function W. At low temperatures, the energy of the most easily removed particle is just $V_0 + E_F$. The corresponding work function is W_0.

Example 10.6

If the density of silver is 10.50 g cm^{3} and its atomic mass 107.9 u, what is its Fermi energy?

Since the predominant valence of silver is +1, we consider that there is one free electron for each atom. Consequently, the number of such electrons in unit volume is

$$N = \frac{(10.50\,\text{g cm}^{-3})(6.022\times10^{23}\,\text{electrons mol}^{-1})(10^6\,\text{cm}^3\,\text{m}^{-3})}{107.9\text{ g mol}^{-1}}$$
$$= 5.86\times10^{28}\,\text{electrons m}^{-3}.$$

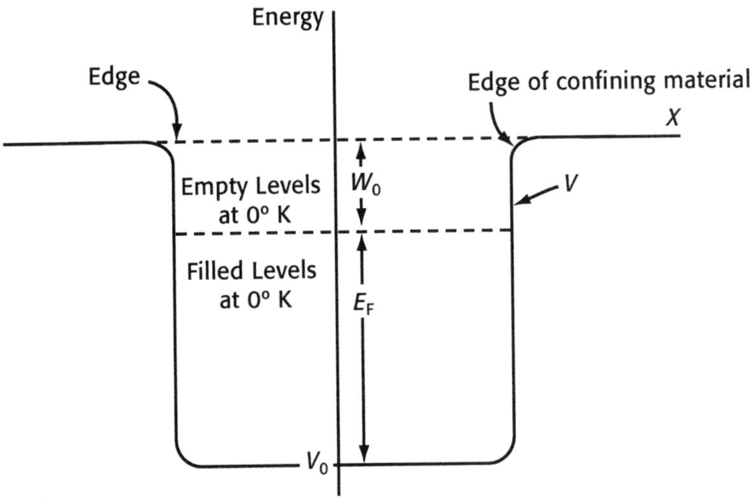

FIGURE 10.2 Relationship of the potential depth -V0 to the Fermi energy EF and to the work function W0 for a system of confined translating fermions.

Introducing this density and the constant from example 10.4 into formula (10.59) yields

$$E_F = \frac{1.5041 \times 10^{-18}\ \text{eV m}^2}{4}\left(\frac{3 \times 5.86 \times 10^{28}\,\text{m}^{-3}}{\pi}\right)^{2/3}$$
$$= 5.5\ \text{eV}.$$

In table 10.1, work functions and Fermi energies are listed for common metals.

10.9 Normalization and Orthogonality Conditions

Pure states of a given kind are subject to both normalization and orthogonality conditions. Here we will see how these arise for freely traveling particles.

A state function for translation in the x direction has form (10.17):

$$X(x) = Ae^{ik_x x}. \qquad [10.60]$$

But from formula (10.6), the probability that a particle is at a given x, per unit length, is X^*X. So if there is one particle between $x = 0$ and $x = a$, on the average, we have

$$\int_0^a X^* X\,dx = \int_0^a A^* e^{-ik_x x} A e^{ik_x x}\,dx$$
$$= A^* A \int_0^a dx = A^* Aa = 1. \qquad [10.61]$$

The integral of X^*X over x is called the *normalization integral* for function $X(x)$. It determines the magnitude but not the phase of the normalization constant.

If we also take A to be real, we have

$$A = \frac{1}{\sqrt{a}}. \qquad [10.62]$$

If there is a particle between $y = 0$ and $y = b$, on the average, we similarly have

$$B = \frac{1}{\sqrt{b}}. \qquad [10.63]$$

And a particle between $z = 0$ and $z = c$, on the average, leads to

$$C = \frac{1}{\sqrt{c}}. \qquad [10.64]$$

TABLE 10.1 Tanslational Data

Metal	*Work Function, eV*	*Fermi Energy, eV*
Ag	4.7	5.5
Au	4.8	5.5
Ca	3.2	4.7
Cu	4.1	7.1
K	2.1	2.1
Li	2.3	4.7
Na	2.3	3.1

For the complete state function,

$$\Psi = XYZT, \tag{10.65}$$

formula (10.21) now yields

$$N = \frac{1}{\sqrt{abc}}, \tag{10.66}$$

The corresponding complex form of the normalization constant is

$$N = \frac{e^{i\delta}}{\sqrt{abc}}, \tag{10.67}$$

where δ is a real parameter.

In the rectangularly symmetric translational motion, two states with different allowed wavevectors are physically independent. They are disjoint states. Let us suppose that the difference is in k_x. For the two states, we have

$$X_1 = A_1 e^{ik_{x1}x}, \qquad X_2 = A_2 e^{ik_{x2}x}. \tag{10.68}$$

The wavevectors are restricted by condition (10.24) in the forms

$$k_{x1} = n_{x1}\frac{2\pi}{a}, \qquad k_{x2} = n_{x2}\frac{2\pi}{a}. \tag{10.69}$$

Now, construct the integral

$$\begin{aligned} S_{12} &= \int_0^a X_1^* X_2 \, dx = A_1^* A_2 \int_0^a e^{i(k_{x2}-k_{x1})x} dx \\ &= A_1 * A_2 \left. \frac{e^{i(k_{x2}-k_{x1})x}}{i(k_{x2}-k_{x1})} \right|_0^a . \end{aligned} \tag{10.70}$$

The phase angle for the exponential at the upper limit is

$$\alpha_2 = (k_{x2} - k_{x1})a = (n_{x2} - n_{x1})2\pi = (\Delta n)2\pi, \tag{10.71}$$

while this angle at the lower limit is

$$\alpha_1 = (k_{x2} - k_{x1})0 = 0. \tag{10.72}$$

Since Δn is an integer, the exponential has the same value, one, at both limits. Consequently,

$$S_{12} = 0. \tag{10.73}$$

For complete state functions Ψ_i and Ψ_j, corresponding to different allowed wavevectors, we therefore obtain

$$S_{ij} = \int_{\substack{\text{rectangular} \\ \text{unit}}} \Psi_i^* \Psi_j \, dx \, dy \, dz = 0. \tag{10.74}$$

Expression S_{ij} is called the *overlap integral* between Ψ_i and Ψ_j. Whenever the overlap integral between two state functions is zero, the functions are said to be *orthogonal*.

10.10 Standing Wave Translational Functions

When a beam represented by formula (10.15) strikes a smooth reflecting wall, the component perpendicular to the wall is reversed. When the beam is introduced between parallel reflecting walls, successive reflections occur leading to the establishment of a standing wave.

Consider an independent particle confined within the rectangular box of figure 10.3. Suppose that the walls act as smooth planar reflectors. Also suppose that movement inside the box is free, at constant potential.

Let the coordinate axes lie along three intersecting edges, as figure 10.3 shows. Then reflection at either wall perpendicular to the x axis reverses the sign of k_x without altering its magnitude.

By symmetry, half of the time the particle is traveling in the positive direction of x between the two walls. The integral in equation (10.61) should then be set equal to 1/2 and, instead of (10.62), we obtain $1/\sqrt{2a}$ for A. The state function for this part of the movement is

$$X_1 = \frac{1}{\sqrt{2a}} e^{i|k_x|x}, \qquad [10.75]$$

where a is the distance between the reflecting walls. For the part of the movement in the negative direction between the two walls, we have

$$X_2 = \frac{1}{\sqrt{2a}} e^{-i|k_x|x}, \qquad [10.76]$$

In the first instance, the wavevector is $|k_x|$; in the second instance, $-|k_x|$.

Superposing X_1 and X_2 with the given phases leads to

$$\begin{aligned} X_3 &= \frac{1}{\sqrt{2a}} e^{i|k_x|x} + \frac{1}{\sqrt{2a}} e^{-i|k_x|x} \\ &= \sqrt{\frac{2}{a}} \frac{e^{i|k_x|x} + e^{-i|k_x|x}}{2} = \sqrt{\frac{2}{a}} \cos|k_x|x. \end{aligned} \qquad [10.77]$$

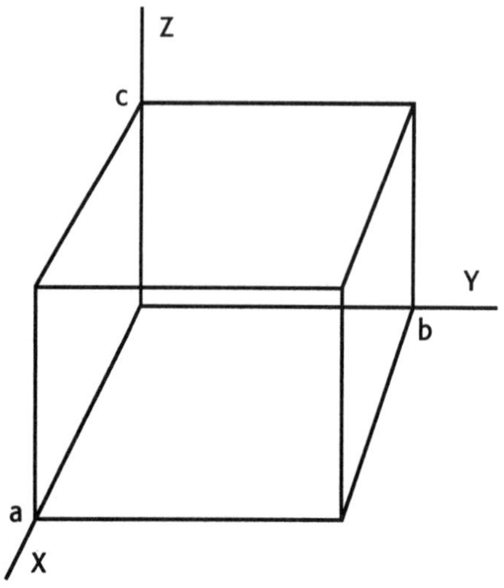

FIGURE 10.3 Coordinate axes for a confining rectangular box.

A real state function orthogonal to X_3 is constructed by multiplying X_1 by $1/i$, X_2 by $-1/i$, and adding:

$$X_4 = \frac{1}{i\sqrt{2a}} e^{i|k_x|x} - \frac{1}{i\sqrt{2a}} e^{-i|k_x|x} = \sqrt{\frac{2}{a}} \frac{e^{i|k_x|x} - e^{-i|k_x|x}}{2i} = \sqrt{\frac{2}{a}} \sin|k_x|x. \quad [10.78]$$

The general real superposition of X_3 and X_4 is

$$X = (\cos\alpha)X_3 + (\sin\alpha)X_4 = \sqrt{\frac{2}{a}}(\cos\alpha\cos|k_x|x + \sin\alpha\sin|k_x|x). \quad [10.79]$$

For wavevector $|k_y|$, the similar function is

$$Y = \sqrt{\frac{2}{b}}(\cos\beta\cos|k_y|y + \sin\beta\sin|k_y|y). \quad [10.80]$$

while for wavevector $|k_z|$, the similar function is

$$Z = \sqrt{\frac{2}{c}}(\cos\gamma\cos|k_z|z + \sin\gamma\sin|k_z|z). \quad [10.81]$$

10.11 Confined Rectangularly Symmetric Motion

Let us continue with the independent particle confined in a rectangular box. Let the axes be chosen as figure 10.3 shows. Also consider the potential function to be zero inside the box and infinite outside.

Each wall then acts as a perfect reflector. Beyond it, the probability density ρ, its factor Ψ, and the pertinent factor of Ψ, such as X, Y, or Z, vanish. Since the function Ψ is presumed to be continuous, it goes to zero inside as a wall is approached.

To make Ψ vanish at the wall where $x = 0$, we set

$$\cos\alpha = 0, \qquad \sin\alpha = 1 \quad [10.82]$$

in formula (10.79). To make Ψ vanish at the opposite wall, at $x = a$, we set

$$|k_x|a = n_x\pi \qquad \text{with} \quad n_x = 1, 2, 3, \ldots \quad [10.83]$$

Quantum number n_x cannot be zero because X must be different from zero within the box for the particle to be there. With these boundary conditions, the x factor in Ψ reduces to

$$X = \sqrt{\frac{2}{a}} \sin\frac{n_x\pi}{a}x \qquad \text{with} \quad n_x = 1, 2, 3, \ldots \quad [10.84]$$

In like manner, we obtain the factors

$$Y = \sqrt{\frac{2}{b}} \sin\frac{n_y\pi}{b}y \qquad \text{with} \quad n_y = 1, 2, 3, \ldots \quad [10.85]$$

$$Z = \sqrt{\frac{2}{c}} \sin\frac{n_z\pi}{c}z \qquad \text{with} \quad n_z = 1, 2, 3, \ldots \quad [10.86]$$

In the Newtonian approximation, the particle energy is

$$E = \frac{p^2}{2m} + 0 = \frac{1}{2m}({p_x}^2 + {p_y}^2 + {p_z}^2). \qquad [10.87]$$

But in quantum mechanics, each component of momentum is related to the corresponding wavevector component by the de Broglie equation

$$\mathbf{p} = \hbar\mathbf{k}. \qquad [10.88]$$

Taking the absolute value of the x component and introducing condition (10.83) leads to

$$\left|p_x\right| = \frac{h}{2\pi}\frac{n_x\pi}{a} = \frac{n_x h}{2a}. \qquad [10.89]$$

Similarly,

$$\left|p_y\right| = \frac{n_y h}{2b}. \qquad [10.90]$$

$$\left|p_z\right| = \frac{n_z h}{2c}. \qquad [10.91]$$

With these momenta, equation (10.87) yields

$$E = \frac{h^2}{8m}\left(\frac{{n_x}^2}{a^2} + \frac{{n_y}^2}{b^2} + \frac{{n_z}^2}{c^2}\right). \qquad [10.92]$$

Here E is the translational energy of the particle, h is Planck's constant, m is the mass of the particle, while a, b, c are the dimensions of the confining box and n_x, n_y, n_z are positive integers, quantum numbers identifying the state.

Note that for Ψ to vanish at opposite walls, the distance between them must be a half integral number of wavelengths. Decreasing this distance without changing the quantum number shortens the corresponding wavelength and increases the magnitude of the corresponding momentum. It thus raises the kinetic energy.

In general, a confined particle has a mean kinetic energy and a mean potential energy. Increasing the confinement without altering the shape of the potential or the shape of the Ψ would shorten the effective de Broglie wavelength in each small interior region and thus raise the kinetic energy there. The mean kinetic energy would be raised and also the mean potential energy. So increasing the confinement of a particle without introducing other changes raises its total energy.

Example 10.7

What does coefficient $h^2/(8m)$ equal when the translating particle is an electron? Divide the results obtained in example 10.4 by 4 to get

$$\frac{h^2}{8m} = 6.0247 \times 10^{-38}\ \text{J m}^2$$

$$= 3.7603 \times 10^{-19}\ \text{eV m}^2$$
$$= 37.603\ \text{eV A}^2.$$

Example 10.8

An electron is confined in a cubic box 8.00 Å^3 in size. What is its kinetic energy in its lowest state?

From the given volume, the length of an edge of the cubic box

$$a = b = c = \sqrt[3]{8.00}\ \text{Å} = 2.00\ \text{Å}.$$

In the ground state, the quantum numbers are as small as they can get:

$$n_x = n_y = n_z = 1.$$

Substituting these numbers and the pertinent one from example 11.7 into formula (10.92) yields

$$E = (37.603\ \text{eV A}^2)\frac{3}{4.00\ \text{A}^2}$$
$$= 28.2\,\text{eV}.$$

Thus, the particle in a box possesses kinetic energy when it is in its lowest state. In general, any confined particle would have a mean kinetic energy and a mean potential energy in this state. As a consequence, motion does not cease at absolute zero (0 K). The energy a system has in its lowest state is called its *zero point energy*.

10.12 Attenuated Motion

The symmetry argument behind construction of equation (10.3) does not require k to be real. It only requires successive elements of space and time, of a given size, to be equivalent. This equivalence is ensured by having the potential energy V of each particle constant and neighboring rays parallel throughout the pertinent region.

Consider a homogeneous beam of independent particles traveling in the positive direction and striking the barrier plotted in figure 10.4. Through the negative x region, potential V is zero; through the positive x region, potential V is constant and greater than the particle energy E; thus

$$V > E. \tag{10.93}$$

In the Newtonian approximation, the particle energy is

$$E = \frac{p^2}{2m} + V. \tag{10.94}$$

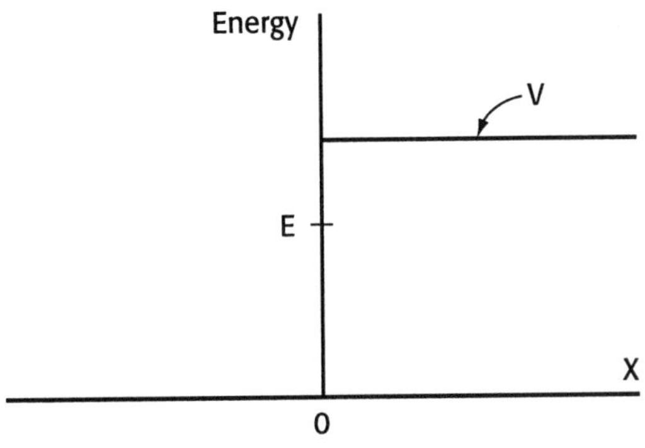

FIGURE 10.4 A step potential function

Solving for p^2,

$$p^2 = 2m(E - V) = -2m(V - E), \tag{10.95}$$

and taking the square root leads to

$$p = i\left[2m(V - E)\right]^{1/2}. \tag{10.96}$$

In classical physics, momentum p cannot be imaginary; the particles cannot move to the right of the origin. But in quantum physics, one can substitute (10.96) into the de Broglie equation to get

$$k = \frac{p}{\hbar} = i\frac{\left[2m(V - E)\right]^{1/2}}{\hbar} = i\kappa. \tag{10.97}$$

Expression κ defined by the last equality is called the *attenuation constant.*

With relationship (10.97), equation (10.5) becomes

$$\Psi = Ae^{i(i\kappa)x}e^{i\omega t} = Ae^{-\kappa x}e^{-i\omega t}. \tag{10.98}$$

Thus the state function can be extended into the positive x region. In addition to a reflected beam, there is penetration into this high potential region.

Formula (10.98) leads to the decreasing particle density

$$\rho = \Psi^*\Psi = A^*Ae^{-2\kappa x}. \tag{10.99}$$

Indeed, the probability that a particle penetrates distance x into the high potential region decreases exponentially with x. This penetration can be observed when V falls again below E, as in figure 10.5.

10.13 Joining Different Constant Potential Regions

But what conditions govern movement of particles across the junction between constant potential regions? What boundary conditions should be imposed?

Suppose that in the first region the potential of a particle is the constant V_1, in the second region its potential is the constant V_2. Let the two regions be connected by a transition zone in which the potential varies smoothly from V_1 to V_2.

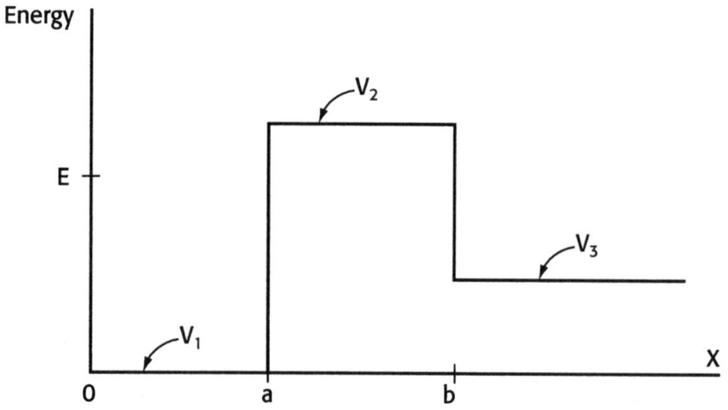

FIGURE 10.5 Piecewise constant potential energy barrier for a beam of particles.

Such variation in potential causes the probability density ρ to vary smoothly. The factors of ρ, state function Ψ and its complex conjugate, should vary smoothly. Furthermore, its first and immediate higher derivatives should also be continuous.

Decreasing the thickness of the transition zone would not destroy these continuities. However, it would cause the higher derivatives to change more abruptly. In the limit, when the thickness becomes zero, we should find that we can still take not only Ψ continuous, as we did at a confining wall, but also its first derivative normal to the boundary.

Consider the barrier in figure 10.5 struck from the left by a homogeneous beam of independent particles. Such a beam is partially reflected at $x = a$. It is attenuated on traveling to $x = b$ and is again partially reflected. A fraction of the beam emerges and travels on.

In the left region, where $x < a$, the forward and backward traveling de Broglie waves superpose:

$$\psi_1 = A_1 e^{ik_1 x} + B_1 e^{-ik_1 x}. \qquad [10.100]$$

The magnitude of the wavevector is given by (10.97) with V_1 zero:

$$k_1 = \frac{(2mE)^{1/2}}{\hbar}. \qquad [10.101]$$

For the central region, where $a < x < b$, we similarly find that

$$\psi_2 = A_2 e^{-\kappa x} + B_2 e^{\kappa x}. \qquad [10.102]$$

with

$$\kappa = \frac{\left[2m(V_2 - E)\right]^{1/2}}{\hbar}. \qquad [10.103]$$

For the right region, the forward traveling de Broglie wave is

$$\psi_3 = A_3 e^{ik_3 x} \qquad [10.104]$$

with

$$k_3 = \frac{\left[2m(E - V_3)\right]^{1/2}}{\hbar}. \qquad [10.105]$$

Equating Ψ and $d\Psi/dx$ from the first and second regions at $x = a$ imposes two conditions on the coefficients. Similarly, equating Ψ and $d\Psi/dx$ from the second and third regions at $x = b$ imposes two more conditions. Since there are five coefficients, one may be chosen arbitrarily. This may be A_1 which is determined by the intensity and phasing of the initial beam. From the physics of the situation, we know that these can be chosen arbitrarily, so no difficulty arises.

A nonzero A_1 leads to a nonzero A_3, regardless of the potential in the intermediate region. When there is a beam incident on the barrier, some of the beam passes on through. In classical physics, such transmission could not occur when V became greater than E unless a tunnel was constructed below the E level, through the barrier. As a consequence, the peculiar quantum mechanical transmission is called *tunneling*.

Inside a high Z nucleus, an alpha particle may form from two protons and two neutrons. As it leaves the nucleus, its bonds to other nucleons are broken and the electrostatic potential energy remains. Initially this is above the total energy of the particle. With increasing distance, it falls below. So there is a potential energy barrier inhibiting the process. But it does occur, as we know. Qualitatively, the situation is similar to that considered here.

In a tunnel diode, a barrier roughly 100 Å thick lies between two doped semiconductors. A small forward bias causes current to flow. But when the band of conducting electrons on one side is raised so that an appreciable part of it coincides with the forbidden band on the other side, the current drops with increasing bias. This negative resistance behavior can be used to amplify signals.

Example 10.9

How are the reflected and the transmitted waves related to an incident de Broglie wave for the potential function in figure 10.4?

To the left of the origin, the forward and the backward traveling waves superpose:

$$\psi_1 = A_1 e^{ikx} + B_1 e^{-ikx}.$$

But in the high potential region, there is no reflected wave; so

$$\psi_2 = A_2 e^{-\kappa x}.$$

The pertinent derivatives are

$$\frac{d\psi_1}{dx} = ikA_1 e^{ikx} - ikB_1 e^{-ikx}$$

and

$$\frac{d\psi_2}{dx} = -\kappa A_2 e^{-\kappa x}.$$

At the boundary, at $x = 0$, the state functions and the first derivatives are equal. And since $e^0 = 1$, we have

$$A_1 + B_1 = A_2,$$

$$ikA_1 - ikB_1 = -\kappa A_2.$$

Solving these equations for A_2 and B_1 yields

$$A_2 = \frac{2A_1}{1 + i(\kappa / k)} = \left(\frac{1 + i(\kappa / k)}{1 + i(\kappa / k)} + \frac{1 - i(\kappa / k)}{1 + i(\kappa / k)}\right) A_1$$

$$= (1 + e^{-2i\alpha}) A_1.$$

$$B_1 = \frac{1 - i(\kappa / k)}{1 + i(\kappa / k)} A_1 = e^{-2i\alpha} A_1.$$

Angle α is defined by the equation

$$\alpha = \tan^{-1}(\kappa / k).$$

Note that the magnitude of B_1 equals that of A_1 when κ is real. Then the reflected wave is of the same intensity as the incident one.

Substituting the expressions for A_2 and B_1 into the original forms and reducing leads to

$$\psi = 2Ae^{-i\alpha}\cos(kx + \alpha) \quad \text{for} \quad x < 0,$$

$$\psi = 2Ae^{-i\alpha}\cos\alpha \; e^{-\kappa x} \quad \text{for} \quad x > 0.$$

Here we have dropped the subscript 1 on A and have dropped the subscripts on the Ψ's.

10.14 A Differential Equation for Multidirectional Propagation

The general de Broglie wave for a given energy state includes components traveling in both the $+x$ and $-x$ directions, both the $+y$ and $-y$ directions, and both the $+z$ and $-z$ directions at each pertinent point in space. So the differential equation previously constructed has to be generalized.

For a homogeneous beam propagating in one direction along a constant potential line, with neighboring rays parallel, formula (10.3) applies. If s measures distance in the propagation direction on the line, we have

$$\frac{\mathrm{d}\Psi_1}{\Psi_1} = ik\,\mathrm{d}s - i\omega\,\mathrm{d}t. \quad [10.106]$$

But from equations (10.12) and (10.14), we obtain the expression

$$k\,\mathrm{d}s = k_x\mathrm{d}x + k_y\mathrm{d}y + k_z\mathrm{d}z, \quad [10.107]$$

which converts (10.106) to

$$\mathrm{d}\Psi_1 = ik_x\Psi_1\,\mathrm{d}x + ik_y\Psi_1\,\mathrm{d}y + ik_z\Psi_1\,\mathrm{d}z - i\omega\Psi_1\,\mathrm{d}t, \quad [10.108]$$

whence

$$\frac{\partial\Psi_1}{\partial x} = ik_x\Psi_1. \quad [10.109]$$

Here Ψ_1 represents the component of Ψ with the wavevector k. But at the same energy, there may be a component propagating in the opposite direction. Its wavevector would be $-k$. Reversing the sign of each wavevector component in (10.108) leads to

$$\mathrm{d}\Psi_2 = -ik_x\Psi_2\,\mathrm{d}x - ik_y\Psi_2\,\mathrm{d}y - ik_z\Psi_2\,\mathrm{d}z - i\omega\Psi_2\,\mathrm{d}t, \quad [10.110]$$

whence

$$\frac{\partial\Psi_2}{\partial x} = -ik_x\Psi_2. \quad [10.111]$$

Differentiating (10.109), (10.111), then substituting for the first derivatives gives

$$\frac{\partial^2\Psi_1}{\partial x^2} = ik_x\frac{\partial\Psi_1}{\partial x} = -k_x^{\;2}\,\Psi_1 \quad [10.112]$$

and

$$\frac{\partial^2\Psi_2}{\partial x^2} = -ik_x\frac{\partial\Psi_2}{\partial x} = -k_x^{\;2}\,\Psi_2. \quad [10.113]$$

But since the state function equals the superposition of its components,

$$\Psi = \Psi_1 + \Psi_2, \quad [10.114]$$

equations (10.112) and (10.113) are added to form

$$\frac{\partial^2\Psi}{\partial x^2} = -k_x^{\;2}\,\Psi. \quad [10.115]$$

In a similar way, one obtains

$$\frac{\partial^2\Psi}{\partial y^2} = -k_y^{\;2}\,\Psi. \quad [10.116]$$

and

$$\frac{\partial^2\Psi}{\partial z^2} = -k_z^{\ 2}\Psi. \qquad [10.117]$$

Since

$$k_x^{\ 2} + k_y^{\ 2} + k_z^{\ 2} = k^2, \qquad [10.118]$$

adding equations (10.115), (10.116), and (10.117) leads to

$$\frac{\partial^2\Psi}{\partial x^2} + \frac{\partial^2\Psi}{\partial y^2} + \frac{\partial^2\Psi}{\partial z^2} = -k^2\Psi. \qquad [10.119]$$

In rectangular coordinates, the *del operator* is represented as

$$\nabla = \mathrm{i}\frac{\partial}{\partial x} + \mathrm{j}\frac{\partial}{\partial y} + \mathrm{k}\frac{\partial}{\partial z}. \qquad [10.120]$$

But since ***i***, ***j***, and ***k*** are mutually perpendicular, del dot del is given by

$$\nabla^2 = \frac{\partial^2}{\partial x^2} + \frac{\partial^2}{\partial y^2} + \frac{\partial^2}{\partial z^2}. \qquad [10.121]$$

So equation (10.119) may be rewritten in the form

$$\nabla^2\Psi + k^2\Psi = 0 \qquad [10.122]$$

From the de Broglie equation and the Newtonian expression for the energy of the particle under consideration, we have

$$k^2 = \frac{p^2}{\hbar^2} = \frac{2m(E-V)}{\hbar^2}. \qquad [10.123]$$

Substituting this expression into equation (10.122) yields

$$\nabla^2\Psi + \frac{2m}{\hbar^2}(E-V)\Psi = 0 \qquad [10.124]$$

Thus we have constructed a single differential equation governing the multidirectional propagation.

From equations (10.108), (10.110), and (10.114), we find that

$$\frac{\partial\Psi}{\partial t} = -i\omega\Psi. \qquad [10.125]$$

Let us recall the Einstein equation (2.35) relating the energy of a photon to the corresponding angular frequency,

$$E = \hbar\omega. \qquad [10.126]$$

Just as the de Broglie equation holds for both photons and particles with a rest mass, we presume that this equation does also.

Multiplying equation (10.125) by $-\hbar/i$ leads to

$$-\frac{\hbar}{i}\frac{\partial}{\partial t}\Psi = \hbar\omega\Psi = E\Psi. \qquad [10.127]$$

In this equation, an operator acting on the Ψ for a *definite* energy *state* yields this energy times Ψ. We call E the *eigenvalue* for the operator. The relationship is called an *eigenvalue equation.*

10.15 The Schrödinger Equations

In the general situation, the potential V varies smoothly with the coordinates. But over a small region, the variation is small. About the limit, when the region becomes infinitesimal, the variation is infinitesimal and the effect on altering Ψ is infinitesimal of higher order.

Neglecting these higher order effects, we consider that equation (10.124) still holds. Thus, we write

$$\nabla^2\Psi + \frac{2m}{\hbar^2}\left[E - V(\mathbf{r})\right]\Psi = 0 \qquad [10.128]$$

a relationship known as the *time-independent Schrödinger equation.*

A rearranged form of (10.128) is

$$\left[-\frac{\hbar^2}{2m}\nabla^2 + V(\mathbf{r})\right]\Psi = E\Psi. \qquad [10.129]$$

This has the eigenvalue equation form

$$H\Psi = E\Psi, \qquad [10.130]$$

in which the *Hamiltonian operator* is

$$H = -\frac{\hbar^2}{2m}\nabla^2 + V(\mathbf{r}). \qquad [10.131]$$

One considers - $(\hbar^2/2m)\,\nabla^2$ to be the kinetic energy part of the operator.

Introducing relationship (10.127) into the right side of (10.129) leads to

$$\left[-\frac{\hbar^2}{2m}\nabla^2 + V(\mathbf{r})\right]\Psi = -\frac{\hbar}{i}\frac{\partial}{\partial t}\Psi, \qquad [10.132]$$

the *time-dependent Schrödinger equation.* Since this equation no longer contains E as a parameter, it may be applied to a system undergoing change. A person can consider that it governs the temporal development of the system.

When a system is in a definite state, the variation of Ψ over space is independent of its variation over time. As a consequence, we have

$$\Psi = \psi(\mathbf{r})T(t). \qquad [10.133]$$

Substitute this form into (10.128), factor out $T(t)$, and cancel $T(t)$ to get

$$\nabla^2\psi + \frac{2m}{\hbar^2}\left[E - V(\mathbf{r})\right]\psi = 0 \qquad [10.134]$$

a convenient form of the time-independent Schrödinger equation. As before, this rearranges to

$$H\psi = E\psi, \qquad [10.135]$$

with the Hamiltonian operator H given by formula (10.131).

10.16 Suitable Solutions

We have seen how a particle, and a system of independent particles, is governed by a state function. Let us now consider what general conditions such a function must satisfy.

In principle, a person can determine the probability density ρ as a function of position and time for a given system. A state of the system is believed to provide a given distribution. As the measurement point varies, and potential V varies smoothly, density ρ presumably also varies smoothly, continuously, with continuous derivatives. The integral of ρ over a region in which one particle is found would equal 1.

But state function Ψ is related to ρ by the factoring

$$\rho = \Psi^{*}\Psi. \quad [10.136]$$

While this determines the magnitude of Ψ, it does not determine the phase of Ψ. Nevertheless, we have seen how the phase involves k^2, ω, and an arbitrary constant. The last appears in the phase for the normalization factor. Integrating the probability density yields the restriction

$$\int_{\substack{\text{volume}\\\text{containing}\\\text{1 particle}}} \Psi^{*}\Psi \mathrm{d}^3\mathbf{r} = 1. \quad [10.137]$$

Here $\mathrm{d}^3\boldsymbol{r}$ is the element of volume. A function that can be normalized to meet condition (10.137) is said to be *quadratically integrable*. Note that this condition does not prevent Ψ from becoming infinite at a point.

We are thus led to impose the following requirements:

(a) State function Ψ is *continuous*.
(b) State function Ψ is *single-valued* insofar as motion over space is concerned.
(c) State function Ψ must be *quadratically integrable*.
(d) State function Ψ must satisfy the imposed *boundary conditions*.

Most of the analytical solutions of a given Schrödinger equation do not satisfy all of these conditions. While the Schrödinger equation for the system can be solved for any value of the parameter E, only the properly quantized values of E yield suitable Ψ functions.

The different acceptable solutions for a given potential $V(\mathbf{r})$ can be superposed. The result is a function still meeting the above requirements. However, there is *no* other condition on it. Thus, the eigenfunctions form a *complete* basis set.

10.17 Expressing ∇^2 in Orthogonal Generalized Coordinates

In treating atoms and molecules, a person finds that the Laplacian ∇^2 in the Schrödinger equation needs to be expressed in coordinates fitting the symmetry or the approximate symmetry of the physical system. Very useful is a representation in orthogonal coordinates.

Let the generalized coordinates replacing the rectangular coordinates be q_1, q_2, q_3. Let the three components of displacement when these numbers increase by infinitesimals $\mathrm{d}q_1$, $\mathrm{d}q_2$, $\mathrm{d}q_3$ be $h_1\, \mathrm{d}q_1$, $h_2\, \mathrm{d}q_2$ and $h_3\, \mathrm{d}q_3$, respectively. Furthermore, take these elements to be mutually perpendicular; as figure 10.6 illustrates. The coordinates are then said to be *orthogonal*.

Let us consider a model conserved effect diffusing in the space under consideration. The amount transported from a small section of one layer to the corresponding section of the next varies directly with the cross sectional area, with the time, and with the driving force. About a given concentration of the effect, this force is proportional to the negative concentration gradient.

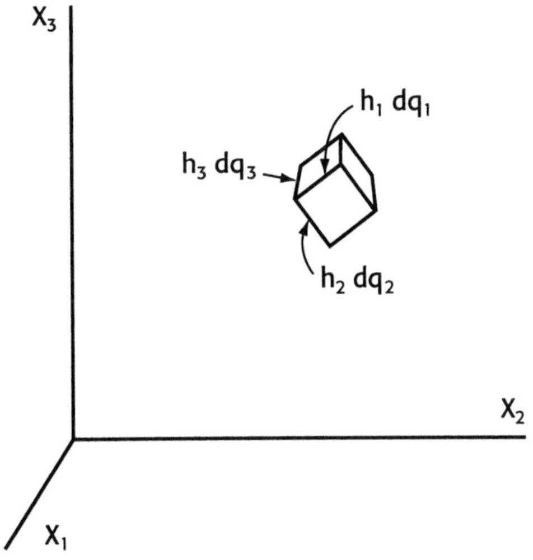

FIGURE 10.6 Volume element bordered by orthogonal line elements.

Let us measure each portion of the effect in a certain unit and report the result as number N. The concentration about a given point is represented by C, an area by S, and the time by t.

For movement through the volume element in figure 10.6 in the direction of increasing q_1, we have the cross sectional area

$$\mathrm{d}S_{q_1} = (h_2\,\mathrm{d}q_2)(h_3\,\mathrm{d}q_3) \qquad [10.138]$$

and the concentration gradient

$$\frac{\partial C}{h_1\,\partial q_1}. \qquad [10.139]$$

With D the coefficient of proportionality, the amount transported through area $\mathrm{d}S_{q_1}$ in time $\mathrm{d}t$ is

$$\mathrm{d}^3N_{q_1} = -D\frac{\partial C}{h_1\,\partial q_1}\mathrm{d}S_{q_1}\mathrm{d}t, \qquad [10.140]$$

whence

$$\frac{\partial \mathrm{d}^2N_{q_1}}{\partial t} = -D\frac{h_2h_3}{h_1}\frac{\partial C}{\partial q_1}dq_2dq_3. \qquad [10.141]$$

The rate of accumulation within the element caused by the excess of movement into the element through the left face over movement out through the right face is given by the negative differential of expression (10.141):

$$\frac{\partial}{\partial q_1}\left(D\frac{h_2h_3}{h_1}\frac{\partial C}{\partial q_1}\right)dq_1\,dq_2\,dq_3. \qquad [10.142]$$

The other two pairs of faces contribute similar expressions. Adding these and dividing by the volume of the element, which is $h_1h_2h_3\,\mathrm{d}q_1\,\mathrm{d}q_2\,\mathrm{d}q_3$, yields the rate of change in the concentration:

$$\frac{\partial C}{\partial t} = \frac{1}{h_1h_2h_3}\left[\frac{\partial}{\partial q_1}\left(D\frac{h_2h_3}{h_1}\frac{\partial C}{\partial q_1}\right) + \frac{\partial}{\partial q_2}\left(D\frac{h_3h_1}{h_2}\frac{\partial C}{\partial q_2}\right) + \frac{\partial}{\partial q_3}\left(D\frac{h_1h_2}{h_3}\frac{\partial C}{\partial q_3}\right)\right]. \qquad [10.143]$$

When coefficient D is constant, equation (10.143) reduces to

$$\frac{\partial C}{\partial t} = D\frac{1}{h_1h_2h_3}\left[\frac{\partial}{\partial q_1}\left(\frac{h_2h_3}{h_1}\frac{\partial C}{\partial q_1}\right)+\frac{\partial}{\partial q_2}\left(\frac{h_3h_1}{h_2}\frac{\partial C}{\partial q_2}\right)+\frac{\partial}{\partial q_3}\left(\frac{h_1h_2}{h_3}\frac{\partial C}{\partial q_3}\right)\right]. \quad [10.144]$$

In rectangular coordinates, equation (10.144) becomes

$$\frac{\partial C}{\partial t} = D\left(\frac{\partial^2 C}{\partial x_1{}^2}+\frac{\partial^2 C}{\partial x_2{}^2}+\frac{\partial^2 C}{\partial x_3{}^2}\right). \quad [10.145]$$

We recognize the expression multiplying D as the Cartesian form of $\nabla^2 C$. Consequently, formula (10.144) tells us that

$$\nabla^2 C = \frac{1}{h_1h_2h_3}\left[\frac{\partial}{\partial q_1}\left(\frac{h_2h_3}{h_1}\frac{\partial C}{\partial q_1}\right)+\frac{\partial}{\partial q_2}\left(\frac{h_3h_1}{h_2}\frac{\partial C}{\partial q_2}\right)+\frac{\partial}{\partial q_3}\left(\frac{h_1h_2}{h_3}\frac{\partial C}{\partial q_3}\right)\right]. \quad [10.146]$$

Here q_1, q_2, q_3 are the orthogonal generalized coordinates while h_1, h_2, h_3 are the multipliers that change dq_1, dq_2, dq_3 into the perpendicular line elements.

Example 10.10

Express $\nabla^2\Psi$ in spherical coordinates.

The spherical coordinates of a point are defined with respect to a center and Cartesian axes based on the center as figure 10.7 shows. The line element traced out when only r changes is dr. When only θ increases, the line element is $r\,d\theta$. When only ϕ increases, it is $r \sin\theta\, d\phi$.

So with $q_1 = r$, $q_2 = \theta$, and $q_3 = \phi$, we have

$$q_1 = r, \qquad q_2 = \theta, \qquad \text{and} \quad q_3 = \varphi,$$
$$h_1 = 1, \qquad h_2 = r, \qquad h_3 = r\sin\theta.$$

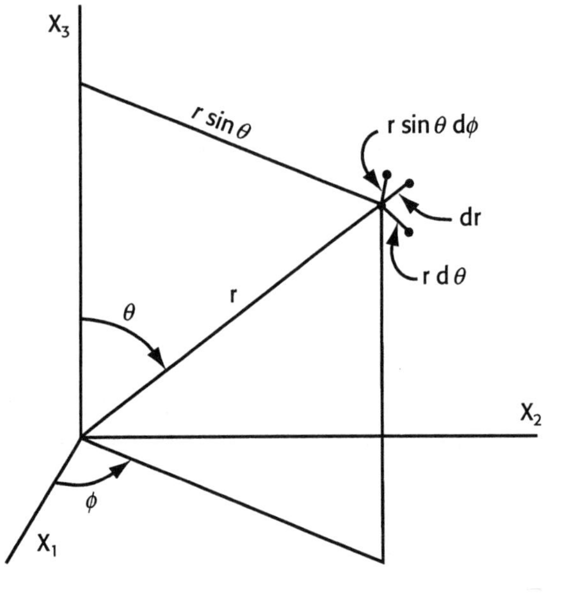

FIGURE 10.7 Three line elements traced out when each spherical coordinate increases infinitesimally by itself from a given point.

If we also replace C with Ψ, equation (10.146) becomes

$$\nabla^2\psi = \frac{1}{r^2 \sin\theta}\left[\frac{\partial}{\partial r}\left(\frac{r^2 \sin\theta}{1}\frac{\partial\psi}{\partial r}\right)+\frac{\partial}{\partial\theta}\left(\frac{r\sin\theta}{r}\frac{\partial\psi}{\partial\theta}\right)+\frac{\partial}{\partial\varphi}\left(\frac{r}{r\sin\theta}\frac{\partial\psi}{\partial\varphi}\right)\right].$$

$$= \frac{1}{r^2}\frac{\partial}{\partial r}\left(r^2\frac{\partial\psi}{\partial r}\right)+\frac{1}{r^2\sin\theta}\frac{\partial}{\partial\theta}\left(\sin\theta\frac{\partial\psi}{\partial\theta}\right)+\left(\frac{1}{r^2\sin^2\theta}\frac{\partial^2\psi}{\partial\varphi^2}\right).$$

Example 10.11

Construct a differential volume element for spherical coordinates.

From figure 10.7, we see that the line elements dr, $r\,d\,\theta$, $r\sin\theta\,d\phi$ are mutually perpendicular. Furthermore, these form the edges of an infinitesimal volume element such as figure 11.6 shows. So we have

$$\mathrm{d}^3\mathbf{r} = (\mathrm{d}r)(r\,\mathrm{d}\theta)(r\sin\theta\,\mathrm{d}\varphi)$$
$$= r^2\sin\theta\,\mathrm{d}r\,\mathrm{d}\varphi.$$

Introducing the element of solid angle

$$\mathrm{d}\Omega = \sin\theta\,\mathrm{d}\theta\,\mathrm{d}\varphi$$

yields the alternate form

$$\mathrm{d}^3\mathbf{r} = r^2\mathrm{d}\Omega\,\mathrm{d}r,$$

which can also be constructed from the description of $\mathrm{d}\Omega$ in section 8.6.

Questions

10.1 What is wrong with the classical description of a particle?
10.2 How can one give the uncertainty in a particle's position and momentum (or velocity) a fundamental role?
10.3 Explain how a complex function is flexible enough to describe a particle.
10.4 Why are observable properties derivable from the state function Ψ?
10.5 What is the principle of continuity? Do nonlinear systems obey this principle?
10.6 What is an ideal homogeneous beam?
10.7 What symmetries appear in the homogeneous beam?
10.8 How do these symmetries affect the state function?
10.9 Show how the state function for a homogeneous beam is constructed.
10.10 How is this state function normalized?
10.11 When does coherence between the state functions for different particles prevail?
10.12 When is incoherence between the state functions for different particles introduced?
10.13 How do the spatial factors for independent motions within a pure state combine? Explain.
10.14 Define wavevector, angular frequency, phase velocity, wavelength, conventional frequency, de Broglie wave.
10.15 How is the momentum of a particle related to its wavevector? What direct evidence do we have for this relationship?
10.16 How do the waves for alternative motions combine?
10.17 Discuss the behavior of (a) the molecules in a gas, (b) the conduction electrons in a metal.
10.18 Justify the use of periodic boundary conditions in a rectangularly symmetric system.
10.19 Construct a formula for the translational energy levels of a rectangularly symmetric system.
10.20 What is degeneracy? When does it appear in the rectangularly symmetric system?
10.21 Define the partition function Z. What distribution law does its use imply?
10.22 Explain how the internal energy E depends on Z.
10.23 Deduce a formula for the translational partition function.

10.24 What is the Pauli exclusion principle?

10.25 How does one determine the number of translational states below a certain energy? How are these states filled?

10.26 Construct a formula for the Fermi energy.

10.27 Show how a state function for translation is normalized.

10.28 Show that states with different allowed wavevectors are orthogonal.

10.29 From the normalized traveling-wave translational functions, construct normalized standing-wave functions.

10.30 Show how these are fitted to a rectangular box.

10.31 Then deduce the corresponding allowed energy levels.

10.32 Why does increasing the confinement of a particle without altering the shape of the confining potential raise both components of its energy?

10.33 What is zero-point energy?

10.34 Why can the state function for a particle extend into a region where $V > E$?

10.35 Explain how the state function behaves at the boundary between two different constant potential regions.

10.36 What is tunneling?

10.37 For particles moving in a particular direction in a constant potential region, symmetry conditions yielded the equation

$$\frac{d\Psi}{\Psi} = ik\,ds - i\omega\,dt.$$

Why is this not adequate for multidirectional movement in the region?

10.38 Show how this equation is generalized to allow for such movement.

10.39 Define operator, eigenvalue, eigenvalue equation, Hamiltonian operator.

10.40 Justify the two Schrödinger equations.

10.41 What general conditions must a state function satisfy?

10.42 Why are most analytical solutions of the Schrödinger equation rejected?

10.43 What is an orthogonal coordinate system? Give examples.

10.44 By considering diffusion of a hypothetical conserved effect, deduce the form for ∇^2 in orthogonal generalized coordinates.

10.45 How is ∇^2 expressed in spherical coordinates?

Problems

10.1 A beam of electrons is accelerated from rest by a potential increment of 120 V. Calculate the resulting (a) speed v of an electron, (b) wavevector k, (c) wavelength λ.

10.2 The beam in problem 10.1 carried, on the average, 10.0 electrons per second through each 1.00 Å^2 cross section. Construct the real normalization constant so that $\Psi^*\Psi$ yields (a) the probability density ρ, (b) the intensity at a stationary point in the beam.

10.3 Calculate ω for the beam in problem 10.1.

10.4 Construct the state function Ψ for the beam in problem 10.1.

10.5 Resolve the state function $A(\cos ax)e^{-i\omega t}$ into unidirectional translational functions.

10.6 If the density of copper is 8.92 g cm^{-3} and its atomic mass 63.546 u, what is its Fermi energy?

10.7 An electron confined in a cubic box has 7.02 eV translational energy. What is the size of the box if the electron is in its lowest state?

10.8 An electron is confined in a cubic box 8.00 Å^3 in size. How much energy is needed to raise it from its lowest state to its first excited state?

10.9 For a particle confined at constant potential within a cubic box, determine the quantum numbers of (a) the lowest excited state that is not degenerate, and (b) the lowest energy level with a degeneracy of 6. Neglect spin.

10.10 An electron is confined in a rectangular box with edges 2.00 Å, 1.00 Å, and 3.00 Å long. Calculate the normalization factor for the state function.

10.11 A particle moves freely back and forth between the planes $x = -a/2$ and $x = a/2$. Sketch the forms of the state function for the lowest five levels. Then construct $X(x)$ for the third excited state.

10.12 Each electron in a homogeneous beam striking a barrier of height 10.0 eV has 1.00 eV kinetic energy. If the barrier is essentially infinite in depth, at what distance is the particle density reduced to e^{-1} of its value at the surface?

10.13 For the system in problem 10.12, what fraction of the maximum probability density in the free region is ρ at the surface of the barrier?

10.14 For the system in problem 10.12, how far from the barrier is the first maximum in the state function?

10.15 The state function to the right of the first maximum of problem 10.14 can be reflected to the left to fit the potential in figure 10.8. Then a is the distance calculated in problem 10.14, Determine the corresponding energy if impenetrable walls were at $x = -a$ and at $x = a$. By how much is the energy lowered by making the height of the barrier 10.0 eV, as we have done?

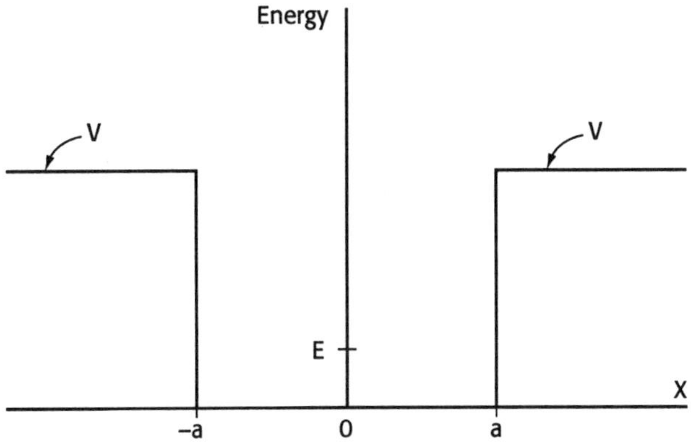

FIGURE 10.8 Symmetrical rectangular potential well with the bottom at zero energy.

— — —

10.16 A beam of protons is accelerated from rest by a potential decrement of 12 5 keV. Calculate the resulting (a) speed v of a proton, (b) wavevector k, (c) wavelength λ, (d) angular frequency ω.

10.17 The beam in problem 10.16 carried, on the average, 10.0 protons per second through each 1.00 Å^2 cross section. Calculate the normalization constant A.

10.18 Resolve the state function $A(\cosh ax)e^{-i\omega t}$ into unidirectional translational functions.

10.19 Construct expressions for wavevector k and angular frequency ω at a point in a region of varying potential where $\Psi = A \exp [if(x,t)]$.

10.20 Show that in a metal the average translational energy of a conduction electron is $(3/5)E_F$.

10.21 If the density of gold is 19.3 g cm^{-3} and its atomic mass 196.97 u, what is its Fermi energy?

10.22 An electron confined in a cubic box has 5.53 eV translational energy. What is the size of the box if the electron is in its first excited state?

10.23 An electron confined in a cubic box 5.00 Å^3 in volume is in its third excited level. How much energy would it lose on falling to its first excited level?

10.24 For a particle confined at constant potential within a cubic box, find the quantum numbers associated with the lowest level represented by state functions that cannot all be reorientations of each other.

10.25 An electron is confined in a rectangular box with edges 1.25 Å, 2.75 Å, 2.10 Å long. Determine the maximum in the state function. What multiple of the average particle density is the density at the maximum?

10.26 Each electron in a homogeneous beam striking the barrier shown in figure 10.4 possesses 1.00 eV kinetic energy. If the probability density ρ fell by one half for each 1.00 Å penetration, what is the height of the barrier?

10.27 For the system in problem 10.26, what fraction of the maximum probability density in the free region is ρ at the surface of the barrier?

10.28 For the system in problem 10.26, how far from the barrier is the first maximum in the state function?

10.29 The state function to the right of the first maximum for problem 10.28 can be reflected to the left to fit the potential in figure 10.8, Then a is the distance calculated in problem 10.14, Determine the corresponding energy if impenetrable walls were at x = -a and at x = a . By how much is the energy lowered by reducing the height of the barriers from infinity to the number calculated in problem 10.26?

10.30 Show that

$$\frac{1-i(\kappa / k)}{1+i(\kappa / k)} = e^{-2i\alpha}$$

when

$$\alpha = \tan^{-1}(\kappa / k).$$

References

Books

Borowitz, S.: 1967, *Fundamentals of Quantum Mechanics*, W. A. Benjamin, New York, pp 1-169.

Borowitz reviews classical wave theory and particle dynamics in detail. Then he constructs the wave equation much as Schrödinger did. A comparison with the Bohr theory is made and the wave function is interpreted following Born and Jordan.

Bunge, M.: 1985, *Treatise on Basic Philosophy*, vol. 7, part I, *Formal and Physical Sciences*, Reidel, Dordrecht, Holland, pp. 165-219.

While there is general agreement on the mathematics of quantum mechanics, there is considerable disagreement on its interpretation. Bunge criticizes the various extant views in detail, finally arriving at a tenable position.

Duffey, G. H.: 1984, *A Development of Quantum Mechanics Based on Symmetry Considerations*, Reidel, Dordrecht, Holland, pp. 1-37, 113-120.

More details on the approach followed in this chapter are presented here.

Duffey, G. H.: 1992, *Quantum States Processes*, Prentice- Hall, Englewood Cliffs, NJ, pp. 1-105.

The mathematical representation of quantum states by functions and then by vectors in Hilbert space is considered in chapter 1. In chapters 2 and 3, how the possible symmetries over physical space and time, and the assumed localizability of particles, lead to the commutation relations and to the form of the Hamiltonian operator is developed.

Articles

Baggott, J.: 1990, "Quantum Mechanics and the Nature of Physical Reality," *J. Chem. Educ.* **67**, 638-642.

Ballentine, L. E.: 1986, "Probability Theory in Quantum Mechanics," *Am. J. Phys.* **54**, 883-889.

Bardou, F.: 1991, "Transition between Particle Behavior and Wave Behavior," *Am. J. Phys.* **59**, 458-461.

Cereceda, J. L.: 1996, "An Apparent Paradox at the Heart of Quantum Mechanics," *Am. J. Phys.* **64**, 459-456.

El-Issa, B. D.: 1936, "The Particle in a Box Revisited," *J. Chem. Educ.* **63**, 761-764.

Fucaloro, A. F.: 1986, "Some Characteristics of Approximate Wave Functions," *J. Chem. Educ.* **63**, 579-581.

Gellene, G. I.: 1995, "Resonant States of a One-Dimensional Piecewise Constant Potential," *J. Chem. Educ.* **72**, 1015-1018.

Gutierrez, G., and Yanez, J. M.: 1997, "Can an Ideal Gas Feel the Shape of its Container?" *Am. J. Phys.* **65**, 739-743.

Jordan, T. F.: 1991, "Assumptions Implying the Schrödinger Equation," *Am. J. Phys.* **59**, 606-608.

Kash, M. M., and Shields, G. C.: 1994, "Using the Franck- Hertz Experiment to Illustrate Quantization," *J. Chem. Educ.* **71**, 466-468.

Leming, C. W., and Smith, A. F.: 1991, "A Numerical Study of Quantum Barrier Penetration in One Dimension," *Am. J. Phys.* **59**, 441-443.

Li, W. -K., and Blinder, S. M.: 1987, "Particle in an Equilateral Triangle: Exact Solution of a Non-separable Problem," *J. Chem. Educ.* **64**, 130-132.

Liang, Y. Q., and Dardenne, Y. X.: 1995, "Momentum Distributions for a Particle in a Box," *J. Chem. Educ.* **72**, 148- 151.

Mermin, N. D.: 1994, "Quantum Mysteries Refined," *Am. J. Phys.* **62**, 880-887.

Rioux, F.: "Numerical Methods for Finding Momentum Space Distributions," *J. Chem. Educ.* **74**, 605-606.

Styer, D. F.: 1996, "Common Misconceptions Regarding Quantum Mechanics," *J. Chem. Educ.* **64**, 31-34.

Torre, A. C. de la, and Dotson, A. C.: 1996, "An Entangled Opinion on the Interpretation of Quantum Mechanics," *Am. J. Phys.* **64**, 174.

Volkamer, K., and Lerom, M. W.: 1992, "More about the Particle- in-a-Box System: The confinement of Matter and the Wave-Particle Dualism," *J. Chem. Educ.* **69**, 100-107.

11

Free Rotational and Angular Motion

11.1 Introduction

IN A GAS PHASE THAT IS NOT TOO DENSE, molecules not only translate but also rotate freely. Along one or more disjoint coordinates, each moves at constant potential. Furthermore, a particle in a central field, traveling a given distance r from the center, moves at constant potential.

For such behavior, the state functions are analogous. Similar to the wave functions for free translation are those for free rotation about a given axis. Similar to the wave function factors for rotation are those for motion at a given r in a central field. Thus, the angular eigenfunctions for a hydrogen-like atom have the same form as those for the angular movement of a linear rotator.

But the eigenenergies behave differently. For a rotating molecule, these are related to the quantized angular momenta. But for an electron in the hydrogen-like atom, the state energy depends only on the quantization of the radial motion. On introducing more electrons, the interelectronic repulsion varies with the rotational state and so splittings are introduced.

As preparation for the development, we need to construct the pertinent classical formulas.

Example 11.1

Show that a rigid linear rotator is represented by a single particle of mass μ traveling on a sphere at distance r from the center-of-mass position.

Consider point masses

$$m_1, m_2, \ldots, m_n$$

located at fixed positions

$$r_1, r_2, \ldots, r_n$$

along a straight line. Let the origin of a spherical coordinate system be located at the center of mass, as figure 11.1 indicates. We then have

$$m_1 r_1 + m_2 r_2 + \ldots + m_n r_n = 0.$$

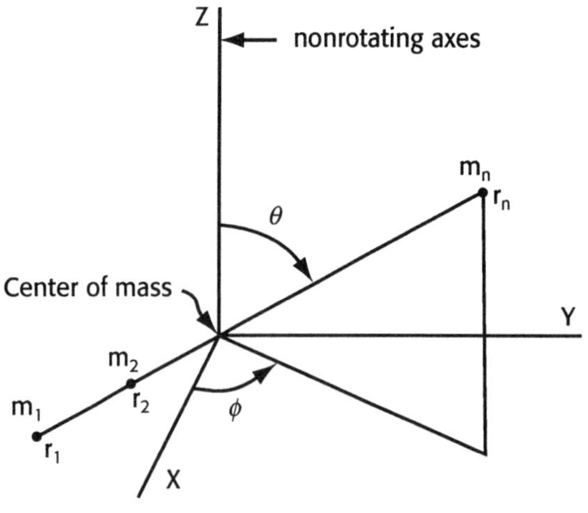

FIGURE 11.1 Coordinates for the atoms in a rotating linear molecule. While the center of mass and the angles θ and φ may vary with time, the axes do not turn or rotate.

The square of an element of path traced out by point mass m_j in time dt is

$$ds_j^2 = r_j^2\, d\theta^2 + r_j^2 \sin^2\theta\, d\phi^2.$$

For the kinetic energy of the system, we obtain

$$T = \sum \frac{1}{2} m_j v_j^2 = \frac{1}{2}\sum m_j r_j^2 \left[\left(\frac{d\theta}{dt}\right)^2 + \sin^2\theta \left(\frac{d\phi}{dt}\right)^2\right] = \frac{I}{2}\left[\dot{\theta}^2 + \left(\sin^2\theta\right)\dot{\phi}^2\right] = \frac{1}{2} I\omega^2,$$

where a dot over a symbol indicates differentiation with respect to t, the second summation is identified as the moment of inertia,

$$I = m_1 r_1^2 + m_2 r_2^2 + \ldots + m_n r_n^2,$$

and ω is the net angular velocity.

Proceeding in the same way for the model in figure 11.2, we find that

$$T = \frac{1}{2}\mu v^2 = \frac{1}{2}\mu r^2\left[\dot{\theta}^2 + \left(\sin^2\theta\right)\dot{\phi}^2\right] = \frac{1}{2} I\omega^2$$

if we set

$$I = \mu r^2.$$

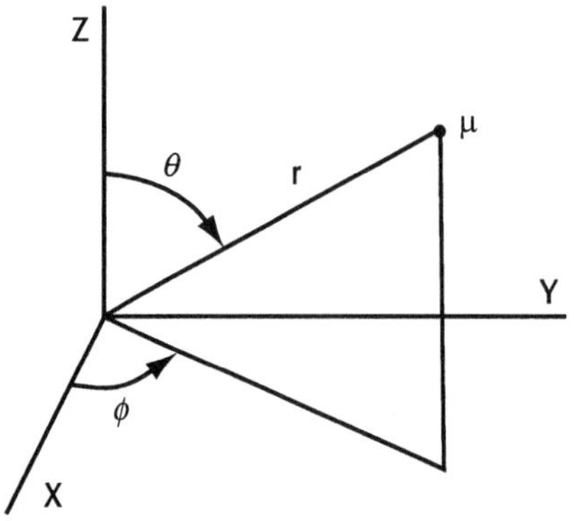

FIGURE 11.2 Model for representing the rotation of a linear molecule, The axes and the angles are the same as in figure 11.1.

The kinetic energy T is taken as that of the linear rotator. To complete the definition of the model, we then choose the reduced mass μ by the formula

$$\frac{1}{\mu}=\frac{1}{m_1}+\frac{1}{m_2}+\ldots+\frac{1}{m_n}.$$

Example 11.2

Show that for a diatomic molecule, r in figure 11.2 equals the interatomic distance.
To locate the origin at the center of mass, we set

$$m_1 r_1 = -m_2 r_2.$$

For the reciprocal of the reduced mass, we have

$$\frac{1}{\mu}=\frac{1}{m_1}+\frac{1}{m_2}=\frac{m_1+m_2}{m_1 m_2}.$$

But the moment of inertia is

$$\mu r^2 = m_1 r_1^2 + m_2 r_2^2.$$

Divide this by μ and introduce the expression for $1/\mu$:

$$r^2=\frac{1}{\mu}\left(m_1 r_1^2+m_2 r_2^2\right)=\frac{m_1}{m_2}r_1^2+r_1^2+r_2^2+\frac{m_2}{m_1}r_2^2.$$

The equation determining the center of mass tells us that

$$\frac{m_1 r_1^2}{m_2}=-r_1 r_2, \qquad \frac{m_2 r_2^2}{m_1}=-r_1 r_2.$$

These relations reduce the preceding equation to

$$r^2=r_1^2-2r_1 r_2+r_2^2=\left(r_2-r_1\right)^2,$$

whence

$$r=r_2-r_1.$$

Since r_1 is negative, this is the interatomic distance.

11.2 Classical Rotational Kinetic Energy

In a given vibrational and rotational state, a molecule may be considered to rotate as a rigid body. This rotational motion may be referred to nonrotating axes based on the center of mass. Over an infinitesimal interval when these axes coincide with the principal axes, the classical energy expression assumes a simple form.

Consider a rigid body with the moments of inertia I_1, I_2, I_3 about its principal axes 1, 2, 3. Let the instantaneous angular velocities about these axes be ω_1, ω_2, ω_3. The kinetic energy is then

$$T=\frac{1}{2}I_1\omega_1^2+\frac{1}{2}I_2\omega_2^2+\frac{1}{2}I_3\omega_3^2=\frac{p_1^2}{2I_1}+\frac{p_2^2}{2I_2}+\frac{p_3^2}{2I_3}. \qquad [11.1]$$

In the last equality, the components of angular momentum

$$I_1\omega_1=p_1, \qquad I_2\omega_2=p_2, \qquad I_3\omega_3=p_3 \qquad [11.2]$$

have been introduced. Note the similarity between the final forms in lines (11.1) and (10.31).

In a cylindrically symmetric rotator, two of the principal moments of inertia are equal. Then formula (11.1) reduces to

$$T = \frac{1}{2} I_1 \omega_1^{\ 2} + \frac{1}{2} I_2 \left(\omega_2^{\ 2} + \omega_3^{\ 2} \right) = \frac{p_1^2}{2I_1} + \frac{1}{2I_2} \left(p_2^2 + p_3^2 \right) = \frac{p_\alpha^2}{2I_2} + p_1^2 \left(\frac{1}{2I_1} - \frac{1}{2I_2} \right), \quad [11.3]$$

where

$$p_\alpha^2 = p_1^2 + p_2^2 + p_3^2. \quad [11.4]$$

We call p_α the total angular momentum.

For a linear rotator, where

$$I_1 = 0, \qquad I_2 = I, \quad [11.5]$$

equation (11.3) reduces to

$$T = \frac{1}{2} I \left(\omega_2^2 + \omega_3^2 \right) = \frac{1}{2} I \omega^2 = \frac{p_\phi^2}{2I}. \quad [11.6]$$

In example 11.1, we saw that this rotator is modeled by a particle of reduced mass μ moving freely at distance r from the center-of-mass point with

$$I = \mu r^2. \quad [11.7]$$

For a particle of mass m_1 bound to a particle of mass m_2 we have

$$\frac{1}{\mu} = \frac{1}{m_1} + \frac{1}{m_2}, \quad [11.8]$$

with r equaling the distance between the centers of the two particles.

11.3 Quantized Angular Momentum States of a Linear Rotator

With a linear molecule, the average positions of the nuclei lie on a moving straight line. The movements of the nuclei out of this line are made up of vibrational modes. The movements of the electrons about the line are formed from the electronic modes. The movements of the center of mass are made up of translational modes. One is left with the line's turning about the center of mass, two equivalent rotational modes.

A model for this rotation is provided by the particle of reduced mass μ traveling around at distance r from the molecule's center of mass, as we saw in section 11.2. The moment of inertia I is given by expression (11.7); the reduced mass for a two-atom molecule, by (11.8).

Consider the model in a definite angular momentum state. Let the axis for this momentum be the z axis. Now, a possible path for the point mass μ lies at a given r and θ.

Along this path, successive points are at the same potential V. Furthermore, neighboring rays at slightly different θ's are parallel. As a consequence, one can consider successive elements ds along the path to be equivalent for changes in the state function. Condition (10.3) applies in the form

$$\frac{d\Psi}{\Psi} = ik \, ds - i\omega \, dt \qquad \text{with} \qquad r = \text{const}, \quad \theta = \text{const}. \quad [11.9]$$

When in addition, time t is constant, we replace symbol Ψ with $\Phi(\phi)$. But the line element is

$$ds = r \sin\theta \, d\phi. \quad [11.10]$$

Thus condition (11.9) leads to the formula

$$\frac{d\Phi}{\Phi} = ikr \sin\theta \, d\phi = iM \, d\phi \tag{11.11}$$

in which

$$M = kr \sin\theta. \tag{11.12}$$

Integration of equation (11.11) yields

$$\Phi = e^{iM\phi}. \tag{11.13}$$

Since $\Phi(\phi)$ describes motion over space, it must be single valued. This is ensured if M is an integer. By symmetry, the positive and negative limits on M must be the same. We have

$$M = -J, -J+1, \ldots, J. \tag{11.14}$$

Note that for a given J, there are

$$g = 2J + 1 \tag{11.15}$$

disjoint states.

One may also consider that half the time the model particle is traveling in the positive ϕ direction and the other half the time in the negative ϕ direction with the same magnitude of angular momentum. For the former part of the motion, we write

$$\Phi_+ = \frac{1}{\sqrt{2}} e^{i|M|\phi}, \tag{11.16}$$

and for the latter part,

$$\Phi_- = \frac{1}{\sqrt{2}} e^{-i|M|\phi}. \tag{11.17}$$

Superposing Φ_+ and Φ_- with the given phases leads to

$$\Phi_1 = \sqrt{2}\,\frac{e^{i|M|\phi} + e^{-i|M|\phi}}{2} = \sqrt{2}\cos|M|\phi. \tag{11.18}$$

A real state function orthogonal to Φ_1, and so disjoint from it, is

$$\Phi_2 = \sqrt{2}\,\frac{e^{i|M|\phi} - e^{-i|M|\phi}}{2i} = \sqrt{2}\sin|M|\phi. \tag{11.19}$$

When $M = 0$, there is only one disjoint state. From (11.13), it is

$$\Phi_0 = e^0 = 1 \tag{11.20}$$

For a given J, there are $2J + 1$ real orthogonal forms.

Since all the angular momentum is about the z axis in the formulation of traveling wave (11.13), the motion over θ at a given ϕ and r is described by a standing wave $\Theta(\theta)$. Pertinent functions will be constructed later.

11.4 Energy Levels for the Linear Rotator

When a molecule possesses a given amount of rotational energy, various quantized amounts may be associated with the allowed angular momenta and with the corresponding standing wave states. But an expression for the total can be derived on assuming that each disjoint state in a complete set of standing wave states is equally probable.

Consider the model for a linear rotator in a definite angular momentum state. Label the axis for the angular momentum the z axis. By symmetry, equation (11.11) then applies.

Multiply equation (11.12) by $\hbar$. Then identify $\hbar k$ as linear momentum p. We obtain

$$M\hbar = \hbar kr\sin\theta = pr\sin\theta = p_\phi. \qquad [11.21]$$

In the last step, the linear momentum times the lever arm is identified as angular momentum p_ϕ. The energy associated with this motion is

$$E_z = \frac{p_\phi^2}{2I} = \frac{M^2\hbar^2}{2I} \qquad [11.22]$$

with

$$M = -J, -J+1, \ldots, J. \qquad [11.23]$$

For a given J, we have $2J + 1$ disjoint states. By symmetry, we take each of these to be equally probable. So the average energy associated with the motion is

$$\overline{E_z} = \frac{\sum_{M=-J}^{M=J} M^2\hbar^2}{2I(2J+1)} = \frac{2\hbar^2\sum_{1}^{J} M^2}{2I(2J+1)}. \qquad [11.24]$$

Weighting each traveling wave solution for a given J equally corresponds to weighting each standing wave solution for the J equally. So the same result applies to the standing wave solutions about the z axis.

In example 11.3, we find that

$$\sum_{1}^{J} M^2 = \frac{J(J+1)(2J+1)}{6}. \qquad [11.25]$$

Substituting this expression into (11.24) leads to

$$\overline{E_z} = J(J+1)\frac{\hbar^2}{6I}. \qquad [11.26]$$

Since result (11.26) applies to standing wave as well as to traveling wave solutions about the z axis, it applies to the accompanying standing wave motions about the x and y axes, which make up $\Theta(\theta)$. But by symmetry, there is no fundamental difference between the orthogonal axes for standing wave Ψ's. So we have

$$\overline{E_z} = \overline{E_x} = \overline{E_y}. \qquad [11.27]$$

Furthermore, we expect the total rotational energy to be the sum of these; so

$$E = 3\overline{E_z} = J(J+1)\frac{\hbar^2}{2I}. \qquad [11.28]$$

Here E is the energy of a linear rotator with moment of inertia I. Integer J is called the *rotational quantum number* for the state. Integer M is referred to as the *magnetic quantum number*.

Example 11.3

Verify formula (11.25). When $J = 1$, $\Sigma M^2 = 1^2 = 1$ and

$$\frac{J(J+1)(2J+1)}{6} = \frac{(1)(2)(3)}{6} = 1.$$

When $J = 2$, $\Sigma M^2 = 1^2 + 2^2 = 5$ and

$$\frac{J(J+1)(2J+1)}{6} = \frac{(2)(3)(5)}{6} = 5.$$

Thus, the equation is true for the two lowest values of J.

Let us suppose that the equation has been verified for numbers up to and including $J - 1$. Then for number J, we have

$$\sum_{1}^{J} M^2 = \sum_{1}^{J-1} M^2 + J^2 = \frac{(J-1)(J)(2J-1)}{6} + J^2 = \frac{1}{6}\left(2J^3 - 3J^2 + J + 6J^2\right) = \frac{1}{6}\left(2J^3 + 3J^2 + J\right)$$
$$= \frac{1}{6}J(J+1)(2J+1).$$

Thus, the equation is true when the sum goes to J. By iteration, one can proceed from the two lowest values, where we know that the formula holds, to any desired finite value of J, by unit steps.

11.5 Rotational Spectrum of a Linear Molecule

A linear molecule with definite rotational energy need not exhibit a definite M or $|M|$. But each of the M's with respect to a particular axis may be present with a certain probability. Transitions to differing mixtures of these M's and to a different J may occur during collisions and interactions with other molecules. An analysis of such inelastic collisions is beyond the scope of this text.

When the molecule is polar, or polarizable, it can also interact with the electromagnetic field, emitting or absorbing photons, or altering photons that are passing by: The former process is observed in emission and absorption spectra; the latter, in Raman spectra.

Light may be either circularly polarized or plane polarized. The latter type is a superposition of oppositely turning circularly polarized waves. A circularly polarized wave may be broken down into photons carrying one unit, $\hbar$, of angular momentum. The energy of a photon is given by the Einstein equation

$$E = \hbar\omega = h\nu, \qquad [11.29]$$

in which ω is the angular frequency, ν the conventional frequency, of the light.

Commonly, one photon is absorbed or emitted at a time and no angular momentum is associated with relative motion of the photon and the molecule. So in the elementary process of absorption or emission, the magnitude of angular momentum of the molecule jumps by $\hbar$ with respect to some axis. With the linear molecule, this action requires the maximum angular momentum about the axis to change by one unit and we have

$$\Delta J = \pm 1. \qquad [11.30]$$

When a molecule is exposed to a relatively stable external field, quantization with respect to the direction of this field occurs. Then during absorption or emission of a photon, we also have

$$\Delta M = 0 \qquad \text{or} \qquad \Delta M = \pm 1. \tag{11.31}$$

Conditions (11.30) and (11.31) are called *selection rules.*

When the rotational quantum number changes from J_0 to $J_0 \pm 1$, the energy of the molecule changes by

$$\Delta E = \left(J_0 \pm 1\right)\left(J_0 \pm 1 + 1\right)\frac{\hbar^2}{2I} - J_0\left(J_0 + 1\right)\frac{\hbar^2}{2I} = \frac{\hbar^2}{2I}\left(\pm 2J_0 \pm 1 + 1\right) = Bhc\left(\pm 2J_0 \pm 1 + 1\right) \tag{11.32}$$

where

$$B = \frac{\hbar}{4\pi Ic}. \tag{11.33}$$

The corresponding photon wave number is

$$\bar{k} = \frac{1}{\lambda} = \frac{h\nu}{hc} = \frac{\pm\Delta E}{hc} = \pm B\left(\pm 2J_0 \pm 1 + 1\right). \tag{11.34}$$

For absorption of a photon, one employs the upper sign; for emission, the lower sign.

In the approximation that the moment of inertia of the linear molecule is constant, the rotational levels plot as figure 11.3 shows. Then the separation between successive spectral lines equals wave number $2B$. Actually, the rotation introduces centrifugal effects so that I increases with J. As a result, the spacing between successive lines decreases as J_0 increases. Furthermore, the lines are broadened by the effects described in the next section.

Example 11.4

Evaluate the constant $\hbar/(4\pi c)$.

Substitute accepted values of the fundamental constants into the expression and carry out the indicated operations:

$$\frac{\hbar}{4\pi c} = \frac{1.05457 \times 10^{-27}\,\text{g cm}^2\,\text{s}^{-1}}{4\pi\left(2.99792 \times 10^{10}\,\text{cm s}^{-1}\right)} = 2.7993 \times 10^{-39}\,\text{g cm}.$$

Example 11.5

The absorption spectrum of HCl has maxima spaced about 20.68 cm^{-1} apart in the far infrared. What is the corresponding moment of inertia I?

In the approximation that the molecule is rigid, equation (11.34) applies, the spacing equals $2B$, and

$$B = 10.34\ \text{cm}^{-1}.$$

Solving formula (11.33) for I and employing the result from example 11.4 leads to

$$I = \frac{\hbar}{4\pi cB} = \frac{2.7993 \times 10^{-39}\,\text{g cm}}{10.34\ \text{cm}^{-1}} = 2.707 \times 10^{-40}\,\text{g cm}^2.$$

Example 11.6

From this moment of inertia, determine the bond length in HCl.

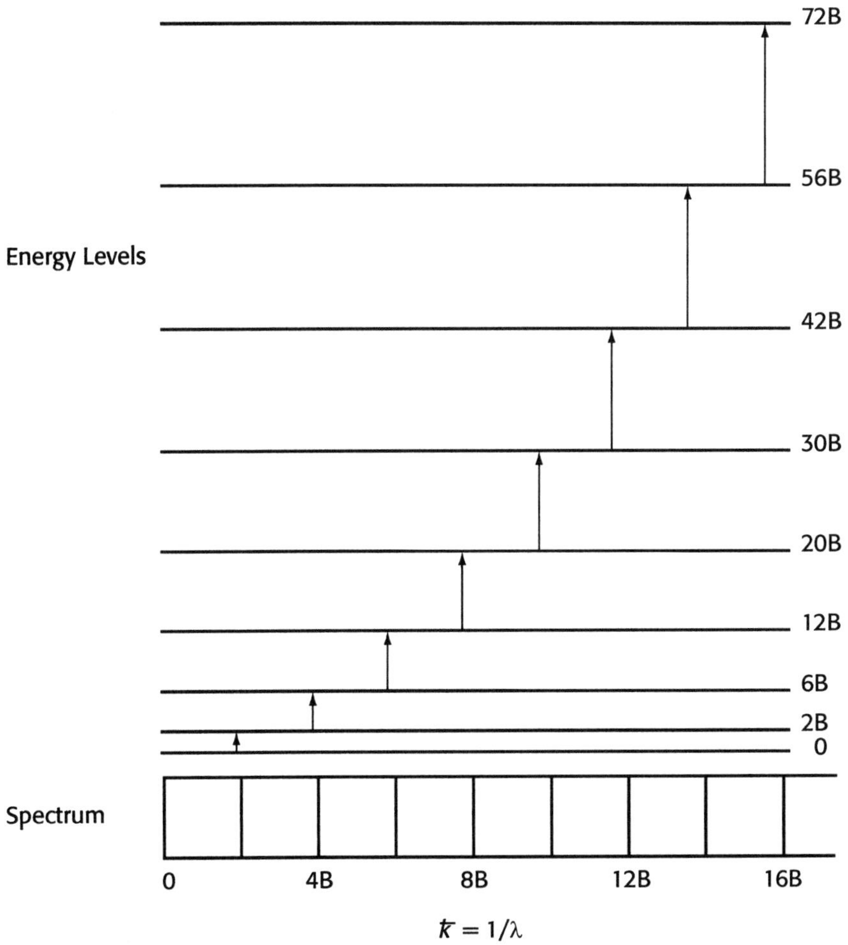

FIGURE 11.3 The lower rotational energy levels of a rigid linear molecule and the allowed transitions that occur between these in absorption.

Solve equation (11.8) for the reduced mass, insert the atomic masses of H and Cl, and carry out the indicated operations:

$$\mu = \frac{m_1 m_2}{m_1 + m_2} = \frac{(1.00797\ \mathrm{u})(35.453\ \mathrm{u})}{36.461\ \mathrm{u}} = 0.9801\ \mathrm{u}.$$

Convert to grams using Avogadro's number:

$$\mu = \frac{0.9801\ \mathrm{u}}{6.0221 \times 10^{23}\,\mathrm{u\ g^{-1}}} = 1.628 \times 10^{-24}\,\mathrm{g}.$$

Solve equation (11.7) for r and insert the values of I and μ:

$$r = \left(\frac{I}{\mu}\right)^{1/2} = \left(\frac{2.707 \times 10^{-40}\,\mathrm{g\ cm^2}}{1.628 \times 10^{-24}\,\mathrm{g}}\right)^{1/2} = 1.29 \times 10^{-8}\,\mathrm{cm} = 1.29\ \text{Å}.$$

11.6 Broadening of Spectral Lines

Since the states of a molecule are discrete, one might expect the observed spectrum of wave numbers to be discrete, to consist of lines. However, a given transition generally appears spread over a range of wave numbers, as a distribution. There are several causes.

Because of *instrumental defects* a particular wavelength in an incident beam is recorded over a range of wavelengths. Systematic errors arise from the width of a slit, unwanted diffraction effects, graininess in the receiver. Random errors result from fluctuations in the line voltage, vibrations in the apparatus.

In addition, the molecules in a macroscopic system lurch about randomly with respect to an observing instrument. This agitation produces a Doppler effect that varies with the relative velocity of each molecule. The result is a *Doppler broadening* that increases with increasing temperature. However, it does not go to zero as the temperature is lowered to 0 K; any confinement causes random motion even at absolute zero. Nevertheless, this is referred to as temperature broadening,

Thirdly, *interaction broadening* occurs. The energy levels of a molecule are shifted by any neighbors close enough to interact with it. In a phase that is not too diffuse, the effect is appreciable, varying with the distribution around the emitting or absorbing molecule. The result may be called a pressure or a concentration broadening.

When the preceding effects are made small enough, a *natural width* is observed. In each emission or absorption step, the photon may be described by a wave packet, a limited compact electromagnetic wave. A Fourier analysis of this reveals it as a superposition of sinusoidal waves. The distribution over wavelengths varies with the mean life of the excited state(s).

11.7 Quantization of Cylindrically Symmetric Rotators

A linear rotator is a cylindrically symmetric rotator with the I_1 of section 11.2 zero. Increasing I_1 from zero while keeping $I_2 = I_3$ introduces an additional angular momentum p_1, which must be quantized.

But motion of the elements of the body around its unique axis is like motion of the model particle around the z axis. So we expect a similar quantization; the angular momentum about the principal axis 1 must be an integral multiple of $\hbar$. Letting K be the integer gives us

$$p_1 = K\hbar \qquad \text{with} \qquad K = -J, -J+1, \ldots, J. \qquad [11.35]$$

We also expect the total angular momentum to be quantized as before. So in the quantum version of formula (11.3), the first term is given by expression (11.28). With p_1 given by formula (11.35), we obtain

$$E = J(J+1)\frac{\hbar^2}{2I_2} + K^2\hbar^2\left(\frac{1}{2I_1} - \frac{1}{2I_2}\right) = hc\left[J(J+1)B + K^2(A-B)\right]. \qquad [11.36]$$

For each principal moment, we have a rotational constant

$$A = \frac{\hbar}{4\pi I_1 c}, \qquad B = \frac{\hbar}{4\pi I_2 c}, \qquad C = \frac{\hbar}{4\pi I_3 c}. \qquad [11.37]$$

In formula (11.36), we had $I_2 = I_3$ and $B = C$. Dividing (11.36) by hc yields the energy in wave numbers

$$E_{\text{rot}} = J(J+1)B + K^2(A-B). \qquad [11.38]$$

A particular example of these levels appears in figure 11.4.

Transitions among the various rotational states are induced by collisions and interactions with neighboring molecules. They also result from interactions with the electromagnetic field. Let us consider the latter.

A molecule with $I_2 = I_3$ acts as if it is cylindrically symmetric about its first principal axis. Thus it cannot exhibit an unbalanced shift of charge away from this axis. Rotation about the axis is like that of a nonpolar linear molecule. There is no dipole moment turning about this axis for the electromagnetic field to interact with and we have the selection rule

$$\Delta K = 0 \qquad [11.39]$$

for emission and absorption of photons.

But when the molecule exhibits a dipole moment *along* the axis, the total angular momentum does interact with the field. As before, we then have the selection rule

$$\Delta J = \pm 1 \qquad [11.40]$$

for emission and absorption of photons.

Since K does not change, the separation between maxima in the spectrum is still $2B$ in the approximation that the molecule is rigid.

11.8 The Asymmetric Rotator

In the general asymmetric molecule, all three principal moments of inertia are different. The state function for rotation then depends on the three angles that determine the rotator's orientation with respect to a nonrotating set of axes based on the center of mass. The complete analysis is too complicated to present here.

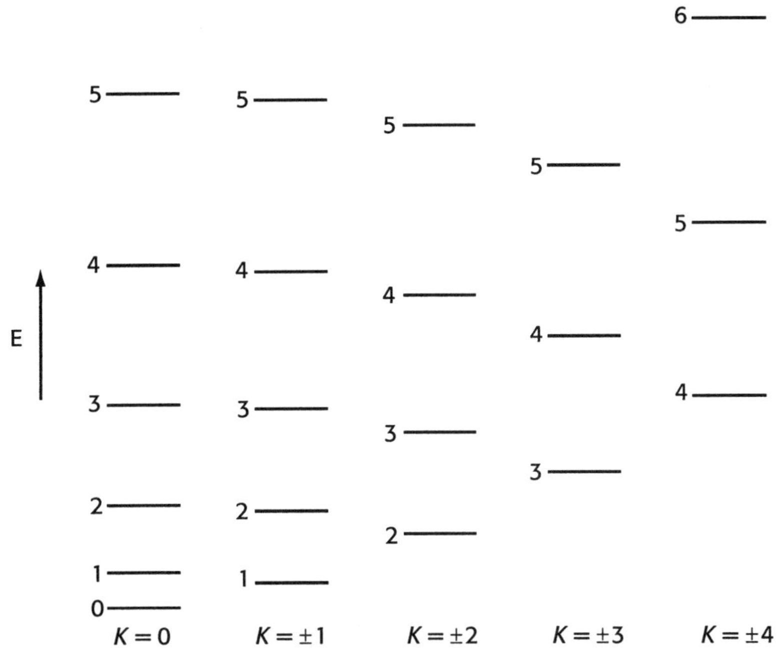

FIGURE 11.4 Energy levels of an oblate rotator for which B = 2A. The number at left of each line is the J for that level.

However, solving the Schrödinger equation for the motion yields the energy levels listed in table 11.1. These can be correlated with the levels of two related cylindrically symmetric rotators.

Consider a polyatomic molecule with the principal moments I_1, I_2, I_3. Rotational constants A, B, C are related to these by equations (11.37). When $I_2 = I_3$, the energy levels are given by equation (11.38). On the other hand, for $I_1 = I_2$, we would rewrite (11.38) in the form

$$E_{\text{rot}} = J(J+1)A + K^2(C-A). \tag{11.41}$$

Let us label the principal axes so that

$$A \geq B \geq C. \tag{11.42}$$

The cylindrically symmetric rotator where

$$A > B = C \tag{11.43}$$

TABLE 11.1 Energy Levels of an Asymmetric Rotator

J_{po}	E_{rot}
0_{00}	0
1_{01}	$B+C$
1_{11}	$A+C$
1_{10}	$A+B$
2_{02}	$2A+2B+2C-2\left[(B-C)^2+(A-C)(A-B)\right]^{1/2}$
2_{12}	$A+B+4C$
2_{11}	$\mathrm{A+4B+C}$
2_{21}	$4A+B+C$
2_{20}	$2A+2B+2C+2\left[(B-C)^2+(A-C)(A-B)\right]^{1/2}$
3_{03}	$2A+5B+5C-2\left[4(B-C)^2+(A-B)(A-C)\right]^{1/2}$
3_{13}	$5A+2B+5C-2\left[4(A-C)^2+(A-B)(B-C)\right]^{1/2}$
3_{12}	$5A+5B+2C-2\left[4(A-B)^2+(A-C)(B-C)\right]^{1/2}$
3_{22}	$4A+4B+4C$
3_{21}	$2A+5B+5C+2\left[4(B-C)^2+(A-B)(A-C)\right]^{1/2}$
3_{31}	$5A+2B+5C+2\left[4(A-C)^2+(A-B)(B-C)\right]^{1/2}$
3_{30}	$5A+5B+2C+2\left[4(A-B)^2+(A-C)(B-C)\right]^{1/2}$

is said to be *prolate* (p). The cylindrically symmetric rotator where

$$A = B > C. \quad [11.44]$$

is said to be *oblate* (o).

Each state of the asymmetric rotator is labeled by its J and by the $|K|$'s of the prolate and oblate states to which it is related. See figure 11.5 for an example.

The dipole moment of an asymmetric molecule interacts with the electromagnetic field, causing absorption or emission of photons. In each step, the angular momentum $\hbar$ of a photon either increases or decreases J by 1, or merely rearranges the total angular momentum. We have the selection rule

$$\Delta J = \pm 1 \quad \text{and} \quad \Delta J = 0. \quad [11.45]$$

Rules also exist for combining the subscripts, but these will not be considered here.

11.9 Rotational Partition Functions and the Resulting Energies

Knowing how the rotational states are quantized for a given molecule, a person can construct the corresponding partition function and from it various thermodynamic properties.

Consider a pure homogeneous thermodynamic system at equilibrium at temperature T. We suppose that T is high enough so that the probability for a molecule to occupy any one disjoint state is small. Then the Boltzmann distribution law applies as in section 10.7.

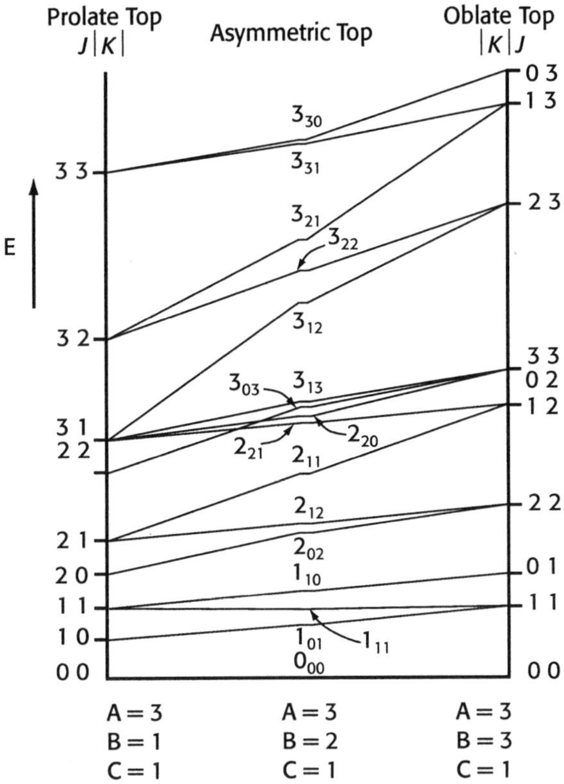

FIGURE 11.5 Correlation of low lying asymmetric rotator energy levels with levels of the corresponding prolate and oblate rotators.

The energy of a molecule equals the sum of the different energy contributions. We have

$$E_i = E_j^{tr} + E_k^{rot} + E_l^{vib} + E_m^{el}, \tag{11.46}$$

where the first term on the right is the contribution from translation, the second term that from rotation, the third term that from vibration, and the fourth term that from the electron states. These different contributions are independent only as a first approximation. To this approximation, one can express the partition function as a product:

$$Z = \Sigma e^{-E_i/kT} = \Sigma e^{-E_j^{tr}/kT}\ \Sigma e^{-E_k^{rot}/kT}\ \Sigma e^{-E_l^{vib}/kT}\ \Sigma e^{-E_m^{el}/kT} = Z_{tr}Z_{rot}Z_{vib}Z_{el}. \tag{11.47}$$

In section 10.7, a calculation of Z_{tr} was presented. Here our concern is with Z_{rot}.

Let us first consider a linear molecule where J may take on any integral value. At the Jth level, the energy is

$$E = J(J+1)B, \tag{11.48}$$

while the degeneracy is

$$g = 2J + 1. \tag{11.49}$$

The corresponding partition function is

$$Z_{\text{rot}} = \sum_{J=0}^{\infty}(2J+1)e^{-J(J+1)B/kT}. \tag{11.50}$$

When B/kT is small, one may replace the summation with an integration. Then we have

$$Z_{\text{rot}} = \int_0^{\infty}(2J+1)e^{-J(J+1)B/kT}\,\mathrm{d}J = -\frac{kT}{B}\int_0^{\infty}e^{-J(J+1)B/kT}\,\mathrm{d}\left(-\frac{J(J+1)B}{kT}\right) = -\frac{kT}{B}e^{-J(J+1)B/kT}\Big|_0^{\infty} = \frac{kT}{B} \tag{11.51}$$

When C'_2, rotation by 1/2 turn about an axis perpendicular to the unique principal axis, converts the molecule into an equivalent molecule, there are only one half as many disjoint states. Then result (11.51) is replaced with

$$Z_{\text{rot}} = \frac{kT}{2B}. \tag{11.52}$$

This formula applies to symmetrical linear molecules, such as I_2, CO_2, and HC≡CH.

In equation (11.33), parameter B appears as a reciprocal distance. To convert it to the same unit of energy as we have in h, we multiply it by hc, following the relation in line (11.34):

$$B = \frac{\hbar}{4\pi Ic}hc = \frac{h^2}{8\pi^2 I}. \tag{11.53}$$

Inserting this expression into (11.51) and (11.52) yields

$$Z_{\text{rot}} = \frac{8\pi^2 IkT}{\sigma h^2}, \tag{11.54}$$

where σ is the symmetry number 1 or 2, respectively.

In general, the number of physical operations that rotate a molecule into an equivalent molecule is called the *symmetry number* σ. And the number of disjoint rotational states is $1/\sigma$ times that for the same structure without the equivalence. Thus, σ equals the order of the rotational subgroup for the molecule.

With a nonlinear molecule, one can proceed similarly. For the cylindrically symmetric rotator when there is no limitation on J or K, we have

$$Z_{\text{rot}} = \sum_{J=0}^{\infty}\sum_{K=-J}^{J}(2J+1)e^{-\left[J(J+1)B+K^2(A-B)\right]/kT} \tag{11.55}$$

For small B/kT, one may replace the summations with integrations. With some further approximations, we then find that

$$Z_{\text{rot}} \cong \left(\frac{\pi kT}{A}\right)^{1/2} \frac{kT}{B}. \qquad [11.56]$$

For an asymmetric rotator, there is no general formula describing the energy levels. So a person cannot proceed as above. However, a reasonable approximation is obtained on replacing B with $(BC)^{1/2}$ in equation (11.56). We thus construct the formula

$$Z_{\text{rot}} \cong \left(\frac{\pi}{ABC}\right)^{1/2} (kT)^{3/2}. \qquad [11.57]$$

Into this, introduce the expressions for A, B, and C, converted to the same unit of energy as we have in h, and bring in the appropriate symmetry number σ to get

$$Z_{\text{rot}} = \left[\frac{\pi\left(8\pi^2\right)^3 I_1 I_2 I_3}{h^6}\right]^{1/2} \frac{(kT)^{3/2}}{\sigma} = \frac{8\pi^2\left(8\pi^3 I_1 I_2 I_3\right)^{1/2}(kT)^{3/2}}{\sigma h^3}. \qquad [11.58]$$

Here I_1, I_2, I_3 are the principal moments of inertia of the molecule, k is the Boltzmann constant, h the Planck constant, and T the absolute temperature.

To derive the corresponding energies, one substitutes into formula (10.39). Partition function (11.54) for the linear rotator yields

$$\ln Z_{\text{rot}} = \ln T + \text{constant} \qquad [11.59]$$

whence

$$E_{\text{rot}} = NkT^2 \frac{1}{T} = NkT. \qquad [11.60]$$

Similarly, partition function (11.58) for the nonlinear rotator gives

$$\ln Z_{\text{rot}} = \frac{3}{2} \ln T + \text{constant} \qquad [11.61]$$

whence

$$E_{\text{rot}} = NkT^2 \frac{3}{2T} = \frac{3}{2} NkT. \qquad [11.62]$$

These energy results agree with the classical forms constructed in chapter 3.

At temperatures so low that B/kT is not small enough, the continuum approximation breaks down. Then one has to return to the summation form for the partition function.

Example 11.7

Express the Boltzmann constant in the reciprocal centimeter energy unit.

From equations (11.34), we see that dividing the photon energy by hc yields its energy in reciprocal distance. So we set

$$k = \frac{1.38066 \times 10^{-23}\,\text{J K}^{-1}}{\left(6.62608 \times 10^{-34}\,\text{J s}\right)\left(2.99792 \times 10^{10}\,\text{cm s}^{-1}\right)} = 0.69504\ \text{cm}^{-1}\,\text{K}^{-1}$$

11.10 Independence of the Angular Factor $Y(\theta,\phi)$

We have seen how the rotational behavior of a linear molecule is modeled by a single particle in a central field. When only two particles are involved, as in a diatomic molecule or a hydrogenlike atom, the radial motion of the model particle also models the vibrational motion or oscillatory motion of the system. Furthermore, the radius vector r locating model μ equals the displacement of one particle from the other.

Consider a particle of mass m_1 joined to a particle of mass m_2 by the potential $V(r)$. The model is then a single particle of mass

$$\mu = \frac{m_1 m_2}{m_1 + m_2}. \qquad [11.63]$$

at distance r from the origin of an inertial frame.

Schrödinger equation (10.134) for the system has the form

$$\nabla^2\psi + 2\mu/\hbar^2[E - V(r)]\psi = 0 \qquad [11.64]$$

or

$$\nabla^2\psi + k^2(r)\psi = 0. \qquad [11.65]$$

With the result from example 10.10, we rewrite this equation as

$$\frac{1}{r^2}\frac{\partial}{\partial r}\left(r^2\frac{\partial\psi}{\partial r}\right) + \frac{1}{r^2\sin\theta}\frac{\partial}{\partial\theta}\left(\sin\theta\frac{\partial\psi}{\partial\theta}\right) + \frac{1}{r^2\sin^2\theta}\frac{\partial^2\psi}{\partial\phi^2} + k^2(r)\psi = 0. \qquad [11.66]$$

When the model particle is in a definite energy state, we expect the radial motion to be independent of the angular motion. As a result, the state function should then factor:

$$\psi = R(r)Y(\theta,\phi). \qquad [11.67]$$

Indeed substituting form (11.67) into equation (11.66) yields

$$\frac{Y}{r^2}\frac{\mathrm{d}}{\mathrm{d}r}\left(r^2\frac{\mathrm{d}R}{\mathrm{d}r}\right) + \frac{R}{r^2\sin\theta}\frac{\partial}{\partial\theta}\left(\sin\theta\frac{\partial Y}{\partial\theta}\right) + \frac{R}{r^2\sin^2\theta}\frac{\partial^2 Y}{\partial\phi^2} + k^2(r)RY = 0. \qquad [11.68]$$

Multiplying equation (11.68) by r^2/RY and rearranging leads to

$$\frac{1}{R}\frac{\mathrm{d}}{\mathrm{d}r}\left(r^2\frac{\mathrm{d}R}{\mathrm{d}r}\right) + r^2k^2(r) = -\frac{1}{Y\sin\theta}\frac{\partial}{\partial\theta}\left(\sin\theta\frac{\partial Y}{\partial\theta}\right) - \frac{1}{Y\sin^2\theta}\frac{\partial^2 Y}{\partial\phi^2} = \text{const.} \qquad [11.69]$$

Since the left side of (11.69) can vary only when r varies while the middle section can vary only when θ or ϕ vary, the equality between them means that they do not vary. For the linear rotator, the constant is $J(J + 1)$; for the hydrogenlike atom, it is $l(l + 1)$. Parameter J is the rotational quantum number; parameter l is called the *azimuthal quantum number*.

Note that the separated differential equation for Y, in line (11.69), does not involve k. As a consequence, the form of $Y(\theta, \phi)$ is independent of k. For simplicity in determining Y, one may take k to be 0. The Schrödinger equation then reduces to Laplace's equation

$$\nabla^2\psi = 0. \qquad [11.70]$$

Example 11.8

Show that $1/r$ is a solution of the Laplace equation.

When Ψ depends only on r, the angular terms in the representation of $\nabla^2\Psi$ from example 10.10 drop out. Equation (11.70) then reduces to

$$\frac{1}{r^2}\frac{\mathrm{d}}{\mathrm{d}r}\left(r^2\frac{\mathrm{d}\psi}{\mathrm{d}r}\right)=0.$$

But when

$$\psi=\frac{1}{r},$$

we have

$$\frac{\mathrm{d}\psi}{\mathrm{d}r}=-\frac{1}{r^2}$$

and

$$\frac{\mathrm{d}}{\mathrm{d}r}\left(r^2\frac{\mathrm{d}\psi}{\mathrm{d}r}\right)=\frac{\mathrm{d}}{\mathrm{d}r}\left(-1\right)=0.$$

So $1/r$ satisfies the Laplace equation.

Example 11.9

Let $\partial x/\partial s = a$, $\partial y/\partial s = b$, $\partial z/\partial s = c$, with a, b, and c constant. Then show that whenever Ψ is a solution to the Laplace equation, then $\partial\psi/\partial s$ is also.

Here a, b, and c are the direction cosines for a straight line passing through the origin, while s is the distance to a point on this line. We have

$$\frac{\partial}{\partial s}=\frac{\partial x}{\partial s}\frac{\partial}{\partial x}+\frac{\partial y}{\partial s}\frac{\partial}{\partial y}+\frac{\partial z}{\partial s}\frac{\partial}{\partial z}=a\frac{\partial}{\partial x}+b\frac{\partial}{\partial y}+c\frac{\partial}{\partial z}.$$

Let this operator act on equation (11.70) with a, b and c fixed:

$$\frac{\partial}{\partial s}\nabla^2\psi=\left(a\frac{\partial}{\partial x}+b\frac{\partial}{\partial y}+c\frac{\partial}{\partial z}\right)\left(\frac{\partial^2\psi}{\partial x^2}+\frac{\partial^2\psi}{\partial y^2}+\frac{\partial^2\psi}{\partial z^2}\right)=\frac{\partial^2}{\partial x^2}\left(a\frac{\partial\psi}{\partial x}+b\frac{\partial\psi}{\partial y}+c\frac{\partial\psi}{\partial z}\right)$$
$$+\frac{\partial^2}{\partial y^2}\left(a\frac{\partial\psi}{\partial x}+b\frac{\partial\psi}{\partial y}+c\frac{\partial\psi}{\partial z}\right)+\frac{\partial^2}{\partial z^2}\left(a\frac{\partial\psi}{\partial x}+b\frac{\partial\psi}{\partial y}+c\frac{\partial\psi}{\partial z}\right)=\nabla^2\left(\frac{\partial\psi}{\partial s}\right)=0.$$

We see that the spatial derivative $\partial\psi/\partial s$ is a solution whenever ψ is.

11.11 A Multiaxial Formula for Y(θ, φ)

A general real expression for the angular factor can now be constructed.

We consider the model particle to be in a definite energy state. The radial motion then separates from the angular motion. Since the latter is independent of the variable wavevector k, one may set k equal to zero without affecting $Y(\theta, \phi)$. So we look at the angular behavior of solutions of Laplace's equation.

In example 11.8, we saw that $1/r$ is a solution of equation (11.70). In example 11.9, we saw that any spatial derivative of a solution is a solution. By induction, therefore, any multiple spatial derivative of $1/r$ is a solution. To eliminate variable r from the solution, one multiplies the derivative by r raised to the appropriate power. A phasing factor and a normalization factor are also introduced.

Following James Clerk Maxwell, we thus construct

$$Y(\theta,\phi) = N(-1)^{l} r^{l+1} \frac{\partial^{l}}{\partial s_1 \partial s_2 \ldots \partial s_l} \frac{1}{r} \tag{11.71}$$

with the direction cosines for ds_1, ds_2, . . . , ds_l all constant. These direction cosines define l axes. As it stands, formula (11.71) describes atomic orbitals. For the linear rotator, we replace l with J.

A set of angular functions, complete for l equal to 1, 2, and 3, appears in table 11.2. The axes have been chosen so that the functions yield definite $|m|$'s. Those with nonzero $|m|$'s combine in pairs to give the standard traveling wave functions in table 11.3. For the linear rotator $|M|$ and M replace $|m|$ and m.

The general analytic form for a solution to the Laplace equation is a power series in the rectangular coordinates. But the Laplace operator

$$\nabla^2 = \frac{\partial^2}{\partial x^2} + \frac{\partial^2}{\partial y^2} + \frac{\partial^2}{\partial z^2} \tag{11.72}$$

reduces the degree of *each* term by 2. So equation (11.70) relates only terms of the same degree. A typical set of terms of degree l is

$$\psi = \sum_{i,j} A_{ij} x^{l-i-j} y^{i} z^{j}. \tag{11.73}$$

When the exponent on x is l, l - 1,..., 2, 1, 0, the number of terms forms the sequence

$$1 + 2 + \ldots + (l-1) + l + (l+1). \tag{11.74}$$

But after the action of ∇^2, the number of terms forms the sequence

$$1 + 2 + \ldots + (l-1), \tag{11.75}$$

since the degree of each term has been reduced by 2. Sum (11.74) minus sum (11.75) equals $2l + 1$. This is the number of independent polynomials of degree l satisfying (11.70). In table 11.2, such polynomials appear in the numerators of the expressions.

In general, the letters s, p, d, f, g,... stand for the quantum number l (or J) being 0, 1, 2, 3, 4,..., respectively. On each symbol for the real functions, the pertinent letter appears as a subscript. The pertinent polynomial factor appears as a subscript to the letter. On each symbol for the complex functions, number l (or J) appears as the subscript, number m (or M) as the superscript.

11.12 Directional Behavior of the Standard Rotational State Functions

In tables 11.2 and 11.3, angular factor Y appears as a function of x/r, y/r, and z/r for particular Maxwell axes and quantum numbers. Let us now consider the angular dependence of each Y.

In figure 11.2, the coordinates of the model particle may be taken as either x, y, z or r, θ, ϕ . With the definitions of the trigonometric functions, these are related by the equations

$$z = r\cos\theta, \tag{11.76}$$

$$x = r\sin\theta\cos\phi, \tag{11.77}$$

$$y = r\sin\theta\sin\phi. \tag{11.78}$$

TABLE 11.2 Standing Wave Rotational Eigenfunctions

Symbol	*Maxwell Axes* s_1	s_2	s_3	*Formula*
Y_{s}				$\sqrt{\frac{1}{4\pi}}$
Y_{p_z}	z			$\sqrt{\frac{3}{4\pi}}\frac{z}{r}$
Y_{p_x}	x			$\sqrt{\frac{3}{4\pi}}\frac{x}{r}$
Y_{p_y}	y			$\sqrt{\frac{3}{4\pi}}\frac{y}{r}$
$Y_{\mathrm{d}_{3z^2-r^2}}$	z	z		$\sqrt{\frac{5}{16\pi}}\frac{3z^2-r^2}{r^2}$
$Y_{\mathrm{d}_{zx}}$	z	x		$\sqrt{\frac{15}{4\pi}}\frac{zx}{r^2}$
$Y_{\mathrm{d}_{yz}}$	y	z		$\sqrt{\frac{15}{4\pi}}\frac{yz}{r^2}$
$Y_{\mathrm{d}_{xy}}$	x	y		$\sqrt{\frac{15}{4\pi}}\frac{xy}{r^2}$
$Y_{\mathrm{d}_{x^2-y^2}}$	$\frac{1}{\sqrt{2}}(\mathrm{x}-\mathrm{y})$	$\frac{1}{\sqrt{2}}(\mathrm{x}+\mathrm{y})$		$\sqrt{\frac{15}{16\pi}}\frac{x^2-y^2}{r^2}$
$Y_{\mathrm{f}_{z(5z^2-3r^2)}}$	z	z	z	$\sqrt{\frac{7}{16\pi}}\frac{z(5z^2-3r^2)}{r^3}$
$Y_{\mathrm{f}_{x(5z^2-r^2)}}$	x	z	z	$\sqrt{\frac{21}{32\pi}}\frac{x(5z^2-r^2)}{r^3}$
$Y_{\mathrm{f}_{y(5z^2-r^2)}}$	y	z	z	$\sqrt{\frac{21}{32\pi}}\frac{y(5z^2-r^2)}{r^3}$
$Y_{\mathrm{f}_{xyz}}$	x	y	z	$\sqrt{\frac{105}{4\pi}}\frac{xyz}{r^3}$
$Y_{\mathrm{f}_{(x^2-y^2)z}}$	$\frac{1}{\sqrt{2}}(\mathrm{x}-\mathrm{y})$	$\frac{1}{\sqrt{2}}(\mathrm{x}+\mathrm{y})$	z	$\sqrt{\frac{105}{16\pi}}\frac{(x^2-y^2)z}{r^3}$

Continued on next page.

TABLE 11.2 Standing Wave Rotational Eigenfunctions

Symbol	*Maxwell Axes* s_1	s_2	s_3	*Formula*
$Y_{\mathrm{f}_{x(x^2-3y^2)}}$	$\frac{1}{2}\left(\mathrm{x}-\sqrt{3}\mathrm{y}\right)$	$\frac{1}{2}\left(\mathrm{x}+\sqrt{3}\mathrm{y}\right)$	x	$\sqrt{\frac{35}{32\pi}}\frac{x\left(x^2-3y^2\right)}{r^3}$
$Y_{\mathrm{f}_{y(y^2-3x^2)}}$	$\frac{1}{2}\left(\sqrt{3}\mathrm{x}-\mathrm{y}\right)$	$\frac{1}{2}\left(\sqrt{3}\mathrm{x}+\mathrm{y}\right)$	y	$\sqrt{\frac{35}{32\pi}}\frac{y\left(y^2-3x^2\right)}{r^3}$

TABLE 11.3 Traveling Wave Rotational Eigenfunctions

Symbol	*Quantum Numbers* l	m	*Formula*
Y_0^0	0	0	$\sqrt{\frac{1}{4\pi}}$
Y_1^0	1	0	$\sqrt{\frac{3}{4\pi}}\frac{z}{r}$
$Y_1^{\pm1}$	1	±1	$\mp\sqrt{\frac{3}{8\pi}}\frac{x\pm iy}{r}$
Y_2^0	2	0	$\sqrt{\frac{5}{16\pi}}\frac{3z^2-r^2}{r^2}$
$Y_2^{\pm1}$	2	±1	$\mp\sqrt{\frac{15}{8\pi}}\frac{(x\pm iy)z}{r^2}$
$Y_2^{\pm2}$	2	±2	$\sqrt{\frac{15}{32\pi}}\frac{(x\pm iy)^2}{r^2}$
Y_3^0	3	0	$\sqrt{\frac{7}{16\pi}}\frac{z\left(5z^2-3r^2\right)}{r^3}$
$Y_3^{\pm1}$	3	±1	$\mp\sqrt{\frac{21}{64\pi}}\frac{(x\pm iy)\left(5z^2-r^2\right)}{r^3}$
$Y_3^{\pm2}$	3	±2	$\sqrt{\frac{105}{32\pi}}\frac{(x\pm iy)^2 z}{r^3}$
$Y_3^{\pm3}$	3	±3	$\mp\sqrt{\frac{35}{64\pi}}\frac{(x\pm iy)^3}{r^3}$

Furthermore, we have

$$x \pm iy = r\sin\theta\left(\cos\phi \pm i\sin\phi\right) = r\sin\theta\, e^{\pm i\phi} \tag{11.79}$$

and

$$\left(x \pm iy\right)^{|m|} = r^{|m|}\sin^{|m|}\theta\, e^{im\phi}. \tag{11.80}$$

For table 11.2, the axes have been chosen so that each angular function equals a product of a function of θ and a function of ϕ:

$$Y = \Theta(\theta)\Phi(\phi). \tag{11.81}$$

But clearly, a general choice of the axes does not allow such factoring. Nevertheless, introducing relations (11.76) - (11.78) into the orbitals of table 11.2 yields the forms in table 11.4.

To see how these vary with direction from the center of mass, a person may plot the magnitude $|Y|$. One thus obtains the pictures in figures 11.6, 11.7, and 11.8. Perturbations of these standing wave orbitals are employed in forming covalent bonds between or among atoms.

Substitutions (11.76) and (11.80) transform the functions in table 11.3 to those in table 11.5. With relation (11.81), one sees that they have the form

$$Y = N\Theta(\theta)e^{im\phi}. \tag{11.82}$$

Let us differentiate (11.82) with respect to ϕ and multiply by $\hbar/i$. We find that

$$\frac{\hbar}{i}\frac{\partial Y}{\partial\phi} = \frac{\hbar}{i}N\Theta(\theta)\left(ime^{im\phi}\right) = m\hbar Y. \tag{11.83}$$

The complete state function is

$$\Psi = R(r)Y(\theta,\phi)T(t), \tag{11.84}$$

so that we have

$$\frac{\hbar}{i}\frac{\partial}{\partial\phi}\Psi = m\hbar\Psi = p_\phi\Psi. \tag{11.85}$$

So when the angular factor Y has form (11.82), the state function is an eigenfunction of the operator

$$\frac{\hbar}{i}\frac{\partial}{\partial\phi} \tag{11.86}$$

with the eigenvalue p_ϕ.

From equation (11.21), we recognize that eigenvalue p_ϕ is the angular momentum of the system. This is the angular momentum that interacts with an imposed magnetic field. The back and forth motion over θ does not contribute to this interaction.

With formula (11.71), a person can construct angular functions for quantum numbers

$$l = 0, 1, 2, \ldots. \tag{11.87}$$

These can be normalized.

For each l, one can pick out angular functions of type

$$Y = \Theta(\theta)\Phi(\phi). \tag{11.88}$$

TABLE 11.4 Angular Dependence of the Eigenfunctions

Symbol	*Quantum Numbers* *J or l*	*|M| or |m|*	*Formula*
Y_{s}	0	0	$\sqrt{\frac{1}{4\pi}}$
Y_{p_z}	1	0	$\sqrt{\frac{3}{4\pi}}\cos\theta$
Y_{p_x}	1	1	$\sqrt{\frac{3}{4\pi}}\sin\theta\cos\phi$
Y_{p_y}	1	1	$\sqrt{\frac{3}{4\pi}}\sin\theta\sin\phi$
$Y_{\mathrm{d}_{3z^2-r^2}}$	2	0	$\sqrt{\frac{5}{16\pi}}\left(3\cos^2\theta-1\right)$
$Y_{\mathrm{d}_{zx}}$	2	1	$\sqrt{\frac{15}{4\pi}}\sin\theta\cos\theta\cos\phi$
$Y_{\mathrm{d}_{yz}}$	2	1	$\sqrt{\frac{15}{4\pi}}\sin\theta\cos\theta\sin\phi$
$Y_{\mathrm{d}_{xy}}$	2	2	$\sqrt{\frac{15}{16\pi}}\sin^2\theta\sin 2\phi$
$Y_{\mathrm{d}_{x^2-y^2}}$	2	2	$\sqrt{\frac{15}{16\pi}}\sin^2\theta\cos 2\phi$
$Y_{\mathrm{f}_{z\left(5z^2-3r^2\right)}}$	3	0	$\sqrt{\frac{7}{16\pi}}\left(5\cos^2\theta-3\right)\cos\theta$
$Y_{\mathrm{f}_{x\left(5z^2-r^2\right)}}$	3	1	$\sqrt{\frac{21}{32\pi}}\left(5\cos^2\theta-1\right)\sin\theta\cos\phi$
$Y_{\mathrm{f}_{y\left(3z^2-r^2\right)}}$	3	1	$\sqrt{\frac{21}{32\pi}}\left(5\cos^2\theta-1\right)\sin\theta\sin\phi$
$Y_{\mathrm{f}_{xyz}}$	3	2	$\sqrt{\frac{105}{16\pi}}\sin^2\theta\cos\theta\sin 2\phi$
$Y_{\mathrm{f}_{\left(x^2-y^2\right)z}}$	3	2	$\sqrt{\frac{105}{16\pi}}\sin^2\theta\cos\theta\cos 2\phi$
$Y_{\mathrm{f}_{y\left(3x^2-y^2\right)}}$	3	3	$\sqrt{\frac{35}{32\pi}}\sin^3\theta\sin 3\phi$
$Y_{\mathrm{f}_{x\left(x^2-3y^2\right)}}$	3	3	$\sqrt{\frac{35}{32\pi}}\sin^3\theta\cos 3\phi$

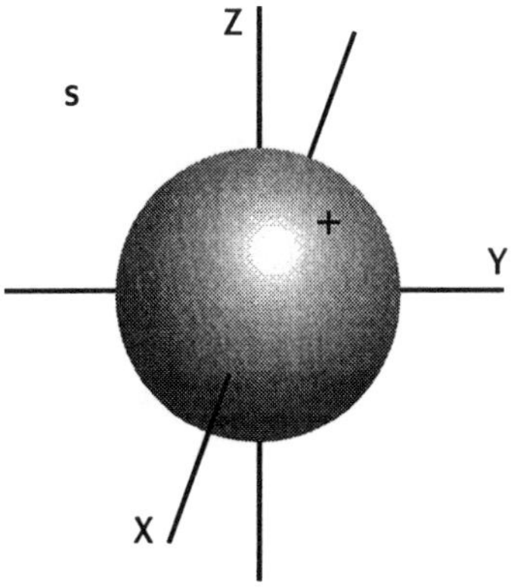

FIGURE 11.6 Plot of the angular dependence of an s orbital.

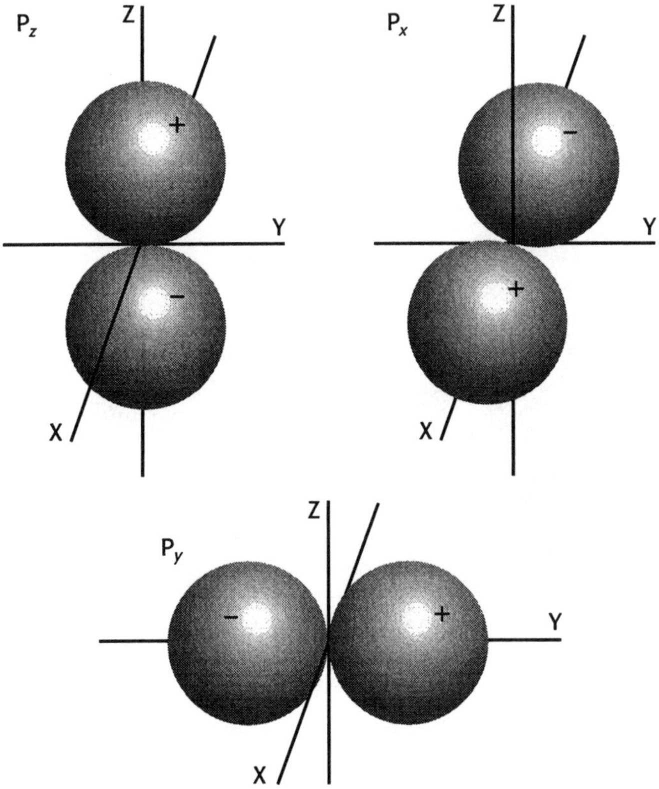

FIGURE 11.7 Plot of the angular dependence of the standard real p orbitals.

From these, one can pick out those for which

$$\Phi = \sqrt{2}\cos|m|\phi \qquad [11.89]$$

and

$$\Phi = \sqrt{2}\sin|m|\phi \qquad [11.90]$$

with

$$m = 0, 1, 2..., l. \qquad [11.91]$$

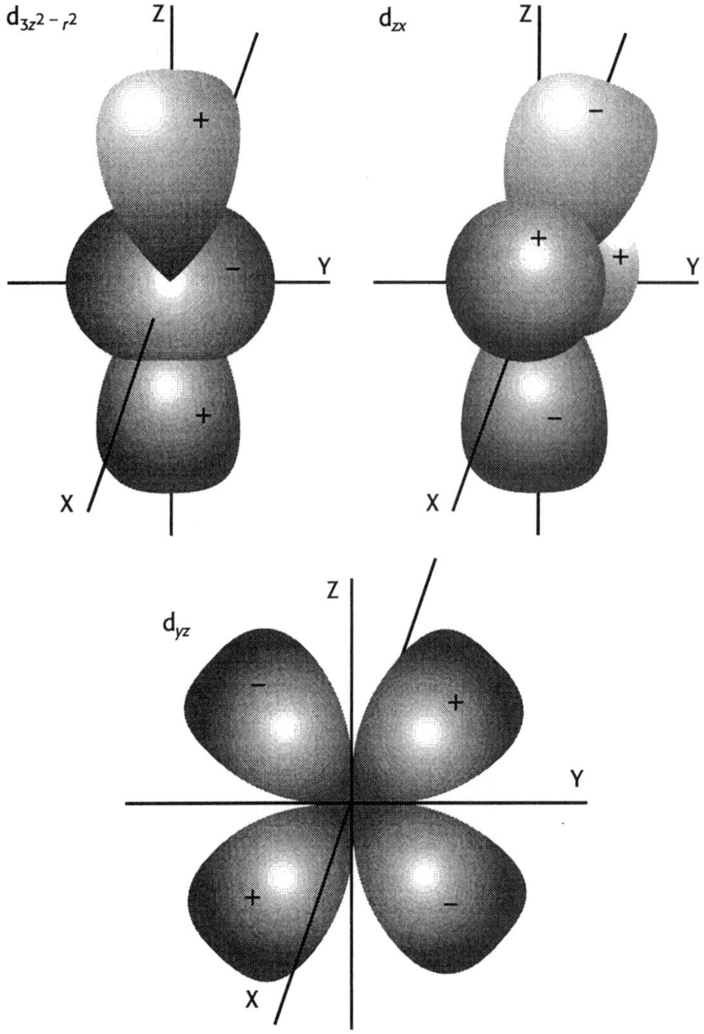

FIGURE 11.8A Plot of the angular dependence of the standard real d orbitals.

These can be combined to form complex normalized *Y*'s with

$$\Phi = e^{im\phi} \tag{11.92}$$

and

$$m = -l, -l+1, \dots, l. \tag{11.93}$$

The *Y*'s with ϕ factors (11.89), (11.90) form a real basis set; those with ϕ factors (11.92) form a complex basis set. Every possible rotational state can be expressed as a linear combination of the *Y*'s in either set. So any normalizable single valued continuous function of θ and ϕ can be expressed as such a superposition. Either set of eigenfunctions forms a *complete set* over the variables θ and ϕ.

11.13 Normalization and Orthogonality Conditions Revisited

The state of a model particle is represented by a function Ψ. The probability density about any given point is symbolized by expression ρ. Now, the integral of the probability

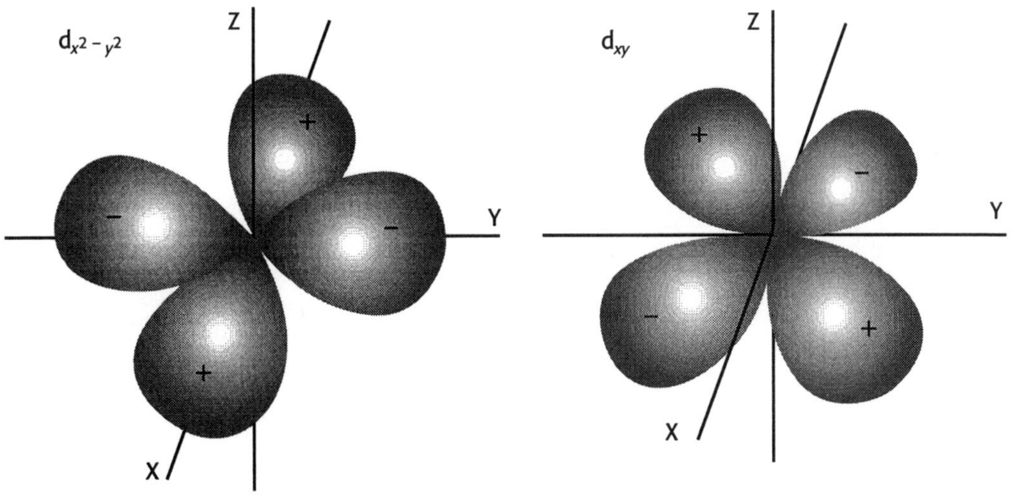

FIGURE 11.8B

density over the space R in which the particle may be found equals unity. So setting ρ equal to $\Psi^*\Psi$ gives us

$$\int_R \Psi^*\Psi \, d^3\mathbf{r} = 1. \tag{11.94}$$

If Ψ_i and Ψ_j are two possible state functions, we have

$$\int_R \Psi_i^* \Psi_i \, d^3\mathbf{r} = 1 \quad \text{and} \quad \int_R \Psi_j^* \Psi_j \, d^3\mathbf{r} = 1. \tag{11.95}$$

When the state functions superpose,

$$\Psi = c_i\Psi_i + c_j\Psi_j, \tag{11.96}$$

we find that

$$\int_R \Psi^*\Psi \, d^3\mathbf{r} = c_i^* c_i \int_R \Psi_i^*\Psi_i \, d^3\mathbf{r} + c_i^* c_j \int_R \Psi_i^*\Psi_j \, d^3\mathbf{r} + c_i c_j^* \int_R \Psi_j^*\Psi_i \, d^3\mathbf{r} + c_j^* c_j \int_R \Psi_j^*\Psi_j \, d^3\mathbf{r}. \tag{11.97}$$

With the functions normalized to 1 and the overlap integral represented as

$$\int_R \Psi_i^* \Psi_j \, d^3\mathbf{r} = S_{ij}. \tag{11.98}$$

equation (11.97) becomes

$$1 = c_i^* c_i + c_i^* c_j S_{ij} + c_i c_j^* S_{ij}^* + c_j^* c_j. \tag{11.99}$$

When the ith and jth states are disjoint, so that Ψ_j contains no part of Ψ_i and vice versa, we have the relation

$$\text{Total probability} = \text{probability of } i\text{th state} + \text{probability of } j\text{th state}. \tag{11.100}$$

We expect the first term on the right to depend only on c_i, the second term on the right to depend only on c_j. This requires

$$S_{ij} = 0 \tag{11.101}$$

TABLE 11.5 Standard Complex Rotational State Functions

Symbol	Quantum Numbers *J* or *l*	*M* or *m*	Formula
Y_0^0	0	0	$\sqrt{\frac{1}{4\pi}}$
Y_1^0	1	0	$\sqrt{\frac{3}{4\pi}}\cos\theta$
$Y_1^{\pm 1}$	1	± 1	$\mp\sqrt{\frac{3}{8\pi}}\sin\theta\, e^{\pm i\phi}$
Y_2^0	2	0	$\sqrt{\frac{5}{16\pi}}\left(3\cos^2\theta - 1\right)$
$Y_2^{\pm 1}$	2	± 1	$\mp\sqrt{\frac{15}{8\pi}}\sin\theta\cos\theta\, e^{\pm i\phi}$
$Y_2^{\pm 2}$	2	± 2	$\sqrt{\frac{15}{32\pi}}\sin^2\theta\, e^{\pm 2i\phi}$
Y_3^0	3	0	$\sqrt{\frac{7}{16\pi}}\left(5\cos^2\theta - 3\right)\cos\theta$
$Y_3^{\pm 1}$	3	± 1	$\mp\sqrt{\frac{21}{64\pi}}\left(5\cos^2\theta - 1\right)\sin\theta\, e^{\pm i\phi}$
$Y_3^{\pm 2}$	3	± 2	$\sqrt{\frac{105}{32\pi}}\sin^2\theta\cos\theta\, e^{\pm 2i\phi}$
$Y_3^{\pm 3}$	3	± 3	$\mp\sqrt{\frac{35}{64\pi}}\sin^3\theta\, e^{\pm 3i\phi}$

in equation (11.99).

Conditions (11.95) and (11.100) are summarized in the equation

$$S_{ij} = \int_R \Psi_i^* \Psi_j \, d^3\mathbf{r} = \delta_{ij}, \qquad [11.102]$$

where δ_{ij} is the Kronecker delta, which is 1 when $i = j$ and 0 when $i \neq j$. The functions in tables 11.2 - 11.5 satisfy this normalization and orthogonality condition.

11.14 Magnetic Energy Associated with Angular Momentum

When the model particle is in a definite angular momentum state and carries charge, its motion generates a magnetic moment. Furthermore, a fundamental particle with spin possesses an intrinsic magnetic moment. These moments combine vectorially and interact with a nonhomogeneous magnetic field.

In classical electromagnetic theory, the force acting on a particle carrying charge q at velocity $\mathbf{v}$ is

$$\mathbf{F} = q(\mathbf{E} + \mathbf{v} \times \mathbf{B}). \quad [11.103]$$

Here $\mathbf{E}$ is the electric field intensity and $\mathbf{B}$ the magnetic induction.

Let us consider the particle following a circular path of radius r about the origin, in the xy plane, as figure 11.9 shows. Let us suppose that $\mathbf{E}$ vanishes and that the lines of magnetic induction exhibit cylindrical symmetry about the z axis.

Let us introduce cylindrical coordinates r, ϕ, z by the equations

$$x = r\cos\phi, \quad [11.104]$$

$$y = r\sin\phi, \quad [11.105]$$

$$z = z. \quad [11.106]$$

The component of the magnetic field in the direction of increasing ϕ is zero by design. So we have

$$\mathbf{B} = B_r\hat{\mathbf{r}} + B_z\hat{\mathbf{z}}. \quad [11.107]$$

Since magnetic lines have no source or sink, the divergence of $\mathbf{B}$ vanishes everywhere:

$$\nabla \cdot \mathbf{B} = 0. \quad [11.108]$$

For the cylindrically symmetric field, equation (11.108) becomes

$$\frac{1}{r}\frac{\partial}{\partial r}(rB_r) + \frac{\partial B_z}{\partial z} = 0, \quad [11.109]$$

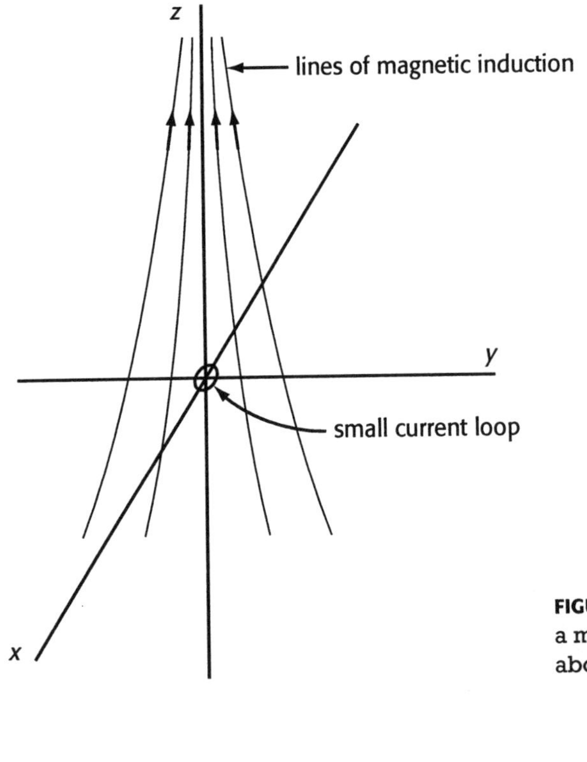

FIGURE 11.9 Orbit of a charged particle in a magnetic field cylindrically symmetric about the z axis.

whence

$$d\left(rB_r\right) = -\frac{\partial B_z}{\partial z} r\,dr. \qquad [11.110]$$

Over the current loop, one may consider that $\partial B_z/\partial z$ is approximately constant. Then (11.110) integrates to

$$rB_r = -\frac{\partial B_z}{\partial z}\frac{r^2}{2} \qquad [11.111]$$

or

$$B_r = -\frac{r}{2}\frac{\partial B_z}{\partial z}. \qquad [11.112]$$

Consider the particle to be moving at speed v on its orbit. Then since the vector **v** is perpendicular to vector $B_r\hat{\mathbf{r}}$, the z component of the-magnetic force from formula (11.103) is

$$F_z = -qvB_r = \frac{1}{2}qvr\frac{\partial B_z}{\partial z} = \boldsymbol{m}_z\frac{\partial B_z}{\partial z}. \qquad [11.113]$$

On the other hand, the average radial component of the force over an integral number of cycles is zero.

In line (11.113), we have introduced the *magnetic moment*

$$\boldsymbol{m}_z = \frac{1}{2}qvr = \frac{qv}{2\pi r}\pi r^2 = I\pi r^2. \qquad [11.114]$$

Note that it equals the effective current I times the area within the loop.

The magnetic energy of an orbiting particle equals the force opposing the effect of the magnetic field times displacement of the loop, integrated over a path extending from a point where the field vanishes. Here we find that

$$\Delta E = -\int \mathbf{F}\cdot d\mathbf{s} = -\int \boldsymbol{m}_z\frac{\partial B_z}{\partial z}\,dz = -\boldsymbol{m}_z B_z. \qquad [11.115]$$

In general, one would have

$$\Delta E = -\boldsymbol{m}\cdot\mathbf{B}. \qquad [11.116]$$

Because angular momentum is quantized in units of $\hbar$,

$$\mu vr = M\hbar, \qquad [11.117]$$

the magnetic moment is quantized in the following manner:

$$\boldsymbol{m}_z = \frac{q}{2\mu}\mu vr = M\frac{q\hbar}{2\mu}. \qquad [11.118]$$

Putting relation (11.118) into (11.115) yields the energy of a rotator in a magnetic field B:

$$E = E_0 - M\frac{q\hbar}{2\mu}B. \qquad [11.119]$$

For an electron circulating in an atom, M is replaced with m, q with $-e$, and μ by m_e. The magnetic moment becomes

$$\boldsymbol{m}_z = -m\frac{e\hbar}{2m_e}, \qquad [11.120]$$

while the energy relationship becomes

$$E = E_0 + m\frac{e\hbar}{2m_e}B. \quad [11.121]$$

The effect on spin energy is added to this.

The unit of magnetic moment,

$$\frac{e\hbar}{2m_e} = 9.2740 \times 10^{-24}\ \mathrm{J\,T^{-1}} \quad [11.122]$$

is called the *Bohr magneton.*

For the electron in an atom, we have the selection rules

$$\Delta l = \pm 1 \quad [11.123]$$

and

$$\Delta m = 0 \quad \text{or} \quad \Delta m = \pm 1 \quad [11.124]$$

following the argument in section 11.5. When m changes, the perturbation of the energy levels due to field B changes and the corresponding spectral line is shifted. Such a shift in a spectral line is referred to as the *Zeeman effect.*

11.15 Observing Spatial Quantization in a Beam

With respect to the direction of a magnetic field B, the magnetic moment of a rotator is quantized following formula (11.118), that of an electron in an atom, following formula (11.120). As a result, the force acting on the rotator or atom is quantized, through formula (12.113). If a homogeneous beam of the units passes through a nonhomogeneous magnetic field, the deflection of particles is similarly quantized. In the apparatus introduced by Otto Stern and Walther Gerlach, this effect was studied.

A typical setup is depicted in figure 11.10. The chosen material is vaporized, collimated by slits, and passed through a velocity filter. The beam then enters the magnetic field. Nonhomogeneity is achieved by employing shaped poles on the magnet.

In the field, force (11.113) acts on each particle. But factor M_z obeys equation (11.118) or (11.120). As a consequence, Newton's second law applies in the form

$$\frac{d^2z}{dt^2} = kM \quad [11.125]$$

with k a constant. Integration over the time t the particle spends in the field yields

$$z = \frac{1}{2}kMt^2. \quad [11.126]$$

Here z is the vertical displacement produced in the field. This is projected onto a screen as shown, to form narrow bands.

For a given J, we expect M to assume values one unit apart, arranged symmetrically about zero. There are thus $2J + 1$ different values. When J is an integer, the number of bands is odd. But when J is half integral, the number of bands is even.

A beam of silver or hydrogen atoms yields two bands. But in each atom, there is no net angular momentum. There is just the spin angular momentum of an odd electron. We presume that this spin is governed by a quantum number s analogous to J and l. The angular momentum in the direction of the field is governed by an m_s analogous to M and m.

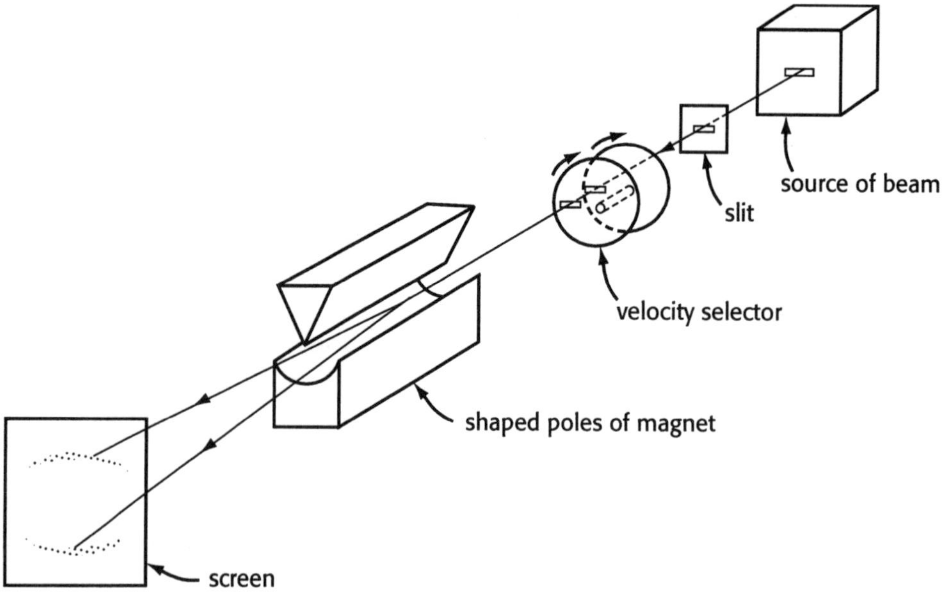

FIGURE 11.10 Arrangement of parts for a Stern Gerlach experiment.

Since neighboring angular momenta differ by $\hbar$, the appearance of only two bands shows that the odd electron may have only the two momenta

$$\frac{1}{2}\hbar \quad \text{and} \quad -\frac{1}{2}\hbar. \qquad [11.127]$$

Zeeman studies show that the first state possesses the magnetic moment

$$-\frac{e\hbar}{2m_e}, \qquad [11.128]$$

while the second state possesses the magnetic moment

$$\frac{e\hbar}{2m_e}. \qquad [11.129]$$

These are twice what one would expect for a charge - e with angular momenta given by (11.127). This fact suggests that interpreting particle spin as rotation of the particle is wrong. One has to consider the intrinsic angular momentum of the electron and its magnetic moment as properties.

11.16 Spin States

Besides electrons, other fundamental fermions exhibit two spin states. These may be represented by introducing a spin coordinate τ and a spin function $\chi(\tau)$ for each such particle.

Consider a single fermion with two disjoint spin states. About the z axis, the spin angular momentum may be either of the two values (11.127). Let the spin function χ for

$1/2\hbar$ about the z axis be α, the spin function χ for - $1/2\hbar$ about the z axis be β. Then we have the eigenvalue equations

$$\hat{S}_z\alpha = \frac{1}{2}\hbar\alpha, \tag{11.130}$$

$$\hat{S}_z\beta = -\frac{1}{2}\hbar\beta, \tag{11.131}$$

in which S_z is the spin operator analogous to (11.86).

For a linear rotator, the energy is given by formulas (11.6) and (11.28). From these, the angular momentum squared is represented by

$$p_\phi^2 = J(J+1)\hbar^2. \tag{11.132}$$

Analogous to this, we have the spin angular momentum squared

$$S^2 = \frac{1}{2}\left(\frac{1}{2}+1\right)\hbar^2. \tag{11.133}$$

The corresponding eigenvalue equations are

$$\hat{S}^2\alpha = \frac{1}{2}\left(\frac{1}{2}+1\right)\hbar^2\alpha, \tag{11.134}$$

$$\hat{S}^2\beta = \frac{1}{2}\left(\frac{1}{2}+1\right)\hbar^2\beta, \tag{11.135}$$

in which $\hat{S}^2$ is the operator for S^2.

A complete Ψ function for the electrons in an atom must indicate the spin state of each electron in each contributing state. For a single electron, we have

$$\Psi = \Psi(\mathbf{r},\tau,t). \tag{11.136}$$

In the approximation that magnetic interaction can be neglected, the spin is independent of the orbital motion and the state function factors as follows:

$$\Psi = \psi(\mathbf{r})\chi(\tau)T(t). \tag{11.137}$$

Questions

11.1 How is a linear rotator modeled by a single particle?
11.2 What does the mass μ of this particle equal? What is its distance from the actual center of mass?
11.3 What symmetries does this model rotator see?
11.4 How do these symmetries affect the state function for rotation?
11.5 Show how the traveling-wave state functions for the linear rotator are constructed.
11.6 What is the source of the quantization of these state functions?
11.7 How are these state functions normalized?

11.8 From the normalized traveling-wave rotational functions, construct the standard normalized standing-wave functions.
11.9 Show what angular momentum states are allowed.
11.10 How is the rotational quantum number J defined?
11.11 What energy is associated with a single angular momentum state?
11.12 How does one go from this energy to the average energy associated with motion around the z axis for a given J?
11.13 How then do we get the total rotational energy for a given J?
11.14 Justify the selection rules

$$\Delta J = \pm 1 \quad \text{and} \quad \Delta M = 0, \pm 1.$$

11.15 How does the resulting spectrum appear?
11.16 What cause spectral lines to be broadened?
11.17 Explain how the cylindrically symmetric rotator is quantized.
11.18 How are the energy levels of an asymmetric rotator correlated with those of cylindrically symmetric rotators?
11.19 Deduce formulas for the rotational partition function of (a) a linear rotator, (b) a cylindrically symmetric rotator, (c) an asymmetric rotator.
11.20 Show that for two particles bound by a potential varying only with the interparticle distance r the state function for a definite energy state has the form $R(\mathrm{r})Y(\theta, \phi)$.
11.21 Why is the form of $Y(\theta, \phi)$ independent of the chosen energy E and the potential function $V(r)$?
11.22 Why does the Laplace equation surrounding a singularity at the origin yield the same angular dependence as Schrodinger's equation with $V = V(r)$?
11.23 Show that spatial derivatives of $1/r$ along lines passing through the origin are solutions to the Laplace equation.
11.24 Why do we multiply the lth derivative of $1/r$ by r^{l+1} in calculating a $Y(\theta, \phi)$?
11.25 How are the directions of the axes along which differentiation proceeds chosen for table 11.2?
11.26 What guides the construction of the traveling- wave functions in table 11.3?
11.27 Summarize the directional properties of the low lying standard rotational state functions.
11.28 Why does one normalize state function Ψ to 1?
11.29 When Ψ is expressed as a linear combination of Ψ_i and Ψ_j, what is accomplished by choosing Ψ_i orthogonal to Ψ_j?
11.30 How are the electric field intensity and the magnetic induction defined by the Lorentz force equation?
11.31 Why does the divergence of the magnetic induction vanish throughout space?
11.32 How is the magnetic moment of a current loop defined? What force acts on such a loop?
11.33 Explain how this magnetic moment is quantized.
11.34 How is the quantization of the magnetic moment observed?
11.35 What direct evidence do we have that an electron exhibits two disjoint spin states?
11.36 How may one formally represent a spin state?
11.37 What eigenvalue equations does the formal representative satisfy?

Problems

11.1 Calculate the separation between successive maxima in the rotational absorption band for $^1H^{79}Br$. The interatomic distance is 1.414 Å while the masses of ^1H and ^{79}Br are 1.007825 and 78.9183 u, respectively.
11.2 Determine the separation between successive maxima in the rotational absorption band for $^2H^{35}Cl$ from that for $^1H^{35}Cl$, which is 20.68 cm . The masses of ^1H, ^2H, and ^{35}Cl are 1.007825, 2.014102, and 34.96885 u, respectively.
11.3 Where are the first three.maxima in the rotational spectrum of $^{127}I^{35}Cl$? The bond length is 2.32 Å while the masses of ^{127}I and ^{35}Cl are 126.9044 and 34.96885 u.

11.4 Calculate the rotational constants for $^{11}B^{19}F_3$. In this molecule, each fluorine is at the corner of an equilateral triangle while the boron is at the center, with each B - F bond 1.29 Å long. The masses of ^{11}B and ^{19}F are 11.00931 and 18.99840 u.

11.5 With the results from problem 11.4, locate the lowest rotational levels of $^{11}B^{19}F_3$ and plot these.

11.6 At what B does the 3_{03} level cross (a) the 2_{20} level, (b) the 2_{21} level, when $A = 3C$ in an asymmetric rotator?

11.7 Construct an equation for the J corresponding to the most highly populated rotational level of a diatomic molecule at temperature T. Solve the equation for ICl, for which $B = 0.1142$ cm^{-1} at 25° C.

11.8 What relationship must exist among the coefficients in

$$\Psi = ax^2 + by^2 + cz^2 + dxy + eyz + fzx$$

for this function to satisfy the Laplace equation? Show that the homogeneous polynomial factor in each standard d orbital obeys this condition.

11.9 Rotate the d_{xy} eigenfunction around the z axis by angle α and express the result as a linear combination of the standard d orbitals.

11.10 By integration, show that Y_0^0, Y_1^0, and Y_1^{+1} in table 11.5 are normalized.

11.11 By integration, show that Y_0^0, Y_1^0, and Y_1^{+1} in table 11.5 are mutually orthogonal.

11.12 Determine the overlap integral S_{12} for the superpositions
(a) $\Psi = 3/5\Psi_1 + 3/5\Psi_2$, (b) $\Psi = 1/2\,\Psi_1 + 3/4\Psi_2$ if functions Ψ, Ψ_1 and Ψ_2 are real and normalized to 1.

11.13 A doubly degenerate energy level is represented by the normalized real functions Ψ_1 and Ψ_2 for which S_{12} equals 1/2. Letting $\Psi_I = \Psi_1$, construct a normalized $\Psi_{II} = a\Psi_1 + b\Psi_2$ such that it is orthogonal to Ψ_I.

— — —

11.14 Calculate the separation between successive maxima in the rotational absorption band for $^1H^{127}I$. The interatomic distance is 1.604 Å while the masses of ^2H and ^{127}I are 1.007825 and 126.9044 u, respectively.

11.15 Determine the separation between successive maxima in the rotational absorption band for $^2H^{127}I$ from the separation calculated in problem 11.14. The mass of ^2H is 2.014102 u.

11.16 Where are the first four maxima in the rotational spectrum of $^7Li^1H$? The bond length is 1.60 Å while the mass of ^7Li is 7.01600 u.

11.17 If the rotational constant A for CH_3Br is 5.08 cm^{-1} , while $B = C = 0.31$ cm^{-1}, where are the lowest rotational energy levels? What transitions are allowed and what spectrum do these allowed transitions produce?

11.18 The H_2O molecule is nonlinear with an HOH angle of 104° 27' and an O - H bond distance of 0.958 Å. Calculate its rotational constants. The mass of ^{16}O is 15.994915 u.

11.19 With the results from problem 11.17, locate the lowest rotational levels of the H_2O molecule.

11.20 Construct an equation for the J corresponding to the most highly populated rotational level of a molecule for which $A = B = C$. Here each level has the degeneracy $(2J + 1)^2$. Solve the equation for CH_4. for which $B = 5.24$ cm^{-1}, at 25° C.

11.21 What relationships must exist among the coefficients in

$$\psi = ax^3 + by^3 + cz^3 + dx^2y + ex^2z + fy^2z + gy^2x + hz^2x + jz^2y + kxyz$$

for this function to satisfy the Laplace equation? Show that the homogeneous polynomial factor in each Y_3^m meets these conditions.

11.22 Rotate the eigenfunction $N(yz/r^2)$ 45° counterclockwise around the x axis.

11.23 Determine how homogeneous polynomial (11.73) and the corresponding Y behave on inversion through the origin.

11.24 Superpose the standard real f eigenfunctions to form

$$\sqrt{\frac{105}{16\pi}}\frac{x\left(y^2 - z^2\right)}{r^3}.$$

11.25 Determine the probability density distribution over angles θ and ϕ for (a) a p_x orbital, (b) a p_y orbital, (c) a p_z orbital. What angular distribution arises from placing one electron in a p_x orbital, one electron in a p_y orbital, and one electron in a p_z orbital, each with the same energy?

11.26 The normalized orbital for bond A - B has the form

$$\psi_1 = \frac{1}{2}\chi_A + \frac{\sqrt{3}}{2}\chi_B$$

where χ_A is a valence orbital of A and χ_B a valence orbital of B. Construct the antibonding orbital ψ_2 formed from χ_A and χ_B, which is orthogonal to Ψ_1.

References

Books

Barrow, G. M.: 1962, *Molecular Spectroscopy*, McGraw-Hill, New York, pp. 47-114.

Barrow takes the quantum mechanical results on faith and discusses rotational spectra in some detail.

Duffey, G. H.: 1984, *A Development of Quantum Mechanics Based on Symmetry Considerations*, Reidel, Dordrecht, Holland, pp. 38-75, 212-218.

In the first section cited, suitable azimuthal factors for *Y* are obtained as in this chapter. But then, the corresponding dependence of *Y* on θ is derived by a symmetry argument. In the second section, the multiaxial formula for *Y* is constructed.

Möller, K. D., and Rothschild, W. G.: 1971, *Far-Infrared Spectroscopy*, John Wiley & Sons, New York, pp. 1-759.

The first four chapters of this reference describe in considerable detail the instrumentation used. The next seven chapters deal with vapor and liquid phase spectra. The following two chapters discuss crystal spectra. Throughout, the theory is directed to the subject under consideration.

Articles

Allendoerfer, R. D.: 1990, "Teaching the Shapes of the Hydrogen-like and Hybrid Atomic Orbitals," *J. Chem. Educ.* **67**, 37- 39.

Boeyens, J. C. A.: 1995, "Understanding Electron Spin," *J. Chem. Educ.* **72**, 412-415.

Chattaraj, P. K., and Sannigrahi, A. B.: 1990, "A Simple Group Theoretical Derivation of the Selection Rules for Rotational Transitions," *J. Chem. Educ.* **67**, 653-655.

Compaan, A., and Wagoner, A.: 1994, "Rotational Raman Scattering in the Instructional Laboratory," *Am. J. Phys.* **62**, 639- 645.

Gauerke, S. J., and Campbell, M. L.: 1994, "A Simple, Systematic Method for Determining J Levels for jj Coupling," J. Chem. Educ. **71**, 457-463.

Henderson, G., and Lcysdon, B.: 1995, "Stark Effects on Rigid- Rotor Wavefunctions," *J. Chem. Educ.* **72**, 1021-1024.

Kikuchi, O., and Suzuki, K.: 1985, "Orbital Shape Representations," *J. Chem. Educ.* **62**, 206-209.

Levine, I. N.: 1985, "Thermodynamic Internal Energy of an Ideal Gas of Rigid Rotors," *J. Chem. Educ.* **62**, 53-54.

Rioux, F.: 1994, "Quantum Mechanics, Group Theory, and C_{60}," *J. Chem. Educ.* **71**, 464-465.

Waite, B. A.: 1989, "An Aufbau Methodology for the Modeling of Rotational Fine Structure of Infrared Spectral Bands," *J. Chem. Educ.* **66**, 805-809.

Woods, R., and Henderson, G.: 1987, "FTIR Rotational Spectroscopy," *J. Chem. Educ.* **64**, 921-924.

12

Vibrational and Radial Motion

12.1 Pertinent Potentials

IN A HYDROGENLIKE ATOM, the electron is bound to the nucleus by a Coulomb potential. This has the form

$$V(r) = -\frac{Ze^2}{4\pi\varepsilon r}, \tag{12.1}$$

where r is the interparticle distance, Ze the charge on the nucleus, $-e$ the charge on the electron, and ε the permittivity of free space. The Schrödinger equation has form (11.64). So the angular motion separates from the radial motion. The former we have considered; the latter we will take up in the second part of this chapter.

In a molecule, the nuclei are surrounded by orbitals containing electrons. These fall into two classes. There are the core orbitals about individual nuclei. Then there are the molecular orbitals enclosing two or more nuclei. These serve to bind the atoms together.

In the classical picture, the valence electrons circulate about the atomic cores. Since an electron is 1/1836 as massive as a proton, 1/1839 as massive as a neutron, the electron is 3- 4 orders of magnitude lighter than any nucleus. So it would execute a complete orbit while the nuclei moved only slightly.

In the corresponding quantum picture, the orbitals are little affected by motions of the nuclei. In the *Born - Oppenheimer approximation*, this effect is neglected. The energy of the molecule is calculated for representative positions of the nuclei. A function is fitted to the resulting points and it is considered to be the potential function for relative motions of the nuclei, that is, for vibrations of the molecule.

Classically, the vibrations can be separated into modes that are independent when the amplitudes are small. Quantum mechanically, these modes are independent to a first approximation.

Now, a mode of vibration is modeled by a particle of reduced mass μ confined in a one - dimensional potential well. In the *harmonic approximation*, this well is parabolic, with the form

$$V = \frac{1}{2}fx^2. \tag{12.2}$$

Here f is the force constant for displacement x. With a diatomic molecule, this displacement is given by

$$x = r - r_e, \qquad [12.3]$$

where r is the internuclear distance and r_e is the equilibrium r.

Example 12.1

Construct a one particle model for the oscillation of a diatomic molecule.

The electron mass associated with an atom in a molecule is very small with respect to the nuclear mass. So even if the center of its electron mass is displaced considerably from the nucleus, the center of the atom's mass is practically at the nuclear center. So in our discussions we will consider the mass of each atom to reside at the center of its nucleus.

Consider the diatomic molecule depicted in figure 12.1. Suppose that the masses of the two atoms, m_1 and m_2, are at positions r_1 and r_2 along an axis at time t. Let us locate the origin at the center of mass, so

$$m_1 r_1 + m_2 r_2 = 0$$

whence

$$r_1 = -\frac{m_2}{m_1} r_2.$$

Since the masses are constant, the derivative of r_1 with respect to time t is given by

$$\dot{r}_1 = -\frac{m_2}{m_1} \dot{r}_2$$

Note that a dot over a symbol represents differentiation with respect to t. With this notation, the vibrational kinetic energy is

$$T = \frac{1}{2} m_1 \dot{r}_1^{\,2} + \frac{1}{2} m_2 \dot{r}_2^{\,2} = \frac{1}{2} m_1 \frac{m_2^{\,2}}{m_1^{\,2}} \dot{r}_2^{\,2} + \frac{1}{2} m_2 \dot{r}_2^{\,2} = \frac{1}{2}\left(\frac{1}{m_1} + \frac{1}{m_2}\right) m_2^2 \dot{r}_2^{\,2} = \frac{1}{2} \mu \frac{m_2^{\,2}}{\mu^2} \dot{r}_2^{\,2}.$$

In the last step, the reduced mass μ has been introduced by the equation

$$\frac{1}{\mu} = \frac{1}{m_1} + \frac{1}{m_2}.$$

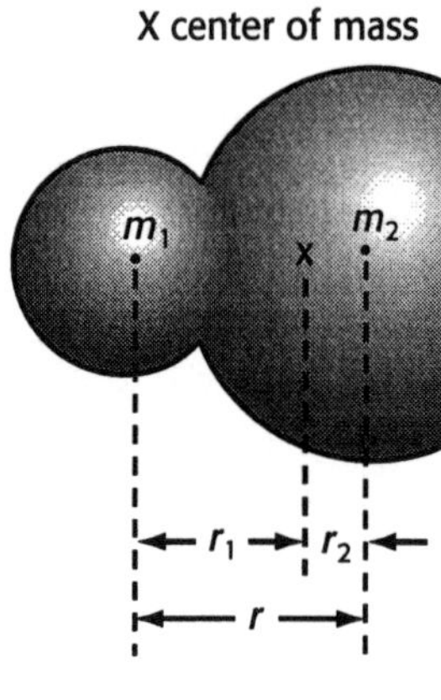

FIGURE 12.1 Two nuclei held together by an enveloping electron cloud.

The internuclear distance is given by

$$r = r_2 - r_1 = r_2 + \frac{m_2}{m_1} r_2 = \left(\frac{1}{m_1} + \frac{1}{m_2} \right) m_2 r_2 = \frac{m_2 r_2}{\mu},$$

whence

$$\dot{r} = \frac{m_2 \dot{r}_2}{\mu}.$$

Substituting this into the expression for T yields

$$T = \frac{1}{2} \mu \dot{r}^2.$$

But differentiating equation (12.3),

$$\dot{x} = \dot{r},$$

and inserting the result into the kinetic energy form leads to

$$T = \frac{1}{2} \mu \dot{x}^2.$$

Thus, a particle of mass μ in the potential

$$V = V(x) = V(r - r_e)$$

has the same kinetic energy and potential energy as the diatomic molecule. As a model for the vibrational motion, it represents these properties faithfully.

Example 12.2

If the two atom system in figure 12.1 obeyed Newtonian mechanics, what would its frequency of oscillation be?

The internuclear distance in the system is

$$r = r_2 - r_1.$$

So differentiating (12.3) yields

$$dx = dr = dr_2 - dr_1.$$

In the harmonic approximation, equation (12.2) holds. Differentiating it and inserting the above relationship yields

$$dV = fx \, dx = fx dr_2 - fx dr_1.$$

But we have

$$dV = \frac{\partial V}{\partial r_2} dr_2 + \frac{\partial V}{\partial r_1} dr_1 = -F_2 dr_2 - F_1 dr_1,$$

where F_2 and F_1 are the forces acting on masses m_2 and m_1

Applying Newton's second law to each mass yields

$$m_2 \frac{d^2 r_2}{dt^2} = F_2 = -fx, \qquad m_1 \frac{d^2 r_1}{dt^2} = F_1 = fx.$$

These equations combine to give

$$\frac{d^2x}{dt^2} = \frac{d^2r_2}{dt^2} - \frac{d^2r_1}{dt^2} = -\left(\frac{f}{m_2} + \frac{f}{m_1}\right)x = -\frac{f}{\mu}x$$

or

$$\frac{d^2x}{dt^2} + \frac{f}{\mu}x = 0.$$

Integrating this equation leads to the result

$$x = A\cos\left[\left(\frac{f}{\mu}\right)^{1/2} t + \alpha\right] = A\cos(\omega t + \alpha)$$

whence

$$\omega = \left(\frac{f}{\mu}\right)^{1/2} \quad \text{and} \quad \nu = \frac{\omega}{2\pi} = \frac{1}{2\pi}\left(\frac{f}{\mu}\right)^{1/2}.$$

We call ω the angular frequency and ν the conventional frequency.

12.2 Schrödinger Equation for the Harmonic Oscillator

In the harmonic approximation, each vibrational mode is modeled by a particle of mass μ confined by a parabolic potential

$$V = \frac{1}{2}fx^2 \qquad [12.4]$$

The parameters μ and f may vary from mode to mode.

Substituting these into the single independent variable form of equation (10.134) yields

$$\frac{d^2\psi}{dx^2} + \frac{2\mu}{\hbar^2}\left(E - \frac{1}{2}fx^2\right)\psi = 0. \qquad [12.5]$$

The parabolic potential rises without limit as x goes to $\pm\infty$. So as long as the mode carries a finite amount of energy, the probability of finding particle μ at $x = \pm\infty$ is zero. Consequently, ρ and its factor Ψ must go asymptotically to zero as x increases or decreases without limit. Both of these conditions can be met only for certain discrete E's. We need to determine these allowed energies.

One may remove the reference to a particular oscillator by introducing the substitutions

$$\frac{f\mu}{\hbar^2} = a^2, \qquad [12.6]$$

$$\frac{2\mu E}{\hbar^2} = ab, \qquad [12.7]$$

$$w = a^{1/2}x. \qquad [12.8]$$

These convert equation (12.5) to the dimensionless form

$$\frac{d^2\psi}{dw^2} + \left(b - w^2\right)\psi = 0. \qquad [12.9]$$

If we let operator D be given by

$$D = \frac{d}{dw}, \qquad [12.10]$$

equation (12.9) becomes

$$\left(D^2 - w^2\right)\psi = -b\psi. \qquad [12.11]$$

But combining (12.7), (12.6) and solving for E yields

$$E = b\frac{\hbar^2}{2\mu}\left(\frac{f\mu}{\hbar^2}\right)^{1/2} = \frac{1}{2}b\hbar\left(\frac{f}{\mu}\right)^{1/2} = \frac{1}{2}b\hbar\omega. \qquad [12.12]$$

In the last step, we have introduced the classical angular frequency of the oscillator from example 12.2.

We next turn to determining the values of b that yield acceptable state functions.

Example 12.3

Evaluate (a) $(D - w)(D + w)\psi$, (b) $(D + w)(D - w)\psi$.

Since D is the differentiating operator (12.10), it has transforming as well as algebraic properties. Employing the rule for differentiating a product, we find that

$$\left(D - w\right)\left(D + w\right)\psi = \left(D - w\right)\left(D\psi + w\psi\right) = D^2\psi + wD\psi + \psi - wD\psi - w^2\psi = \left(D^2 - w^2 + 1\right)\psi$$

and

$$\left(D + w\right)\left(D - w\right)\psi = \left(D + w\right)\left(D\psi - w\psi\right) = D^2\psi - wD\psi - \psi + wD\psi - w^2\psi = \left(D^2 - w^2 - 1\right)\psi.$$

12.3 Quantization of the Harmonic Oscillator

A universal equation for the harmonic oscillator is (12.11). Introducing it into the right sides of the expressions found in example 12.3 leads to

$$\left(D - w\right)\left(D + w\right)\psi = -\left(b - 1\right)\psi \qquad [12.13]$$

and

$$\left(D + w\right)\left(D - w\right)\psi = -\left(b + 1\right)\psi. \qquad [12.14]$$

If we let

$$b = 2v + 1 \qquad [12.15]$$

and identify each v and corresponding ψ with the same subscript, these equations become

$$\left(D - w\right)\left(D + w\right)\psi_2 = -2v_2\psi_2, \qquad [12.16]$$

$$\left(D + w\right)\left(D - w\right)\psi_1 = -\left(2v_1 + 2\right)\psi_1. \qquad [12.17]$$

Now act on (12.16) with operator $(D + w)$, on (12.17) with $(D - w)$, to construct

$$\left(D+w\right)\left(D-w\right)\left[\left(D+w\right)\psi_2\right]=-2v_2\left[\left(D+w\right)\psi_2\right], \qquad [12.18]$$

$$\left(D-w\right)\left(D+w\right)\left[\left(D-w\right)\psi_1\right]=-\left(2v_1+2\right)\left[\left(D-w\right)\psi_1\right]. \qquad [12.19]$$

On comparing (12.19) with (12.16), we see that

$$\left(D-w\right)\psi_1=c_+\psi_2 \qquad [12.20]$$

and

$$v_1+1=v_2. \qquad [12.21]$$

And on comparing (12.18) with (12.17), we see that

$$\left(D+w\right)\psi_2=c_-\psi_1 \qquad [12.22]$$

with

$$v_2=v_1+1. \qquad [12.23]$$

Substitute (12.22) and (12.20) into the left side of equation (12.16) and set the result equal to the right side:

$$\left(D-w\right)\left(D+w\right)\psi_2=\left(D-w\right)c_-\psi_1=c_+c_-\psi_2=-2v_2\psi_2. \qquad [12.24]$$

To satisfy the last equality, one may take

$$c_-=\left(2v_2\right)^{1/2} \qquad [12.25]$$

and

$$c_+=-\left(2v_2\right)^{1/2}. \qquad [12.26]$$

Then (12.20) would become

$$\psi_v=-\frac{D-w}{\left(2v\right)^{1/2}}\psi_{v-1}, \qquad [12.27]$$

and (12.22) would become

$$\psi_{v-1}=\frac{D+w}{\left(2v\right)^{1/2}}\psi_v, \qquad [12.28]$$

for number v replacing 2 and v_2. A person can show that choices (12.25) and (12.26) do preserve normalization.

The operator acting on ψ_{v-1} in equation (12.27) is called a *step-up operator* because it increases quantum number v by 1. Similarly, the operator acting on ψ_v in equation (12.28) is called a *step-down operator*. It decreases quantum number v by 1.

When the harmonic oscillator is in its lowest state, action of the step-down operator on its ψ should yield nothing. We would have

$$\left(D+w\right)\psi_0==0. \qquad [12.29]$$

Substituting this condition into equation (12.16) makes the eigenvalue, and v, zero. On the other hand, each action of the step-up operator on ψ increases v by one unit. So substituting relation (12.15) into formula (12.12) leads to

$$E = \left(v + \frac{1}{2}\right)\hbar\omega = \left(v + \frac{1}{2}\right)h\nu, \qquad v = 0,1,2,.... \tag{12.30}$$

For transitions in which v changes by 1, we have

$$\Delta E_{\text{vib}} = \pm\hbar\omega_{\text{vib}} = \pm h\nu_{\text{vib}}. \tag{12.31}$$

The corresponding photon wave number is

$$\bar{k}_0 = \frac{h\nu_{\text{vib}}}{hc} = \bar{k}_{\text{vib}}. \tag{12.32}$$

A spectral line involving no other change in the molecule is called a Q line or *branch*.

12.4 Nodeless Solutions

The simplest solutions to the harmonic oscillator Schrödinger equation will now be constructed. These also satisfy the Schrödinger equation asymptotically at all energies.

When $v = 0$, parameter $b = 1$ and equation (12.13) is satisfied when condition (12.29) holds. This equation has the form

$$\frac{d\psi}{dw} + w\psi = 0, \tag{12.33}$$

whence

$$\frac{d\psi}{\psi} = -w\,dw. \tag{12.34}$$

Integration leads to

$$\ln\psi = \ln A - \frac{1}{2}w^2 \tag{12.35}$$

and

$$\psi = Ae^{-w^2/2}. \tag{12.36}$$

From example 12.4, the normalized form is

$$\psi = \left(\frac{a}{\pi}\right)^{1/4} e^{-w^2/2}. \tag{12.37}$$

But when v = -1, parameter b = - 1 and equation (12.14) is satisfied by

$$(D - w)\psi = 0 \tag{12.38}$$

or

$$\frac{d\psi}{dw} - w\psi = 0, \tag{12.39}$$

whence

$$\frac{d\psi}{\psi} = w\,dw. \tag{12.40}$$

Integration gives us

$$\ln\psi = \ln B + \frac{1}{2}w^2 \quad [12.41]$$

and

$$\psi = Be^{w^2/2}. \quad [12.42]$$

This function is not normalizable. Instead, it makes the probability density ρ increase without limit as $|w|$ and $|x|$ increase. So it is not an acceptable solution.

Example 12.4

Normalize function (12.36).

This solution yields the probability density

$$\rho = \psi^*\psi = \left(A^* e^{-w^2/2}\right)\left(Ae^{-w^2/2}\right) = A^* Ae^{-w^2} = A^* Ae^{-ax^2}$$

Let us take A to be real and set the total probability equal to 1:

$$\int_{-\infty}^{\infty} \rho \, \mathrm{d}x = A^2 \int_{-\infty}^{\infty} e^{-ax^2} \, \mathrm{d}x = 1.$$

Inserting the value of the integral from example 10.5 leads to

$$A = \left(\frac{a}{\pi}\right)^{\frac{1}{4}}.$$

Example 12.5

Show that a sum of the two nodeless solutions satisfies the Schrödinger equation asymptotically at any energy E. Construct the form

$$\psi = Ae^{-w^2/2} + Be^{w^2/2}.$$

Differentiate twice

$$\frac{\mathrm{d}\psi}{\mathrm{d}w} = -Awe^{-w^2/2} + Bwe^{w^2/2},$$

$$\frac{\mathrm{d}^2\psi}{\mathrm{d}w^2} = Aw^2e^{-w^2/2} + Bw^2e^{w^2/2} - Ae^{-w^2/2} + Be^{w^2/2}.$$

As $w \to \pm\infty$, the last two terms can be neglected and

$$\frac{\mathrm{d}^2\psi}{\mathrm{d}w^2} \cong w^2\psi.$$

Furthermore, b can be neglected with respect to w^2 in equation (12.9). Thus, it reduces to

$$\frac{\mathrm{d}^2\psi}{\mathrm{d}w^2} \cong w^2\psi,$$

the same as the preceding approximate equation. Since the ψ above contains two arbitrary constants A and B and since the differential equation is only second order, this is the most general asymptotic solution for either $w = \infty$ or $w = -\infty$.

For a solution to be normalizable, B must be zero at both $w = \infty$ and $w = -\infty$. Thus, one must have an asymptotic solution $\psi = Ae^{-w^2/2}$ at $w = \infty$ join by a wavy path to $\psi = \pm Ae^{-w^2/2}$ at $w = -\infty$. For the even solutions the upper sign prevails; for the odd solutions, the lower sign.

12.5 Acceptable Wavy Solutions

By applying the step-up operator v times to the ground state function, one generates the function for the vth level. Each action of the operator introduces more waviness and an additional node.

For the general harmonic oscillator, we start with expression (12.37). Putting it into formula (12.27) and iterating v times yields

$$\psi_v = \left(-1\right)^v \frac{\left(D-w\right)^v}{\left(2^v v!\right)^{1/2}} \left(\frac{a}{\pi}\right)^{1/4} e^{-w^2/2}. \qquad [12.43]$$

By differentiation and mathematical induction, one finds that

$$\left(D-w\right)^v e^{-w^2/2} = e^{w^2/2} D^v e^{-w^2}. \qquad [12.44]$$

Substituting into (12.43) gives us

$$\psi_v = \frac{1}{\left(2^v v!\right)^{1/2}} \left(\frac{a}{\pi}\right)^{1/4} e^{-w^2/2} \left(-1\right)^v e^{w^2} \frac{\mathrm{d}^v}{\mathrm{d}w^v} e^{-w^2} = N e^{-w^2/2} H_v\left(w\right). \qquad [12.45]$$

In the last step, we have set

$$H_v\left(w\right) = \left(-1\right)^v e^{w^2} \frac{\mathrm{d}^v}{\mathrm{d}w^v} e^{-w^2}. \qquad [12.46]$$

This expression is the *Hermite polynomial* of the vth degree. The factor $(-1)^v$ serves to make the highest term in $H_v(w)$ positive.

In table 12.1, the state functions for the lowest levels are listed. Plots of the eigenfunctions and of the resulting probability density distributions appear in figure 12.2.

12.6 Partition Function for a Vibrational Mode

From how a vibrational mode is quantized, a person can calculate the corresponding partition function and the mode's contribution to the thermodynamic properties of the given material.

Consider a pure homogeneous thermodynamic system at equilibrium at temperature T. Suppose that T is high enough so that the probability for a molecule to reside in any one disjoint state is small. Then the Boltzmann distribution law applies as in section 11.9.

Also suppose that the submicroscopic units of the system possess one or more vibrational modes. A typical mode is represented approximately by the particle of mass μ in the potential $1/2\, fx$. For such a model, the energy is given by equation (12.30),

$$E = \left(v + \frac{1}{2}\right)\hbar\omega, \qquad v = 0,1,2,\ldots. \qquad [12.47]$$

TABLE 12.1 Normalized Harmonic Oscillator Eigenfunctions

Quantum Number v	*Formula*
0	$\left(\frac{a}{\pi}\right)^{1/4} e^{-ax^2/2}$
1	$\sqrt{2}\left(\frac{a}{\pi}\right)^{1/4} a^{1/2} x e^{-ax^2/2}$
2	$\frac{1}{\sqrt{2}}\left(\frac{a}{\pi}\right)^{1/4}\left(2ax^2 - 1\right)e^{-ax^2/2}$
3	$\frac{1}{\sqrt{3}}\left(\frac{a}{\pi}\right)^{1/4}\left(2a^{3/2}x^3 - 3a^{1/2}x\right)e^{-ax^2/2}$
4	$\frac{1}{2\sqrt{6}}\left(\frac{a}{\pi}\right)^{1/4}\left(4a^2x^4 - 12ax^2 + 3\right)e^{-ax^2/2}$
5	$\frac{1}{2\sqrt{15}}\left(\frac{a}{\pi}\right)^{1/4}\left(4a^{5/2}x^5 - 20a^{3/2}x^3 + 15a^{1/2}x\right)e^{-ax^2/2}$
6	$\frac{1}{4\sqrt{45}}\left(\frac{a}{\pi}\right)^{1/4}\left(8a^3x^6 - 60a^2x^4 + 90ax^2 - 15\right)e^{-ax^2/2}$

[a] Here $a = (f\mu)^{1/2}/\hbar$, with f = force constant and μ = reduced mass.

Since there is no degeneracy at any level, the corresponding partition function is

$$Z_{\text{vib}} = \sum_{v=0}^{\infty} e^{-\left(v+\frac{1}{2}\right)\hbar\omega/kT} = e^{-\hbar\omega/2kT}\left(1 + e^{-\hbar\omega/kT} + e^{-2\hbar\omega/kT} + \ldots\right). \quad [12.48]$$

The series in parenthesis has the form

$$S = 1 + a + a^2 + \ldots. \quad [12.49]$$

But since

$$aS = a + a^2 + a^3 + \ldots, \quad [12.50]$$

we have

$$\left(1 - a\right)S = 1 \quad [12.51]$$

and

$$S = \frac{1}{1-a}. \quad [12.52]$$

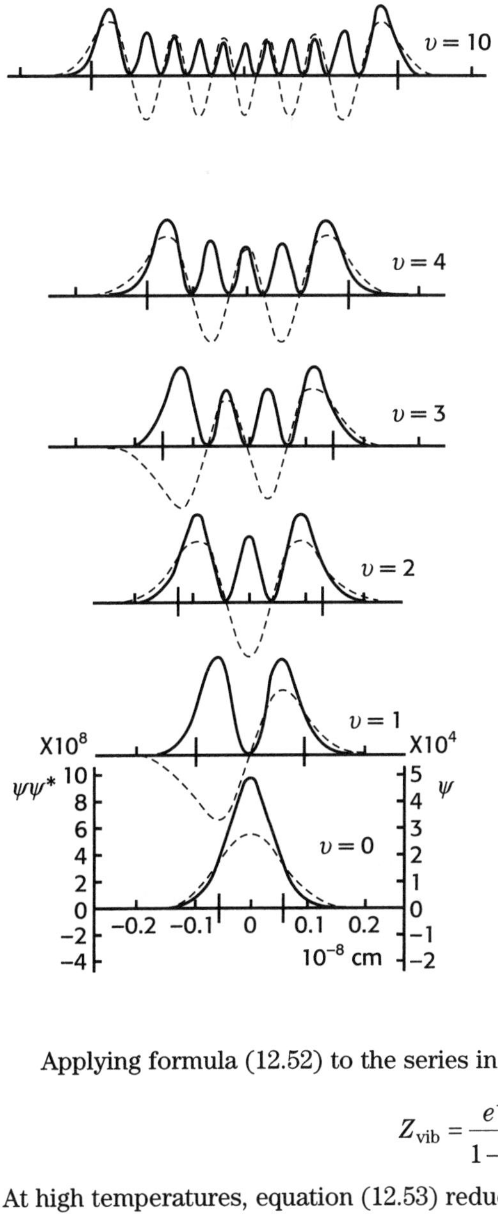

FIGURE 12.2 Probability density distributions (solid lines) and state functions (dashed lines) for low levels of the harmonic oscillator. For the scale, μ = 10 u and $\tilde{k}$= 1000 cm^{-1}.

Applying formula (12.52) to the series in (12.48) leads to

$$Z_{\text{vib}} = \frac{e^{-\hbar\omega/2kT}}{1 - e^{-\hbar\omega/kT}}. \qquad [12.53]$$

At high temperatures, equation (12.53) reduces to

$$Z_{\text{vib}} \cong \frac{1}{1 - \left(1 - \hbar\omega / kT\right)} = \frac{kT}{\hbar\omega}, \qquad [12.54]$$

whence

$$\ln Z_{\text{vib}} \cong \ln T + \ln\left(k / \hbar\omega\right). \qquad [12.55]$$

Substituting into the last expression in line (10.39) leads to

$$E_{\text{vib}} = NkT^2\left(\frac{1}{T}\right) = NkT = nRT. \qquad [12.56]$$

Note that this result agrees with equation (3.48) for a classically excited vibrational mode.

12.7 Selection Rules

An internal mode of motion of a molecule may gain or lose energy (a) by interacting with other nontranslational modes in the same molecule, (b) by interacting with other modes in neighboring molecules, and (c) by interacting with the electromagnetic field. Processes (a) and (b) are of interest in chemical kinetics but will not be considered here. Process (c) is involved in producing absorption, emission, and Raman spectra.

As noted in section 11.5, each photon carries one unit of angular momentum directed along its line of travel. In an infrared interaction, this may involve a change in the rotational quantum number of the participating molecule by 1:

$$\Delta J = \pm 1. \qquad [12.57]$$

For a nonlinear molecule, or a linear one with net electronic angular momentum, the absorption or emission may merely alter the direction of the angular momentum and

$$\Delta J = 0. \qquad [12.58]$$

Furthermore, degenerate modes may couple to produce rotation. Then a rearrangement can yield condition (12.58) also.

In a Raman interaction, the angular momentum of the photon may or may not be reoriented. But since the photon angular momentum is either $\pm\hbar$ along its ray, we have the rule that

$$\Delta J = 0 \quad \text{or} \quad \Delta J = \pm 2 \qquad [12.59]$$

for the molecule. For a nonlinear molecule, we may also have

$$\Delta J = \pm 1 \quad \text{and} \quad \Delta K = 0, \pm 1, \pm 2. \qquad [12.60]$$

In the classical approximation, a polar molecule interacts with the electromagnetic field through its fluctuating dipole moment. This variation is produced both by rotation and by vibration. A molecular change is accompanied by creation or annihilation of an electromagnetic wave segment, or by alteration of it.

In quantum mechanics, the interaction involves the initial and final state functions, ψ_{i} and ψ_{f} through the *transition dipole moment*

$$\mu_{\text{if}} = q \int \psi_{\text{i}}^{*} \mathbf{r} \psi_{\text{f}} \, \text{d}^3\mathbf{r}. \qquad [12.61]$$

For an electron transition, q would equal $-e$ and r would locate the electron. For a vibrational mode, a fractional charge q would reside on the model particle and charge $-q$ would be at the origin. Number q would be taken so that qr would equal the dipole moment for position r.

Absorption of radiation occurs at the rate

$$\frac{\text{d}P}{\text{d}t} = B\rho, \qquad [12.62]$$

where P is the probability and ρ is the energy density of the radiation per unit frequency. The coefficient is given by

$$B = \frac{\left|\mu_{\text{if}}\right|^2}{2\varepsilon\hbar^2}. \qquad [12.63]$$

We will not derive these expressions here.

Integral (12.61) differs from zero for a harmonic oscillator only when

$$\Delta v = \pm 1. \qquad [12.64]$$

So for absorption or emission to occur, quantum number v must change by 1. But when appreciable anharmonicity is present, overtones for which

$$\Delta v = \pm 2, \pm 3, \ldots, \qquad [12.65]$$

appear.

A Raman transition involves a net change in polarizability of the molecule while the photon is passing. This occurs with

$$\Delta v = \pm 1. \qquad [12.66]$$

12.8 Vibration-Rotation Spectra

Because of the size of the parameters involved, the energy needed to effect a vibrational transition is generally much larger than that needed to effect a rotational transition. As a consequence, rotational branches on both sides of a Q branch appear in both infrared and Raman spectra. Here we will consider the former.

When the angular momentum for a photon annihilated or created is supplied by a rearrangement of angular momenta in the molecule, a Q branch at wave number

$$k = k_0 \qquad [12.67]$$

appears. As examples, we have the stretching mode of the NO molecule and the bending mode of the CO_2 molecule.

An absorbed (or emitted) photon excites (or de-excites) both vibration and rotation when the signs of both Δv and ΔJ are the same:

$$\Delta v = \pm 1 \quad \text{and} \quad \Delta J = \pm 1. \qquad [12.68]$$

Then the energy needed to increase (or decrease) v combines additively with the energy needed to increase (or decrease) J. Since the defined wave numbers are proportional to the corresponding energies, expressions (12.67) and (11.34) combine additively.

We obtain

$$k = k_0 + 2B(J_0 + 1) \quad \text{with} \quad J_0 = 0, 1, 2, \ldots \qquad [12.69]$$

for absorption, and

$$k = k_0 + 2BJ_0 \quad \text{with} \quad J_0 = 1, 2, \ldots \qquad [12.70]$$

for emission. These lines form the R *branch* of the spectrum.

When the energy required to increase (or decrease) v is made up of energy from the photon and from a rotational change, the signs on Δv and ΔJ are opposite and

$$\Delta v = \pm 1 \quad \text{while} \quad \Delta J = \mp 1. \qquad [12.71]$$

Governing absorption, we then have

$$k_0 = k + 2BJ_0, \qquad [12.72]$$

whence

$$k = k_0 - 2BJ_0 \quad \text{with} \quad J_0 = 1, 2, 3, \ldots \qquad [12.73]$$

Similarly, the photon wave number for emission is

$$k = k_0 - 2B(J_0 + 1) \quad \text{with} \quad J_0 = 0, 1, 2, \ldots \qquad [12.74]$$

These lines form the P branch of the spectrum.

If there were no variation in the moment of inertia with J and v, the energy levels and the resultant absorption spectrum would appear as figure 12.3 shows. Actually, such variations are always present. Furthermore, each line spreads over a range of wave numbers, for the reasons noted in section 11.6. See figure 12.4. The relative intensity of a line depends on the extent to which the initial rotational level is populated at the prevailing temperature.

Example 12.6

After correcting for anharmonicity, one finds that $\bar{k}_0$ for HCl is 2990 cm^{-1} . From this number, determine the force constant for the molecule.

Solving the final equation in example 12.2 for the force constant yields

$$f = \mu\left(2\pi\nu_{\text{vib}}\right)^2.$$

But from equation (12.32), we have

$$\nu_{\text{vib}} = c\bar{k}_0.$$

From example 11.6, the reduced mass is

$$\mu = 1.628 \times 10^{-24}\ \text{g}.$$

So the force constant is given by

$$f = \mu\left(2\pi c\bar{k}_0\right)^2 = \left(1.628 \times 10^{-27}\ \text{kg}\right)\left[2\pi\left(2.9979 \times 10^{10}\,\text{cm s}^{-1}\right)\left(2990\ \text{cm}^{-1}\right)\right]^2 = 516\ \text{N m}^{-1}.$$

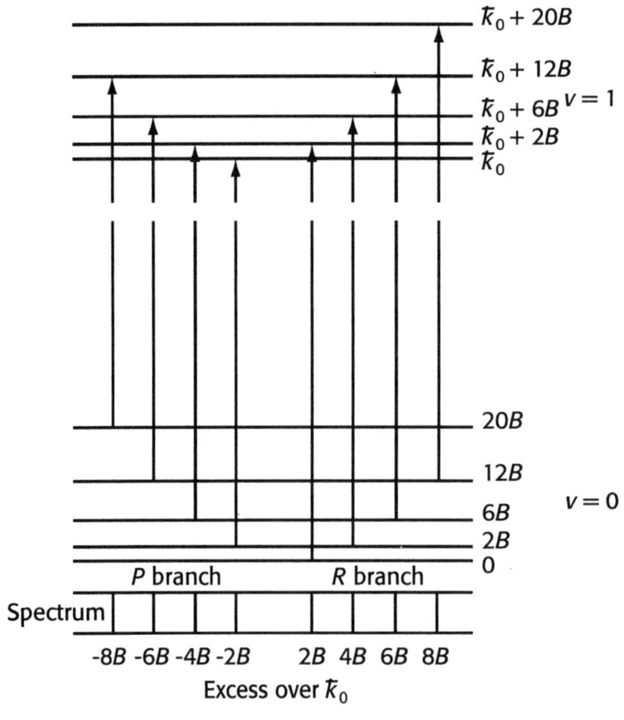

FIGURE 12.3 Energy levels and permissable transitions of an ideal linear molecule, with the corresponding spectrum.

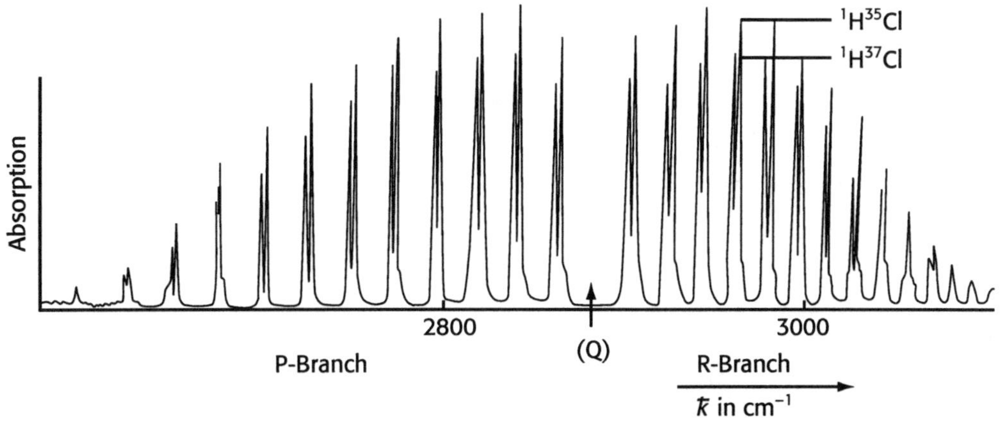

FIGURE 12.4 A high-resolution vibration-rotation spectrum of HCl. Note how $^1H^{35}Cl$ and $^1H^{37}Cl$ both contribute. Their abundance ratio is approximately 3 to 1.

Example 12.7

What is the force constant of a spring that extends by 1.00 cm when a 100 g mass is attached to it?

$$f = \frac{(0.100\ \text{kg})(9.81\ \text{N kg}^{-1})}{0.0100\ \text{m}} = 98.1\ \text{N m}^{-1}.$$

A person can easily construct a spring with the force constant of a typical chemical bond.

Example 12.8

What would the wave number for a given mode be if the reduced mass for its model were 1 u and the force constant 10^2 N m^{-1} = 10^2 kg s^{-2} ?

Since each atomic mas unit (u) contributes 1 g to the mass of a mole of pure substance and 1 mol contains Avogadro's number N of particles, the mass in grams of a particle of mass μ' u is

$$\mu = \frac{\mu'}{N}.$$

But from equation (12.32) and the final equation in example 12.2, we have

$$k_0 = \frac{\nu_{\text{vib}}}{c} = \frac{1}{2\pi c}\left(\frac{f}{\mu}\right)^{1/2} = \frac{1}{2\pi c}\left(\frac{fN}{\mu'}\right)^{1/2}$$

Substitute in the given numbers to obtain:

$$k_0 = \frac{\left[(10^5\ \text{g s}^{-2})(6.02214\times10^{23}\ \text{g}^{-1})\right]^{1/2}}{2\pi(2.99792\times10^{10}\ \text{cm s}^{-1})} = 1302.79\ \text{cm}^{-1}.$$

So from

$$f = \frac{\mu'}{N}(2\pi c k_0)^2,$$

we can write

$$f = \mu'\left(\frac{k_0}{1302.79\ \text{cm}^{-1}}\right)^2 \times 10^2\ \text{N m}^{-1}.$$

12.9 Raman Spectra

On passing a molecule, a photon may be deflected elastically, as in the Compton effect. Or it may transfer energy to or accept energy from the molecule, as in the process first reported by Chandrasekhara V. Raman.

When near a molecule, a photon acts through the electromagnetic field, polarizing the structure. If the polarizability of the molecule then changes, a transfer of energy takes place. The energy lost or gained by the photon equals the energy gained or lost by the molecule.

In general, the direction of polarization differs from the direction of the electric field causing the polarization. As a consequence, the polarizability $\boldsymbol{\alpha}$ is a dyad. Indeed, it behaves as the permittivity $\boldsymbol{\varepsilon}$ and the permeability $\boldsymbol{\mu}$ do in electromagnetism. There are consequently nine components of polarizability. These combine as follows:

$$\alpha = \alpha_{xx}\mathbf{ii} + \alpha_{xy}\mathbf{ij} + \alpha_{xz}\mathbf{ik} + \alpha_{yx}\mathbf{ji} + \alpha_{yy}\mathbf{jj} + \alpha_{yz}\mathbf{jk} + \alpha_{zx}\mathbf{ki} + \alpha_{zy}\mathbf{kj} + \alpha_{zz}\mathbf{kk} \quad [12.75]$$

The dipole moment induced by electric field intensity $\mathbf{E}$ is the polarization

$$\mathbf{P} = \alpha \cdot \mathbf{E}. \quad [12.76]$$

Thus, component P_x is given by

$$P_x = \alpha_{xx}E_x + \alpha_{xy}E_y + \alpha_{xz}E_z. \quad [12.77]$$

In quantum mechanics, the interaction involves the initial and final state functions, ψ_i and ψ_f, through the *transition polarization*

$$\mathbf{P}_{if} = \int \psi_i^* \mathbf{P} \psi_f \, d^3\mathbf{r}. \quad [12.78]$$

This plays the same role as expression (12.61); the rate of transition is proportional to its square.

For a vibrational mode, the polarizability varies with the displacement

$$r - r_e = x. \quad [12.79]$$

To a first approximation, one may set

$$\alpha = \alpha_{v0} + \alpha_{v1}x, \quad [12.80]$$

dropping the nonlinear terms. Then we have

$$P_{if} = \alpha_{v0}\int \psi_i^* \psi_f \, dx + \alpha_{v1}\int \psi_i^* x \psi_f \, dx. \quad [12.81]$$

Because ψ_i and ψ_f are orthogonal, the first term vanishes. The integral in the second term is nonzero only when

$$\Delta v = \pm 1. \quad [12.82]$$

Anharmonicity may bring in an appreciable contribution from

$$\Delta v = \pm 2. \quad [12.83]$$

Furthermore, when changes in two different modes occur at nearly the same wave numbers, they interact and move apart.

Besides selection rule (12.82), we have rules (12.59) and (12.60) for the accompanying rotational transitions. For a diatomic molecule, the changes depicted in figure 12.5 occur with the results diagramed in figure 12.6. Raman lines produced by photons that have lost energy in their interactions with molecules are called *Stokes lines*; those of photons that have gained energy, *anti-Stokes lines*. The branches are designated O, Q, S as shown.

12.10 Bond Anharmonicity and its Effect

Starting with the equilibrium configuration, compressing a chemical bond increases the electric repulsion between the nuclei and increases the confinement of the electrons. If nuclear interaction did not intervene, the potential function would increase without limit as the internuclear distance r went to zero. On the other hand, stretching the bond decreases the overlap of the valence orbitals and decreases the freedom of the occupying electrons. As r increases, the system would move towards the separated atom situation. Thus, the bonding would decrease progressively to zero and the potential function would approach a constant value.

Parabolic potential (12.2) does not allow for either of these effects. Expression $1/2fx^2$ fails to increase fast enough as x decreases from zero and it fails to level off as x increases away from zero. A function that does behave properly is sketched in figure 12.7, together with the parabolic potential that approximates this function around the point of equilibrium.

The parabolic potential leads to state function (12.45) and energy (12.47). The vibrational quantum number v equals the number of nodes in ψ_v in the region where

$$x = r - r_e \qquad [12.84]$$

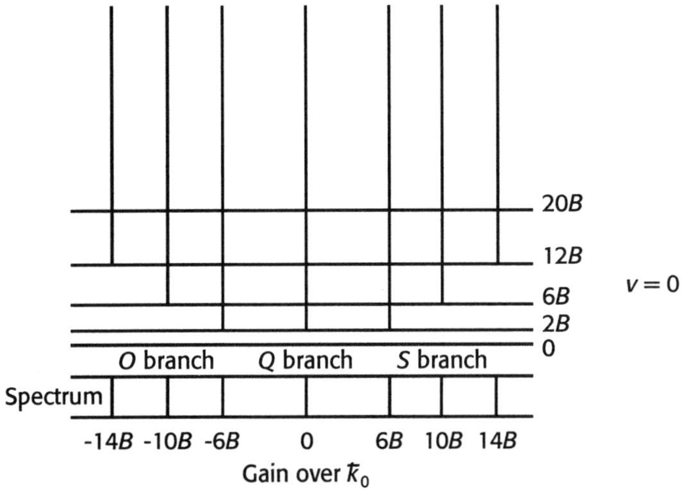

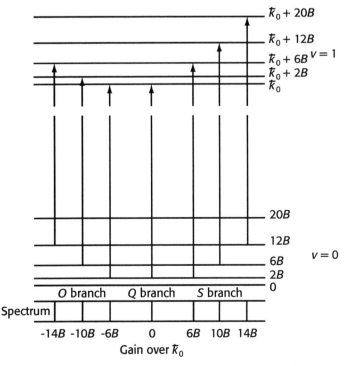

FIGURE 12.5 Energy levels and permissable Raman transitions for an ideal linear molecule, with the corresponding spectrum.

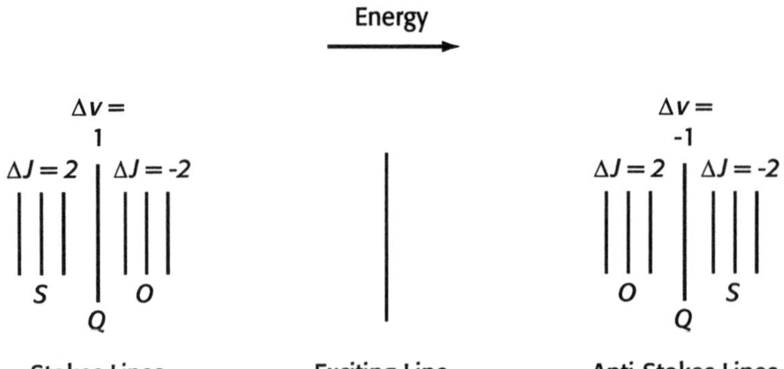

FIGURE 12.6 Schematic picture of a Raman spectrum for a diatomic molecule.

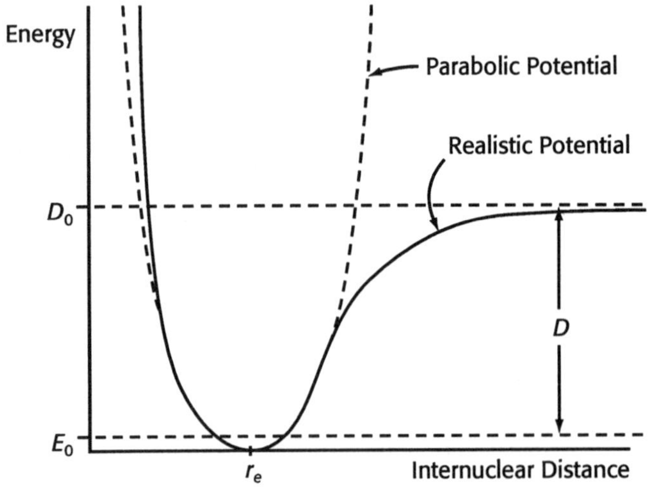

FIGURE 12.7 An approximate parabolic potential and the actual potential binding two atoms together.

is finite. Distorting the parabolic potential to the realistic potential does not affect the curve in the immediate neighborhood of r_e. Motion largely confined to this neighborhood because energy E is low would not be much affected. If $v + 1/2$ in the energy equation

$$E = \left(v + \frac{1}{2}\right)\hbar\omega \qquad [12.85]$$

could approach zero, the expression would presumably become asymptotically valid for the realistic potential.

Correlating with each of the low lying harmonic oscillator state functions is a solution for the realistic potential with the same number of nodes in the finite region about r_e and an asymptotic approach to zero on each side beyond this region. A *vibrational quantum number* v equal to this number of nodes is assigned to the solution. Increasing v from zero shortens the effective wavelength at each position, raises the energy, and causes the asymmetric nature of the potential to become more significant. Since the flattening of the potential on the right has more effect than the steepening on the left, the model particle becomes much less confined than it would be in the parabolic potential. At energy D_0 it becomes essentially free. So we expect successive levels to come closer together until they meet when the D_0 level is reached.

An analytic function that reproduces this behavior is a power series in $v + 1/2$, with the first term given by (12.85). In practice, one generally needs employ only one additional term:

$$E = \left(v+\frac{1}{2}\right)\hbar\omega_0 - \left(v+\frac{1}{2}\right)^2 b\hbar\omega_0 = \left[\left(v+\frac{1}{2}\right)-\left(v+\frac{1}{2}\right)^2 b\right]\hbar\omega_0. \qquad [12.86]$$

In wave number units, this equation becomes

$$\bar{\kappa} = \left[\left(v+\frac{1}{2}\right)-\left(v+\frac{1}{2}\right)^2 b\right]\bar{\kappa}_0 \quad \text{with} \quad v = 0, 1, 2, ... \qquad [12.87]$$

Parameter b is called the *anharmonicity constant.*

The steepening of the potential at small r and the flatening of it at large r, as figure 12.7 shows, compresses the state function on the left and expands it on the right. Nevertheless, integral (12.61) is still large for

$$\Delta v = \pm 1. \qquad [12.88]$$

However, it becomes appreciable for

$$\Delta v = \pm 2, \pm 3, ... \qquad [12.89]$$

Since the additional frequencies observed are approximate multiples of ω_0, they are called overtones or *harmonics.*

At energy D_0 successive levels come together, increasing v no longer increases E, and we have

$$\frac{dE}{dv} = 0. \qquad [12.90]$$

But differentiating (12.86) yields

$$\frac{dE}{dv} == \hbar\omega_0 - 2\left(v+\frac{1}{2}\right)b\hbar\omega_0. \qquad [12.91]$$

Setting this equal to zero, solving for $v + 1/2$

$$v+\frac{1}{2} = \frac{1}{2b}, \qquad [12.92]$$

and substituting back into equation (12.86) yields

$$D_0 = \frac{1}{2b}\hbar\omega_0 - \frac{1}{4b^2}b\hbar\omega_0 = \frac{\hbar\omega_0}{4b}. \qquad [12.93]$$

Subtracting the ground state energy

$$E = \left(\frac{1}{2}-\frac{1}{4}b\right)\hbar\omega_0 \qquad [12.94]$$

gives the dissociation energy

$$D = \frac{\hbar\omega_0}{4b} - \left(\frac{1}{2}-\frac{1}{4}b\right)\hbar\omega_0. \qquad [12.95]$$

An analytic function that approximates the actual potential is the closed form suggested by Philip M. Morse:

$$V = D_0\left[1 - e^{-a(r - r_e)}\right]^2. \qquad [12.96]$$

Example 12.9

Removing the rotational shifts from the vibrational spectrum of HCl leaves a fundamental at 2886 cm^{-1} and an overtone at 5668 cm^{-1}. Calculate the harmonic oscillator wave number $\bar{k}_0$ and the anharmonicity constant for HCl.

For the transition when v goes from 0 to 1, formula (12.87) yields the wave number

$$\bar{k}_1 = \left[\frac{3}{2} - \left(\frac{3}{2}\right)^2 b - \frac{1}{2} + \left(\frac{1}{2}\right)^2 b\right]\bar{k}_0 = (1 - 2b)\bar{k}_0.$$

And when v goes from 0 to 2, the photon wave number is

$$\bar{k}_2 = \left[\frac{5}{2} - \left(\frac{5}{2}\right)^2 b - \frac{1}{2} + \left(\frac{1}{2}\right)^2 b\right]\bar{k}_0 = (2 - 6b)\bar{k}_0.$$

Substitute the given data into these equations,

$$2886 \text{ cm}^{-1} = (1 - 2b)\bar{k}_0,$$
$$5668 \text{ cm}^{-1} = (2 - 6b)\bar{k}_0,$$

multiply the first equation by 2, and subtract the second from it:

$$104 \text{ cm}^{-1} = 2b\bar{k}_0.$$

Use this result to eliminate $2b\bar{k}_0$ from the first equation:

$$2886 \text{ cm}^{-1} = \bar{k}_0 - 104 \text{ cm}^{-1},$$

whence

$$\bar{k}_0 = 2990 \text{ cm}^{-1}$$

and

$$b = \frac{104 \text{ cm}^{-1}}{2(2990 \text{ cm}^{-1})} = 0.0174.$$

Example 12.10

How may the dissociation energy of a diatomic molecule be obtained from the variation of its allowed vibrational transitions with the mean quantum number?

As an approximation, employ equation (12.87). For quantum numbers v_1 and $v_2 = v_1 + 1$, the energy levels are

$$\bar{k}_1 = \left[v_1 + \frac{1}{2} - \left(v_1^2 + v_1 + \frac{1}{4}\right)b\right]\bar{k}_0,$$

$$\bar{k}_2 = \left[v_1 + \frac{3}{2} - \left(v_1^2 + 3v_1 + \frac{9}{4}\right)b\right]\bar{k}_0.$$

For a transition between these levels, the photon wave number equals the difference $k_2 - k_1$, which is

$$k_{v_1+1/2} = \left[1 - 2b - 2bv_1\right]k_0 = \left[1 - b - \left(v_1 + \frac{1}{2}\right)2b\right]k_0,$$

a *linear* function of $v_1 + 1/2$

In practice, the photon wave numbers from the vibrational transitions are plotted against $v_1 + 1/2$. A line is drawn through the points and extrapolated as a straight line to zero k. But adding the energies leading to dissociation produces

$$\frac{D}{hc} = k_{1/2} + k_{3/2} + k_{5/2} + \ldots = \sum k_{i+1/2}.$$

The ith term equals the area under the curve for $i \le v_1 + 1/2 < i + 1$. Consequently, the triangular area under the complete curve is identified as the dissociation energy.

As an example, consider the observed vibrational transitions in H_2^+. These are plotted in figure 12.8. The line through the experimental points intersects the k axis at 2240 cm^{-1}, the $v + 1/2$ axis at 19.1. So the triangular area under the line is

$$\frac{D}{hc} = \frac{1}{2}\left(2240\ cm^{-1}\right)\left(19.1\right) = 21{,}400\ cm^{-1}.$$

This procedure is referred to as a *Birge-Sponer extrapolation*. For most substances, the dissociation energy obtained in this way is somewhat high. So the true curve bends down near the end, as the dashed one in figure 12.8 does.

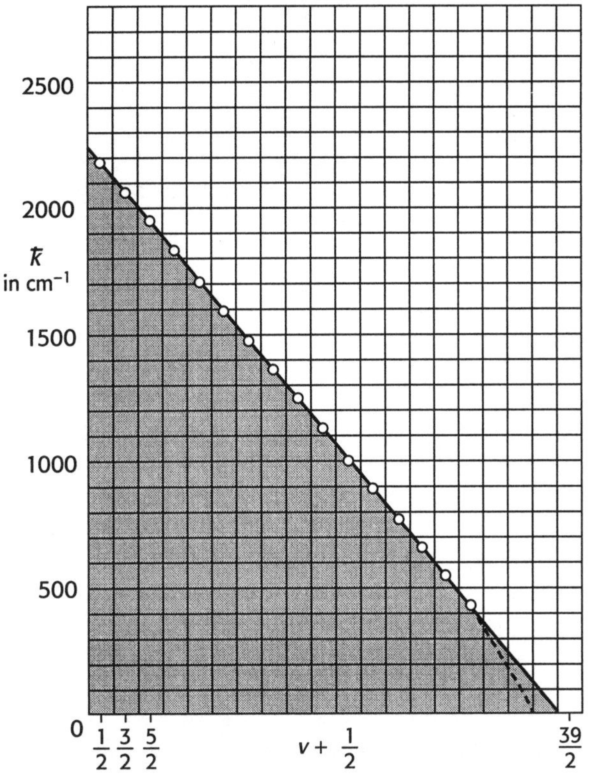

FIGURE 12.8 Observed vibrational energy level separations for H_2^+ plotted against $v_1 + 1/2$.

12.11 Asymptotic Solutions for the Central Field

For the particle of reduced mass μ, changing the potential from the parabolic to the realistic form has a profound effect on the asymptotic behavior as r increases without limit and as r goes to zero. Nevertheless, results for a Coulomb potential and for a shielded Coulomb potential, as we have for an electron in an atom, are similar to those for the realistic internuclear potential in a molecule.

Consider the particle of reduced mass μ in a central field $V(r)$. Its states are governed by Schrödinger equation (11.64). When the radial motion is independent of the angular motion, as in a definite energy state, we have

$$\psi = R(r)Y(\theta,\phi). \tag{12.97}$$

From equation (11.69), factor $R(r)$ must satisfy

$$\frac{1}{r^2}\frac{d}{dr}\left(r^2\frac{dR}{dr}\right)+\left[k^2(r)-\frac{J(J+1)}{r^2}\right]R=0. \tag{12.98}$$

For the electron in an atom, J is replaced by l. But the square of the local wavevector is

$$k^2=\frac{2\mu}{\hbar^2}\left[E-V(r)\right], \tag{12.99}$$

while the rotational energy is

$$T_{\text{rot}}=\frac{J(J+1)\hbar^2}{2\mu r^2}. \tag{12.100}$$

So we can rewrite (12.98) as

$$\frac{d^2R}{dr^2}+\frac{2}{r}\frac{dR}{dr}+\frac{2\mu}{\hbar^2}\left[E-V(r)-T_{\text{rot}}\right]R=0. \tag{12.101}$$

For both the diatomic molecule and the electron in an atom, potential $V(r)$ becomes constant as r increases without limit. If we also consider the energy associated with perpendicular motion, T_{rot}, to be constant, then the increment

$$\Delta E=E-V(r)-T_{\text{rot}}=E_{\text{rad}}-V(r) \tag{12.102}$$

is constant. Here E_{rad} is the energy associated with the radial motion.

At large r, the second term in (12.101) also becomes small. The equation reduces to

$$\frac{d^2R}{dr^2}+\frac{2\mu\Delta E}{\hbar^2}R=0. \tag{12.103}$$

Integration of this equation yields

$$R=Ae^{-\kappa r}+Be^{\kappa r} \tag{12.104}$$

with

$$\kappa=\frac{\left(-2\mu\Delta E\right)^{1/2}}{\hbar}. \tag{12.105}$$

Parameter κ is the attenuation constant defined in equation (10.97).

As long as we have a bound system, ΔE is negative and κ is real. Then for the particle density ρ to go to zero as r increases without limit, we must have

$$B = 0. \tag{12.106}$$

When the energy associated with radial motion exceeds $V(r)$ at $r = \infty$, the system is not bound. Then κ is imaginary and the asymptotic form is sinusoidal, as for free translation.

The behavior as r approaches zero is determined by the form for $V(r)$. Again, one solution is rejected. The criterion here is that of normalizability. Function $R(r)$ must not become infinite at $r = 0$ so strongly that it cannot be normalized.

The acceptable solutions at $r = 0$ and $r = \infty$ must then be joined by a function satisfying the Schrödinger equation. For the bound states, one finds that this is possible, by wavy functions, only at discrete energies E. These are the eigenvalues for the Schrödinger equation.

12.12 Radial State Functions for a Hydrogenlike Atom

The states of complicated systems may be correlated with those of simpler systems. A suitable reference system for atoms and nuclei is the hydrogenlike atom. The angular behavior of this atom has already been discussed. But to make further progress on the radial behavior, we need to introduce the Coulomb field explicitly.

Consider a nucleus of mass m_1 and charge Ze attracting a single electron of mass m_2 and charge $-e$. The potential energy for the electron at distance r from the nucleus is

$$V = -\frac{Ze^2}{4\pi\varepsilon r}, \tag{12.107}$$

if ε is the permittivity of space. Equation (12.101) for the radial factor $R(r)$ in ψ becomes

$$\frac{d^2R}{dr^2} + \frac{2}{r}\frac{dR}{dr} + \left[\frac{2\mu}{\hbar^2}\left(E + \frac{Ze^2}{4\pi\varepsilon r}\right) - \frac{l(l+1)}{r^2}\right]R = 0. \tag{12.108}$$

A suitable unit of distance is the *Bohr radius*

$$a = \frac{4\pi\varepsilon\hbar^2}{\mu Ze^2}. \tag{12.109}$$

As a unit of energy, one may introduce

$$b = \frac{Ze^2}{2(4\pi\varepsilon a)}. \tag{12.110}$$

The system energy is then written as

$$E = \mp\frac{b}{n^2}. \tag{12.111}$$

The negative sign is for bound states; the positive sign for free states. A dimensionless coordinate is

$$\rho = \frac{2r}{na}. \tag{12.112}$$

With these substitutions, the radial equation becomes

$$\frac{d^2R}{d\rho^2}+\frac{2}{\rho}\frac{dR}{d\rho}+\left[\mp\frac{1}{4}+\frac{n}{\rho}-\frac{l(l+1)}{\rho^2}\right]R=0. \quad [12.113]$$

Suitable solutions are obtained for

$$n=1,2,3.... \quad [12.114]$$

These have the form

$$R=N\rho^l L_{n+1}^{2l+1}(\rho)e^{-\rho/2}. \quad [12.115]$$

Function $L_{n+l}^{2l+1}(\rho)$ is an *associated Laguerre polynomial*. These polynomials are listed in standard tables. The normalization factor is

$$N=-\left\{\left(\frac{2}{na}\right)^3\frac{(n-l-1)!}{2n[(n+1)!]^3}\right\}^{1/2}. \quad [12.116]$$

Results for the low lying levels are listed in table 12.2.

Parameter n is called the *principal quantum number*. The quantized energy is

$$E=-\frac{Ze^2}{2(4\pi\varepsilon a)n^2}. \quad [12.117]$$

TABLE 12.2 Hydrogenlike Radial State Functions

Quantum Numbers		$Rnl(\rho)$ with $\rho = 2r/na$
n	l	
1	0	$\frac{2}{a^{3/2}}e^{-\rho/2}$
2	0	$\frac{1}{2\sqrt{2}a^{3/2}}(2-\rho)e^{-\rho/2}$
2	1	$\frac{1}{2\sqrt{6}a^{3/2}}\rho e^{-\rho/2}$
3	0	$\frac{1}{9\sqrt{3}a^{3/2}}(6-6\rho+\rho^2)e^{-\rho/2}$
3	1	$\frac{1}{9\sqrt{6}a^{3/2}}(4-\rho)\rho e^{-\rho/2}$
3	2	$\frac{1}{9\sqrt{30}a^{3/2}}\rho^2 e^{-\rho/2}$

Example 12.11

Evaluate the Bohr radius, considering the nucleus to be fixed, as if it were infinitely massive.

With accepted values of the fundamental constants and the mass of the electron, we find that

$$\frac{4\pi\varepsilon\hbar^2}{\mu e^2} = \frac{\hbar^2}{c^2 \times 10^{-7}\mu e^2} = \frac{\left(1.05457\times10^{-34}\right)^2 (\text{J s})(\text{kg m}^2\ \text{s}^{-1})}{\left(2.99792\times10^{8}\right)^2 \left(10^{-7}\ \text{C}^{-2}\ \text{J m}\right)\left(9.10939\times10^{-31}\ \text{kg}\right)\left(1.60218\times10^{-19}\ \text{C}\right)^2}$$

$$= 0.52918\times10^{-10}\ \text{m} = 0.52918\ \text{Å}$$

Formula (12.109) becomes

$$a = \frac{0.52918}{Z}\ \text{Å},$$

where Z is the number of protons in the nucleus.

Example 12.12

How does the electronic energy in a hydrogenlike system vary with the pertinent quantum numbers?

Combining (12.110) and (12.109) leads to

$$b = \frac{\mu Z^2 e^4}{2\left(4\pi\varepsilon\right)^2 \hbar^2} = \frac{1}{2}\mu\left(\frac{c^2\times10^{-7}e^2}{\hbar}\right)^2 Z^2.$$

Then substituting in accepted values of the fundamental constants, letting μ_r be μ/m_e, yields

$$b = \frac{1}{2}\left(9.10939\times10^{-31}\ \text{kg}\right)\left(\frac{\left(2.99792\times10^{8}\right)^2\left(10^{-7}\ \text{C}^{-2}\ \text{J m}\right)\left(1.60218\times10^{-19}\ \text{C}\right)^2}{1.05457\times10^{-34}}\right)\mu_r Z^2$$

$$= 2.1799\times10^{-18}\mu_r Z^2\ \text{J} = \frac{2.1799\times10^{-18}\mu_r Z^2}{1.60218\times10^{-19}\ \text{J}\left(\text{eV}\right)^{-1}} = 13.606\mu_r Z^2\ \text{eV}.$$

Formula (12.117) becomes

$$E = -13.606\frac{Z^2}{n^2}\mu_r\ \text{eV}.$$

The energy of an electron in the field of a single particle carrying positive charge is independent of the azimuthal and magnetic quantum numbers (l and m_l).

Example 12.13

Express the electronic energy of a hydrogenlike system in reciprocal centimeters.

Since the photon wavenumber is given by

$$\bar{k} = \frac{1}{\lambda} = \frac{\nu}{c} = \frac{h\nu}{hc} = \frac{E}{hc},$$

we divide the b obtained in example 12.12 by hc:

$$R = \frac{b}{hc} = \frac{2.17989 \times 10^{-18}\ \text{J}}{\left(6.62608 \times 10^{-34}\ \text{J s}\right)\left(2.99792 \times 10^{10}\ \text{cm s}^{-1}\right)} = 1.0974 \times 10^{5}\ \text{cm}^{-1}.$$

The formula for energy then becomes

$$k = -R\frac{Z^2}{n^2}\mu_r = -1.0974 \times 10^{5}\frac{Z^2}{n^2}\mu_r\ \text{cm}^{-1}.$$

The most accurate value for coefficient R comes directly from spectroscopic measurements. From these, one finds that

$$R = 109{,}737.3153\ \text{cm}^{-1}.$$

This parameter is known as the *Rydberg constant.*

12.13 Electronic Spectrum of a Hydrogenlike Atom

Compared to an ordinary oscillator, a hydrogenlike atom is highly anharmonic. Furthermore, parameter n acts as quantum number v; quantity n - 1 equals the number of nodal surfaces in each standard real state function $\psi_{nl\,ml}$ just as v equals the number of nodes in ψ_v.

As a consequence, for absorption or emission of a photon, there is no limitation on how n may change. Recall how condition (12.89) arose. Only, since n is integral, we have

$$\Delta n = \text{an integer}. \qquad [12.118]$$

But to take care of the unit of angular momentum that a photon carries, we have the selection rule

$$\Delta l = \pm 1. \qquad [12.119]$$

See figure 12.9 for a cascade of transitions accompanying the emission of successive photons.

During collisions, azimuthal quantum number l may change by any amount. These we will not analyze here.

Note that the 2s state cannot go to the 1s state without violating (12.119). As a consequence, this state is metastable. At low densities where collisions are sufficiently infrequent, the transition may be found as a fluorescence. In general, emissions and absorptions violating rule (12.119) occur at a very low rate.

In a magnetic field, the states for different m's with respect to the direction of the field appear at different energies. We then have the additional selection rule

$$\Delta m = 0 \quad \text{and} \quad \Delta m = \pm 1. \qquad [12.120]$$

The resulting splitting of the spectral lines is called the *Zeeman effect.*

12.14 Multiparticle Systems

Each particle in a multiparticle system is a separate entity. As a result, its state function combines with the state functions for the other particles as the state functions for independent orthogonal coordinates of a single particle combine to produce a complete state function. But when the particles are indistinguishable, a correction must be introduced.

If a first choice can be made in n_1 equivalent ways, a second choice in n_2 equivalent ways, and so forth, the number of ways in which all the choices can be made jointly is

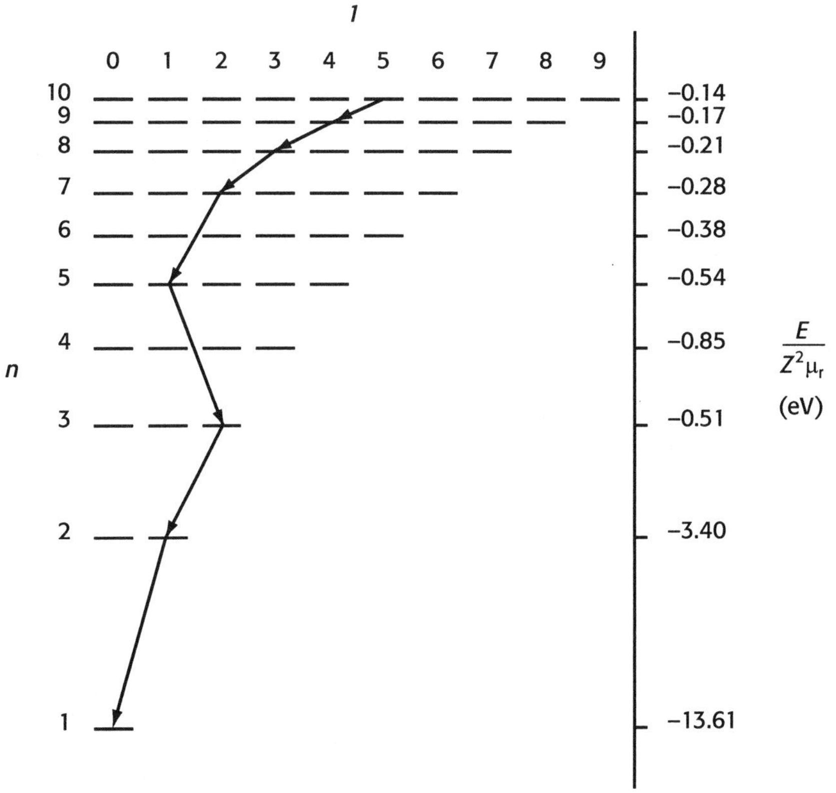

FIGURE 12.9 Low lying electronic levels of a hydrogenlike atom and an allowed series of transitions.

the product $n_1 n_2 \ldots$ But with the given symmetry among ways, the probability of a given outcome is proportional to the number of ways. Then the probability that the independent events all occur equals the product of their respective probabilities.

If each particle in an n-particle system is described by a state function,

$$\Psi_1(\mathbf{r}_1, \tau_1, t), \Psi_2(\mathbf{r}_2, \tau_2, t), \ldots, \Psi_n(\mathbf{r}_n, \tau_n, t), \qquad [12.121]$$

the probabilities that the first particle is in element $d^3\mathbf{r}_1$, the second in $d^3\mathbf{r}_2$, . . . , the nth in $d^3\mathbf{r}_n$ are

$$\Psi_1^* \Psi_1 \, d^3\mathbf{r}_1, \Psi_2^* \Psi_2 \, d^3\mathbf{r}_2, \ldots, \Psi_n^* \Psi_n \, d^3\mathbf{r}_n. \qquad [12.122]$$

But since independent probabilities combine multiplicatively, the overall probability is

$$\begin{aligned} \Psi_1^* \Psi_1 \, d^3\mathbf{r}_1 \Psi_2^* \Psi_2 \, d^3\mathbf{r}_2 \ldots \Psi_n^* \Psi_n \, d^3\mathbf{r}_n &= \left[\Psi_1(\mathbf{r}_1, \tau_1, t)\Psi_2(\mathbf{r}_2, \tau_2, t)\ldots\Psi_n(\mathbf{r}_n, \tau_n, t)\right]^* \\ &\times \left[\Psi_1(\mathbf{r}_1, \tau_1, t)\Psi_2(\mathbf{r}_2, \tau_2, t)\ldots\Psi_n(\mathbf{r}_n, \tau_n, t)\right] d^3\mathbf{r}_1 d^3\mathbf{r}_2 \ldots d^3\mathbf{r}_n \\ &= \Psi^*(\mathbf{r}_1, \mathbf{r}_2, \ldots, \mathbf{r}_n, \tau_1, \tau_2, \ldots, \tau_n, t)\Psi(\mathbf{r}_1, \mathbf{r}_2, \ldots, \mathbf{r}_n, \tau_1, \tau_2, \ldots, \tau_n, t) d^{3n}\mathbf{r}. \end{aligned} \qquad [12.123]$$

With Ψ the state function for the system, we have

$$\Psi(\mathbf{r}_1, \mathbf{r}_2, \ldots, \mathbf{r}_n, \tau_1, \tau_2, \ldots, \tau_n, t) = \Psi_1(\mathbf{r}_1, \tau_1, t)\Psi_2(\mathbf{r}_2, \tau_2, t)\ldots\Psi_n(\mathbf{r}_n, \tau_n, t). \qquad [12.124]$$

In the approximation that magnetic interactions can be neglected, this expression factors into

$$\psi(\mathbf{r}_1,\mathbf{r}_2,\ldots,\mathbf{r}_n)\chi(\tau_1,\tau_2,\ldots,\tau_n)e^{i\omega t} = \psi_1(\mathbf{r}_1)\chi_1(\tau_1)\psi_2(\mathbf{r}_2)\chi_2(\tau_2)\ldots\psi_n(\mathbf{r}_n)\chi_n(\tau_n)e^{i\omega t}. \quad [12.125]$$

So far, we have implied that the particles are distinguishable. But in nature, particles of the same species are indistinguishable. Furthermore, interchanging any two such particles must leave the probability density ρ unchanged.

The state function itself may be left unchanged. It is then said to be *symmetric* with respect to the various interchanges. Or, the state function may change sign in each exchange. It is then said to be *antisymmetric* with respect to the various interchanges. In principle, the orbital factor and the spin factor may also exhibit various mixed symmetries. For bosons, the complete state function is symmetric. For fermions, the complete state function is antisymmetric.

For formula (12.125), some permutations of particles produce a change. Let P_k be the kth permutation that does this. The superposition over all distinct results,

$$\Psi = N\sum_k P_k\psi_1(\mathbf{r}_1)\chi_1(\tau_1)\psi_2(\mathbf{r}_2)\chi_2(\tau_2)\ldots\psi_n(\mathbf{r}_n)\chi_n(\tau_n)e^{i\omega t}, \quad [12.126]$$

is then symmetric in the possible interchanges; it would be employed for a boson. Here N is the normalization factor.

A suitable state function for identical fermions must be antisymmetric with respect to all possible interchanges. Then one employs the superposition

$$\psi = N\begin{vmatrix} \psi_1(\mathbf{r}_1)\chi_1(\tau_1) & \psi_2(\mathbf{r}_1)\chi_2(\tau_1) & \cdot & \psi_n(\mathbf{r}_1)\chi_n(\tau_1) \\ \psi_1(\mathbf{r}_2)\chi_1(\tau_2) & \psi_2(\mathbf{r}_2)\chi_2(\tau_2) & \cdot & \psi_n(\mathbf{r}_2)\chi_n(\tau_2) \\ \cdot & \cdot & \cdot & \cdot \\ \psi_1(\mathbf{r}_n)\chi_1(\tau_n) & \psi_2(\mathbf{r}_n)\chi_2(\tau_n) & \cdot & \psi_n(\mathbf{r}_n)\chi_n(\tau_n) \end{vmatrix} \quad [12.127]$$

for a spatial - spin part of Ψ. Expression (12.127) is called a *Slater determinant.*

One may take the different states that the particles occupy as (a) completely disjoint, (b) partially the same, or (c) completely the same. For bosons, all three possibilities yield suitable overall state functions. For fermions, only the first possibility yields a nonzero determinant (12.127). We thus obtain a general form for the *Pauli exclusion principle.* When the single particle states are distinguished by quantum numbers, we see that no two of these states can have the same set of quantum numbers.

Example 12.14

How do two 1s orbitals and a 2s orbital combine to form a ground state for Li?

Let 1s(i), 2s(i), $\alpha(i)$, and $\beta(i)$ represent the ith electron in the appropriate 1s orbital, 2s orbital, + $1/2\hbar$ spin state, and - $1/2\hbar$ spin state, respectively. Then Slater determinant (12.127) may be either

$$\psi = \frac{1}{\sqrt{6}}\begin{vmatrix} 1s(1)\alpha(1) & 1s(1)\beta(1) & 2s(1)\alpha(1) \\ 1s(2)\alpha(2) & 1s(2)\beta(2) & 2s(2)\alpha(2) \\ 1s(3)\alpha(3) & 1s(3)\beta(3) & 2s(3)\alpha(3) \end{vmatrix} = \frac{1}{\sqrt{6}}[1s(1)\alpha(1)1s(2)\beta(2)2s(3)\alpha(3)$$

$$+1s(2)\alpha(2)1s(3)\beta(3)2s(1)\alpha(1) + 1s(3)\alpha(3)1s(1)\beta(1)2s(2)\alpha(2) - 1s(3)\alpha(3)1s(2)\beta(2)2s(1)\alpha(1)$$

$$-1s(2)\alpha(2)1s(1)\beta(1)2s(3)\alpha(3) - 1s(1)\alpha(1)1s(3)\beta(3)2s(2)\alpha(2)]$$

or

$$\psi = \frac{1}{\sqrt{6}}\begin{vmatrix} 1s(1)\alpha(1) & 1s(1)\beta(1) & 2s(1)\beta(1) \\ 1s(2)\alpha(2) & 1s(2)\beta(2) & 2s(2)\beta(2) \\ 1s(3)\alpha(3) & 1s(3)\beta(3) & 2s(3)\beta(3) \end{vmatrix} = \frac{1}{\sqrt{6}}[1s(1)\alpha(1)1s(2)\beta(2)2s(3)\beta(3)$$
$$+1s(2)\alpha(2)1s(3)\beta(3)2s(1)\beta(1) + 1s(3)\alpha(3)1s(1)\beta(1)2s(2)\beta(2) - 1s(3)\alpha(3)1s(2)\beta(2)2s(1)\beta(1)$$
$$-1s(2)\alpha(2)1s(1)\beta(1)2s(3)\beta(3) - 1s(1)\alpha(1)1s(3)\beta(3)2s(2)\beta(2)].$$

Since two disjoint forms for ψ exist, the ground state of Li is doubly degenerate. This degeneracy is associated with the two orthogonal choices for the last column in the determinant. These correspond to the net spin S_z equal to + $1/2\hbar$; and to - $1/2\hbar$.

A zeroth order approximation would employ hydrogenlike atomic orbitals in the Slater determinant. Then for the first order approximation, one could argue that the average electron cloud in which a given electron moves has the result of reducing the effective charge on the nucleus. The other electrons are said to screen part of the nuclear charge.

To allow for this effect, one can replace the Z for each ls orbital with a variable Z_1. For the 2s orbital, one can replace Z with a variable Z_2. One finds the lowest energy, -201.2 eV, for $Z_1 = 2.69$ and $Z_2 = 1.78$. But the experimental energy is -203.48 eV.

Actually, the screening should vary with r. Thus, the hydrogenlike radial function with altered Z is not adequate.

12.15 Multielectron Uninuclear Structures

In the self- consistent field method developed by Douglas R. Hartree, improved radial functions are constructed.

In the simplest form of the procedure, the state function is considered to be a product of one-electron orbitals:

$$\psi = \psi_1(\mathbf{r}_1)\psi_2(\mathbf{r}_2)...\psi_n(\mathbf{r}_n). \quad [12.128]$$

The quantum numbers for each orbital are related to the independent nodal surfaces present. The Pauli principle is taken into account by placing no more than two electrons in any one orbital.

Each electron is considered to move in the field of the nucleus and the average field of all other electrons. Then with product form (12.128), the Schrodinger equation separates into n one-electron equations

$$\nabla^2\psi_i + \frac{2\mu}{\hbar^2}\left[E_i - V_i(\mathbf{r}_i)\right]\psi_i = 0. \quad [12.129]$$

Orbital ψ_i is expressed as a product,

$$\psi_i = R_i(r_i)Y_i(\theta_i, \phi_i), \quad [12.130]$$

in which the angular factor is the same as in the corresponding hydrogenlike atom. The radial factor satisfies the equation

$$\frac{1}{r^2}\frac{d}{dr}\left(r^2\frac{dR_i}{dr}\right) + \left[\frac{2\mu}{\hbar^2}\left(E_i - V_i(r)\right) - \frac{l(l+1)}{r^2}\right]R_i = 0. \quad [12.131]$$

A distribution for the other electrons is assumed, the corresponding potential $V_i(r)$ calculated, and equation (12.131) is solved numerically for each electron. From the results and the initial assumptions, one constructs an improved electron distribution and repeats the calculations. Iterations are continued until consistency between the assumed electron distributions and those calculated is reached. The energy is taken as the sum of the single electron energies:

$$E = \sum E_i. \quad [12.132]$$

This procedure does not take into account correlations among movements of the electrons. But the correction amounts to only about 1% of the total energy. Configuration interaction calculations provide an estimate of this correction.

In multielectron atoms, the primary determinant of an electron's energy is its principal quantum number n, the number of independent nodes in its state function minus one. A secondary determinant is the azimuthal quantum number l. Electrons with the same n are said to form a *shell*. Those with the same n and l are said to form a subehell.

12.16 Combining Angular Momenta

In a multiparticle system, the orbital and the spin angular momenta combine vectorially. Just as the individual momenta are quantized, particular resultants are also quantized.

In our discussion, let us employ units for which $\hbar$= 1. Also, let us first consider atoms early in the periodic table, where the magnetic interaction between orbital motion and spin is small with respect to the interaction correlating with the relative orbital motions. This stronger interaction is electrostatic, arising from the interelectronic potentials.

From formula (11.132), the apparent angular momentum of the ith electron is

$$p_\phi = \sqrt{l_i\left(l_i + 1\right)}\,\hbar. \quad [12.133]$$

One may consider this the magnitude of a vector that precesses around the imposed direction, the z axis. This vector would combine with those of the other valence electrons as figure 12.10 shows. The projections on the z axis must be quantized in $\hbar$ units. The net angular momentum from closed shells is zero, so they need not be considered.

But since

$$l_i < \sqrt{l_i\left(l_i + 1\right)} < l_i + 1, \quad [12.134]$$

one may replace the square root of $l_i(l_i + 1)$ by l_i, and likewise for similar square roots, in the discussion. For two electrons, a person can define a resultant angular momentum quantum number L with the possibilities

$$L = l_1 + l_2,\ l_1 + l_2 - 1, \ldots, \left|l_1 - l_2\right|. \quad [12.135]$$

An additional electron would couple with each of these resultants in turn, and so on. Representing each final L, a capital letter is employed as follows:

$$\begin{aligned} L = 0, 1, 2, 3, 4, \ldots \\ \mathrm{S, P, D, F, G, \ldots} \end{aligned} \quad [12.136]$$

In a similar way, the spins combine. For two electrons, one obtains two resultants:

$$S = s_1 + s_2,\ s_1 + s_2 - 1. \quad [12.137]$$

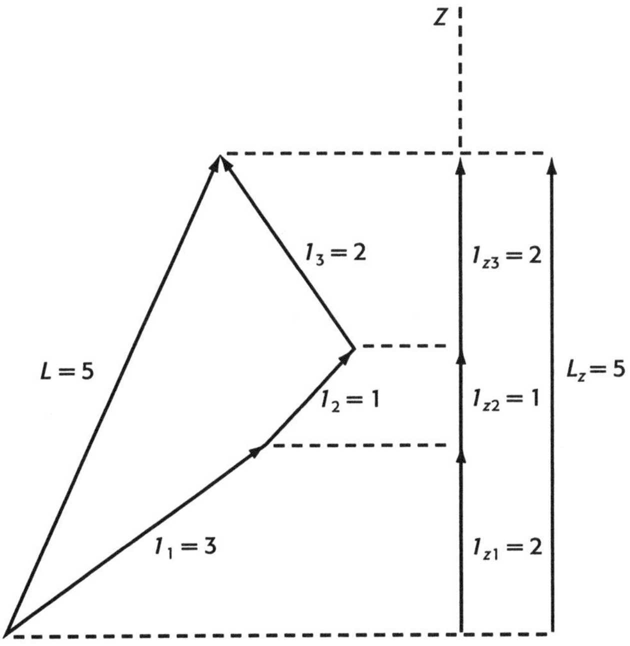

FIGURE 12.10 Vector diagram indicating how three orbital angular momenta may combine. Each projection on the z axis is an integral number of units.

An additional electron would couple with each of these resultants in turn, and so on. The multiplicity due to spin equals $2S + 1$. This multiplicity conventionally appears as a superscript preceding the letter for the state.

Finally for a light atom, the orbital and spin angular momenta combine. By analogy with (12.135), we obtain the possible resultants

$$J = L + S, L + S - 1, \ldots, |L - S|. \quad [12.138]$$

When needed, this number appears as a subscript on the letter for a state. The resulting combination is called the *term symbol* for the state.

With high - Z atoms, those late in the periodic table, the magnetic interaction between orbital motion and spin becomes large. Then one must first combine the $\mathbf{l}_i$ of a valence electron with its spin $\mathbf{s}_i$ to get $\mathbf{j}_i$. The total angular momentum is then obtained by adding the $\mathbf{j}_i$'s vectorially.

For intermediate atoms, neither the *LS* coupling nor the *jj* coupling is adequate. Then the situation is more complicated. The state function is still an eigenfunction of J, however. Here we will focus our attention on the *LS* scheme.

A given set of occupied orbitals is called an *electron configuration.* It may be described by listing the orbitals together with the number of electrons in each as a superscript. The filled inner shells and subshells may be designated by the symbol for the corresponding atom in brackets. Thus, the ground state of carbon has the configuration

$$1s^2 2s^2 2p^2 = [\text{He}] 2s^2 2p^2 = [\text{Be}] 2p^2. \quad [12.139]$$

A completed subshell is spherically symmetric, with $L = 0$ and $S = 0$. As a consequence, the terms of a configuration x^q are the same as those of x^{r-q}, where r is $2(2l + 1)$. For their **L**'s, **S**'s, and **J**'s have to cancel. Thus, the terms for p^5 are the same as those for p; the terms for d^7 are the same as those for d^3.

Furthermore, for a multiply occupied subshell that is not equivalent to a singly occupied one, the Pauli exclusion principle rules out some of the terms we have obtained. To find the remaining possibilities, one may tabulate the sums

$$\sum m_l = M_L \quad \text{and} \quad \sum m_s = M_S \qquad [12.140]$$

and assign them to terms.

Example 12.15

What terms arise when the electrons outside closed shells and subshells have the configuration (a) 3d4d, (b) 2p3p4p?

Sequence (12.136) is employed to find the possible L's, sequence (12.137) for the possible S's, and sequence (12.138) for the possible J's, for successive couplings, until the configuration outside the closed shells and subshells is covered.

(a) Here $l_1 = 2$ and $l_2 = 2$. With sequence (12.135), we obtain

$$L = 4, 3, 2, 1, 0.$$

With sequence (12.137), we find that

$$S = 1, 0 \quad \text{and} \quad 2S + 1 = 3, 1.$$

The corresponding terms are

$$^3\text{G}, ^3\text{F}, ^3\text{D}, ^3\text{P}, ^3\text{S}, ^1\text{G}, ^1\text{F}, ^1\text{D}, ^1\text{P}, ^1\text{S}.$$

For each of these, the possible J's can be readily determined and added as subscripts. Thus for the first term, we obtain

$$^3\text{G}_5, ^3\text{G}_4, ^3\text{G}_3$$

while for the sixth term, we have only

$$^1\text{G}_4.$$

(b) Here $l_1 = 1$, $l_2 = 1$, $l_3 = 1$ From l_1 and l_2, we obtain

$$L' = 2, 1, 0 \quad \text{and} \quad S' = 1, 0.$$

Combining these results with l_3, we find that

$$L = 3, 2, 1, 2, 1, 0, 1 \quad \text{and} \quad S = 3/2, \frac{1}{2}, \frac{1}{2}.$$

These yield the terms

$$^4\text{F}, ^4\text{D}, ^4\text{P}, ^2\text{F}, ^2\text{F}, ^2\text{D}, ^2\text{D}, ^2\text{P}, ^2\text{P}, ^4\text{D}, ^4\text{P}, ^4\text{S}, ^2\text{D}, ^2\text{D}, ^2\text{P}, ^2\text{P}, ^2\text{S}, ^2\text{S}, ^4\text{P}, ^2\text{P}, ^2\text{P}.$$

Again, the J's could be added when needed in a detailed analysis.

Example 12.16

Derive the terms for an $n\text{p}^2$ configuration.

From the results for L' and S' in the (b) part of example 12.15, we see that if the Pauli exclusion principle did not intervene, the terms would be

$$^3\text{D}, ^3\text{P}, ^3\text{S}, ^1\text{D}, ^1\text{P}, ^1\text{S}.$$

TABLE 12.3 Allowed Combinations of Quantum Numbers for an np^2 Configuration

m_l					
+1	*0*	*-1*	M_L	M_S	*Assignments*
↑↓			2	0	^1D
↑	↑		1	1	^3P
↑	↓		1	0	^1D
↓	↑		1	0	^3P
↓	↓		1	-1	^3P
↑		↑	0	1	^3P
↑		↓	0	0	^1D
↓		↑	0	0	^3P
↓		↓	0	-1	^3P
	↑↓		0	0	^1S
	↑	↑	-1	1	^3P
	↑	↓	-1	0	^1D
	↓	↑	-1	0	^3P
	↓	↓	-1	-1	^3P
		↑↓	-2	0	^1D

But half of these are forbidden on ruling out doubly occupied states.

For a p electron, l is 1 and m_l may be +1, 0, or -1. Furthermore, s is 1/2 and m_s may be + 1/2 or - 1/2. Let us list the possibilities with ↑ representing an electron with m_s = 1/2, ↓ an electron with m_s = - 1/2. The Pauli exclusion principle then allows the states appearing in the first three columns of table 12.3.

These yield the sums appearing in the fourth and fifth columns. There is only one state with $M_L = 2$, only one with $M_L = -2$. So the ^3D term is ruled out. Going with the extreme states are one with $M_L = 1$, one with $M_L = 0$, one with $M_L = -1$, for term ^1D. That leaves three states with $M_L = 1$, three with $M_L = -1$. These belong to the ^3P term. Also with these are three states with $M_L = 0$. Left is a $M_L = 0$, $M_S = 0$ state. This forms the S term.

Thus, the allowed terms are

$$^1D,\ ^3P,\ ^1S.$$

Questions

12.1 What is the Born-Oppenheimer approximation? How is it justified?

12.2 Is the energy of a molecule calculated for fixed positions of the nuclei, potential energy? Explain.

12.3 Under what circumstances can we consider this energy to be potential energy?

12.4 What is a mode of motion?

12.5 Why can a vibrational mode be represented by the motion of a single particle in a one-dimensional potential?

12.6 What is the harmonic approximation?

12.7 Construct the Schrödinger equation for a harmonic oscillator.

12.8 How does operator $(D - w)(D + w)$ differ from operator $(D + w)(D - w)$?
12.9 Why is $-N(D - w)$ a step-up operator while $N(D + w)$ is a step-down operator?
12.10 Derive the two nodeless solutions for the harmonic oscillator Schrödinger equation. Why is one of these rejected?
12.11 How are all acceptable solutions for the harmonic oscillator derived from the acceptable nodeless solution?
12.12 What energies do these solutions yield?
12.13 Construct the partition function for a vibrational mode.
12.14 What is the selection rule for a vibrational transition? Explain.
12.15 How do the P, Q, and R branches of the vibrational bands arise?
12.16 What are the selection rules for a Raman transition?
12.17 How do the O, Q, and S branches arise?
12.18 How does a realistic binding potential differ from a parabolic potential?
12.19 In the energy level formula, how does one allow for anharmonicity?
12.20 How may one approximate the dissociation energy of a diatomic molecule from observed vibrational transitions?
12.21 For the model particle of mass μ in the central field $V(r)$, when does the state function factor into a radial function $R(r)$ and an angular function $Y(\theta, \phi)$?
12.22 Show how the rotational energy T_{rot} enters the Schrodinger equation for the radial factor $R(r)$.
12.23 How does this equation reduce when r becomes large? Integrate the reduced equation to obtain an asymptotic solution for $R(r)$
12.24 Why must one arbitrary constant in this solution vanish?
12.25 What determines the behavior of $R(r)$ as r approaches zero? Again, why is one solution rejected?
12.26 Describe qualitatively the radial state functions for a hydrogenlike atom.
12.27 How does the energy of a hydrogenlike atom involve the Rydberg constant and the principal quantum number?
12.28 What selection rules govern the electron transitions in a hydrogenlike atom?
12.29 How do the state functions for equivalent bosons combine to form a suitable composite state function? Explain.
12.30 How do the state functions for equivalent fermions combine to form a suitable composite state function? Explain.
12.31 How does this result lead to a general form of the Pauli exclusion principle?
12.32 Describe the self-consistent field method for treating multielectron structures.
12.33 What restrictions determine how orbital and spin angular momenta combine?
12.34 How are spectroscopic terms described?

Problems

12.1 The state function for a harmonic oscillator in its first excited level has the form

$$\psi = Nwe^{-w^2/2}.$$

By an appropriate integration and manipulation, determine N.

12.2 Determine the acceptable nodeless solution to the Schrödinger equation for a model particle moving in the potential

$$V = \frac{1}{2}fx^2 \quad \text{when} \quad x < 0$$

and

$$V = \frac{1}{2}gx^2 \quad \text{when} \quad x > 0.$$

12.3 Employ formula (12.43) to construct the state function for a harmonic oscillator in its first excited state.

12.4 To what region along the x axis would a classical harmonic oscillator with $1/2\hbar\omega$ energy be restricted? Determine the probability that the corresponding quantum oscillator would be found in this region.

12.5 Gaseous $^1H^{35}Cl$ exhibits maximum absorption of infrared radiation at the wave numbers (in cm^{-1}):
3014, 2998, 2981, 2963, 2945, 2926, 2906, 2865, 2844, 2821.
From these numbers determine the rotational constant B and the vibrational wave number κ_0 for the molecule.

12.6 Calculate the force constants for H_2 and D_2 from the corresponding vibrational wave numbers 4159.2 cm^{-1} and 2990.3 cm^{-1}. The masses of ^1H and ^2H are 1. 007825 u and 2.014102 u, respectively.

12.7 Gaseous HCl was exposed to a mercury arc. About the 2536.5 Å line, weak small Raman displacements, in cm^{-1}, +222.2, +183.3, +143.8, . . ., -101.1, -142.7, -187.5, -229.4, -271.0 were observed. A strong Raman displacement 2886.0 cm^{-1} from the exciting line, surrounded by weak O and S lines, was also seen. Deduce (a) the vibrational wave number κ_0 and (b) the rotational constant B for the HCl molecule.

12.8 From the data in example 12.9, estimate the number of bound levels in oscillating HCl.

12.9 The observed vibrational energy level separations for HgH are 1203.7, 965.6, 632.4, and 172 cm^{-1}. From these numbers, determine the dissociation energy of HgH.

12.10 Substitute function $\psi = Ae^{-r/a}$ into Schrödinger equation (11.66) with the potential V given by formula (12.106). Then from the resulting identity in r, determine a and E for the ls orbital.

12.11 For the 1s orbital $\psi = Ae^{-r/a}$, determine the normalization constant A.

12.12 If a hydrogenlike atom is in its ground state, what is the probability that its electron is out where $V > E$ at a given time?

12.13 What transition in H is associated with the emission of (a) a 97,492 cm^{-1} photon, (b) a 20,565 cm^{-1} photon?

12.14 Construct four independent Slater determinants for the 1s2s configuration of helium. Multiply these out and where possible factor the result.

12.15 Combine the two unfactored functions from problem 12.14 to obtain orthogonal results that do factor into a spatial function and a spin function. Classify the different orbital and spin factors. Which of the spin factors go with (a) $S = 1$, (b) $S = 0$?

— — —

12.16 Substitute the form

$$\psi = A\left(cx^2 - 1\right)e^{-ax^2/2}$$

into the Schrödinger equation for the harmonic oscillator. Then evaluate E and c.

12.17 Employ the pertinent step-up operator to the harmonic oscillator state function for $v = 1$ to obtain this function for $v = 2$.

12.18 If a harmonic oscillator is in its first excited state, what is the probability that it would be found with $V > E$?

12.19 If a particle moves in the potential $V = 1/2 fx^4$, how does its state function behave out where $|x|$ is very large?

12.20 Gaseous HBr exhibits maximum absorption of infrared radiation at the wave numbers (in cm^{-1}): 2671, 2658, 2645, 2631, 2617, 2602, 2587, 2571, 2539, 2523, 2506, 2488, 2470, 2451. From these numbers determine the rotational constant B and the vibrational wave number κ_0 for the molecule.

12.21 Calculate the force constants for CO and NO if the corresponding vibrational wave numbers are 2143.3 cm^{-1} and 1876,0 cm^{-1}. The masses of ^{14}N and ^{16}O are 14.00307 u and 15.99491 u, respectively.

12.22 If the force constant for vibration of $^{23}Na^{35}Cl$ is 118.0 N m^{-1}, what is its wave number κ_0? The masses of ^{23}Na and ^{35}Cl are 22.9898 u and 34.96885 u, respectively.

12.23 The vibrational energy levels of NaI are at wavenumbers 142.81, 427.31, 710.31, 991.81 cm^{-1}. Plot $\kappa/(v + 1/2)$ against $v + 1/2$ and determine κ_0 and b.

12.24 Estimate the anharmonicity constant b and energy D_0 when the number of bound levels of an oscillator drops to (a) 2, (b) 1.

12.25 Substitute function $\psi = A(1 - br)e^{-r/a}$ into Schrödinger equation (11.66) with the potential V given by formula (12.106), Then from the resulting identity in r, determine a, b, and E for the 2s orbital.

12.26 Normalize the 2s orbital obtained in problem 12.25.

12.27 If the electron in a hydrogenlike atom is in its 2s orbital, what is the probability that it is out where $V > E$ at a given time?

12.28 Determine the most probable distance of the electron from the nucleus in the 2s state.

12.29 What terms arise from the electron configuration [Mg] $3p^5 3d$?

12.30 Consider a system of two identical particles, each with the possible spin quantum numbers $m_s = -s, -s + 1, \ldots, s$. Determine how many orthogonal symmetric spin states and antisymmetric spin states occur and calculate the ratio between these numbers.

References

Books

Duffey, G. H.: 1984, *A Development of Ouantum Mechanics Based on Symmetry Considerations*, Reidel, Dordrecht, Holland, pp. 75-142, 157-164.

Further details on the procedures employed in the present chapter are given. Furthermore, the differential equation for the harmonic oscillator is obtained by operating on the expression e^{-w^2}. This confirms the form of solution that we have constructed.

Nakamoto, K.: 1986, *Infrared and Raman Spectra of Inorganic and Coordination Compounds*, 4th edn., John Wiley & Sons, Inc., New York, pp 1-478.

Nakamoto devotes the first 100 pages of his text to the theory of normal modes of vibration. Then he applies this theory to many inorganic, coordination, organometallic, and bioinorganic compounds. The coverage is broad and balanced. Numerous references to the original literature are cited.

Svanberg, S.: 1992, *Atomic and Molecular Spectroscopy*, 2nd edn., Springer-Verlag, New York, pp 1-393.

Svanberg reviews the basic theory using simple mathematics. Rotational transitions, vibrational transitions, and electronic transitions are all considered. The important spectroscopic procedures are described, together with representative results.

Articles

Bacic, Z., and Light, J. C.: 1989, "Theoretical Methods for Rovibrational States of Floppy Molecules," *Annul Rev. Phys. Chem.* **40**, 469-498.

Blaise, P., Henri-Rousseau, O., and Merad, N.: 1984, "Some Further Comments about the Stability of the Hydrogen Atom," *J. Chem. Educ.* **61**, 957-960.

Boulil, B., Henri-Rousseau, O., and Deumie, M.: 1988, "Born-Oppenheimer and Pseudo-Jahn-Teller Effects as Considered in the Framework of the Time-Dependent Adiabatic Approximation," *J. Chem. Educ.* **65**, 395-399.

Boulil, B., and Henri-Rousseau, 0.: 1989, "From Quantum Mechanical Harmonic Oscillators to Classical Ones through Maximization of Entropy," *J. Chem. Educ.* **66**, 467-470.

Bozlee, B. J., Luther, J. H., and Buraczewski, M.: 1992, "The Infrared Overtone Intensity of a Simple Diatomic: Nitric Oxide," *J. Chem. Educ.* **69**, 370-373.

Chem, J. -H.: 1989, "Atomic Term Symbols by Group Representation Methods," *J. Chem. Educ.* **66**, 893-898.

Clark, D. B.: 1991, "Very Large Hydrogen Atoms in Interstellar Space," *J. Chem. Educ.* **68**, 454-455.

David, C. W.: 1996, "IR Vibration-Rotation Spectra of the Ammonia Molecule," J. Chem. Educ. **73**, 46-50.

Dykstra, C. E.: 1988, "Electrical Polarization in Diatomic Molecules," *J. Chem. Educ.* **65**, 198-200.

Fitts, D. D.: 1995, "Ladder-Operator Treatment of the Radial Equation for the Hydrogenlike Atom," *J. Chem. Educ.* **72**, 1066-1069.

Fogarasi, G., and Pulay, P.: 1984, "Ab Initio Vibrational Force Fields," *Annul Rev. Phys. Chem.* **35**, 191-213.

Geldard, J. F., and Pratt, L. R.: 1987, "Statistical Determination of Normal Modes," *J. Chem. Educ.* **64**, 425-426.

Goodfriend, P. L.: 1987, "Diatomic Vibrations Revisited," *J. Chem Educ.* **64**, 753-756.

Grunwald, E., Herzog, J., and Steel, C.: 1995, "Using Fourier Transforms to Understand Spectral Line Shapes," *J. Chem. Educ.* **72**, 210- 214.

Guofan, L., and Ellzey Jr., M. L.: 1987, "Finding the Terms of Configurations of Equivalent Electrons by Partitioning Total Spins," *J. Chem. Educ.* **64**, 771-772.

Kellman, M. E.: 1995, "Algebraic Methods in Spectroscopy," *Annul Rev. Phys. Chem.* **46**, 395-421.

Kiremire, E. M. R.: 1987, "A Numerical Algorithm Technique for Deriving Russell-Saunders (R-S) Terms," *J. Chem. Educ.* **64**, 951-953.

Laane, J.: 1994, "Vibrational Potential Energy Surfaces and Conformations of Molecules in Ground and Excited Electronic States," *Annul Rev. Phys. Chem.* **45**, 179-211.

Lacey, A. R.: 1987, "A Student Introduction to Molecular Vibrations," *J. Chem. Educ.* **64**, 756-761.

Lessinger, L.: 1994, "Morse Oscillators, Birge-Sponer Extrapolation, and the Electronic Absorption Spectrum of I_2," *J. Chem. Educ.* **71**, 388-391.

Lewis, E. L., Palmer, C. W. P., and Cruickshank, J. L.: 1994, "Iodine Molecular Constants from Absorption and Laser Fluorescence," *Am. J. Phys.* **62**, 350-356.

LipRowitz, K. B.: 1995, "Abuses of Molecular Mechanics: Pitfalls to Avoid," *J. Chem. Educ.* **72**, 1070-1075.

Melrose, M. P., and Scerri, E. R.: 1996, "Why the 4s Orbital is Occupied before the 3d," *J. Chem. Educ.* **73**, 498-503.

Milner, J. D., and Peterson, C.: 1985, "Notes on the Factorization Method for Quantum Chemistry," *J. Chem. Educ.* **62**, 567- 568.

Norrby, L. J.: 1991, "Why is Mercury Liquid?" *J. Chem. Educ.* **68**, 110-113.

Pisani, L., Andre, J. -M, Andre, M. -C., and Clementi, E.: 1993, "Study of Relativistic Effects in Atoms and Molecules by the Kinetically Balanced LCAO Approach," *J. Chem. Educ.* **70**, 894-901.

Prais, M. G.: 1986, "Analysis of the Vibrational-Rotational Spectrum of Diatomic Molecules," *J. Chem. Educ.* **63**, 747-752.

Royer, A.: 1996, "Why are the Energy Levels of the Quantum Harmonic Oscillator Equally Spaced?" *Am. J. Phys.* **64**, 1393-1399.

Schor, H. H. R., and Teixeira, E. L.: 1994, "The Fundamental Rotational-Vibrational Band of CO and NO," *J. Chem. Educ.* **71**, 771- 774.

Sen, Z., Kemin, A., Fenglin, Q., and Yinze, Z.: 1991, "A New Method of Deducing Atomic Spectroscopic Terms," *J. Chem. Educ.* **68**, 205-207.

Sheehan, W. F.: 1983, "Wave Functions in One Dimension," *J. Chem. Educ.* **60**, 50-52.

Simons, J.: 1992, "Why Equivalent Bonds Appear as Distinct Peaks in Photoelectron Spectra," *J. Chem. Educ.* **69**, 522-528.

Whetten, R. 1,., Ezra G S.. and Grant, E. R.: 1985, "Molecular Dynamics beyond the Adiabatic Approximation: New Experiments and Theory " *Annul. Rev. Phys. Chem.* **36**, 277-320.

13

Analyzing Organized Structures

13.1 The General Behavior of Bases

THE SYMMETRIES (OR NEAR-SYMMETRIES) OF A MOLECULE or solid unit are embodied in the operations that change it to an equivalent (or nearly equivalent) system. Each such operation transforms a local physical attribute. This may be a displacement associated with an oscillation. It may be an orbital for electrons or nuclei. It may be an angular momentum or spin. It may be electric charge or color charge.

The result may be expressed as a linear combination of standard expressions for the pertinent attribute, The standard expressions are then referred to as *bases*. The bases may be arranged in a row or column matrix. The operation is then represented by a square matrix acting on the row or column matrix.

Two operations applied in succession are equivalent to a single operation. Correspondingly, the matrices for the two operations multiply to yield the matrix for the overall single operation. The operations, and the matrices that represent them, form a closed set.

Included in the set is the identity operation, which leaves the system unaltered Also included is the inverse of each operation, which reverses the effect of the operation. Furthermore, two successive operations can be replaced by their equivalent in a series of operations.

A set with these abstract properties is called a *group*. Some useful aspects of group theory will be considered in this chapter.

13.2 Representing Nondistortive Operations

Geometric operations that do not distort a given molecule or crystal are represented by particular linear transformations of the Cartesian coordinates of a typical point in the system. When the transformation is homogeneous, it may be represented by a matrix equation. The point is represented by a row matrix; the operation, by a square matrix.

For the given system, choose a central point as origin and on it erect convenient mutually perpendicular axes. Choose a point in the system with coordinates x, y, z

before and x', y', z' after the given operation. A linear homogeneous transformation is then defined by

$$x' = xA_{11} + yA_{21} + zA_{31}, \quad [13.1]$$

$$y' = xA_{12} + yA_{22} + zA_{32}, \quad [13.2]$$

$$z' = xA_{13} + yA_{23} + zA_{33}, \quad [13.3]$$

If we replace x, y, z and x', y', z' with x_1, x_2, x_3 and x_1', x_2', x_3', we have

$$x_k' = \sum_j x_j A_{jk}. \quad [13.4]$$

Introducing the matrix rule for combining arrays, we express (13.4) as

$$\begin{pmatrix} x_1' & x_2' & x_3' \end{pmatrix} = \begin{pmatrix} x_1 & x_2 & x_3 \end{pmatrix} \begin{pmatrix} A_{11} & A_{12} & A_{13} \\ A_{21} & A_{22} & A_{23} \\ A_{31} & A_{32} & A_{33} \end{pmatrix} \quad [13.5]$$

or

$$\mathbf{r}' = \mathbf{rA}. \quad [13.6]$$

The simplest operation is the one involving no change, the *identity* operation I, for which

$$\begin{pmatrix} x' & y' & z' \end{pmatrix} = \begin{pmatrix} x & y & z \end{pmatrix} \begin{pmatrix} 1 & 0 & 0 \\ 0 & 1 & 0 \\ 0 & 0 & 1 \end{pmatrix} \quad [13.7]$$

or

$$\mathbf{r}' = \mathbf{rI}. \quad [13.8]$$

Reflection of the system in the xy plane changes the sign of z; then

$$\begin{pmatrix} x' & y' & z' \end{pmatrix} = \begin{pmatrix} x & y & z \end{pmatrix} \begin{pmatrix} 1 & 0 & 0 \\ 0 & 1 & 0 \\ 0 & 0 & -1 \end{pmatrix} \quad [13.9]$$

or

$$\mathbf{r}' = \mathbf{r}\sigma_h, \quad [13.10]$$

if the xy plane is taken to be horizontal. Reflection in a plane containing the z axis is then taken to be relative to a vertical plane. For such a reflection, we write

$$\mathbf{r}' = \mathbf{r}\sigma_v. \quad [13.11]$$

Reflection of the system through the origin changes the sign of all three coordinates; we have

$$\begin{pmatrix} x' & y' & z' \end{pmatrix} = \begin{pmatrix} x & y & z \end{pmatrix} \begin{pmatrix} -1 & 0 & 0 \\ 0 & -1 & 0 \\ 0 & 0 & -1 \end{pmatrix} \quad [13.12]$$

or

$$\mathbf{r}' = \mathbf{ri}. \quad [13.13]$$

This operation is called *inversion.*

Rotation of the system about the z axis by angle ϕ alters the coordinates of a typical point as figure 13.1 shows. Introducing the definitions of cosine and sine and the identities of example 13.1 gives us

$$x' = r\cos(\alpha+\phi) = r\cos\alpha\cos\phi - r\sin\alpha\sin\phi = x\cos\phi - y\sin\phi, \qquad [13.14]$$

$$y' = r\sin(\alpha+\phi) = r\cos\alpha\sin\phi + r\sin\alpha\cos\phi = x\sin\phi + y\cos\phi, \qquad [13.15]$$

We also have

$$z' = z. \qquad [13.16]$$

The corresponding matrix equation is

$$\begin{pmatrix} x' & y' & z' \end{pmatrix} = \begin{pmatrix} x & y & z \end{pmatrix} \begin{pmatrix} \cos\phi & \sin\phi & 0 \\ -\sin\phi & \cos\phi & 0 \\ 0 & 0 & 1 \end{pmatrix} \qquad [13.17]$$

which we abbreviate as

$$\mathbf{r}' = \mathbf{r}\mathbf{C}_n^m. \qquad [13.18]$$

Operation C_n^m consists of rotation of the system by m/n turn about the z axis; thus

$$\phi = \frac{2\pi m}{n}. \qquad [13.19]$$

Example 13.1

Derive the identities employed in equations (13.14) and(13.15).

First construct figure 13.2. Then employ the definitions of cosine and sine to obtain

$$x' = a\cos(\phi+\alpha) = b\cos\alpha - c\sin\alpha = a\cos\phi\cos\alpha - a\sin\phi\sin\alpha \quad \text{and}$$

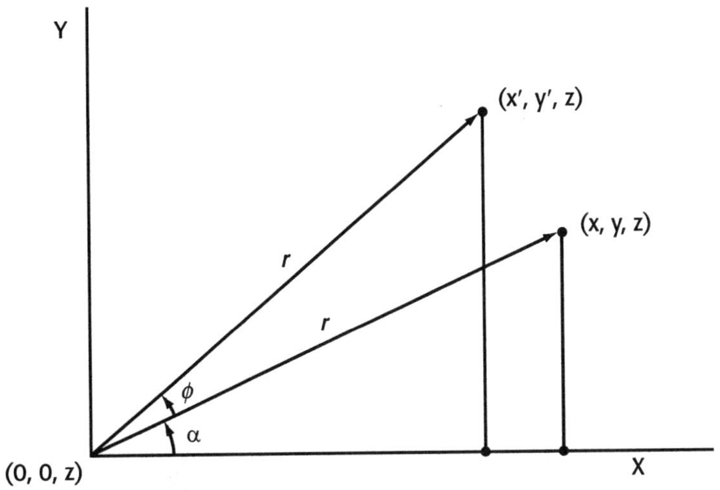

FIGURE 13.1 Rotation of a position vector for a typical point in a system by angle ϕ around the z axis.

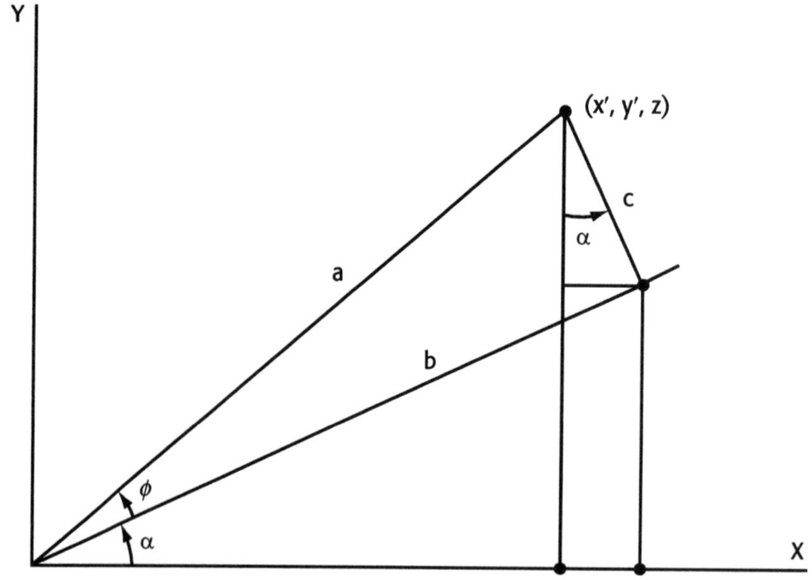

FIGURE 13.2 Right triangles for determining cos (α + ϕ) and sin (α + ϕ).

and

$$y' = a\sin(\phi+\alpha) = b\sin\alpha + c\cos\alpha = a\cos\phi\sin\alpha + a\sin\phi\cos\alpha$$

whence

$$\cos(\alpha+\phi) = \cos\alpha\cos\phi - \sin\alpha\sin\phi,$$
$$\sin(\alpha+\phi) = \cos\alpha\sin\phi + \sin\alpha\cos\phi.$$

13.3 Symmetry Operations

A symmetric structure is characterized by the existence of geometric operations that transform the structure into an equivalent structure. Such an operation is called a *symmetry operation.* In general, the operation occurs with respect to a substructure, such as a point, a line, a plane, a vector, or combinations of these.

The primary symmetry operations for the atoms in molecules and crystals are listed in table 13.1.

Around a given line, the smallest rotation C_n that is a symmetry operation determines the multiplicity of the axis; for C_n, the axis is said to be n-fold. When a system has only one n- fold axis with $n > 2$, this axis is called the principal axis and it is taken to be the z axis. Reflection in a plane perpendicular to the C_n axis is labeled σ_h. Reflection in a plane containing the C_n axis is labeled σ_v or σ_d The letter h stands for horizontal, v for vertical, d for dihedral.

The distinct operations obtained on combining the primary operations are not listed in table 13.1. Since two or more symmetry operations acting in succession transform a system into an equivalent system, the combination is a symmetry operation. And for each symmetry operation, the inverse is also a symmetry operation.

Thus, the presence or absence of certain *key* symmetry operations determines what other operations are present. And one can identify the complete group of operations

Table 13.1 Primary Symmetry Operations

Reference Substructure	*Description of Operation*	*Symbol for Operation*
All space	Identity	I
Mirror plane	Reflection in the plane	σ
Straight line	Rotation by $1/n$ turn about the line	C_n
Point	Inversion	i
Point on straight line	Rotation by $1/n$ turn, then inversion	iC_n
Point on plane	Rotation by $1/n$ turn, then reflection in the plane	S_n
Vector	Translation in ith direction	T_i
Glide plane	Translation, then reflection in the plane	σT_i
Screw axis	Translation, then rotation by $1/n$ turn about the axis	$C_n T_i$

that transform a symmetric system into an equivalent system by answering only a few questions.

A convenient sequence for determining an appropriate group for a molecule appears in figure 13.3. The question at the top is asked first. If the answer is yes, one follows the right branch down to the next question; if it is no, one follows the left branch to the next question. The successive questions are answered in turn till one comes to the end of the line. There a symbol identifying the group appears.

For calculations on a given molecule, a person may not employ the full symmetry group. Instead, it may be simpler or more expedient to employ a subgroup.

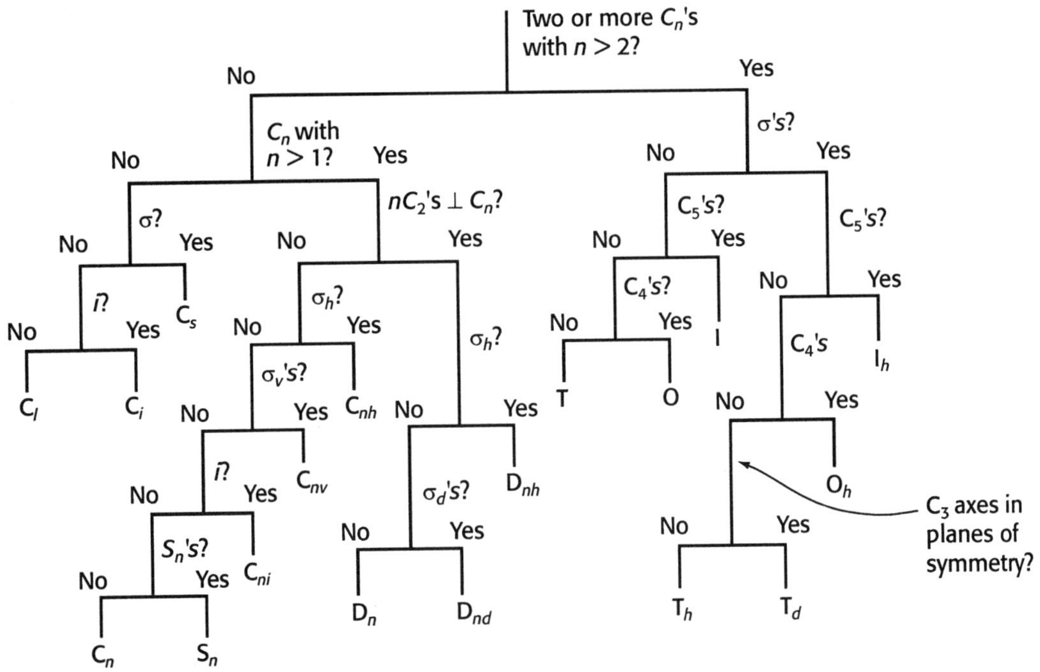

FIGURE 13.3 Successive questions for identifying the group for a molecule.

13.4 Pertinent Permutation Matrices

Whenever a molecule or crystal possesses symmetry, there are operations that permute equivalent sections or parts. Such operations are represented by permutation matrices. The number of such permutations that can be effected by rotations alone is called the *symmetry number* for the system. Additional permutations result from reflections when they are allowed.

Consider a symmetric system such as figure 13.4 illustrates. But suppose that it has n equivalent parts. Let us label these a_1, a_2, ..., a_n before a permutation and a_1', a_2', ..., a_n' after the permutation. Let us also construct the row matrices

$$\mathbf{a} = \begin{pmatrix} a_1 \; a_2 \ldots a_n \end{pmatrix}, \tag{13.20}$$

$$\mathbf{a}' = \begin{pmatrix} a_1' \; a_2' \ldots a_n' \end{pmatrix}. \tag{13.21}$$

After the operation, the jth part is at the position initially occupied by the kth part:

$$a_j' = a_k. \tag{13.22}$$

Thus, the operation is described by n homogeneous linear equations. These are described by a single matrix equation of the form

$$\begin{pmatrix} a_1' \; a_2' \ldots a_n' \end{pmatrix} = \begin{pmatrix} a_1 \; a_2 \ldots a_n \end{pmatrix} \begin{pmatrix} \cdot & \cdot & \cdot & 0 & 1 & 0 & \cdot & \cdot & \cdot \\ \cdot & \cdot & \cdot & \cdot & \cdot & \cdot & \cdot & \cdot & \cdot \\ \cdot & \cdot & \cdot & \cdot & \cdot & \cdot & \cdot & \cdot & \cdot \\ 1 & 0 & \cdot & \cdot & \cdot & \cdot & \cdot & \cdot & 0 \\ \cdot & \cdot & \cdot & \cdot & \cdot & \cdot & \cdot & \cdot & \cdot \\ \cdot & \cdot & \cdot & \cdot & \cdot & \cdot & \cdot & \cdot & \cdot \end{pmatrix}. \tag{13.23}$$

In each row or column of the square matrix, there is only one nonzero element, This element equals 1.

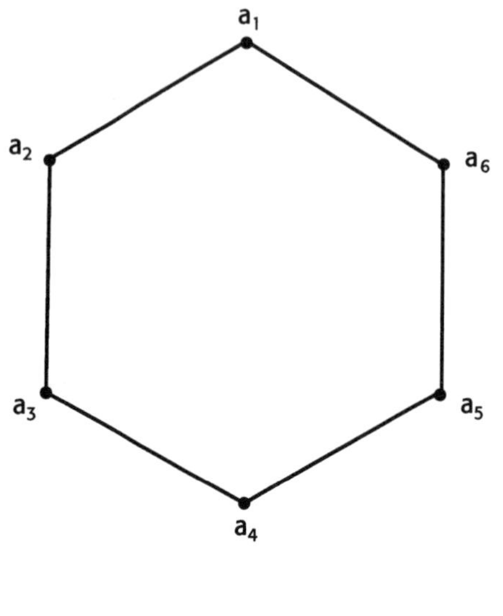

FIGURE 13.4 Hexagonal ring structure with six equivalent parts, as in benzene.

In a particular permutation, some elements may not change; for them

$$a_j' = a_j. \qquad [13.24]$$

Also, some pairs of elements may merely interchange positions; for them

$$a_j' = a_k \quad \text{and} \quad a_k' = a_j \qquad [13.25]$$

On the other hand, a set of m elements may move around a cycle. If we number these in order from 1 to m, then

$$a_j' = a_{j+1} \quad \text{with} \quad m+1 = 1. \qquad [13.26]$$

A general permutation can be broken down into permutations of types (13.24),(13.25), and (13.26).

Example 13.2

Determine the symmetry number for benzene.

Perpendicular to the plane of figure 13.4 through the center is a sixfold axis. The operations I, C_6, C_3, C_2, C_3^{-1}, C_6^{-1} about it yield 6 permutations from the initial orientation. Perpendicular to the sixfold axis are three twofold axes passing through opposite vertices of the hexagon. These yield 3 additional permutations. Also perpendicular to the sixfold axis are three twofold axes passing through the middles of opposite edges of the hexagon. These yield 3 more permutations. So we have

$$\sigma = 6 + 3 + 3 = 12.$$

13.5 Other Transformation Matrices

With atoms and molecules, we are concerned with the behavior of orbitals under the various symmetry operations of the given system. With molecules, we are concerned with the behavior of atomic displacements, the behavior of bond stretchings, the behavior of bond angle variations.

Some symmetry operations act to permute these. Others act to project them in various ways. Then the square matrix in equation (13.23) must be modified to reflect these projections. In applications, we will not be concerned with all the elements, but only with those on the principal diagonal.

Example 13.3

Subject figure 13.5 to the C_2, S_4, and σ_d operations indicated and determine how the base vectors on the carbon atom transform.

Under the operation C_2, we have

$$\mathbf{i}' = C_2\mathbf{i} = -\mathbf{i}, \quad \mathbf{j}' = C_2\mathbf{j} = -\mathbf{j}, \quad \mathbf{k}' = C_2 k = \mathbf{k}, \quad \text{and}$$

and

$$\begin{pmatrix}\mathbf{i}' & \mathbf{j}' & \mathbf{k}'\end{pmatrix} = \begin{pmatrix}\mathbf{i} & \mathbf{j} & \mathbf{k}\end{pmatrix}\begin{pmatrix}-1 & 0 & 0\\ 0 & -1 & 0\\ 0 & 0 & 1\end{pmatrix}.$$

The sum of the elements on the principal diagonal is - 1.

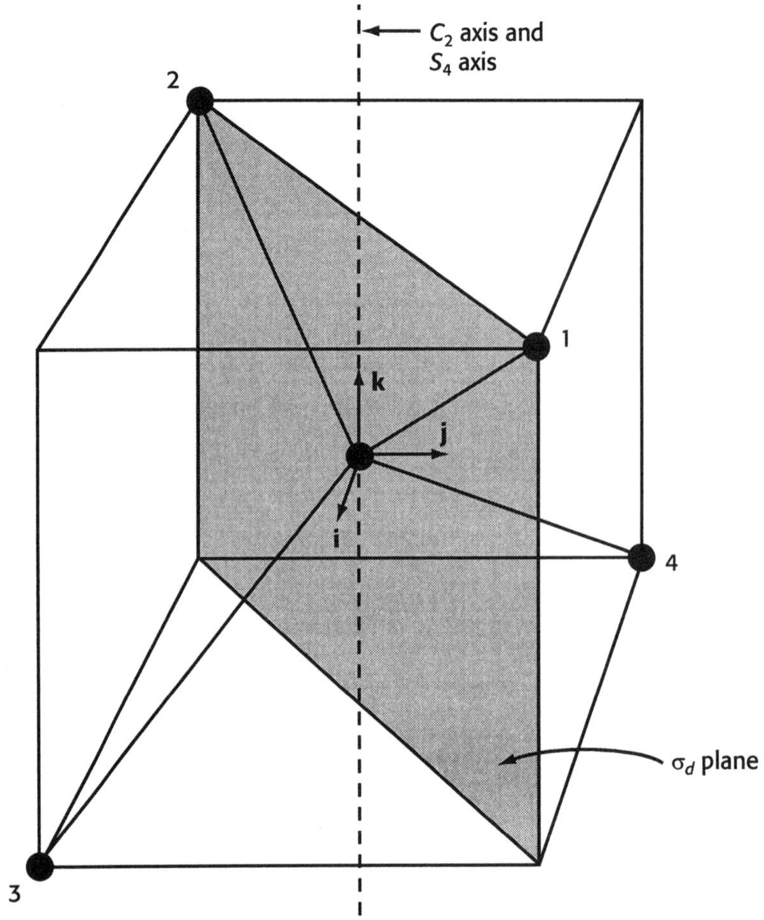

FIGURE 13.5 Orthogonal unit displacement vectors for the carbon atom in tetrahedral methane.

Under the operation S_4, we have

$$\mathbf{i}' = S_4\mathbf{i} = \mathbf{j}, \quad \mathbf{j}' = S_4\mathbf{j} = -\mathbf{i}, \quad \mathbf{k}' = S_4 k = -\mathbf{k}, \quad \text{and}$$

and

$$\begin{pmatrix}\mathbf{i}' \ \mathbf{j}' \ \mathbf{k}'\end{pmatrix} = \begin{pmatrix}\mathbf{i} \ \mathbf{j} \ \mathbf{k}\end{pmatrix}\begin{pmatrix}0 & -1 & 0 \\ 1 & 0 & 0 \\ 0 & 0 & -1\end{pmatrix}.$$

The sum of elements on the principal diagonal is -1.

Under the operation σ_d, we have

$$\mathbf{i}' = \sigma_d\mathbf{i} = \mathbf{j}, \quad \mathbf{j}' = \sigma_d\mathbf{j} = \mathbf{i}, \quad \mathbf{k}' = \sigma_d k = \mathbf{k}, \quad \text{and}$$

and

$$\begin{pmatrix}\mathbf{i}' \ \mathbf{j}' \ \mathbf{k}'\end{pmatrix} = \begin{pmatrix}\mathbf{i} \ \mathbf{j} \ \mathbf{k}\end{pmatrix}\begin{pmatrix}0 & 1 & 0 \\ 1 & 0 & 0 \\ 0 & 0 & 1\end{pmatrix}.$$

The sum of elements on the principal diagonal is 1.

Example 13.4

Subject figure 13.6 to the C_2, S_4, and σ_d operations indicated and determine the elements on the principal diagonal of the representation matrix.

Under the C_2 and S_4 operations, all the base vectors are permuted and we have all zeros on the principal diagonal.

Under the σ_d operation, we find that

$$\mathbf{k}_1{}' = \sigma_d \mathbf{k}_1 = \mathbf{k}_1, \quad \mathbf{k}_2{}' = \sigma_d \mathbf{k}_2 = \mathbf{k}_2,$$

while all other base vectors are permuted. We are left with two ones on the principal diagonal of the representation matrix. These add to give 2.

Example 13.5

Subject figure 13.7 to the C_3 and σ_d operations indicated and determine how the stretching vectors transform.

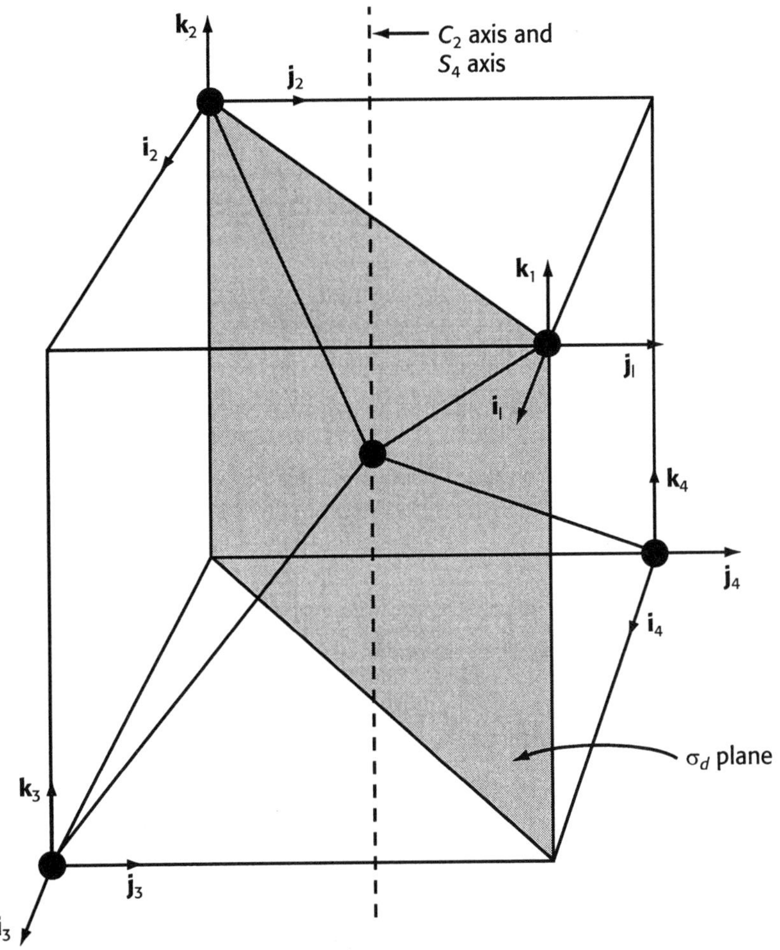

FIGURE 13.6 Orthogonal unit displacement vectors for the hydrogen atoms in tetrahedral methane.

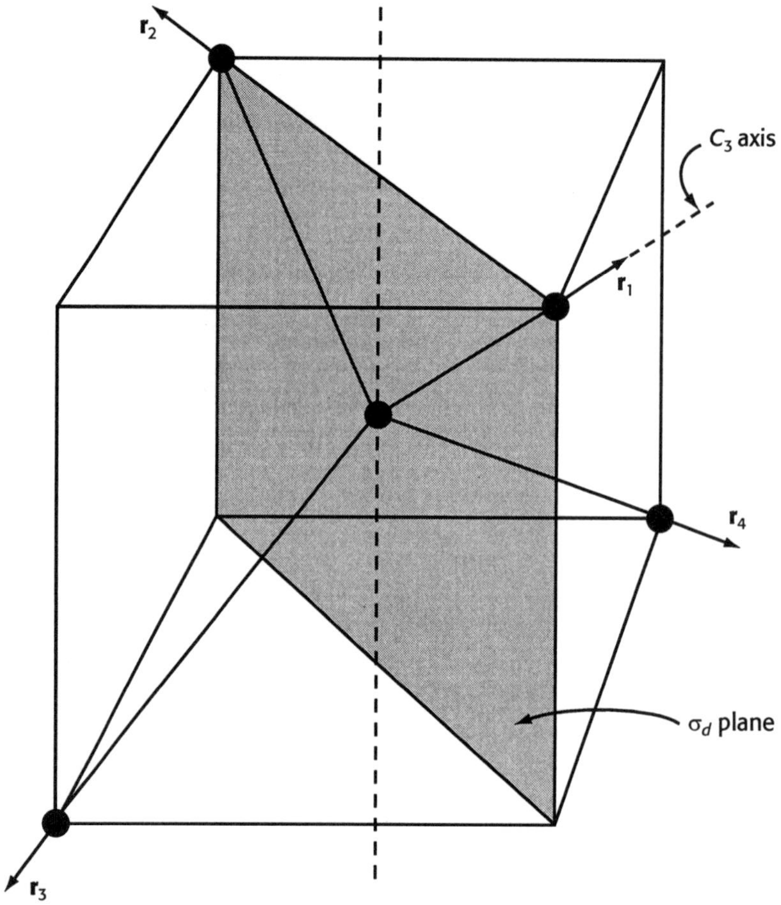

FIGURE 13.7 Stretching displacement vectors for the C-- H bonds in methane.

Under the operation C_3, we have

$$\mathbf{r}_1' = \mathbf{r}_1, \quad \mathbf{r}_2' = \mathbf{r}_3, \quad \mathbf{r}_3' = \mathbf{r}_4, \quad \mathbf{r}_4' = \mathbf{r}_2,$$

and

$$\begin{pmatrix} \mathbf{r}_1' & \mathbf{r}_2' & \mathbf{r}_3' & \mathbf{r}_4' \end{pmatrix} = \begin{pmatrix} \mathbf{r}_1 & \mathbf{r}_2 & \mathbf{r}_3 & \mathbf{r}_4 \end{pmatrix} \begin{pmatrix} 1 & 0 & 0 & 0 \\ 0 & 0 & 0 & 1 \\ 0 & 1 & 0 & 0 \\ 0 & 0 & 1 & 0 \end{pmatrix}.$$

The sum of the elements on the principal diagonal is 1.

Under the operation σ_d, we have

$$\mathbf{r}_1' = \mathbf{r}_1, \quad \mathbf{r}_2' = \mathbf{r}_2, \quad \mathbf{r}_3' = \mathbf{r}_4, \quad \mathbf{r}_4' = \mathbf{r}_3,$$

and

$$\begin{pmatrix} \mathbf{r}_1' & \mathbf{r}_2' & \mathbf{r}_3' & \mathbf{r}_4' \end{pmatrix} = \begin{pmatrix} \mathbf{r}_1 & \mathbf{r}_2 & \mathbf{r}_3 & \mathbf{r}_4 \end{pmatrix} \begin{pmatrix} 1 & 0 & 0 & 0 \\ 0 & 1 & 0 & 0 \\ 0 & 0 & 0 & 1 \\ 0 & 0 & 1 & 0 \end{pmatrix}.$$

The sum of the elements on the principal diagonal is 2.

Example 13.6

Subject figure 13.8 to the C_2 and σ_d operations indicated and determine how the bond angles transform.

Under the C_2 operation, atom 1 moves to the 2 position, atom 2 to the 1 position, atom 3 to the 4 position, and atom 4 to the 3 position. So the angles between these transform as follows:

$$\alpha_{12}' = \alpha_{12}, \quad \alpha_{13}' = \alpha_{24}, \quad \alpha_{14}' = \alpha_{23}, \quad \alpha_{23}' = \alpha_{14}, \quad \alpha_{24}' = \alpha_{13}, \quad \alpha_{34}' = \alpha_{34},$$

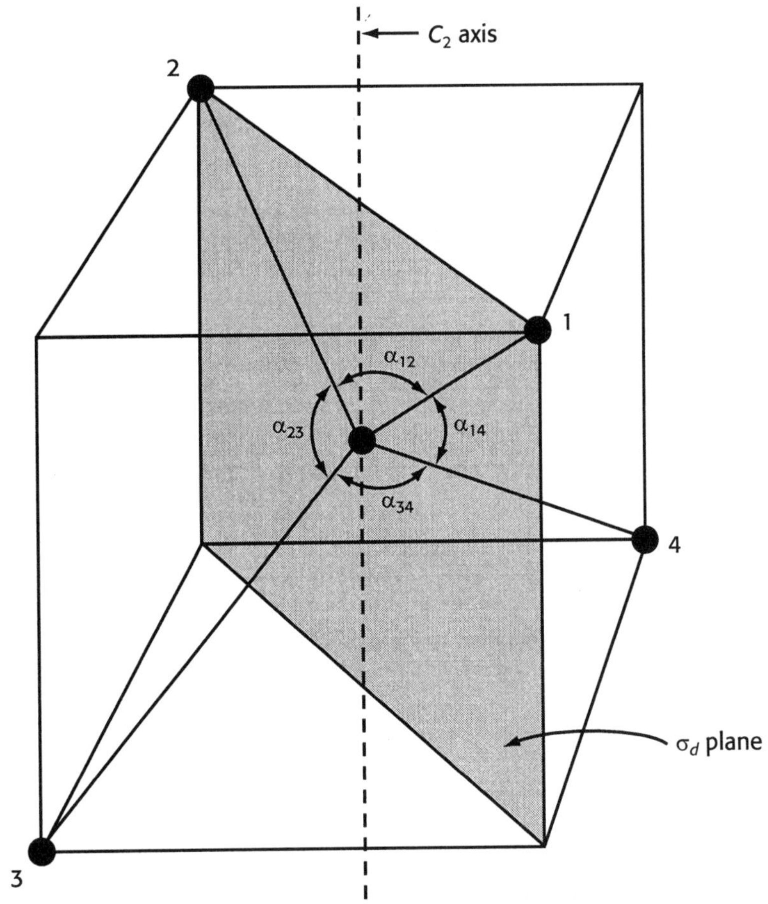

FIGURE 13.8 Bond angles for the CH bonds in methane.

whence

$$\begin{pmatrix}\alpha_{12}' & \alpha_{13}' & \alpha_{14}' & \alpha_{23}' & \alpha_{24}' & \alpha_{34}'\end{pmatrix} = \begin{pmatrix}\alpha_{12} & \alpha_{13} & \alpha_{14} & \alpha_{23} & \alpha_{24} & \alpha_{34}\end{pmatrix}\begin{pmatrix}1 & 0 & 0 & 0 & 0 & 0\\ 0 & 0 & 0 & 0 & 1 & 0\\ 0 & 0 & 0 & 1 & 0 & 0\\ 0 & 0 & 1 & 0 & 0 & 0\\ 0 & 1 & 0 & 0 & 0 & 0\\ 0 & 0 & 0 & 0 & 0 & 1\end{pmatrix}.$$

The sum of the elements on the principal diagonal is 2.

Under the σ_d operation, only atoms 3 and 4 are interchanged. So we have

$$\begin{pmatrix}\alpha_{12}' & \alpha_{13}' & \alpha_{14}' & \alpha_{23}' & \alpha_{24}' & \alpha_{34}'\end{pmatrix} = \begin{pmatrix}\alpha_{12} & \alpha_{13} & \alpha_{14} & \alpha_{23} & \alpha_{24} & \alpha_{34}\end{pmatrix}\begin{pmatrix}1 & 0 & 0 & 0 & 0 & 0\\ 0 & 0 & 1 & 0 & 0 & 0\\ 0 & 1 & 0 & 0 & 0 & 0\\ 0 & 0 & 0 & 0 & 1 & 0\\ 0 & 0 & 0 & 1 & 0 & 0\\ 0 & 0 & 0 & 0 & 0 & 1\end{pmatrix}.$$

The sum of the elements on the principal diagonal is 2.

13.6 Classifying Operations

Rotations by the same fraction of a turn may be made equivalent by the presence of a symmetry operation. Similarly, reflections in different planes may be made equivalent. The process is depicted in figure 13.9.

Consider a system for which P and P^{-1} are symmetry operations. Let the system be subjected to operation A, which transforms row matrix $\mathbf{r}$ to row matrix $\mathbf{s}$:

$$\mathbf{s} = A\mathbf{r} = \mathbf{r}\mathbf{A}. \qquad [13.27]$$

Here $\mathbf{A}$ is the matrix representing operation A.

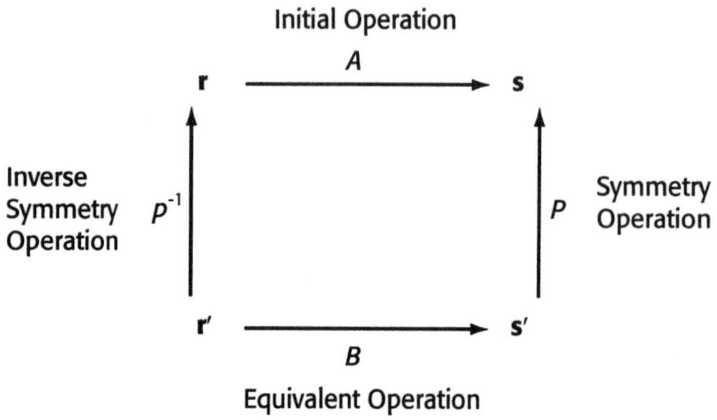

FIGURE 13.9 Relating different operations through a symmetry operation.

But the effects of the symmetry operation P are

$$\mathbf{r}' = P\mathbf{r} = \mathbf{r}\mathbf{P}, \quad [13.28]$$

$$\mathbf{s}' = P\mathbf{s} = \mathbf{s}\mathbf{P}, \quad [13.29]$$

and of its inverse

$$\mathbf{r} = P^{-1}\mathbf{r}' = \mathbf{r}'\mathbf{P}^{-1}. \quad [13.30]$$

So the action of operation PAP^{-1} on $\mathbf{r}'$ is given by

$$\mathbf{s}' = PAP^{-1}\mathbf{r}' = PA\mathbf{r}'\mathbf{P}^{-1} = P\mathbf{r}'\mathbf{A}\mathbf{P}^{-1} = \mathbf{r}'\mathbf{P}\mathbf{A}\mathbf{P}^{-1}. \quad [13.31]$$

On comparing this with

$$\mathbf{s}' = B\mathbf{r}' = \mathbf{r}'\mathbf{B}, \quad [13.32]$$

we see that

$$\mathbf{B} = \mathbf{P}\mathbf{A}\mathbf{P}^{-1} \quad [13.33]$$

and

$$B = PAP^{-1}. \quad [13.34]$$

Since operation P leaves the system in a configuration equivalent to the original one, operation B is equivalent to operation A. Change(13.34) of A is called a *similarity transformation* of A with respect to P. Whenever symmetry operations B and A are linked by a similarity transformation with another symmetry operation, they are said to be in the same *class*.

Example 13.7

Show that C_3 and C_3^{-1} belong to the same class when σ_v is also a symmetry operation.

Consider a molecule whose symmetry operations belong to the $\mathbf{C}_{3v}$ group. Break it down into three equivalent members and label these as in the left triangle of figure 13.10. Note the positions of the C_3 axis and the σ_v plane. Then apply operations σ_v^{-1}, C_3, and σ_v in succession. The resulting configuration is the same as that obtained on applying C_3^{-1} alone. We thus find that

$$\sigma_v C_3 \sigma_v^{-1} = C_3^{-1}.$$

13.7 Character Vectors

One may represent the operations of a group with square matrices of a particular size. The elements in the matrices are generally not related in a simple way. However, those within any given class are.

Let us suppose that $\mathbf{A}$ and $\mathbf{B}$ are matrices from a representation R of a group. Let us furthermore suppose that they are linked by the similarity transformation

$$\mathbf{B} = \mathbf{P}\mathbf{A}\mathbf{P}^{-1} \quad [13.35]$$

for which $\mathbf{P}$ also belongs to the representation. This transformation generally alters the elements in $\mathbf{A}$. However, some functions of the elements are not altered.

The *trace* of a square matrix is defined as the sum of elements on the principal diagonal. For $\mathbf{A}$ and $\mathbf{B}$, the traces are

$$\mathrm{Tr}\,\mathbf{A} = \sum_k A_{kk} \quad \text{and} \quad \mathrm{Tr}\,\mathbf{B} = \sum_j B_{jj}. \quad [13.36]$$

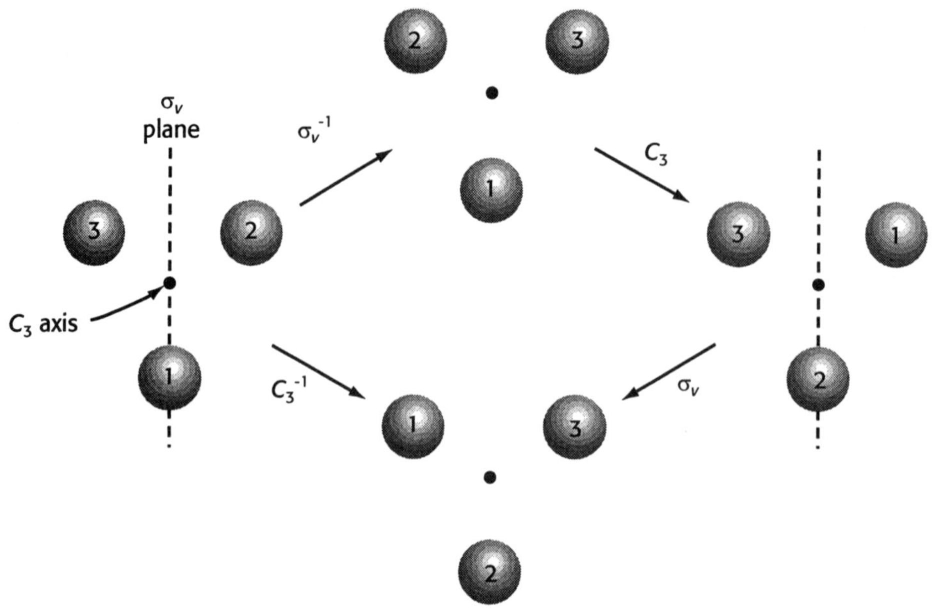

FIGURE 13.10 How reflection operation σ_v puts C_3^{-1} in the same class as C_3.

But from relation(13.35), we find that

$$\text{Tr } \mathbf{B} = \text{Tr } \mathbf{PAP}^{-1} = \sum_j \sum_k \sum_l P_{jk} A_{kl} P_{lj}^{-1} = \sum_j \sum_k \sum_l P_{lj}^{-1} P_{jk} A_{kl} = \sum_l \sum_k \delta_{lk} A_{kl} = \sum_k A_{kk} = \text{Tr } \mathbf{A}. \quad [13.37]$$

Thus, transformation (13.35) does not alter the trace of the matrix.

A closed set of symmetry operators for a given system may be partitioned into classes. Furthermore, it may be represented by square matrices of a particular size.

The trace one obtains for each class then depends on the class and on the representation. The set of traces for a given representation R forms a vector called the *character* $\chi(R)$. A group is distinguished by a set of basis vectors in terms of which all character vectors can be expressed. Components of these basis vectors are listed in character tables.

How a table is organized is illustrated by table 13.2. The heading identifies each class by a typical operation preceded by the number of such operations in the class. Each basis vector is identified by a letter with a subscript in the left column. Letter A is employed for a numerical representation, letter E for a 2×2 matrix representation, letter F for a 3×3 matrix representation, and so on. In the body of the table are the components

Table 13.2 Basis Character Vectors for the $\mathbf{T}_d$ Group

	I	$8C_3$	$3C_2$	$6S_4$	$6\sigma_d$	
A_1	1	1	1	1	1	$x^2 + y^2 + z^2$
A_2	1	1	1	-1	-1	
E	2	-1	2	0	0	$(2z^2 - x^2 - y^2, x^2 - y^2)$
F_1	3	0	-1	1	-1	(R_x, R_y, R_z)
F_2	3	0	-1	-1	1	$(x, y, z), (yz, zx, xy)$

corresponding to each class. In the right column are basis functions that generate a representation for the corresponding row. The symbols R_x, R_y, R_z stand for unit rotations about the x, y, and z axes. They may be represented by the cross products $\mathbf{j} \times \mathbf{k}$, $\mathbf{k} \times \mathbf{i}$, $\mathbf{i} \times \mathbf{j}$.

Example 13.8

Determine the components of the character vector for the representation based on the Cartesian atomic displacements in tetrahedral methane and in similar molecules.

The unit displacement vectors are described by figures 13.5 and 13.6. For the identity operation, the number of vectors left unchanged is $3 \times 5 = 15$. So the entry under I in the character table is 15. With each C_3, all the vectors are permuted; the entry under $8C_3$ is 0. With each C_2 and S_4 , the vectors on the central atom yield -1, as example 13.3 indicated, while those on the attached atoms yield 0, as example 13.4 concluded. The sum equals -1. With each σ_d, the vectors on the central atom yield 1 and those on the attached atoms yield 2, for a total of 3. The results are collected in table 13.3.

Example 13.9

Determine the components of the character vector for the representation based on the bond stretchings in tetrahedral methane and in similar molecules.

The radial unit displacement vectors are described in figure 13.7. For the identity operation, the number of vectors left unchanged is 4. So the entry under I is 4. From example 13.5, the component for the C_3 class is 1; the component for the σ_d class is 2. In the C_2 and S_4 operations, all the $\mathbf{r}_j$'s are permuted; so the corresponding components are 0. The results appear in table 13.3.

Example 13.10

Determine the components of the character vector for the representation based on the bond angle variations in tetrahedral methane and in similar molecules.

There are 6 bond angles. In the identity operation, the number left unchanged is 6. So the entry under I is 6. In the C_3 and S_4operations, all 6 are permuted; so the corresponding components are 0. From example 13.6, the component for the C_2 class is 2; the component for the σ_d class is also 2. The results are listed in table 13.3.

13.8 Symmetry Species

The various attributes of a symmetric system may be characterized by how they behave under the symmetry operations of the different classes. Indeed, each mode of

Table 13.3 Character vectors for atomic displacements, bond stretchings, and bond angle variations in a T_d AB_4 molecule

	I	$8C_3$	$3C_2$	$6S_4$	$6\sigma_d$	
$x(\mathbf{i}_i, \mathbf{j}_i, \mathbf{k}_i)$	15	0	-1	-1	3	$(\mathbf{i}_i, \mathbf{j}_i, \mathbf{k}_i)$ $i = 0, 1, 2, 3, 4$
$x(\mathbf{r}_i)$	4	1	0	0	2	$(\mathbf{r}_1, \mathbf{r}_2, \mathbf{r}_3, \mathbf{r}_4)$
$x(\alpha_{ij})$	6	0	2	0	2	$(\alpha_{12}, \alpha_{13}, \alpha_{14}, \alpha_{23}, \alpha_{24}, \alpha_{34})$

motion (translation, rotation, vibration) of a given molecule belongs to a symmetry species that can be determined. Furthermore, each electron orbital of an atom or molecule can be similarly assigned.

The symmetry operations from a closed set acting on an attribute generate a representation R for the species. The traces for the different classes yield components of the corresponding character vector $\boldsymbol{\chi}(R)$.

Suppose a given group has d distinct classes. A character vector then has d independent components. And it may be expressed as a linear function of d basis vectors. Thus, we write

$$\chi = \sum_{k=1}^{d} a^{(k)} \chi^{(k)} \qquad [13.38]$$

where $\boldsymbol{\chi}^{(k)}$ is the basis vector and $a^{(k)}$ is the number of times it contributes to $\boldsymbol{\chi}$. In general, the basis vectors are normalized so that the coefficients $a^{(k)}$ are integers. A symmetry species that generates a basis vector is referred to as a *primitive symmetry species.*

Example 13.11

Determine the primitive symmetry species that contribute to motions of the atoms in tetrahedral AB_4.

The first row in table 13.3 lists the components of the character vector for these motions. We see that these are obtained by adding the components for A_1, E, F_1, and $3F_2$ from table 13.2. Thus

$$\chi\left(\mathbf{i}_i, \mathbf{j}_i, \mathbf{k}_i\right) = A_1 + E + F_1 + 3F_2.$$

Now the translation of the center of mass behaves as (x, y, z); the rotation about this center, as (R_x, R_y, R_z). Subtracting out the corresponding characters leaves

$$\chi\left(\text{vibration}\right) = A_1 + E + 2F_2.$$

Tetrahedral AB_4 thus has only four fundamental vibrational frequencies, one nondegenerate, one doubly degenerate, and two triply degenerate.

The vibrational F_2 modes involve the central atom moving oppositely to the center of mass of the attached atoms, so that the center of mass of the whole molecule does not move. This central atom motion is neglected in some of the following discussion. Taking it into account mixes some bond bending motion into the bond stretching motion and vice versa.

Example 13.12

Determine the primitive symmetry species that contribute to (a) the radial motions and (b) the angular motions in tetrahedral AB_4.

The components in the second row of table 13.3 are obtained on adding the components for A_1 and F_2. Thus

$$\chi\left(\mathbf{r}_i\right) = A_1 + F_2.$$

The components in the third row of table 13.3 are obtained on adding the components for A_1 , E, and F_2. Thus

$$\chi\left(\alpha_{ij}\right) = A_1 + E + F_2.$$

From example 13.11, only one A_1 vibration exists. Since it involves the molecule expanding and contracting radially (a breathing motion), it does not involve any change in the angles. Only the E and the second F_2 motions are legitimate angular motions.

13.9 Selection Rules

A classical mode of motion is characterized by definite displacements at a given time. The generalized coordinate describing these changes at a rate that varies with time. For a vibration, the variation is described by a sinusoidal function with a particular amplitude and phase. Interactions tend to alter these. Excitation always involves increasing the amplitude; deexcitation, decreasing the amplitude.

In a quantum mechanical mode of motion, the generalized coordinate satisfies a wave function. In the ground state and in even excited states, this function is approximately symmetric; a given classical displacement array is accompanied by an equivalent classical array, in phase, directed oppositely. In the first excited state and in odd higher excited states, the wave function is antisymmetric in the generalized coordinate; a given classical displacement array is accompanied by an oppositely directed classical array 180° out of phase.

As a consequence, an excitation by one step has the symmetry of a classical phase of the motion. This symmetry is described by the basis functions for the motion's symmetry species.

An electromagnetic wave acts on the nuclei in molecules in its path. When the photon energy matches the energy needed to excite a motion by one step, that change may occur. The periodic transverse motion of the electric vector in the wave makes it belong to the same symmetry species as a Cartesian coordinate measured along the vector. It effects transitions with this symmetry.

We therefore have the selection rule: Photons with enough energy will cause a transition and be absorbed (or emitted) by a normal mode of motion when the mode belongs to the same primitive symmetry species as a Cartesian coordinate x, y, or z.

In Raman scattering, passing photons interact with molecules. Energy is transferred when polarizability of a molecule changes. This property behaves as one or more of the quadratic combinations of Cartesian coordinates. Thus, we have the selection rule: In scattering, high energy photons will exchange energy with a mode of motion when the mode belongs to the same symmetry species as a homogeneous quadratic function of the Cartesian coordinates.

There is, however, a very important limitation. When a molecule has a center of symmetry, transitions that are allowed in absorption or emission are forbidden in scattering.

Since their energies of transition are very small, rotational transitions appear in the far infrared absorption and emission spectra. Vibrational transitions appear in the nearer infrared regions. Electronic transitions appear in the optical and near ultraviolet regions.

Example 13.13

For tetrahedral AB_4, how many fundamental transitions are active (a) in infrared absorption and emission, (b) in Raman interaction?

From example 13.11 , the vibrations belong to A_1, E, and two F_2 primitive symmetry types. But according to table 13.2, only an F_2 array behaves as a Cartesian coordinate. So there are two active infrared (IR) transitions. From example 13.12, one of these involves stretching of the bonds, the other bending of the bonds.

Also according to table 13.2, the rotational arrays generate the F_1 representation. So there is one triply degenerate rotationsl band.

The molecule here has a center of symmetry. So the F_2 transitions do not appear in Raman spectra. However, the A_1 and E arrays also behave as homogeneous quadratic functions of the Cartesian coordinates. The corresponding transitions are active in Raman transitions.

Example 13.14

In the absorption spectra of methane, two intense bands centered at 1306.2 and 3020.3 cm^{-1} appear. Assign these.

The force constant for stretching a bond is about ten times as large as that for bending it. So the higher wave number (and frequency) vibration involves this stretching. We assign the 3020.3 cm^{-1} band to the F_2 stretching mode, the 1306.2 cm^{-1} band to the F_2 bending mode.

Strictly, the former frequency has been altered somewhat by the mixing in of some bending motion; the latter frequency, by the mixing in of some stretching motion. Recall the comments in example 13.11.

Example 13.15

How do spectral studies support the tetrahedral structure for methane?

The tetrahedral structure for AB_4 is the most symmetric possible. Reducing the symmetry splits primitive symmetry species and so increases the number of fundamentals. In the problem set, group theoretical calculations on square pyramidal and square planar AB_4 are introduced. The results are summarized in table 13.4 Since methane has only two intense absorption bands, spectral measurements support the tetrahedral structure.

13.10 The Orthogonality Condition

The basis character vectors for a given group are linearly independent. And when the components are properly weighted, they are mutually orthogonal.

Let $\chi^{(j)}(R_i)$ be the character component for operation R_i and primitive symmetry species j. We then find that

$$\sum_i \chi^{(j)}(R_i)\chi^{(k)*}(R_i) = g\delta_{jk}, \qquad [13.39]$$

where the summation is over *all* elements of the group, g is the number of such elements, and δ_{jk} is the Kronecker delta. This is equal to zero when $j \neq k$ and equal to one when $j = k$. Also, the asterisk indicates taking the complex conjugate of the preceding number.

If h_j is the number of elements in the jth class while $\chi_j^{(k)}$ is the character component for that class for the kth symmetry species, then equation (13.39) can be rewritten as

$$\sum_j h_j \chi_j^{(k)} \chi_j^{(l)*} = g\delta_{kl}, \qquad [13.40]$$

Table 13.4 Number of fundamentals active in the infrared for various AB_4 structures

Symmetry	*Fundamental Frequencies*
$\mathbf{T}_d$	2
$\mathbf{D}_{4h}$	3
$\mathbf{C}_{4v}$	4
$\mathbf{C}_1$	9

We can now derive a formula for coefficient $a^{(i)}$ in expansion (13.38). The jth component of equation (13.38) is

$$\chi_j = \sum_m a^{(m)} \chi_j^{(m)}. \qquad [13.41]$$

Multiply this result by $h_j \chi_j^{(i)*}$ and sum over the classes,

$$\sum_j h_j \chi_j \chi_j^{(i)*} = \sum_j \sum_m a^{(m)} h_j \chi_j^{(m)} \chi_j^{(i)*} = \sum_m a^{(m)} g \delta_{mi} = a^{(i)} g. \qquad [13.42]$$

Then solve for $a^{(i)}$,

$$a^{(i)} = \frac{1}{g} \sum_j h_j \chi_j \chi_j^{(i)*}. \qquad [13.43]$$

Example 13.16

Show that primitive symmetry species F_2 occurs three times in the atomic motions of tetrahedral AB_4.

Substitute into (13.43) numbers from the first row of table 13.3 and the last row of table 13.2 to get

$$a^{F_2} = \frac{1}{24}\left[(3)(15) + 0 + 3(-1)(-1) + 6(-1)(-1) + 6(1)(3)\right] = \frac{1}{24}(45 + 3 + 6 + 18) = 3.$$

13.11 Generating Primitive Symmetry Species

An attribute of a part of a physical system generally does not belong to a single primitive symmetry species. Furthermore, a complete set of such attributes usually forms a composite of such species. But with the components of the pertinent character vectors, one can generate the contributing species.

Let $\mathbf{u}_j$ be a mathematical expression measuring an attribute of a part of a symmetric system. The expression may be scalar, vector, dyadic,.... Choose a group of symmetry operations for the system. Let R_i be the ith operation in this group while $\boldsymbol{\chi}^{(k)}$ is the character vector for the kth primitive symmetry species.

Construct the form,

$$\mathbf{f}_{kj} = \sum_i \chi^{(k)*}(R_i) R_i \mathbf{u}_j = \sum_i \chi^{(k)}(R_i)(R_i)^{-1} \mathbf{u}_j, \qquad [13.44]$$

in which the summation proceeds over all elements of the group. One can show that $\mathbf{f}_{kj}$ belongs to the kth primitive symmetry species. The general proof is beyond the scope of this text.

When the $\mathbf{u}_j$'s from a complete set yield only linearly independent $\mathbf{f}_{kj}$'s equal in number to the degeneracy of the species, these serve as bases. When a multiple of this number is obtained, then they need to be combined properly to obtain bases. Combinations that mix only among themselves under operation of the group are employed. When vibrations are mixed with rotations and/or translations, they may be separated on these physical grounds.

Example 13.17

Consider a square molecule with unit displacement vectors arranged on nuclear equilibrium positions as figure 13.11 shows. Combine these to make up bases for pure modes of motion. Each of these modes must belong to a row of a primitive symmetry species.

Under operations of the $\mathbf{C}_4$ group, unit vectors $\mathbf{u}_1$, $\mathbf{u}_5$, and $\mathbf{u}_9$ transform as table 13.5 shows. In table 13.6, the components of the character vectors are listed. Let us combine entries as formula (13.44) indicates.

For the A symmetry species, we find that

$$\mathbf{f}_1 = \mathbf{u}_1 + \mathbf{u}_2 + \mathbf{u}_3 + \mathbf{u}_4 \text{ for the breathing motion,}$$

$$\mathbf{f}_2 = \mathbf{u}_5 + \mathbf{u}_6 + \mathbf{u}_7 + \mathbf{u}_8 \text{ for rotation about the } z \text{ axis,}$$

$$\mathbf{f}_3 = \mathbf{u}_9 + \mathbf{u}_{10} + \mathbf{u}_{11} + \mathbf{u}_{12} \text{ for the translation in the } z \text{ direction.}$$

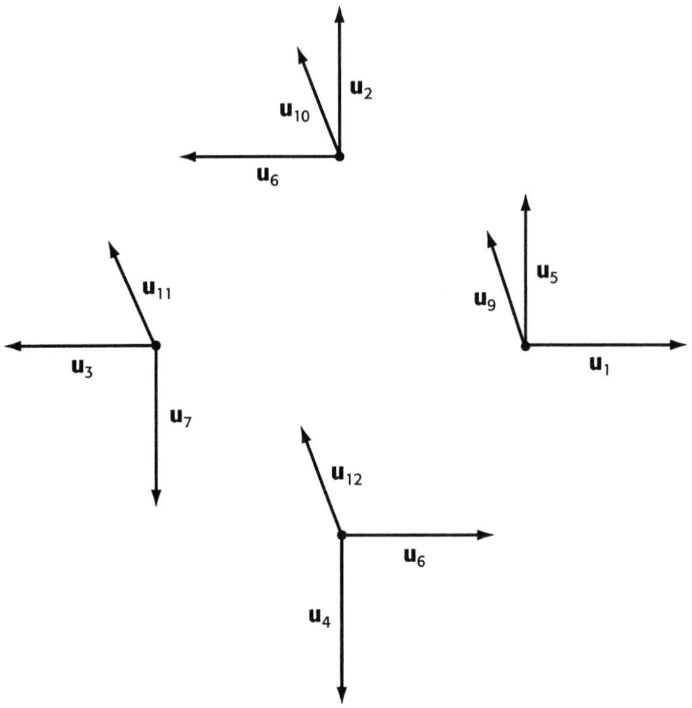

FIGURE 13.11 Unit basis vectors for displacements of the atoms in a square molecule.

Table 13.5 Effect of the Symmetry Operations on Representative Basis Vectors

I	C_4	C_2	C_4^3
$\mathbf{u}_1$	$\mathbf{u}_2$	$\mathbf{u}_3$	$\mathbf{u}_4$
$\mathbf{u}_5$	$\mathbf{u}_6$	$\mathbf{u}_7$	$\mathbf{u}_8$
$\mathbf{u}_9$	$\mathbf{u}_{10}$	$\mathbf{u}_{11}$	$\mathbf{u}_{12}$

Table 13.6 Components of the Primitive Character Vectors of the C_4 Group

	I	C_4	C_2	C_4^3
A	1	1	1	1
B	1	-1	1	-1
C	1	i	-1	$-i$
D	1	$-i$	-1	i

In the B symmetry species, we find that

$$\mathbf{f}_4 = \mathbf{u}_1 - \mathbf{u}_2 + \mathbf{u}_3 - \mathbf{u}_4 \text{ for a squashing motion,}$$

$$\mathbf{f}_5 = \mathbf{u}_5 - \mathbf{u}_6 + \mathbf{u}_7 - \mathbf{u}_8 \text{ for a scissors motion,}$$

$$\mathbf{f}_6 = \mathbf{u}_9 - \mathbf{u}_{10} + \mathbf{u}_{11} - \mathbf{u}_{12} \text{ for a bending in the } z \text{ direction.}$$

For the C and D symmetry species, we obtain the complex forms

$$\mathbf{f}_7 = \mathbf{u}_1 - i\mathbf{u}_2 - \mathbf{u}_3 + i\mathbf{u}_4,$$

$$\mathbf{f}_8 = \mathbf{u}_1 + i\mathbf{u}_2 - \mathbf{u}_3 - i\mathbf{u}_4,$$

$$\mathbf{f}_9 = \mathbf{u}_5 - i\mathbf{u}_6 - \mathbf{u}_7 + i\mathbf{u}_8,$$

$$\mathbf{f}_{10} = \mathbf{u}_5 + i\mathbf{u}_6 - \mathbf{u}_7 - i\mathbf{u}_8,$$

$$\mathbf{f}_{11} = \mathbf{u}_9 - i\mathbf{u}_{10} - \mathbf{u}_{11} + i\mathbf{u}_{12},$$

$$\mathbf{f}_{12} = \mathbf{u}_9 + i\mathbf{u}_{10} - \mathbf{u}_{11} - i\mathbf{u}_{12}.$$

From the real and imaginary parts of $\mathbf{f}_{11}$ and $\mathbf{f}_{12}$, we find that

$$\mathbf{f}_{13} = \text{Re}\,\mathbf{f}_{11} = \text{Re}\,\mathbf{f}_{12} = \mathbf{u}_9 - \mathbf{u}_{11} \text{ for rotation about the } y \text{ axis,}$$

$$\mathbf{f}_{14} = -\text{Im}\,\mathbf{f}_{11} = \text{Im}\,\mathbf{f}_{12} = \mathbf{u}_{10} - \mathbf{u}_{12} \text{ for rotation about the } x \text{ axis.}$$

From the real and imaginary parts of $\mathbf{f}_7$, $\mathbf{f}_8$, $\mathbf{f}_9$, $\mathbf{f}_{10}$, we have

$$\mathbf{f}_{15} = \mathbf{u}_1 - \mathbf{u}_3, \mathbf{f}_{16} = \mathbf{u}_2 - \mathbf{u}_4,$$

$$\mathbf{f}_{17} = \mathbf{u}_5 - \mathbf{u}_7, \mathbf{f}_{18} = \mathbf{u}_6 - \mathbf{u}_8.$$

Then from $\mathbf{f}_{15}$ and $\mathbf{f}_{18}$, we obtain

$$\mathbf{f}_{19} = \mathbf{f}_{15} - \mathbf{f}_{18} = \mathbf{u}_1 - \mathbf{u}_3 - \mathbf{u}_6 + \mathbf{u}_8 \text{ for translation in the } x \text{ direction,}$$

$$\mathbf{f}_{20} = \mathbf{f}_{15} + \mathbf{f}_{18} = \mathbf{u}_1 - \mathbf{u}_3 + \mathbf{u}_6 - \mathbf{u}_8 \text{ for bending in the } x \text{ direction.}$$

Similarly from $\mathbf{f}_{16}$ and $\mathbf{f}_{17}$, we obtain

$$\mathbf{f}_{21} = \mathbf{f}_{16} + \mathbf{f}_{17} = \mathbf{u}_2 - \mathbf{u}_4 + \mathbf{u}_5 - \mathbf{u}_7 \text{ for translation in the } y \text{ direction,}$$

$$\mathbf{f}_{22} = \mathbf{f}_{16} - \mathbf{f}_{17} = \mathbf{u}_2 - \mathbf{u}_4 - \mathbf{u}_5 + \mathbf{u}_7 \text{ for bending in the } y \text{ direction.}$$

Plots of the bases for the various vibrational modes appear on the next page in figure 13.12.

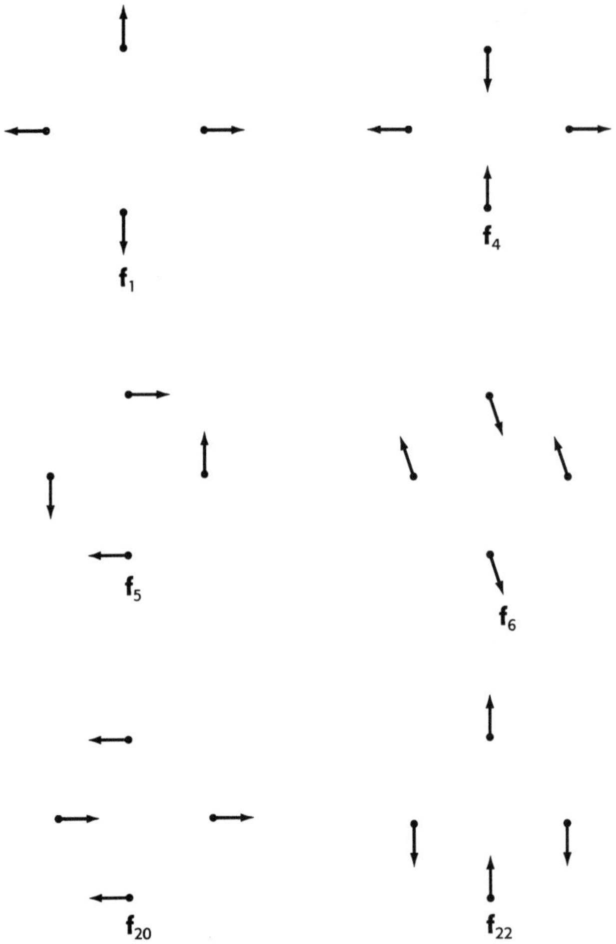

FIGURE 13.12 Arrays of vectors that serve as bases for the various vibrational modes of the square molecule.

Questions

13.1 How are the symmetries in a system expressed?
13.2 What may the bases be for a symmetry discussion?
13.3 How are these bases related?
13.4 Define the term group.
13.5 Why do the relationships among a complete set of bases constitute a group?
13.6 Describe the primary symmetry operations.
13.7 Show how the primary geometric operations are represented by matrices.
13.8 How may one identify the group to which a symmetric molecule belongs?
13.9 Define symmetry number.
13.10 Show how permutations may be represented by matrices.
13.11 How are symmetry operations classified?
13.12 What is a similarity transformation?
13.13 Define the character vector.
13.14 How many independent character vectors does a group possess?
13.15 What is a primitive symmetry species?
13.16 How is a mode of motion of a molecule defined?
13.17 What characterizes the ground state and each even excited state of a mode of motion?
13.18 What characterizes each odd excited state of a mode of motion?
13.19 What is the symmetry of an excitation by (a) one step, (b) two steps, in the state ladder?

13.20 Justify the selection rules for a mode of motion.
13.21 What is the orthogonality condition?
13.22 Show how primitive symmetry species for a molecule are generated.

Problems

13.1 Construct the matrices for rotating a system by 60° and by 120° about the z axis. Show that squaring the C_6 matrix yields the C_3 matrix.

13.2 To what groups do the symmetry operations for (a) H_2O, (b) NH_3, (c) $H_2C = C = CH_2$, (d) C_6H_6, (e) CCl_4, (f) SF_6 belong?

13.3 Construct the permutation matrices generated by rotating a benzene molecule by 60° and by 120° about its C_6 axis.

13.4 Subject figure 13.13 to the I, C_4, C_2, σ_v, and σ_d operations and determine how the displacement vectors transform.

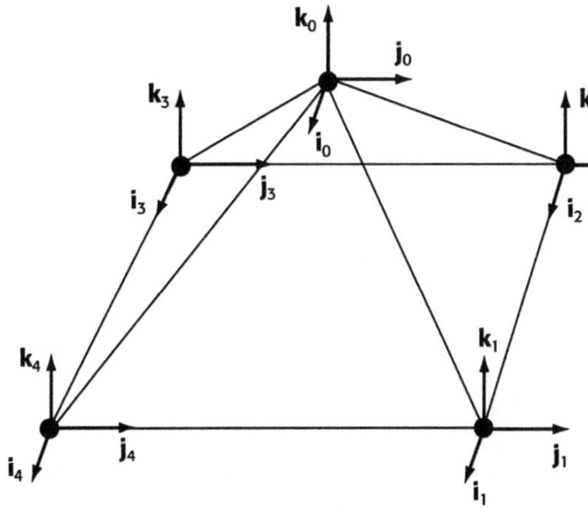

FIGURE 13.13 Orthogonal unit displacement vectors for C_{4v} AB_4.

13.5 With the results from problem 13.4, construct the components of the character vector based on the Cartesian atomic displacements in square pyramidal AB_4.

13.6 With the character table for the C_{4v} group determine the composition of the character vector obtained in problem 13.5. Then subtract out the characters for translation and rotation to obtain those for vibration. Note which of these correspond to active fundamentals in the infrared spectrum.

13.7 Subject figure 13.14 to the I, C_4, C_2, σ_v, σ_d and construct the corresponding representation matrices.

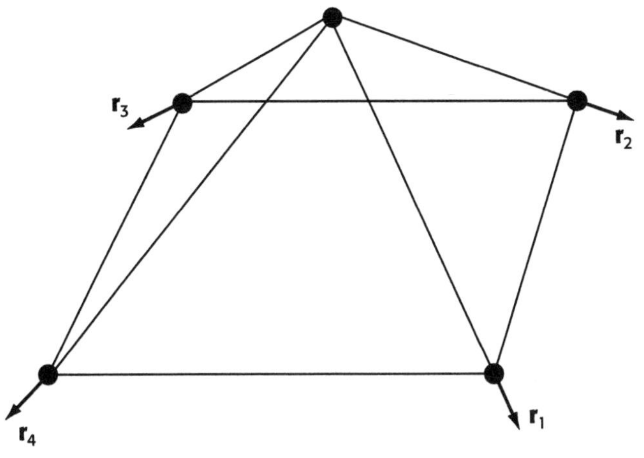

FIGURE 13.14 Stretching displacement vectors for the bonds in C_{4v} AB_4.

13.8 With the results from problem 13.7, construct the components of the character vector based on the bond stretching vectors. Then determine the primitive symmetry species present.

13.9 Subject figure 13.15 to the I, C_4, C_2, σ_v, σ_d operations and construct the corresponding representation matrices.

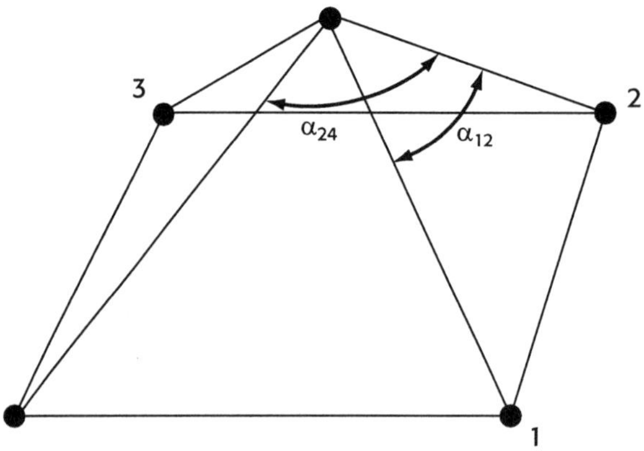

FIGURE 13.15 Representative bond angles in the square pyramidal AB_4 molecule.

13.10 With the results from problem 13.9, construct the components of the character vector based on the angular motions. Then determine the primitive symmetry species present.

13.11 Which of the IR active fundamentals for $\mathbf{C}_{4v}$ AB_4 are principally (a) bond stretching, (b) bond bending?

13.12 What form does equation (13.17) assume when the system is rotated by angle ϕ about the y axis?

13.13 Construct the matrix for subjecting the Cartesian coordinates of a point to operation S_8.

13.14 To what groups do the symmetry operations for functions (a) $x\ R(r)$, (b) $xy\ R(r)$, (c) $xyz\ R(r)$, (d) $(3z^2 - r)\ R(r)$, (e) $(x^2 - y^2)\ R(r)$ belong?

13.15 Subject figure 13.16 to an operation from each class of the $\mathbf{D}_{4h}$ group. Determine the numbers on the principal diagonal of each transformation matrix. Then add these to get each component of the corresponding character vector.

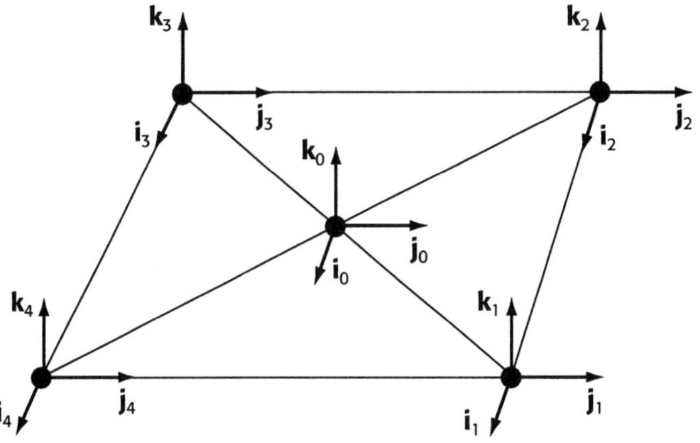

FIGURE 13.16 Orthogonal unit displacement vectors for D_{4h} AB_4.

13.16 With the character table for the $\mathbf{D}_{4h}$ group, determine the composition of the vector obtained in problem 13.15. Then subtract out the characters for translation and rotation to obtain the vibrational species. Note which correspond to active fundamentals in the infrared spectrum.

13.17 Subject figure 13.17 to the operations of the $\mathbf{D}_{4h}$ group. Determine the numbers on the principal diagonals of the transformation matrices. Add these to get the components of the corresponding character vector.

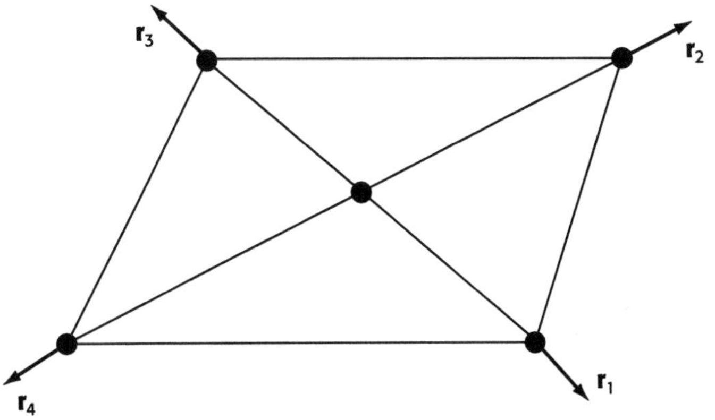

FIGURE 13.17 Londstretching displacement vectors for D_{4h} AB_4.

13.18 With the character table for the $\mathbf{D}_{4h}$ group, determine the primitive symmetry species for bond stretching from the problem 13.17 results. Which are active in the infrared?

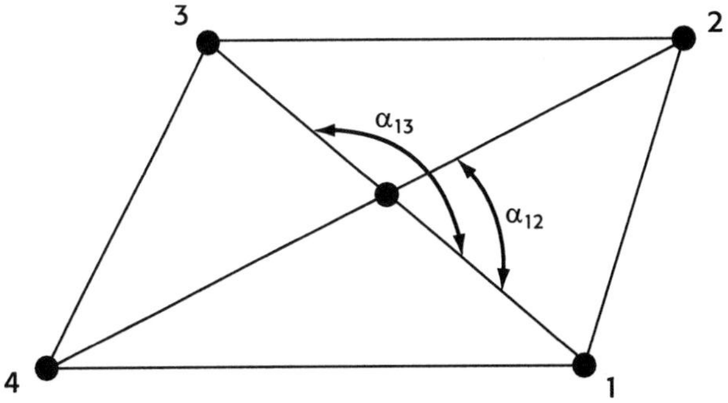

FIGURE 13.18 Representative bond angles in the square planar AB_4 molecule.

13.19 While bond angles α_{12}, α_{23}, α_{34}, α_{41} of figure 13.18 act in the plane of the molecule, subject them to the operations of the $\mathbf{D}_{4h}$ group. Determine the numbers on the principal diagonal of each transformation matrix. Then add these to get each component of the corresponding character vector.

13.20 With the character table for the $\mathbf{D}_{4h}$ group, determine the primitive symmetry species for the in-plane bond bending using the problem 13.19 results. Which are active in the infrared?

13.21 While bond angles α_{13}, α_{24} of figure 13.18 act perpendicular to the plane of the molecule, subject them to the operations of the $\mathbf{D}_{4h}$ group. Determine the numbers on the principal diagonal of each transformation matrix. Then add these to get each component of the corresponding character vector.

13.22 With the character table for the $\mathbf{D}_{4h}$ group, determine the primitive symmetry species for the out-of-plane bond bending using the results from problem 13.21. Which are active in the infrared?

References

Books

Cotton, F. A.: 1990, *Chemical Applications of Group Theory*, 3rd edn, John Wiley & Sons, Inc., New York, pp. 1-295.

Cotton presents group theory simply and illustrates its use by applications to various molecules and complexes. As in my section 3.10, the fundamental orthogonality theorem is stated without proof or much justification.

Duffey, G. H.: 1992, *Applied Group Theory for Physicists and Chemists*, Prentice-Hall, Englewood Cliffs, NJ, pp. 1-105, 161-189.

In the first chapter, the structures of key groups are described with Cayley diagrams. In chapter 2, the basis for character tables is presented algebraicly. Symmetry adapted vibrations and orbitals are constructed using the character tables in chapters 3 and 6.

Hargittai, I., and Hargittai, M.: 1987, *Symmetry through the Eyes of a Chemist*, VCH Publishers, New York, pp. 1-444.

This is a very interesting introductory text. The entire field of chemistry is surveyed from the point of view of symmetry. Considerable group theory is covered in slow stages.

Articles

Baraldi, I., and Carnevali, A.: 1993, "The Relation between Generator Theory and the Theory of Matrix Representations in Symmetry Point Groups of Finite Order," *J. Chem. Educ.* **70**, 964-967.

Baraldi, I., and Vanossi, D.: 1997, "On the Character Tables of Finite Point Groups," *J. Chem. Educ.* **74**, 806-809.

Burrows, E. L., and Clark, M. J.: 1974, "Pictures of Point Groups," *J. Chem. Educ.* **51**, 87-91.

Carter, R. L.: 1991, "The Tabular Method for Reducing Representations," *J. Chem. Educ.* **68**, 373-374.

Condren, S. M.: 1994, "Group Theory Calculations of Molecular Vibrations Using Spreadsheets," *J. Chem. Educ.* **71**, 486-488.

Contreras-Ortega, C., Vera, L., and Quiroz-Reyes, E.: 1991, "How Great is the Great Orthogonality Theorem?" *J. Chem. Educ.* **68**, 200-202.

Contreras-Ortega, C., Vera, L., and Quiroz-Reyes, E.: 1995, "More than One Character Table?" *J. Chem. Educ.* **72**, 821-822.

Ermer, O.: 1990, "Independent Coordinates of Molecular Structures and Group Theory," *J. Chem. Educ.* **67**, 209-210.

Faltynek, R. A.: 1995, "Group Theory in Advanced Inorganic Chemistry," *J. Chem. Educ.* **72**, 20-24.

Fujita, S.: 1986, "Point Groups Based on Methane and Adamantane (T_d) Skeletons," *J. Chem. Educ.* **63**, 744-746.

Gelessus, A., Thiel, W., and Weber, W.: 1995, "Multipoles and Symmetry," *J. Chem. Educ.* **72**, 505-508.

Gutman, I., and Potgieter, J. H.: 1994, "Isomers of Benzene," *J. Chem. Educ.* **71**, 222-224.

Huang, S. O., and Wang, P. G.: 1990, "Further Comment on Infinite Point Groups," *J. Chem. Educ.* **67**, 34-35.

Kettle, S. F. A.: 1989, "How Can xy and $x^2 - y^2$ Have Unique Symmetries if x and y are Not Uniquely Defined?" *J. Chem. Educ.* **66**, 818- 820.

Macomber, R. S.: 1997, "A Unifying Approach to Absorption Spectroscopy at the Undergraduate Level," *J. Chem. Educ.* **74**, 65-67.

McNaught, I. J.: 1997, "Reducible Representations for Linear Molecules," *J. Chem. Educ.* **74**, 809-811.

Smith, D. W.: 1996, "Simple Treatment of the Symmetry Labels for the d-d States of Octahedral Complexes," *J. Chem. Educ.* **73**, 504-507.

Tel, L. M., and Perez-Romero, E.: 1988, "Density of Elements in Continuous Point Groups," *J. Chem. Educ.* **65**, 585- 587.

Thomas, C. H.: 1974, "The Use of Group Theory to Determine Molecular Geometry from IR Spectra," *J. Chem. Educ.* **51**, 91-93.

Verkade, J. G.: 1987, "A Novel Pictorial Approach to Teaching Nfolecular Motions in Polyatomic Molecules," *J. Chem. Educ.* **64**, 411- 416.

14

States of Molecular Electrons

14.1 Introduction

THE CHEMICAL PROPERTIES OF ELEMENTS correlate with their low lying electron configurations. The inertness of the rare gases He, Ne, Ar, Xe, Rn implies that their electron configurations are particularly stable. So when two elements react to form an ionic compound, the resulting electron configurations tend to be those of nearby inert gases. When atoms combine to form covalent bonds, the shared electrons tend to complete the pertinent subshells to also give inert gas configurations.

In covalent structures, we have multielectron multinuclear systems. These are generally more complicated than the single atom or ion systems considered in section 12.15.

Only for the hydrogen molecule ion H_2^+ can a person accurately solve the Schrödinger equation. For molecules in general, approximation methods must be employed. These may be based on the variation theorem.

An orbital encompassing more than one atomic core is called a molecular orbital. Such an orbital may be constructed by superposing atomic orbitals based on the pertinent atomic cores. The extra freedom gained by each valence electron reduces its mean kinetic energy and lowers its mean potential energy. This change implies less pressure on the atomic orbitals. As a result, they shrink. Indeed, much of the bonding energy comes from this shrinkage. Furthermore, some distortion of each atomic orbital, consistent with the overall symmetry, occurs.

Any symmetry or near-symmetry that a molecule possesses may be exploited in constructing its molecular orbitals. For, each disjoint orbital must belong to a primitive symmetry species of the pertinent group and subgroups.

14.2 The Hydrogen Molecule Ion Model

Just as the hydrogen atom serves as a reference structure for multielectron atoms, the hydrogen molecule ion serves as a reference structure for localized chemical bonds.

The system H_2^+ consists of two protons bound together by a single electron. But from the ratio of the electron mass to the reduced mass of the nuclei, we expect that during the traversal of a normal electron orbit, the internuclear distance would change by roughly (1/1000)th the orbit's length. For an average orbit, this movement would be less than 0.01 Å. So in calculating the electronic energy, one may consider the nuclei to be fixed; the

Born - Oppenheimer approximation should be valid. And for each internuclear separation, the energy equals the sum of that arising from the interproton repulsion, the electron - proton attraction, and the electron motion.

For our discussion, we label the two protons A and B. The distance between them is labeled R, while the distance of the electron from A is r_A, the distance from B, r_B. The internuclear repulsion energy is

$$V_{nuc} = \frac{e^2}{4\pi\varepsilon R}, \qquad [14.1]$$

while the electronic potential energy is

$$V = -\frac{e^2}{4\pi\varepsilon}\left(\frac{1}{r_A} + \frac{1}{r_B}\right). \qquad [14.2]$$

The Schrödinger equation for the electron motion becomes

$$H\psi = -\frac{\hbar^2}{2m}\nabla^2\psi + V\psi = E\psi, \qquad [14.3]$$

where m is the electron mass and E the electronic energy, potential plus kinetic. On adding V_{nuc} to E, one obtains total energy E_{tot} for the given R. Results for various bond lengths are sketched in figure 14.1. The total energy E_{tot} is employed as the potential energy for vibration of the molecule.

Fitting the symmetry of the potential and kinetic energies are the elliptic coordinates

$$\mu = \frac{r_A + r_B}{R}, \quad \nu = \frac{r_A - r_B}{R}, \quad \phi, \qquad [14.4]$$

where ϕ is an azimuthal angle measured around the axis passing through the nuclei. Indeed, these variables are separable in equation (14.3), One obtains three first order differential equations.

Integration of these yields functions $M(\mu)$, $N(\nu)$, and exp $(i\,\lambda\,\phi)$ fitting the boundary conditions. These functions combine to yield

$$\psi = M(\mu)N(\nu)e^{i\lambda\phi} \quad \text{with} \quad \lambda = \text{an integer}. \qquad [14.5]$$

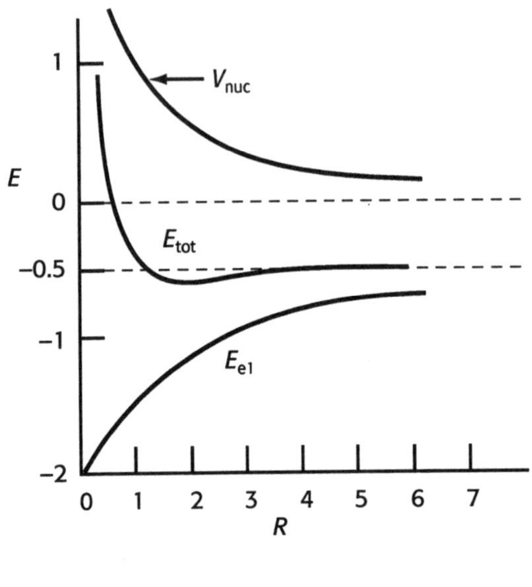

FIGURE 14.1 The internuclear repulsion energy V_{nuc}, electronic energy E_{el}, and total energy E_{tot} of H_2^+ in its ground state. Here the unit of energy is minus twice the H atom ground state energy, the hartree; the unit of distance is the Bohr radius of the H atom.

Expressions M and N depend on λ^2, not on the sign of λ. So the two exponential factors for a given nonzero λ^2 can be replaced with

$$\sqrt{2}\cos|\lambda|\phi \quad \text{and} \quad \sqrt{2}\sin|\lambda|\phi. \qquad [14.6]$$

In the suitable solutions, M has elliptical nodal surfaces, N has hyberbolic nodal surfaces. The functions themselves are quite complicated.

In the symbol for a state, number $|\lambda|$ is represented by a Greek letter as follows:

$$\begin{aligned} |\lambda| &= 0, 1, 2, 3, \ldots \\ \text{Orbital} &= \sigma, \pi, \delta, \phi, \ldots \end{aligned} \qquad [14.7]$$

One of the symmetry operations for the molecule is reflection through the center of mass, inversion i. Since $i^2 = I$, the state function is either left unaltered or changed in sign by the operation i. An orbital that is not altered by the inversion is labeled with the subscript g (for gerade). One that is changed in sign by the inversion is labeled with the subscript u (for ungerade). These symbols are said to specify the *parity* of the orbital.

Substituting two identical atomic cores for the protons yields a similar system. However, explicitly introducing the interelectronic correlations breaks the symmetry that allowed the μ dependence to separate from the v dependence. Adding a second valence electron paired with the first extends this process. But still, the $|\lambda|$ and parity are valid quantum numbers.

14.3 Correlations with Atomic Structures

The disjoint molecular orbitals in a diatomic system can be related to orbitals in the original separated atoms and to orbitals in the pertinent united atom structure.

Consider a homonuclear diatomic system. Let us first consider the atoms well separated, with a common z axis passing through the two nuclei from left to right. Let us superpose equivalent valence orbitals so the result has a definite allowed $|\lambda|$ and parity. The resulting functions are symmetry- adapted molecular orbitals; a person may call them symmetry orbitals (SO) for short.

Then we consider bringing the nuclei close enough so that the atomic orbitals interact strongly. During this process, the symmetry is conserved; both $|\lambda|$ and parity are fixed. Sketches indicating how s and p orbitals may combine appear in figure 14.2

One may imagine that the nuclei are brought closer and closer together until they coalesce. Again, both $|\lambda|$ and parity are fixed. One would end up with united atom structures oriented as figure 14.3 shows.

In correlating the separated atom symmetry orbitals with the united atom atomic orbitals, a person requires them to have the same symmetry (same $|\lambda|$ and parity). To resolve the remaining ambiguities, we introduce the *noncrossing rule*: As a parameter is changed continuously, the energy levels for orbitals with the same symmetry do not cross.

So we match the lowest energy united atom orbital of a given symmetry with the lowest energy separated atom orbital of the same symmetry, and so on up the ladder. In this way, we obtain the correlations sketched in figure 14.4.

In the center of this diagram, the molecular orbitals are numbered according to their energy order within each symmetry type (as in $2\sigma_g$). On the right, the symmetry type is followed by the symbol for the atomic orbitals used (as in $\sigma_g 2s$). On the left, the united atom orbital symbol precedes the symmetry type (as in $2s\sigma_g$).

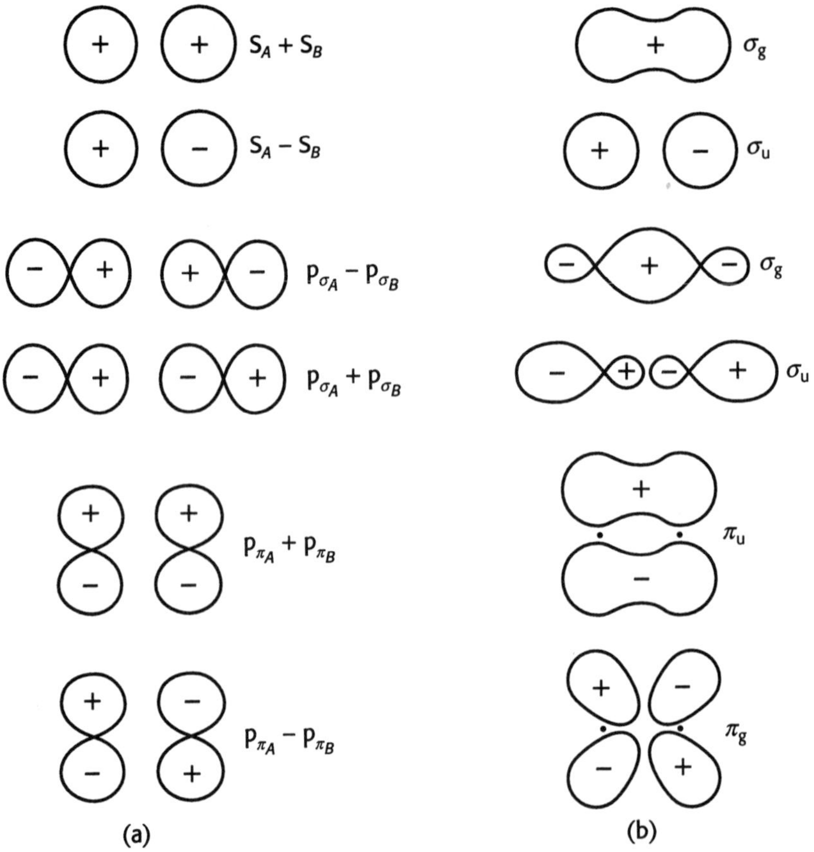

FIGURE 14.2 Symmetry orbitals constructed from s and p atomic orbitals. In column (a) the two atoms are effectively separated; in column (b) they interact strongly.

14.4 Bonding between Two Equivalent Atoms

Valence orbitals arranged appropriately on two equivalent atoms can superpose to form a covalent bond. A single bond lying on the axis employs one or two electrons in a σ_g molecular orbital. Less stable would be one or two electrons in a π_u molecular orbital, or a δ_g molecular orbital. The σ_u, π_g, and δ_u states would be relatively unstable.

Consider two equivalent normalized valence orbitals, ψ_A and ψ_B, on two equivalent atoms, A and B. Let us suppose that each electron moves in the average field of all other electrons as well as in the field of the two nuclei. By symmetry, each valence orbital contributes equally to the molecular orbital. So for it, we may write

$$\psi = \psi_A \pm \psi_B. \quad [14.8]$$

Note that here we have not normalized the molecular orbital.

The Schrödinger equation has the form

$$H\psi = E\psi. \quad [14.9]$$

Multiplying it by ψ^*,

$$\psi^* H\psi = E\psi^*\psi, \quad [14.10]$$

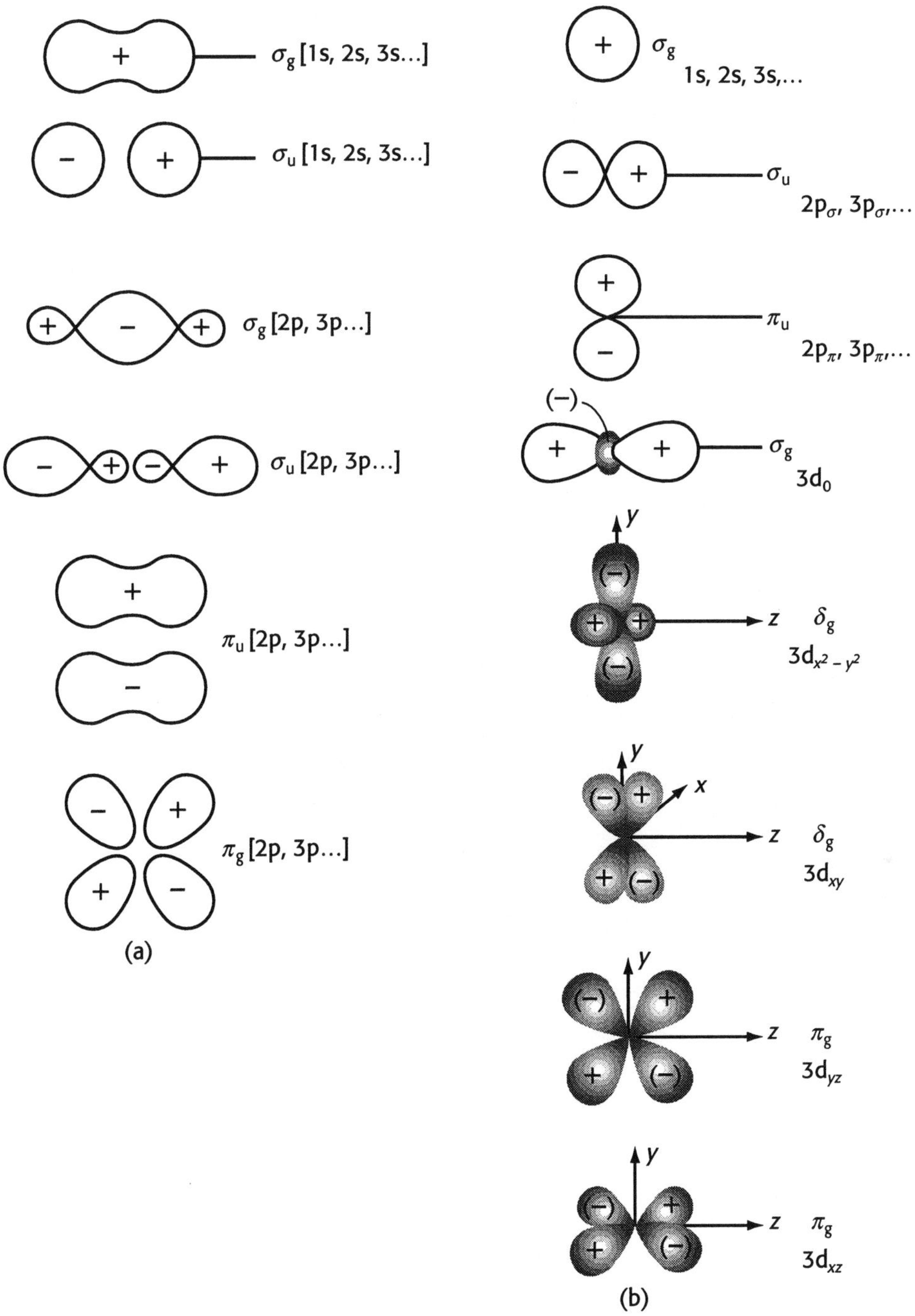

FIGURE 14.3 (a) Molecular symmetry orbitals and (b) united atom atomic orbitals similarly oriented.

integrating over all space, and solving for energy E leads to

$$E = \frac{\int_{\mathrm{R}} \psi^* H \psi \, \mathrm{d}^3\mathbf{r}}{\int_{\mathrm{R}} \psi^* \psi \, \mathrm{d}^3\mathbf{r}}. \qquad [14.11]$$

If ψ had been normalized, the denominator would equal 1.

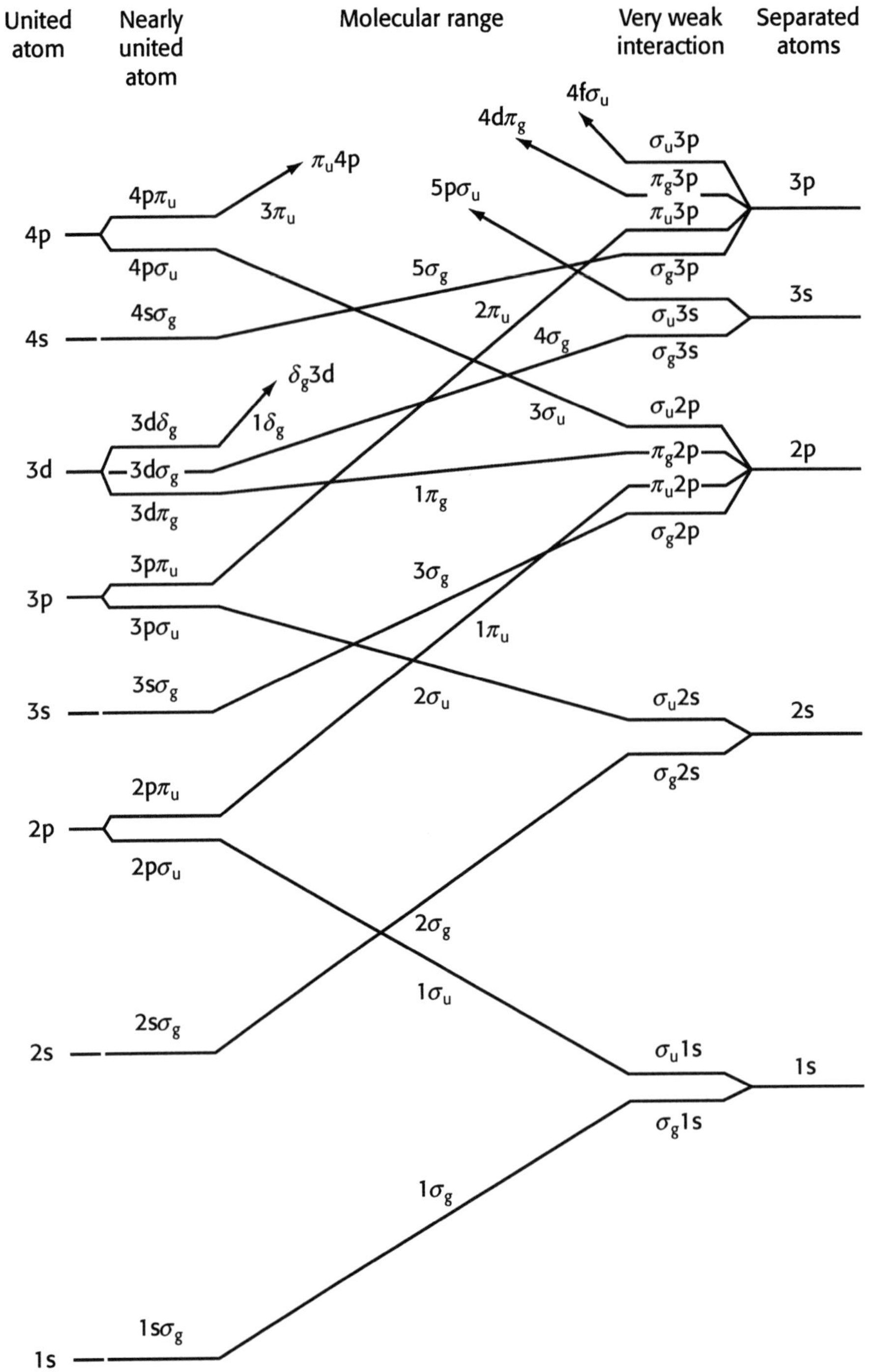

FIGURE 14.4 Approximate variation of energy levels with internuclear separation for homonuclear diatomic structures.

Introduce superposition (14.8) into formula (14.11)

$$E = \frac{\int_R (\psi_A \pm \psi_B)^* H (\psi_A \pm \psi_B)\, d^3\mathbf{r}}{\int_R (\psi_A \pm \psi_B)^* (\psi_A \pm \psi_B)\, d^3\mathbf{r}}. \quad [14.12]$$

By definition and symmetry, we set

$$h_{AA} = \int_R \psi_A^* H \psi_A \, d^3\mathbf{r} = \int_R \psi_B^* H \psi_B \, d^3\mathbf{r}, \quad [14.13]$$

$$h_{AB} = \int_R \psi_A^* H \psi_B \, d^3\mathbf{r} = \int_R \psi_B^* H \psi_A \, d^3\mathbf{r}, \quad [14.14]$$

$$1 == \int_R \psi_A^* \psi_A \, d^3\mathbf{r} = \int_R \psi_B^* \psi_B \, d^3\mathbf{r}, \quad [14.15]$$

$$s_{AB} = \int_R \psi_A^* \psi_B \, d^3\mathbf{r} = \int_R \psi_B^* \psi_A \, d^3\mathbf{r}. \quad [14.16]$$

Equation (14.12) now reduces to

$$E = \frac{2h_{AA} \pm 2h_{AB}}{2 \pm 2s_{AB}} = \frac{h_{AA} \pm h_{AB}}{1 \pm s_{AB}}. \quad [14.17]$$

Integral h_{AA} can be interpreted as the energy of an electron when it occupies σ_A alone, But here B carries 1 unit of charge when the electron is on A. The resulting field causes h_{AA} to vary strongly with the distance R of B from A. A similar remark applies to integral h_{BB}.

Integral s_{AB} varies with the internuclear distance R and with the nature and orientation of the atomic orbitals. Integral h_{AB} varies with the overlap s_{AB}. For s and p orbitals, approximate proportionality exists. But for d orbitals, h_{AB} may be large when s_{AB} is fairly small.

Parameter h_{AA} is called the *Coulomb integral* α. Parameter h_{AB} is called the *resonance integral* or *bond integral* β. Parameter s_{AB} is called the *overlap integral* S. Equation (14.17) can be rewritten as

$$E = \frac{\alpha \pm \beta}{1 \pm S}. \quad [14.18]$$

Since α and β are negative, bonding occurs when the positive sign is chosen.

To simplify calculations, one may employ atomic units (a.u.) in which

$$\hbar = 1, \quad m_e = 1, \quad \frac{e^2}{4\pi\varepsilon} = 1. \quad [14.19]$$

Then formula (12.109) yields

$$a = 1 \quad [14.20]$$

when $Z = 1$. This unit of distance is called the *bohr*. Also, formula (12.111) yields

$$E = -\frac{1}{2n^2} \quad [14.21]$$

for the bound states of H. The corresponding unit of energy is called the *hartree*.

If the hydrogen molecule ion is constructed with ψ_A and ψ_B standard 1s orbitals, the results plotted in figure 14.5 are obtained, when $R = 2$ bohrs. Actually, the atomic orbitals need to be distorted and shrunk somewhat to allow for the nonspherical field and the extra freedom. These effects can be introduced by adding contributions from higher

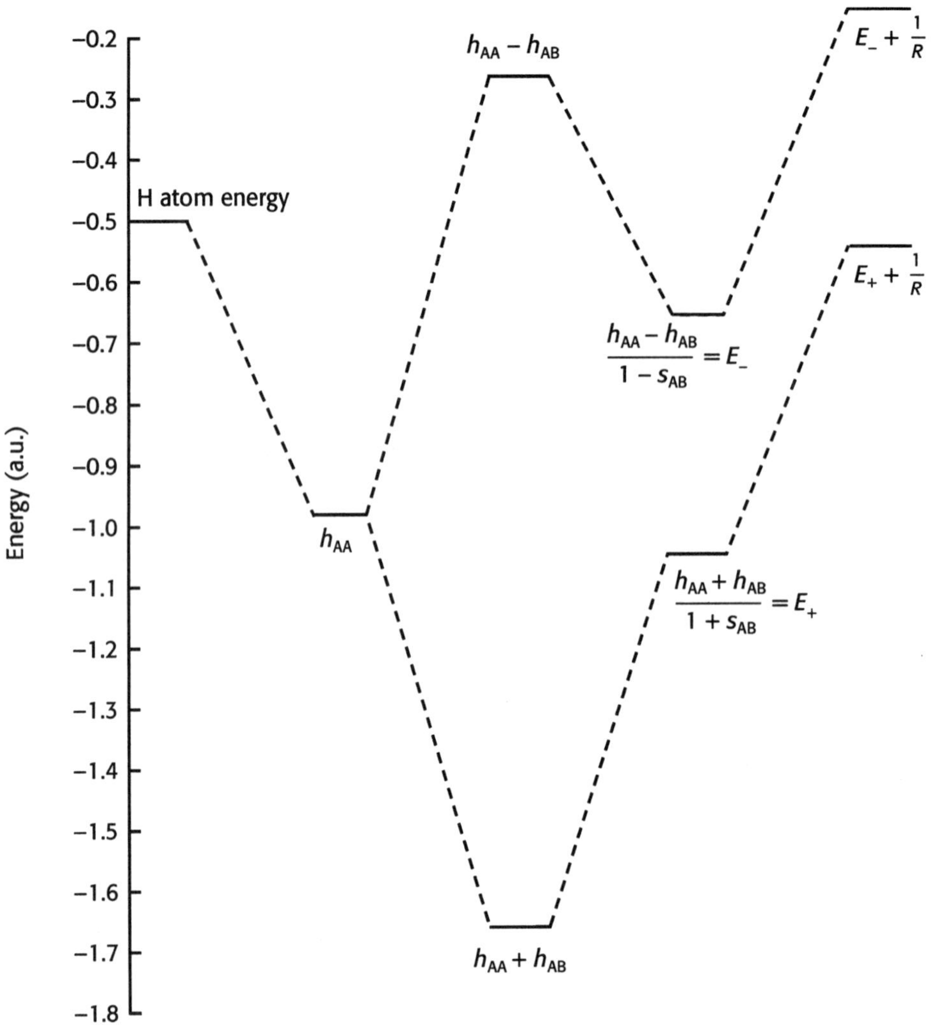

FIGURE 14.5 Approximate contributions to the energy of H_2^+ for R = 2 bohrs.

standard H atom orbitals. The amounts needed and their phases can be determined with the variation theorem to be established later.

In the hydrogen molecule, there are two valence electrons. To the approximation that either electron moves in the field of the nuclei and the average field of the other electron, result (14.17) still applies to each. But now the second atom carries no net charge. As a consequence, h_{AA} and h_{BB} do not vary strongly with the internuclear distance R. Instead, one may consider them to be roughly constant and equal to α. The effect of interelectron correlation can be allowed for by a configuration interaction calculation. Similar considerations apply to localized chemical bonds between any two like atoms.

14.5 Symmetric Three-Center Bonds

In a simple covalent bond, two orbitals from neighboring atoms combine as equation (14.8) implies. Ordinarily, the resulting molecular orbital holds two paired electrons. But in many electron deficient structures, orbitals from three neighboring atoms combine to form three-center bonds.

Such bonds are common in boron hydride structures. Thus in diborane, the 1s orbital of a bridge hydrogen overlaps a tetrahedral orbital from each boron that it links. The other two tetrahedral orbitals on each boron hold the terminal hydrogens. See figure 14.6.

Symmetric three center bonds are found in many cage structures. As examples, we cite $Ta_6Cl_{12}^{+a}$, $Ta_6Br_{12}^{+a}$, $Nb_6Cl_{12}^{+a}$, where each metal atom is surrounded by valence orbitals with tetragonal antiprism symmetry. Four of these point toward the neighboring halogen atoms and four toward the middles of the neighboring faces. About the middle of each face, three atomic orbitals from the three adjacent vertex atoms meet. See figure 14.7.

Consider three equivalent valence orbitals, ψ_1, ψ_2, ψ_3, from three equivalent atoms arranged about a threefold axis. Under operations of the $\mathbf{C}_3$ group, atomic orbital ψ_1 then behaves as table 14.1 shows.

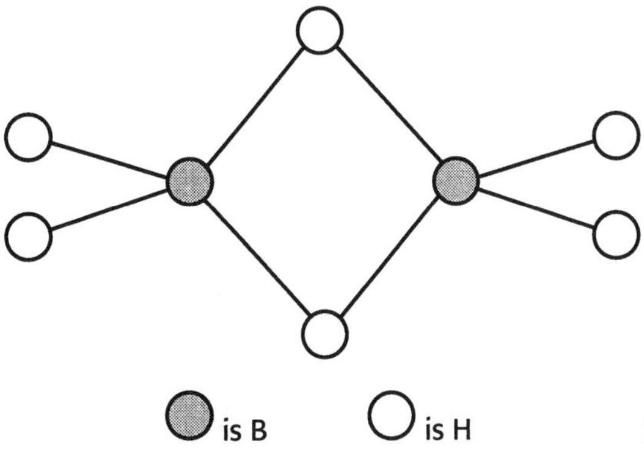

FIGURE 14.6 The B_2H_6 structure.

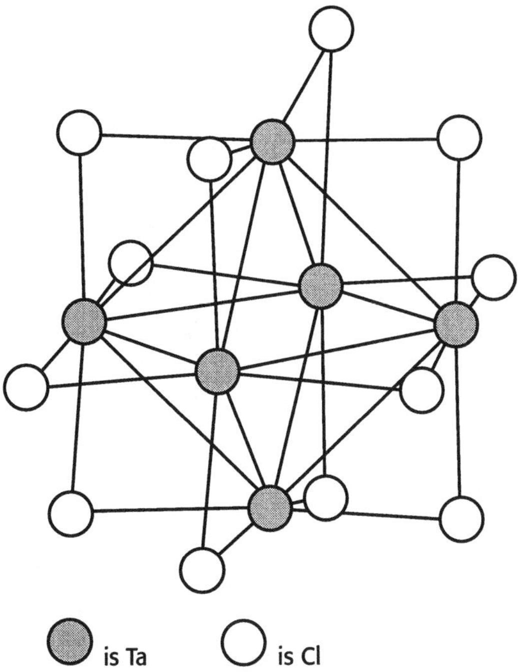

FIGURE 14.7 The $Ta_6Cl_{12}^{+a}$ structure.

But the basic character vectors for the $\mathbf{C}_3$ group have the components listed in table 14.2. Applying formula (13.44) to the expressions in tables 14.1 and 14.2 yields the symmetry-adapted molecular orbitals. For species A, we obtain

$$\psi_{\mathrm{I}} = \psi_1 + \psi_2 + \psi_3. \qquad [14.22]$$

Similarly for species C and D, we find that

$$\psi_{\mathrm{II}} = \psi_1 + \omega^2\psi_2 + \omega\psi_3, \qquad [14.23]$$

$$\psi_{\mathrm{III}} = \psi_1 + \omega\psi_2 + \omega^2\psi_3. \qquad [14.24]$$

Note that with the atomic orbitals real, we have

$$\psi_{\mathrm{I}}^* = \psi_{\mathrm{I}}, \quad \psi_{\mathrm{II}}^* = \psi_{\mathrm{III}}, \quad \psi_{\mathrm{III}}^* = \psi_{\mathrm{II}}. \qquad [14.25]$$

The corresponding energy levels are obtained with formula (14.11). We consider the atomic orbitals to be normalized to 1:

$$\int_{\mathrm{R}} \psi_i^* \psi_i \, \mathrm{d}^3\mathbf{r} = 1. \qquad [14.26]$$

In addition, we have the overlap integral,

$$\int_{\mathrm{R}} \psi_i^* \psi_j \, \mathrm{d}^3\mathbf{r} = s_{ij}, \qquad [14.27]$$

the Coulomb integral,

$$\int_{\mathrm{R}} \psi_i^* H \psi_i \, \mathrm{d}^3\mathbf{r} = h_{ii}, \qquad [14.28]$$

and the bond integral,

$$\int_{\mathrm{R}} \psi_i^* H \psi_j \, \mathrm{d}^3\mathbf{r} = h_{ij}. \qquad [14.29]$$

These can be evaluated for specific atomic orbitals.

Multiplication tables are used in determining how these integrals appear in the energy expressions. List the contributions to ψ^* in the left column, the contributions to ψ or $H\psi$ across the top. Take the integral of each combination and list the result in the body of the table, introducing the equalities that the symmetries demand.

Table 14.1 Transformations of ψ_1

I	C_3	C_3^2
ψ_1	ψ_2	ψ_3

Table 14.2 $\mathbf{C}_3$ Primitive Character Vectors

	I	C_3	C_3^2
A	1	1	1
C	1	ω	ω^2
D	1	ω^2	ω

$\omega = \exp(2\pi i/3)$

For the A symmetry species, we construct tables 14.3 and 14.4, Adding the expressions in the bodies of the tables following formula (14.11) leads to

$$E_1 = \frac{3h_{11} + 6h_{12}}{3 + 6s_{12}} = \frac{h_{11} + 2h_{12}}{1 + 2s_{12}}. \quad [14.30]$$

For the C symmetry species, we construct tables 14.5 and 14.6. Summing the contributions to the integrals in (14.11), with $\omega + \omega^2 = -1$, yields

$$E_2 = \frac{3h_{11} - 3h_{12}}{3 - 3s_{12}} = \frac{h_{11} - h_{12}}{1 - s_{12}}. \quad [14.31]$$

For the D symmetry species, one obtains the same result:

$$E_3 = \frac{h_{11} - h_{12}}{1 - s_{12}}. \quad [14.32]$$

Table 14.3 Parts of $\int \psi^* \psi \, d^3\mathbf{r}$ for the A Species

	ψ_1	ψ_2	ψ_3
ψ_1^*	1	s_{12}	s_{12}
ψ_2^*	s_{12}	1	s_{12}
ψ_3^*	s_{12}	s_{12}	1

Table 14.4 Parts of $\int \psi^* H \psi d^3\mathbf{r}$ for the A Species

	$H\psi_1$	$H\psi_2$	$H\psi_3$
ψ_1^*	h_{11}	h_{12}	h_{12}
ψ_2^*	h_{12}	h_{11}	h_{12}
ψ_3^*	h_{12}	h_{12}	h_{11}

Table 14.5 Parts of $\int \psi^* \psi d^3\mathbf{r}$ for C

	ψ_1	$\omega^2\psi_2$	$\omega\psi_3$
ψ_1^*	1	$\omega^2 s_{12}$	ωs_{12}
$\omega\psi_2^*$	ωs_{12}	1	$\omega^2 s_{12}$
$\omega^2\psi_3^*$	$\omega^2 s_{12}$	s_{12}	1

Table 14.6 Parts of $\int \psi^* H \psi d^3\mathbf{r}$ for C

	$H\psi_1$	$\omega^2 H\psi_2$	$\omega H\psi_3$
ψ_1^*	h_{11}	$\omega^2 h_{12}$	ωh_{12}
$\omega\psi_2^*$	ωh_{12}	h_{11}	$\omega^2 h_{12}$
$\omega^2\psi_3^*$	$\omega^2 h_{12}$	ωh_{12}	h_{11}

Parameter h_{11} is often designated the Coulomb integral α; parameter h_{12}, the bond integral β; and parameter s_{12}, the overlap integral S. Then formulas (14.30), (14.31), and (14.32) reduce to

$$E_1 = \frac{\alpha + 2\beta}{1 + 2S}, \qquad [14.33]$$

$$E_2 = E_3 = \frac{\alpha - \beta}{1 - S}. \qquad [14.34]$$

Integrals α and β are negative while integral S is generally positive. Bonding in a three-center system is usually provided by paired electrons in the A symmetry orbital. The A orbital is said to be bonding; the C and D orbitals, antibonding.

In B_2H_6, each boron supplies 3 valence electrons. Two of these are used in bonding the terminal hydrogens. Each of the two hydrogen bridges is a three-center bond. It holds 1 electron from the H atom and 1 electron from the B atoms.

In $Ta_6Cl_{12}{}^{+a}$, each tantalum atom supplies 5 valence electrons. For the six Ta atoms, this amounts to 30 electrons. But each chlorine bridge takes 1 electron. For the twelve bridges, this requires 12 electrons. The three-center bond over each face takes 2 electrons. For the eight faces, this requires 16 electrons. The difference,

$$30 - 12 - 16 = 2, \qquad [14.35]$$

represents the electrons lost in forming the cage from the atoms. Consequently, the formula for the ion is $Ta_6Cl_{12}{}^{+2}$.

Example 14.1

To the approximation that the atomic orbitals and parameter α do not change in the process, what is the energy increment on forming a single symmetric covalent bond?

Applying formula (14.11) to the separated atoms gives

$$E = h_{AA} + h_{BB} = 2\alpha.$$

But formula (14.18) applies to each electron in the final state. We have

$$\Delta E = \frac{2\alpha + 2\beta}{1 + S} - 2\alpha = \frac{2\beta - 2\alpha S}{1 + S}$$

In the approximation that S is small, we find that

$$\Delta E \cong 2\beta.$$

This form explains why β is called the bond integral.

Example 14.2

How much stability is gained in the formation of a three-center bond from a two-center bond and an empty orbital on the third center?

As a rough approximation, employ formulas (14.18) and (14.33) with α and β constant. Then we have

$$\Delta E = \frac{2\alpha + 4\beta}{1 + 2S} - \frac{2\alpha + 2\beta}{1 + S} = \frac{2\beta - 2\alpha S}{(1 + 2S)(1 + S)}.$$

If $S \cong 0$, then we find that

$$\Delta E \cong 2\beta.$$

14.6 Symmetric Six-Center Bonds

When bonding extends over more than two atoms, it is often said to be *delocalized.* The extra freedom gained by the valence electrons in the resulting molecular orbitals reduces the mean electron kinetic energy and accordingly lowers the mean electron potential energy. There is less pressure on the atomic orbitals than if all bonds were two-center bonds. So the atomic orbitals shrink further. Additional distortions may also occur.

A fundamental unit in aromatic molecules is the benzene ring. In benzene itself, we have six CH units arranged in a regular hexagonal ring. The 2s and two of the 2p real carbon orbitals are used to form the three σ bonds extending out from each C atom. These are all localized bonds. Then the real 2p orbitals directed perpendicular to the ring form a delocalized system. Each C atom contributes one electron to this system.

So let us consider six equivalent valence orbitals, $\psi_1, \psi_2, \psi_3, \psi_4, \psi_5, \psi_6$ from six equivalent atoms arranged about a sixfold axis as figure 14.8 indicates. Under operations of the $\mathbf{C}_6$ group, atomic orbital ψ_1 behaves as table 14.7 shows.

The basic character vectors of the $\mathbf{C}_6$ group have the components listed in table 14.8, Applying formula (13.44) to the expressions in tables 14.7 and 14.8 yields the symmetry-adapted orbitals. For species A, we obtain

$$\psi_{\mathrm{I}} = \psi_1 + \psi_2 + \psi_3 + \psi_4 + \psi_5 + \psi_6. \qquad [14.36]$$

Similarly for species C_1 and D_1, we find that

$$\psi_{\mathrm{II}} = \psi_1 - \omega\psi_2 + \omega^2\psi_3 - \psi_4 + \omega\psi_5 - \omega^2\psi_6, \qquad [14.37]$$

$$\psi_{\mathrm{III}} = \psi_1 - \omega^2\psi_2 + \omega\psi_3 - \psi_4 + \omega^2\psi_5 - \omega\psi_6. \qquad [14.38]$$

For species C_2 and D_2, we obtain

$$\psi_{\mathrm{IV}} = \psi_1 + \omega^2\psi_2 + \omega\psi_3 + \psi_4 + \omega^2\psi_5 + \omega\psi_6, \qquad [14.39]$$

$$\psi_{\mathrm{V}} = \psi_1 + \omega\psi_2 + \omega^2\psi_3 + \psi_4 + \omega\psi_5 + \omega^2\psi_6. \qquad [14.40]$$

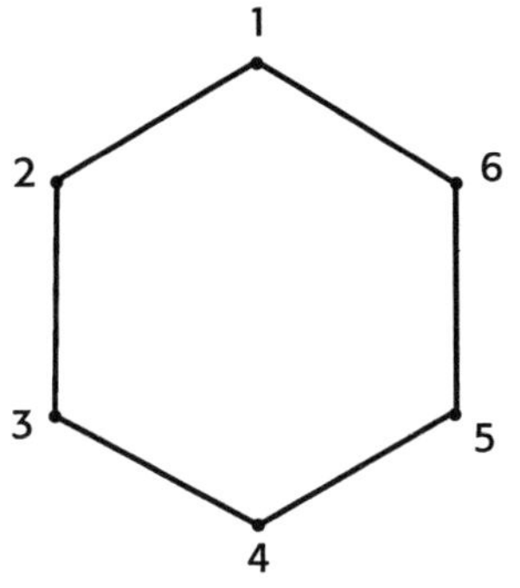

FIGURE 14.8 Numbering of atoms placed at the vertices of a regular hexagon.

Table 14.7 Effect of Each Operation of the $\mathbf{C}_6$ Group on Atomic Orbital ψ_1

I	C_6	C_3	C_2	C_3^2	C_6^5
ψ_1	ψ_2	ψ_3	ψ_4	ψ_5	ψ_6

Table 14.8 Components of the Primitive Character Vectors of the $\mathbf{C}_6$ Group

	I	C_6	C_3	C_2	C_3^2	C_6^5
A	1	1	1	1	1	1
B	1	-1	1	-1	1	-1
C_1	1	$-\omega^2$	ω	-1	ω^2	$-\omega$
D_1	1	$-\omega$	ω^2	-1	ω	$-\omega^2$
C_2	1	ω	ω^2	1	ω	ω^2
D_2	1	ω^2	ω	1	ω^2	ω

$\omega = \exp(2\pi i/3)$

And for species B, we find that

$$\psi_{\mathrm{VI}} = \psi_1 - \psi_2 + \psi_3 - \psi_4 + \psi_5 - \psi_6. \quad [14.41]$$

As before we set

$$\int_{\mathrm{R}} \psi_i^* \psi_j \, \mathrm{d}^3\mathbf{r} = s_{ij} \quad [14.42]$$

and

$$\int_{\mathrm{R}} \psi_i^* H \psi_j \, \mathrm{d}^3\mathbf{r} = h_{ij}. \quad [14.43]$$

Now, the atomic orbitals are equivalent, arranged with sixfold symmetry, as table 14.7 indicates. So we have

$$s_{i\,i+1} = s_{12}, \qquad h_{i\,i+1} = h_{12}, \quad [14.44]$$

$$s_{i\,i+2} = s_{13}, \qquad h_{i\,i+2} = h_{13}, \quad [14.45]$$

$$s_{i\,i+3} = s_{14}, \qquad h_{i\,i+3} = h_{14}, \quad [14.46]$$

and

$$s_{ij} = s_{ji} \qquad h_{ij} = h_{ji}. \quad [14.47]$$

Thus, there is an ortho, a meta, and a para overlap integral, an ortho, a meta, and a para bond integral.

To determine how these integrals appear in the energy expression, we construct multiplication tables. The contributions to ψ^* are placed in the left column, the contributions to ψ or to $H\psi$ across the top. The integral of each combination is taken, equalities (14.47), (14.44) - (14.46) are introduced, and the result is placed where the pertinent row and column meet in the body of the table.

For the A symmetry species, we construct tables 14.9 and 14.10. For the numerator of expression (14.11), we sum the entries in table 14.10; for the denominator, we sum the entries in table 14.9, Dividing both numerator and denominator by six then yields

$$E_1 = \frac{h_{11} + 2h_{12} + 2h_{13} + h_{14}}{1 + 2s_{12} + 2s_{13} + s_{14}}. \quad [14.48]$$

For the C_1 symmetry species, tables 14.11 and 14.12 are constructed. Summing the expressions in the bodies of the tables, as directed by formula (14.11), and reducing leads to

$$E_2 = \frac{h_{11} + h_{12} - h_{13} - h_{14}}{1 + s_{12} - s_{13} - s_{14}}. \quad [14.49]$$

Table 14.9 Contributions to $\int \psi^* \psi \, d^3\mathbf{r}$ for A Species

	ψ_1	ψ_2	ψ_3	ψ_4	ψ_5	ψ_6
ψ_1^*	1	s_{12}	s_{13}	s_{14}	s_{13}	s_{12}
ψ_2^*	s_{12}	1	s_{12}	s_{13}	s_{14}	s_{13}
ψ_3^*	s_{13}	s_{12}	1	s_{12}	s_{13}	s_{14}
ψ_4^*	s_{14}	s_{13}	s_{12}	1	s_{12}	s_{13}
ψ_5^*	s_{13}	s_{14}	s_{13}	s_{12}	1	s_{12}
ψ_6^*	s_{12}	s_{13}	s_{14}	s_{13}	s_{12}	1

Table 14.10 Contributions to $\int \psi^* H \psi \, d^3\mathbf{r}$ for$A$ Species

	$H\psi_1$	$H\psi_2$	$H\psi_3$	$H\psi_4$	$H\psi_5$	$H\psi_6$
ψ_1^*	h_{11}	h_{12}	h_{13}	h_{14}	h_{13}	h_{12}
ψ_2^*	h_{12}	h_{11}	h_{12}	h_{13}	h_{14}	h_{13}
ψ_3^*	h_{13}	h_{12}	h_{11}	h_{12}	h_{13}	h_{14}
ψ_4^*	h_{14}	h_{13}	h_{12}	h_{11}	h_{12}	h_{13}
ψ_5^*	h_{13}	h_{14}	h_{13}	h_{12}	h_{11}	h_{12}
ψ_6^*	h_{12}	h_{13}	h_{14}	h_{13}	h_{12}	h_{11}

Table 14.11 Contributions to $\int \psi^* \psi \, d^3\mathbf{r}$ for the C_1 Species

	ψ_1	$-\omega\psi_2$	$\omega^2\psi_3$	$-\psi_4$	$\omega\psi_5$	$-\omega^2\psi_6$
ψ_1^*	1	$-\omega s_{12}$	$\omega^2 s_{13}$	$-s_{14}$	ωs_{13}	$-\omega^2 s_{12}$
$-\omega^2\psi_2^*$	$-\omega^2 s_{12}$	1	$-\omega s_{12}$	ωs_{13}	$-s_{14}$	ωs_{13}
$\omega\psi_3^*$	ωs_{13}	$-\omega^2 s_{12}$	1	$-\omega s_{12}$	$\omega^2 s_{13}$	$-s_{14}$
$-\psi_4^*$	$-s_{14}$	ωs_{13}	$-\omega^2 s_{12}$	1	$-\omega s_{12}$	$\omega^2 s_{13}$
$\omega^2\psi_5^*$	$\omega^2 s_{13}$	$-s_{14}$	ωs_{13}	$-\omega^2 s_{12}$	1	$-\omega s_{12}$
$-\omega\psi_6^*$	$-\omega s_{12}$	$\omega^2 s_{13}$	$-s_{14}$	ωs_{13}	$-\omega^2 s_{12}$	1

$\omega^2 + \omega = -1$

Table 14.12 Contributions to $\int \psi^* H \psi \, d^3\mathbf{r}$ for C_1 Species

	$H\psi_1$	$-\omega H\psi_2$	$\omega^2 H\psi_3$	$-H\psi_4$	$\omega H\psi_5$	$-\omega^2 H\psi_6$
ψ_1^*	h_{11}	$-\omega h_{12}$	$\omega^2 h_{13}$	$-h_{14}$	ωh_{13}	$-\omega^2 h_{12}$
$-\omega^2\psi_2^*$	$-\omega^2 h_{12}$	h_{11}	$-\omega h_{12}$	$\omega^2 h_{13}$	$-h_{14}$	ωh_{13}
$\omega\psi_3^*$	ωh_{13}	$-\omega^2 h_{12}$	h_{11}	$-\omega h_{12}$	$\omega^2 h_{13}$	$-h_{14}$
$-\psi_4^*$	$-h_{14}$	ωh_{13}	$-\omega^2 h_{12}$	h_{11}	$-\omega h_{12}$	$\omega^2 h_{13}$
$\omega^2\psi_5^*$	$\omega^2 h_{13}$	$-h_{14}$	ωh_{13}	$-\omega^2 h_{12}$	h_{11}	$-\omega h_{12}$
$-\omega\psi_6^*$	$-\omega h_{12}$	$\omega^2 h_{13}$	$-h_{14}$	ωh_{13}	$-\omega^2 h_{12}$	h_{11}

For the D_1 symmetry species, the same result is obtained:

$$E_3 = E_2. \qquad [14.50]$$

For the C_2 symmetry species, one similarly obtains

$$E_4 = \frac{h_{11} - h_{12} - h_{13} + h_{14}}{1 - s_{12} - s_{13} + s_{14}}. \qquad [14.51]$$

The D_2 symmetry species yields the same result:

$$E_5 = E_4. \qquad [14.52]$$

And for the B symmetry species, one finds that

$$E_6 = \frac{h_{11} - 2h_{12} + 2h_{13} - h_{14}}{1 - 2s_{12} + 2s_{13} - s_{14}}. \qquad [14.53]$$

As before, parameter h_{11} is designated the Coulomb integral α; parameter h_{12}, the bond integral β; and parameter s_{12}, the overlap integral S. Parameters h_{13}, h_{14}, s_{13}, and s_{14} involve atomic orbitals farther apart and are often neglected. The three bonding molecular orbitals then appear with

$$E_1 = \frac{\alpha + 2\beta}{1 + 2S}, \qquad [14.54]$$

$$E_2 = E_3 = \frac{\alpha + \beta}{1 + S}, \qquad [14.55]$$

while the three antibonding molecular orbitals have

$$E_4 = E_5 = \frac{\alpha - \beta}{1 - S}, \qquad [14.56]$$

$$E_6 = \frac{\alpha - 2\beta}{1 - 2S}. \qquad [14.57]$$

Two paired electrons may be placed in each orbital. The three bonding orbitals nicely accommodate the six valence electrons. In the approximation that S is small, we have

$$E \cong 2E_1 + 2E_2 + 2E_3 = 2(\alpha + 2\beta) + 4(\alpha + \beta) = 6\alpha + 8\beta. \qquad [14.58]$$

The energy of an electron in a two-center bond is given by (14.18). If the six-center bond were replaced by three two center bonds, with no other change, then we would have

$$E = 6(\alpha + \beta) = 6\alpha + 6\beta. \qquad [14.59]$$

Associated with the extra freedom in the six - center bond is the energy increment

$$\Delta E = (6\alpha + 8\beta) - (6\alpha + 6\beta) = 2\beta. \qquad [14.60]$$

In the $\mathbf{D}_{6h}$ group, which describes the full symmetry of benzene, orbital (14.36) belongs to the A_{2u} species; orbitals (14.37) and (14.38) belong to the E_{1g} species; orbitals (14.39) and (14.40) belong to the E_{2u} species; orbital (14.41) belongs to the B_{2g} species. So in the literature, these are labeled a_{2u}, e_{1g}, e_{2u}, and b_{2g} respectively. Associated with the formation of these orbitals is the splitting sketched in figure 14.9.

In classical chemistry, only two-center bonds were considered. But in many particularly stable structures, the standard valences could be satisfied only by introducing alternating single and double bonds. In such a system, the double bonds were said to be

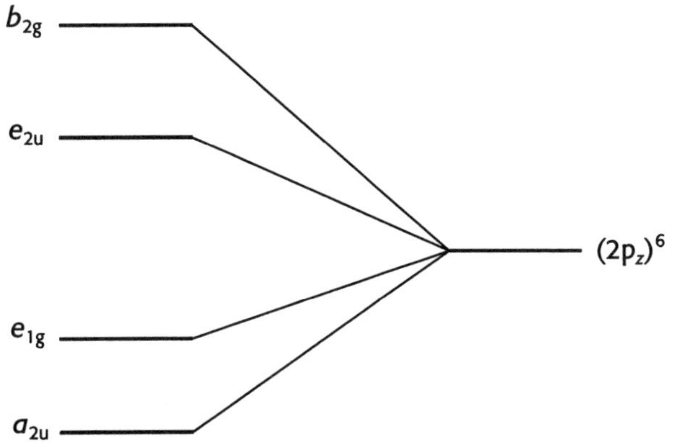

FIGURE 14.9 Splitting of the π energy levels in benzene associated with the formation of the π molecular orbitals.

conjugated. But such conjugation implies that the π electrons enter molecular orbitals extending over the whole chain. The stabilization energy coming from this gain in freedom for the electrons is called *conjugation energy*.

For the benzene ring, the two Kekule structures in figure 14.10 exist. Organic chemists consider that resonance between the two occurs. So the conjugation energy, given approximately by formula (14.60), is also called resonance energy.

14.7 Conventional Resonance Energy and the True Delocalization Energy

By thermochemical measurements, one can determine the energy changes associated with various changes in bonding. From electron diffraction and spectroscopic measurements, a person can determine the equilibrium bond lengths. With vibrational frequencies, one can fit a Morse curve and so estimate how the energy of a bond increases as it is stretched or compressed.

The hydrogenation of benzene follows the equation

$$C_6H_6 + 3H_2 \rightarrow C_6H_{12} + 208.4 \text{ kJ}, \qquad [14.61]$$

while the hydrogenation of cyclohexene involves

$$C_6H_{10} + H_2 \rightarrow C_6H_{12} + 119.5 \text{ kJ}. \qquad [14.62]$$

So removing $3H_2$ from cyclohexane requires the input (a) of 208.4 kJ mol^{-1} if benzene is formed but (b) of 3(119.5) kJ mol^{-1} if cyclohexatriene, with alternating single and double bonds, is formed. See figure 14.11.

From electron diffraction studies, each single carbon-carbon bond is 1.54 Å in length, each double carbon-carbon bond 1.33 Å in length. But in benzene the carbon-carbon bond length is 1.397 Å. To compress the three single bonds and extend the three double bonds in cyclotriene to this uniform distance requires about 113 kJ mol^{-1}.

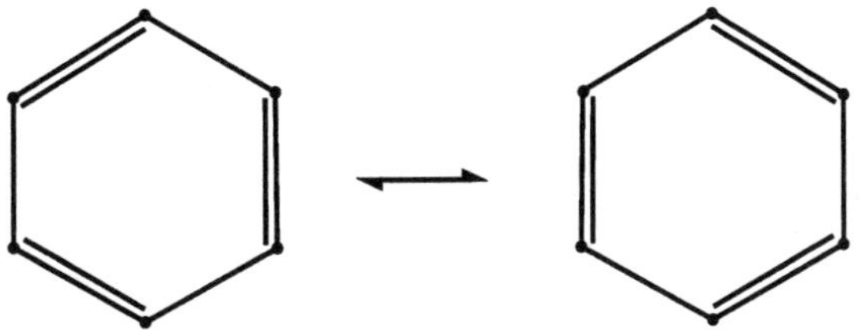

FIGURE 14.10 Alternate two-center bonds in a benzene ring.

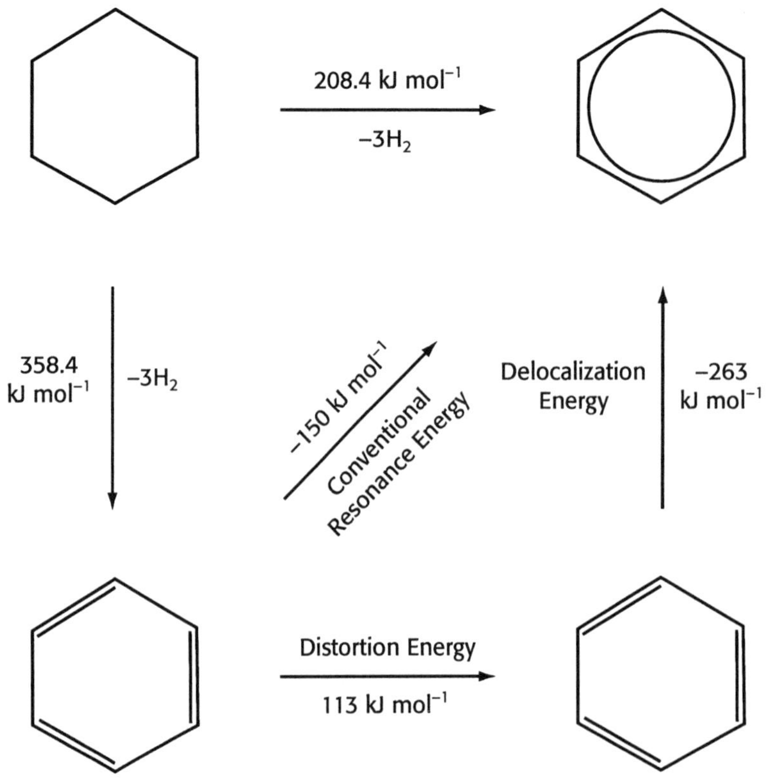

FIGURE 14.11 Reaction energies used to determine the conventional resonance energy and the true delocalization energy of benzene.

From these numbers, the delocalization energy is

$$\left(208.4 - 358.4 - 113\right) \text{ kJ mol}^{-1} = -263 \text{ kJ mol}^{-1}. \qquad [14.63]$$

The bond integral for (C2p , C2p) π overlap is consequently

$$\beta = \frac{1}{2}\left(-263 \text{ kJ mol}^{-1}\right) = -131.5 \text{ kJ mol}^{-1}. \qquad [14.64]$$

But in practice, corrections for bond length changes are often not introduced. For the conventional resonance energy in benzene, we have

$$\left(208.4 - 358.4\right) \text{kJ mol}^{-1} = -150 \text{ kJ mol}^{-1}, \qquad [14.65]$$

as figure 14.11 shows. For the carbon-carbon π bond integral, one then employs

$$\beta = \frac{1}{2}\left(-150 \text{ kJ mol}^{-1}\right) = -75 \text{ kJ mol}^{-1}. \qquad [14.66]$$

Most molecules do not have enough symmetry so that each pertinent molecular orbital belongs to a different symmetry species. Then two or more molecular orbitals belong to the same species and an additional formula is needed to determine the resulting energies and contributions. This is provided by the variation theorem.

14.8 The Variation Theorem

To make further progress, we need to investigate a general form of expression (14.11).

When a submicroscopic system possesses a definite energy E_j, its state function Ψ_j satisfies the Schrödinger equation

$$H\Psi_j = E_j\Psi_j, \tag{14.67}$$

in which the Hamiltonian operator is

$$H = \sum_{k=1}^{n} -\frac{\hbar^2}{2\mu_k}\nabla_k^2 + V\left(\mathbf{r}_1, \mathbf{r}_2, \ldots, \mathbf{r}_n\right), \tag{14.68}$$

as long as magnetic effects can be neglected. Introducing these effects merely involves adding the pertinent terms to operator (14.68). In this expression, the kth particle is located by radius vector $\mathbf{r}_k$, so the corresponding Laplacian operator is

$$\nabla_k^2 = \frac{\partial^2}{\partial x_k^2} + \frac{\partial^2}{\partial y_k^2} + \frac{\partial^2}{\partial z_k^2} \tag{14.69}$$

in rectangular coordinates.

In principle, the possible states are spanned by an infinite number of disjoint eigenfunctions that are normalized and that meet the boundary conditions. Any allowable state is represented by a superposition of these eigenfunctions:

$$\Psi = \sum c_i\Psi_i. \tag{14.70}$$

Within the boundaries, the standing wave eigenfunctions exhibit a complete set of nodal surfaces of each integral number including zero. So any arbitrary function that meets the analytical conditions to be a state function can be expressed as a linear combination of them. This function Ψ may be real or complex. Furthermore, the real eigenfunctions may be combined linearly to form a complete set of disjoint complex eigenfunctions.

Consider a complete set of eigenfunctions for equation (14.67) that are normalized and mutually orthogonal. Furthermore, multiply (14.67) by the complex conjugate of any one of these to form

$$\Psi_i^* H\Psi_j = E_j\Psi_i^*\Psi_j. \tag{14.71}$$

Let us integrate over the coordinates of the system to obtain

$$\int_R \Psi_i^* H\Psi_j \, d^{3n}\mathbf{r} = E_j\int_R \Psi_i^*\Psi_j \, d^{3n}\mathbf{r} = \delta_{ij}E_j, \tag{14.72}$$

where δ_{ij} is the Kronecker delta.

But multiplying (14.70) by Ψ^* and integrating yields

$$\int_R \Psi^*\Psi \, d^{3n}\mathbf{r} = \int_R \sum c_i^*\Psi_i^* \sum c_j\Psi_j \, d^{3n}\mathbf{r} = \sum c_i^* c_j \int_R \Psi_i^*\Psi_j \, d^{3n}\mathbf{r} = \sum c_i^* c_j\delta_{ij} = \sum c_i^* c_i. \tag{14.73}$$

With Ψ normalized to 1, we have

$$\sum c_i^* c_i = 1. \tag{14.74}$$

One may let the Hamiltonian operator act on any normalized candidate function Ψ, multiply by the complex conjugate of this function, and integrate to obtain

$$\int_R \Psi^* H\Psi \, d^{3n}\mathbf{r} = \int_R \left(\sum c_i^*\Psi_i^*\right) H \sum c_j\Psi_j \, d^{3n}\mathbf{r} = \sum c_i^* c_j \int_R \Psi_i^* E_j\Psi_j \, d^{3n}\mathbf{r} = \sum c_i^* c_j E_j\delta_{ij} = \sum c_i^* c_i E_i. \tag{14.75}$$

Let us order the energies so that

$$E_j \geq E_{j-1}. \quad [14.76]$$

Then from (14.75), we find that

$$\int_R \Psi^* H \Psi \, d^{3n}\mathbf{r} \geq \sum c_i^* c_i E_1 = E_1. \quad [14.77]$$

In the last equality, we have used formula (14.74), Since one cannot extract an infinite amount of energy from any given small system, the lower limit on the integral, E_1, is finite.

When the candidate function is not normalized to 1, inequality (14.77) is replaced with

$$E_1 \leq \frac{\int_R \Psi^* H \Psi \, d^{3n}\mathbf{r}}{\int_R \Psi^* \Psi \, d^{3n}\mathbf{r}}. \quad [14.78]$$

If one constructs an analytic function Ψ that satisfies the boundary conditions, this function must satisfy inequality (14.78), with E_1 the lowest energy eigenvalue. When Ψ depends on certain parameters, the best values for the parameters are those that make the expression on the right of (14.78) as low as possible.

If a person subtracts out the contribution of the first eigenfunction to Ψ, then the lowest value of the expression is equal to or greater than E_2. Similarly, if a person subtracts out the contributions of the first j - 1 eigenfunctions to Ψ, then the lowest value is equal to or greater than E_j. As a consequence, the first n minima of the original expression on the right of (14.78) yield approximations to E_1, E_2,..., E_n.

We call this result the *variation theorem.*

14.9 Linear Variation Functions

A molecular orbital is formed by superposing distorted valence orbitals from the participating atoms. Thus, it is a linear combination of these orbitals. So let us examine what the variation theorem implies when the parameters appear in linear positions in the variation function.

We consider that the chosen contributing functions Ψ_1, Ψ_2, ..., Ψ_n are known. The variation theorem then allows us to determine the best values for the coefficients in

$$\psi = c_1\psi_1 + c_2\psi_2 + \ldots + c_n\psi_n = \sum_1^n c_i\psi_i. \quad [14.79]$$

and the corresponding energies.

First, introduce E as a symbol for the ratio on the right of (14.78):

$$E = \frac{\int_R \Psi^* H \Psi \, d^{3n'}\mathbf{r}}{\int_R \Psi^* \Psi \, d^{3n'}\mathbf{r}} \quad [14.80]$$

On varying the coefficients, one finds n stationary values, not necessarily all distinct, for E. With the variation theorem, we identify these with the n low lying energy levels. In a molecule, we expect about half of these to be bonding, about half to be antibonding.

Clearing (14.80) of the fraction and introducing substitution (14.79) leads to

$$E\int_R \sum c_i^* \Psi_i^* \sum c_j \Psi_j \, d^{3n'}\mathbf{r} = \int_R \sum c_i^* \Psi_i^* H \sum c_j \Psi_j \, d^{3n'}\mathbf{r}. \quad [14.81]$$

Now let

$$\int_R \Psi_i^* \Psi_j \, d^{3n'}\mathbf{r} = s_{ij}, \tag{14.82}$$

$$\int_R \Psi_i^* H \Psi_j \, d^{3n'}\mathbf{r} = h_{ij}. \tag{14.83}$$

Then equation (14.81) becomes

$$\sum_{i,j} E c_i^* c_j s_{ij} = \sum_{i,j} c_i^* c_j h_{ij}, \tag{14.84}$$

whence

$$\sum_{i,j} \left(h_{ij} - E s_{ij}\right) c_i^* c_j = 0. \tag{14.85}$$

Variation of c_k generally causes E to vary. But at the desired minima, E acts as a constant. At these points, differentiation of equation (14.85) with respect to the real part of c_k yields

$$\sum \left(h_{kj} c_j + h_{ik} c_i^* - E s_{kj} c_j - E s_{ik} c_i^*\right) = 0, \tag{14.86}$$

differentiation with respect to the imaginary part of c_k yields

$$\sum i\left(-h_{kj} c_j + h_{ik} c_i^* + E s_{kj} c_j - E s_{ik} c_i^*\right) = 0. \tag{14.87}$$

Multiply (14.87) by i, then add the result to (14.86) and divide by 2:

$$\sum \left(h_{kj} - E s_{kj}\right) c_j = 0. \tag{14.88}$$

We have here n simultaneous equations determining the coefficients $c_1, c_2, \ldots, c_n$. For $k = 1, 2, \ldots, n$, these are

$$\begin{aligned} &\left(h_{11} - E s_{11}\right) c_1 + \left(h_{12} - E s_{12}\right) c_2 + \ldots + \left(h_{1n} - E s_{1n}\right) c_n = 0, \\ &\left(h_{21} - E s_{21}\right) c_1 + \left(h_{22} - E s_{22}\right) c_2 + \ldots + \left(h_{2n} - E s_{2n}\right) c_n = 0, \\ &\qquad\vdots \\ &\left(h_{n1} - E s_{n1}\right) c_1 + \left(h_{n2} - E s_{n2}\right) c_2 + \ldots + \left(h_{nn} - E s_{nn}\right) c_n = 0. \end{aligned} \tag{14.89}$$

Cramer's rule yields 0 for all c_j's unless the determinant of the coefficients of the c_j's is zero. But then we have

$$\begin{vmatrix} h_{11} - E s_{11} & h_{12} - E s_{12} & \cdots & h_{1n} - E s_{1n} \\ h_{21} - E s_{21} & h_{22} - E s_{22} & \cdots & h_{2n} - E s_{2n} \\ \vdots & \vdots & \cdots & \vdots \\ h_{n1} - E s_{n1} & h_{n2} - E s_{n2} & \cdots & h_{nn} - E s_{nn} \end{vmatrix} = 0. \tag{14.90}$$

This is known as the *secular equation*. It is of degree n in E. But each of the roots corresponds to a stationary value for expression (14.80). So these are approximations to the n lowest energy levels for the given system.

Functions $\psi_1, \psi_2, \ldots, \psi_n$ are called basis functions. For convenience, they are generally normalized to 1, so that each s_{ii} equals 1, When they are also approximations to valence orbitals, s_{ij} is the ijth overlap integral, h_{ii} the ith Coulomb integral, and h_{ij} the ijth bond integral.

In determining the possible states of a system, with given potential interactions among given particles, one first constructs reasonable basis functions that meet the boundary conditions. Integrals (14.82) and (14.83) are calculated by computer. The results are inserted into the secular determinant and the equation solved by computer.

For molecular orbital calculations, a person employs suitably altered hydrogenlike orbitals. The changes are introduced to expedite the calculation of the needed integrals. First, one may eliminate the radial nodes. The resulting Slater orbitals have radial factors of the form

$$R = Ar^{n-1} \exp\left[-(Z-s)r/n\right]. \qquad [14.91]$$

The angular factors are unchanged. Here n is the principal quantum number while s is the screening constant, determined by standard rules. Eliminating the radial nodes causes some loss of orthogonality. This must be taken into account.

To further expedite calculation of the needed integrals, one may replace the Slater orbitals with Gaussian orbitals. These have radial factors of the form

$$R = A \exp\left(-\alpha r^2\right). \qquad [14.92]$$

To get good results, each Slater orbital may be replaced by a linear combination of three or more Gaussians.

14.10 Bonding between Two Nonequivalent Atoms

When the atoms bound together are different, the symmetry argument employed in section 14.4 breaks down. One then needs to use the secular equation.

Consider two normalized valence orbitals, ψ_A and ψ_B on two neighboring atoms, A and B. A molecular orbital from these has the form

$$\psi = c_A\psi_A + c_B\psi_B = c_1\psi_1 + c_2\psi_2. \qquad [14.93]$$

Coefficients $c_A = c_1$ and $c_B = c_2$ will be chosen so that ψ is also normalized to 1.

Furthermore, let us employ real valence orbitals, so that

$$s_{12} = s_{21} = S. \qquad [14.94]$$

We also set

$$h_{11} = \alpha_A, \qquad h_{22} = \alpha_B, \qquad [14.95]$$

and

$$h_{12} = h_{21} = \beta. \qquad [14.96]$$

The secular equation then becomes

$$\begin{vmatrix} \alpha_A - E & \beta - ES \\ \beta - ES & \alpha_B - E \end{vmatrix} = 0. \qquad [14.97]$$

As an approximation, let us neglect overlap S. Also let us introduce the substitutions

$$E_1 = \alpha_B + \beta \cot\zeta \qquad [14.98]$$

and

$$E_2 = \alpha_A - \beta \cot\zeta. \qquad [14.99]$$

Form E_1 leads to

$$\begin{vmatrix} \dfrac{\alpha_A - \alpha_B}{\beta} - \cot\zeta & 1 \\ 1 & -\cot\zeta \end{vmatrix} = 0, \qquad [14.100]$$

while form E_2 leads to

$$\begin{vmatrix} \cot\zeta & 1 \\ 1 & \dfrac{\alpha_B - \alpha_A}{\beta} + \cot\zeta \end{vmatrix} = 0. \tag{14.101}$$

These yield the condition

$$\frac{\alpha_A - \alpha_B}{2\beta} = \frac{\cot^2\zeta - 1}{2\cot\zeta} = \cot 2\zeta, \tag{14.102}$$

which relates parameter ζ to α_A, α_B, and β.

When the two atoms are equivalent, we have

$$\cot 2\zeta = 0 \tag{14.103}$$

and

$$\zeta = 45°. \tag{14.104}$$

Then equations (14.98) and (14.99) reduce to

$$E_1 = \alpha + \beta, \tag{14.105}$$

and

$$E_2 = \alpha - \beta, \tag{14.106}$$

in agreement with (14.18) when $S = 0$.

From (14.100), one can write down two simultaneous equations for c_1 and c_2. But formula (14.102) makes them consistent. So we need formulate only the simpler one, which is

$$c_1 - (\cot\zeta)c_2 = 0. \tag{14.107}$$

This yields the relations

$$c_1 = \cos\zeta, \qquad c_2 = \sin\zeta. \tag{14.108}$$

Similarly from (14.101), we obtain

$$(\cot\zeta)c_1 + c_2 = 0, \tag{14.109}$$

whence

$$c_1 = \sin\zeta, \qquad c_2 = -\cos\zeta. \tag{14.110}$$

So the molecular orbital for E_1 is

$$\psi_1 = (\cos\zeta)\psi_A + (\sin\zeta)\psi_B, \tag{14.111}$$

while that for E_2 is

$$\psi_2 = (\sin\zeta)\psi_A - (\cos\zeta)\psi_B. \tag{14.112}$$

The approximate energy change associated with the reaction

$$\frac{1}{2}A_2 + \frac{1}{2}B_2 \rightarrow AB \tag{14.113}$$

is

$$\Delta E \cong 2\alpha_B + 2\beta\cot\zeta - \alpha_A - \beta - \alpha_B - \beta = \alpha_B - \alpha_A + 2\beta(\cot\zeta - 1). \tag{14.114}$$

In this expression, the magnitude of α_B - α_A dominates. So it provides the main impetus for reaction (14.113) to proceed.

In a semiempirical calculation,the parameters are constructed from experimental data. The magnitude of a Coulomb integral α is approximated by the average between the ionization energy and the electron affinity for the appropriate valence state. This average is known as the *Mulliken electronegativity*. The magnitude of the bond integral β may be estimated from representative bond energies.

In the so-called ab initio methods, the integrals are calculated on a computer employing the pertinent Hamiltonian operator and the chosen basis functions.

Example 14.3

Construct molecular orbitals for HF and estimate their energies.

For the 1s valence state of hydrogen, the ionization energy I is 13.60 eV and the electron affinity A is 0.75 eV. So we have

$$\alpha_{\mathrm{H}} = -\frac{13.60 + 0.75}{2}\ \mathrm{eV} = -7.175\ \mathrm{eV}.$$

For the 2p valence state of fluorine, I is 20.86 eV and A is 3.50 eV. So

$$\alpha_{\mathrm{F}} = -\frac{20.86 + 3.50}{2}\ \mathrm{eV} = -12.18\ \mathrm{eV}.$$

From standard bond energies, we also estimate that

$$\beta = -2.0\ \mathrm{eV}.$$

Substituting these values into formula (14.102) yields

$$\cot 2\zeta = \frac{5.005\ \mathrm{eV}}{-4.0\ \mathrm{eV}} = -1.251$$

whence

$$\zeta = 70.68^{\circ}.$$

Equations (14.98), (14.111), (14.99), (14.112) now lead to

$$E_1 = -12.88\ \mathrm{eV}, \qquad \psi_1 = 0.331\psi_{\mathrm{H}} + 0.944\psi_{\mathrm{F}},$$
$$E_2 = -6.47\ \mathrm{eV}, \qquad \psi_2 = 0.944\psi_{\mathrm{H}} - 0.331\psi_{\mathrm{F}}.$$

14.11 Pi Bonding in Fulvene

Any symmetry that a molecule possesses may be used to factor the secular equation. An interesting example is provided by an isomer of benzene, fulvene. This consists of a five-membered aromatic ring with a $=CH_2$ side chain. The carbon skeleton is described in figure 14.12.

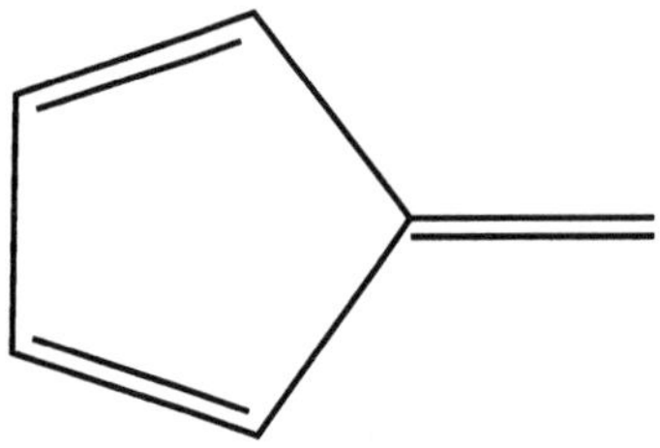

FIGURE 14.12 Conjugated bond description of fulvene.

In fulvene, there is a real 2p orbital directed perpendicular to the plane of the ring on each carbon atom. The six valence electrons in these orbitals form a delocalized system.

Let the perpendicular 2p orbitals be designated $\psi_1, \psi_2, \psi_3, \psi_4, \psi_5, \psi_6$ with the subscript identifying the atom as in figure 14.13, The symmetry operations for the molecule form the $\mathbf{C}_{2v}$ group. Under these operations, the π orbitals behave as table 14.13 indicates.

Applying formula;(13.44) to the expressions in tables 14.13 and 14.14 yields the following results. For species A_2 we find the combinations

$$\frac{1}{\sqrt{2}}\psi_1 - \frac{1}{\sqrt{2}}\psi_4, \qquad [14.115]$$

$$\frac{1}{\sqrt{2}}\psi_2 - \frac{1}{\sqrt{2}}\psi_3. \qquad [14.116]$$

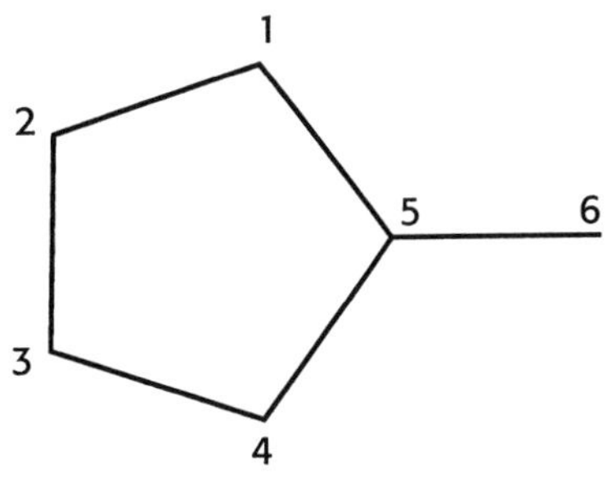

FIGURE 14.13 Numbering of atoms in the fulvene molecule.

Table 14.13 Transformation of Fulvene Orbitals

I	C_2	σ_v	σ_v'
ψ_1	$-\psi_4$	$-\psi_1$	ψ_4
ψ_2	$-\psi_3$	$-\psi_2$	ψ_3
ψ_5	$-\psi_5$	$-\psi_5$	ψ_5
ψ_6	$-\psi_6$	$-\psi_6$	ψ_6

Table 14.14 Primitive Character Vectors of $\mathbf{C}_{2v}$ Group

	I	C_2	σ_v	σ_v'
A_1	1	1	1	1
A_2	1	1	-1	-1
B_1	1	-1	1	-1
B_2	1	-1	-1	1

For species B_2, we find the orthogonal contributions

$$\frac{1}{\sqrt{2}}\psi_1 + \frac{1}{\sqrt{2}}\psi_4, \qquad [14.117]$$

$$\frac{1}{\sqrt{2}}\psi_2 + \frac{1}{\sqrt{2}}\psi_3, \qquad [14.118]$$

$$\psi_5, \qquad [14.119]$$

$$\psi_6. \qquad [14.120]$$

The overlap integrals and the Coulomb integrals between different primitive symmetry species vanish. So we have a separate secular equation for each species. We also set

$$s_{11} = s_{22} = s_{33} = s_{44} = s_{55} = s_{66} = 1, \qquad [14.121]$$

$$h_{11} = h_{22} = h_{33} = h_{44} = h_{55} = h_{66} = \alpha, \qquad [14.122]$$

$$h_{12} = h_{23} = h_{34} = h_{45} = h_{51} = h_{56} = \beta, \qquad [14.123]$$

and take all other *atomic* overlap and Coulomb integrals to be zero.

For the A_2 species, we then obtain the secular equation

$$\begin{vmatrix} \alpha - E & \beta \\ \beta & \alpha - \beta - E \end{vmatrix} = 0, \qquad [14.124]$$

which yields

$$\alpha - E = x = \frac{1 \pm \sqrt{5}}{2}\beta. \qquad [14.125]$$

For the B_2 species, we obtain the secular equation

$$\begin{vmatrix} \alpha - E & \beta & \sqrt{2}\beta & 0 \\ \beta & \alpha + \beta - E & 0 & 0 \\ \sqrt{2}\beta & 0 & \alpha - E & \beta \\ 0 & 0 & \beta & \alpha - E \end{vmatrix} = 0. \qquad [14.126]$$

Again letting α - E be x, we multiply (14.126) out to get

$$x^4 + \beta x^3 - 4\beta^2 x^2 - 3\beta^3 x + \beta^4 = 0 \qquad [14.127]$$

or

$$\left(x + \beta\right)\left(x^3 - 4\beta^2 x + \beta^3\right) = 0. \qquad [14.128]$$

From the first factor, we have

$$x = \alpha - E = -\beta. \qquad [14.129]$$

On solving the cubic equation, we obtain

$$x = \alpha - E = -2.1149\beta,\ 0.2541\beta,\ 1.8608\beta. \qquad [14.130]$$

So the energy levels for the delocalized orbitals are

$$\begin{aligned}&\alpha - 1.8608\beta,\\ &\alpha - 1.6180\beta,\\ &\alpha - 0.2541\beta,\\ &\alpha + 0.6180\beta,\\ &\alpha + 1.0000\beta,\\ &\alpha + 2.1149\beta.\end{aligned} \qquad [14.131]$$

Associated with the extra freedom in the six-center bond is the energy increment

$$\Delta E = 6\alpha + 2(2.1149 + 1.0000 + 0.6180)\beta - 6\alpha - 6\beta = 1.4658\beta. \qquad [14.132]$$

This is 0.733 of the stabilization in benzene.

From the formulas for the molecular orbitals, one finds that position 6 carries a considerable positive charge while the ring is equally negative. Thus, fulvene is a polar hydrocarbon.

Example 14.4

Construct the molecular orbital for the α + 0.61803 β level in fulvene.

The simultaneous equations behind equation (14.124) are

$$(\alpha - E)a_1 + \beta a_2 = 0,$$
$$\beta a_1 + (\alpha - \beta - E)a_2 = 0.$$

Here a_1 is the coefficient multiplying expression (14.115), a_2 the coefficient multiplying expression (14.116), in the molecular orbital.

The specified solution of the secular equation is

$$\alpha - E = -0.61803\beta.$$

Substituting this into the first equation yields

$$-0.61803\beta a_1 + \beta a_2 = 0$$

whence

$$a_2 = 0.61803\,a_1.$$

Parameterize the normalization relation

$$a_1^2 + a_2^2 = 1$$

by setting

$$a_1 = \cos\alpha, \quad a_2 = \sin\alpha.$$

Then we find that

$$\alpha = \tan^{-1} 0.61803 = 31.72^\circ$$

and

$$a_1 = \cos 31.72^\circ = 0.85065,$$

$$a_2 = \sin 31.72^\circ = 0.52573.$$

But since

$$c_1 = \frac{a_1}{\sqrt{2}} = -c_4,$$

$$c_2 = \frac{a_2}{\sqrt{2}} = -c_3,$$

the molecular orbital is

$$\psi_{\text{III}} = 0.6015\psi_1 + 0.3718\psi_2 - 0.3718\psi_3 - 0.6015\psi_4.$$

Questions

14.1 What is a molecular orbital? How may one be constructed?

14.2 On forming an A- B bond, why should the atomic orbitals shrink? Why are the atomic orbitals distorted?

14.3 What is the symmetry of H_2^+? How does this affect the state function ψ?

14.4 How does quantum number λ arise? When is it replaced by $|\lambda|$?

14.5 What do the symbols σ, π, δ, ϕ,... represent?

14.6 What is parity? What do the subscripts g and u stand for?

14.7 What effect does substituting two identical atomic cores for the protons have? What about adding a second valence electron?

14.8 What is a symmetry orbital? How is such an orbital changed on going from the separated atom state to the united atom state?

14.9 How does a person correlate the separated atom levels with the molecular levels and with the united atom levels?

14.10 Construct the formula

$$E = \frac{\int_R \psi^* H \psi \, d^3\mathbf{r}}{\int_R \psi^* \psi \, d^3\mathbf{r}}.$$

14.11 Then derive the equation

$$E = \frac{h_{AA} \pm h_{AB}}{1 \pm s_{AB}} = \frac{\alpha \pm \beta}{1 \pm S}$$

for a valence electron with A and B equivalent.

14.12 What signs do α, β, and S have? Which solution is (a) bonding, (b) antibonding?

14.13 Why is $1/R$ added to these solutions in figure 14.5?

14.14 What effect does adding a second valence electron have?

14.15 Consider the symmetric three-center bond and construct molecular orbitals for species A, C, and D.

14.16 Then determine the corresponding energy levels.

14.17 Consider the symmetric six-center bond and construct molecular orbitals for the different symmetry species.

14.18 Then determine the corresponding energy levels.

14.19 The resonance energy in benzene is often said to equal 2β. How is this result obtained from the expressions found for question 14.18.

14.20 How is the conventional resonance energy related to the delocalization energy in benzene?

14.21 Why can any suitable arbitrary function be expressed as a linear combination of the normalized orthogonal eigenfunctions for a given system?

14.22 Derive the variation theorem.

14.23 What is the physical significance of a linear variation function?

14.24 Derive the secular equation.

14.25 How may one modify the pertinent hydrogenlike orbitals for a molecular orbital calculation?

14.26 Apply the secular equation to a single bond between two nonequivalent atoms.

14.27 Then construct expressions for the corresponding molecular orbitals.
14.28 How may a person estimate (a) a Coulomb integral α, (b) a bond integral β?
14.29 Show how symmetry considerations allow one to factor the secular equation for fulvene.
14.30 How can a pure hydrocarbon be polar?

Problems

14.1 From figure 14.4, write down the electron configurations for He_2 and for Li_2 (a) in the ground state, (b) in the first excited state.
14.2 Determine the resonance energy of the radical

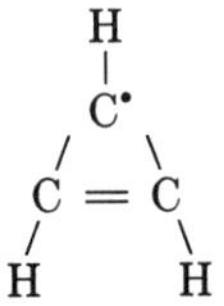

14.3 Construct the symmetry-adapted molecular orbitals for four equivalent π valence orbitals placed at the corners of a square.
14.4 Construct parametric expressions for the energies of the orbitals found in problem 14.3.
14.5 Then determine the resonance energy of square cyclobutadiene.
14.6 Construct symmetry-adapted molecular orbitals for the four π valence electrons (a) in trans-butadiene, (b) in cis-butadiene.
14.7 Construct a separate secular equation for each symmetry species found in problem 14.6.
14.8 Solve the secular equations for butadiene and determine the resonance energy of the molecule.
14.9 From the two bonding energies, construct the corresponding molecular orbitals for butadiene.
14.10 Construct the molecular orbital for the α + 1.0000 level in fulvene.

— — —

14.11 From figure 14.4, formulate the electron configurations for Be_2, B_2, and C_2 (a) in the ground state, (b) in the first excited state.
14.12 Again consider the structure in problem 14.2. Determine the π electron energy and the resonance energy (a) if one electron is removed, giving $C_3H_3^+$, (b) if one electron is added, giving $C_3H_3^-$.
14.13 Construct the symmetry-adapted molecular orbitals for five equivalent π valence orbitals placed at the corners of a regular pentagon.
14.14 Construct parametric expressions for the energies of the orbitals found in problem 14.13.
14.15 Then determine the resonance energy of pentagonal $C_5H_5^{+a}$ carrying (a) one positive charge, (b) no charge, (c) one negative charge.
14.16 Construct symmetry-adapted molecular orbitals for the ten π valence electrons in naphthalene, $C_{10}H_8$.
14.17 Construct a separate secular equation for each symmetry species found in problem 14.16.
14.18 Solve the secular equations for naphthalene and determine the resonance energy of the molecule.
14.19 Construct molecular orbitals for HCl and estimate their energies. For the 3p valence state of chlorine, I is 15.03 eV and A is 3.73 eV.
14.20 Construct the molecular orbital for the α + 2.1149 level β in fulvene.

References

Books

Carsky, P., and Urban, M.: 1980, *Ab Initio Calculations: Methods and Applications in Chemistry*, Springer-Verlag, Berlin, pp. 1-241.

In an ab initio calculation, all significant integrals are calculated by computer. The accuracy is limited by the basis set employed. Commonly, Gaussian functions based on the participating atoms are used. Also, relativistic effects are neglected.

In this text, various practical procedures are described and results summarized. Numerous references are cited.

Del Re, G., Berthier, G., and Serre, J.: 1980, *Electronic States of Molecules and Atom Clusters: Foundations and Prospects of Semiempirical Methods*, Springer-Verlag, Berlin, pp. 1-177.

A simplified model based on classical chemical theory is often employed. For it, a Hamiltonian operator representing the ideal system must be constructable. But the integrals in the formulas may be evaluated from empirical data.

In this text, practical models are described and evaluated. Useful parameters are tabulated and numerous references cited.

Douglas, B. E., and Hollingsworth, C. A.: 1985, *Symmetry in Bonding and Spectra*, Academic Press, Orlando, Fl, pp. 1- 253.

After reviewing the pertinent group theory, the authors turn to applications. Symmetry properties of the Hamiltonian operator, common atomic orbitals, spectral terms, sigma bonding, pi bonding, multicenter bonding, ligand field theory are all considered in detail. Many helpful figures are included.

Hinchliffe, A.: 1988, *Computational Quantum Chemistry*, John Wiley & Sons, Inc., New York, pp. 1-112.

After a survey of the concepts of valence theory, Hinchliffe describes various ab initio calculations, representative results, and their interpretation. The text is quite readable, with much of practical value.

Lowe, J. P.: 1978, *Quantum Chemistry*, Academic Press, New York, pp. 135-571.

This introductory text covers basic quantum mechanics in a conventional manner. But molecular orbital theory is presented and applied in detail. Furthermore, pertinent group theory is developed.

Simons, J., and Nichols, J: 1997, *Quantum Mechanics in Chemistry*, Oxford University Press, Oxford, pp. 123-185.

The authors present the rules of quantum mechanics without justification or explanation. Operator expressions and Schrödinger equations are constructed and applied to a large number of chemical systems with various approximations. Many problems are presented and worked out in detail.

Articles

Baird, N. C.: 1986, "The Chemical Bond Revisited," *J. Chem. Educ.* **63**, 560-664.

Blaise, P., and Henri-Rousseau, O.: 1988, "Variational Energy Lowering May Increase Hamiltonian Dispersion," *J. Chem. Educ.* **65**, 9-11.

Bratsch, S. G.: 1988, "Revised Mulliken Electronegativities I, II," *J. Chem. Educ.* **65**, 34-41, 223-227.

David, C. W.: 1991, "Computing Overlaps between Nonorthogonal Orbitals," *J. Chem. Educ.* **68**, 129-130.

DeKock, R. L., and Bosma, W. B.: 1988, "The Three-Center, Two-Electron Chemical Bond," *J. Chem. Educ.* **65**, 194-197.

Dias, J. R.: 1987, "Facile Calculations of the Characteristic Polynomial and π-Energy Levels of Molecules Using Chemical Graph Theory," *J. Chem. Educ.* **64**, 213-216.

Dias, J. R.: 1992, "An Example Molecular Orbital Calculation Using the Sachs Graph Method," *J. Chem. Educ.* **69**, 695-700.

Duke, B. J., and Leary, B.: 1995, "Non-Koopmans' Molecules," *J. Chem. Educ.* **72**, 501-504.

George, P., Bock, C. W., and Trachtman, M.: 1984, "The Evaluation of Empirical Resonance Energies as Reaction Enthalpies with Particular Reference to Benzene," *J. Chem. Educ.* **61**, 225-227.

Hofmann, H. F.: 1997, "A Dynamical Model of the Chemical Bond: Kinetic Energy Resonances between Atomic Orbitals," *Eur. J. Educ.* **18**, 354-362.

Hollingsworth, C. A.: 1991, "Degeneracies in Separable Systems with 0_h Symmetry," *J. Chem. Educ.* **68**, 23-24.

Karafiloglou, P., and Chanessian, G.: 1991, "Understanding Molecular Orbital Wave Functions in Terms of Resonance Structures," *J. Chem. Educ.* **68**, 583-586.

Keeports, D.: 1986, "A Comparison of Molecular Vibrational Theory to Huckel Molecular Orbital Theory," *J. Chem. Educ.* **63**, 753- 756.

Keeports, D.: 1989, "Application of the Variational Method to the Particle-in-the-Box Problem," *J. Chem. Educ.* **66**, 314-318.

Maitland, A., and Brown, R. D. H.: 1983, "Systematics in the Assignment of Electronic and Vibronic States for Linear Molecules," *J. Chem. Educ.* **60**, 202-206.

Mazo, R. M.: 1990, "Molecular Electronic Terms and Molecular Orbital Configurations," *J. Chem. Educ.* **67**, 135-138.

Pisanty, A.: 1991, "The Electronic Structure of Graphite," *J. Chem. Educ.* **66**, 804-808.

Reed, J. L.: 1992, "Electronegativity and Atomic Charge," *J. Chem. Educ.* **69**, 785-790.

Reed, L. H., and Murphy, A. R.: 1986, "An Investigation of the Quality of Approximate Wave Functions," *J. Chem. Educ.* **63**, 757-759.

Sannigrahi, A. B., and Kar, T.: 1988, "Molecular Orbital Theory of Bond Order and Valency," *J. Chem. Educ.* **65**, 674-676.

Taubmann, G.: 1992, "Calculation of the Hückel Parameter from the Free-Electron Model," *J. Chem. Educ.* **69**, 95-97.

von Nagy-Felsobuki, E. I.: 1989, "Hückel Theory and Photoelectron Spectroscopy," *J. Chem. Educ.* **66**, 821-824.

Vincent, A.: 1996, "An Alternative Derivation of the Energy Levels of the 'Particle on a Ring' System," *J. Chem. Educ.* **73**, 1001- 1003.

Vos, M., and McCarthy, I.: 1997, "Measuring Orbitals and Bonding in Atoms, Molecules, and Solids," *Am. J. Phys.* **65**, 544-553.

Willis, C. J.: 1991, "Describing Electron Distributions in the Hydrogen Molecule," *J. Chem. Educ.* **68**, 743-747.

15

Phenomenological Chemical Kinetics

15.1 The Rate of a Reaction

IN CARRYING OUT CHEMICAL PROCESSES, one is concerned not only with whether a reaction can take place but also with the speed of the reaction. This must generally be neither too slow nor too fast. One may proceed empirically, measuring the rate of the given reaction as a function of time under precisely imposed conditions. Or, one may proceed theoretically, investigating what molecules are involved and how they interact.

In any investigation, a person would first determine what reactants are involved and what products are produced. The main stoichiometry may be represented as the overall reaction

$$aA + bB \rightarrow cC + dD. \qquad [15.1]$$

However, this process may be accompanied by side reactions. If appreciable, these must be taken into account.

Since reaction rates often vary strongly with temperature, this condition must be kept practically constant. The reaction vessel may be immersed in a constant temperature bath or placed in contact with a block of metal maintained at the desired temperature.

The concentration of a key constituent may be determined chemically or physically at known times. Thus, one may withdraw a sample and titrate it. If necessary, the reaction may be quenched by rapid cooling or dilution. Alternatively, one may follow the intensity of absorption in a particular spectral region where one reactant or product predominates. For a gas phase reaction where the number of molecules changes, one may measure the total pressure as a function of time. In condensed phases, one may employ potentiometric or conductimetric procedures.

From the concentration - time data, one may calculate the slope at representative times. This is divided by minus or plus the corresponding coefficient in the stoichiometric equation to avoid ambiguity. Thus for reaction (15.1), the *rate* R is

$$R = -\frac{1}{a}\frac{d[A]}{dt} = -\frac{1}{b}\frac{d[B]}{dt} = \frac{1}{c}\frac{d[C]}{dt} = \frac{1}{d}\frac{d[D]}{dt}. \qquad [15.2]$$

Here [A], [B], [C], and [D] are the concentrations of A, B, C, and D. The standard unit for R is the mole per cubic decimeter per second, abbreviated mol dm^{-3} s^{-1} or M s^{-1}.

Example 15.1

How does the total pressure in the reaction

$$2N_2O_5(g) \rightarrow 4NO_2(g) + O_2(g)$$

vary with the degree of dissociation at a given temperature?

In the approximation that the constituents behave as ideal gases, we have

$$P = \frac{n}{V}RT = \frac{n}{n_0}\frac{n_0}{V}RT = \frac{n}{n_0}P_0.$$

Consider that we start with n_0 moles of pure gaseous reactant. Also let ν be the number of moles of product gases produced from one mole of reactant. Then if α is the fraction dissociating in time t, we have

$$n = \left[1 + (\nu - 1)\alpha\right]n_0.$$

Combining these two equations yields

$$P = \left[1 + (\nu - 1)\alpha\right]P_0.$$

For the decomposition of N_2O_5, parameter $\nu = 5/2$ and

$$P = \left(1 + \frac{3}{2}\alpha\right)P_0.$$

Example 15.2

How is the partial pressure of N_2O_5 related to the total pressure in the system described in example 15.1?

For the reaction

$$A \rightarrow \text{products},$$

the degree of dissociation α satisfies the equation

$$1 - \alpha = \frac{n_A}{n_{A0}}.$$

When the reactant behaves as an ideal gas, we also have

$$P_A = \frac{n_A}{V}RT = \frac{n_A}{n_{A0}}\frac{n_{A0}}{V}RT = (1 - \alpha)P_0.$$

But from the penultimate equation in example 15.1, we obtain

$$(\nu - 1)\alpha = \frac{P}{P_0} - 1.$$

Thus from pressure measurements and knowledge of ν, a person can determine P_A.

For the N_2O_5 decomposition, we find that

$$\frac{P_A}{P_0} = \left[1 - \frac{2}{3}\left(\frac{P}{P_0} - 1\right)\right] = \frac{5}{3} - \frac{2}{3}\frac{P}{P_0},$$

where A is N_2O_5.

15.2 Reaction Order and Molecularity

The rate of a homogeneous reaction generally depends on the concentration of one or more of the reactants. It may also depend on the concentration of a constituent that is not in the stoichiometric equation. This constituent may be an intermediate substance or a catalyst. Furthermore, the concentration of one or more of the products may affect the rate.

In section 3.15, we saw that the rate of bimolecular collisions varies with the product of the concentrations of the participating molecules. For a reaction between A and B, we suppose that only a fraction of the collisions lead to reaction. Furthermore, this fraction may increase strongly with temperature. At a given temperature, we would have

$$R = k[\mathrm{A}][\mathrm{B}]. \qquad [15.3]$$

For reaction (15.1), one might expect a generalization of law (15.3), such as

$$R = k[\mathrm{A}]^m[\mathrm{B}]^n. \qquad [15.4]$$

When this form holds, exponent m is called the *order* of the reaction with respect to constituent A, exponent n the order with respect to constituent B. The overall order is the sum of the exponents, here $m + n$. Coefficient k is the rate per unit [A] and unit [B], the *specific reaction rate.* This generally varies with temperature T. For the reaction at a given temperature, it may be called the rate coefficient or the *rate constant.*

Most reactions do not proceed as the stoichiometric equation suggests but involve more than one step. Each step in which the indicated products are formed directly from the indicated reactants is called an *elementary reaction.*

The *molecularity* of an elementary reaction is the number of reactant molecules involved in the step. When only one molecule is involved, as with

$$\mathrm{A} \rightarrow \text{products}, \qquad [15.5]$$

the reaction is *unimolecular*. When two molecules react, as with

$$\mathrm{A} + \mathrm{B} \rightarrow \text{products}, \qquad [15.6]$$

the reaction is *bimolecular*. When three molecules are involved, as with

$$\mathrm{A} + \mathrm{B} + \mathrm{C} \rightarrow \text{products}, \qquad [15.7]$$

the reaction is *termolecular*. Higher molecularities are not found in the gas phase.

For an elementary reaction, the molecularity and the order are the same. But for the overall reaction, the order depends on the contributing elementary reactions. Rate determinations may yield numbers between -2 and 3. A negative order with respect to a constituent implies that the constituent slows down the reaction. Such a constituent is called an *inhibitor*. A fractional reaction order indicates the participation of two or more elementary reactions.

15.3 Zero-Order Reactions

Under certain circumstances, the rate of a reaction may be independent of the concentration of any reactant or product. If A is a reactant, we then have

$$R = -\frac{d[\mathrm{A}]}{dt} = k[\mathrm{A}]^0 = k, \qquad [15.8]$$

whence

$$[A] = [A]_0 - kt. \qquad [15.9]$$

And a plot of concentration [A] against time t yields a straight line with intercept $[A]_0$ and slope - k.

This behavior occurs when the rate is determined by the concentration of a catalyst. Parameter k may then be proportional to the concentration of the catalyst. It also occurs in a photochemical reaction when the rate is determined by the light intensity. Parameter k would then be proportional to the light intensity.

15.4 First-Order Reactions

Another possibility is that the given reaction is first order in one reactant. Thus for

$$A \rightarrow \text{products}, \qquad [15.10]$$

we may have

$$R = -\frac{d[A]}{dt} = k[A]. \qquad [15.11]$$

Rearranging equation (15.11),

$$\frac{d[A]}{[A]} = -k\,dt, \qquad [15.12]$$

and integrating leads to

$$\ln[A] = \ln[A]_0 - kt \qquad [15.13]$$

or

$$[A] = [A]_0 e^{-kt}. \qquad [15.14]$$

In analyzing data, one may plot ln [A], or log [A], against time t. If a straight line is obtained, the reaction is first order. The slope then yields coefficient -k, or -k log e, respectively.

A person may rewrite result (15.14) in the form

$$[A] = [A]_0 e^{-t/\tau}. \qquad [15.15]$$

Parameter $\tau = 1/k$ is called the *time constant*, or *decay time*, for the reaction.

As examples of first order reactions, we have various isomerizations and rearrangements, some decompositions.

15.5 Second-Order Reactions

A reaction may be (a) second order in one reactant or (b) first order in two different reactants.

When there is only one species of reactant, we have the process

$$A + A \rightarrow \text{products} \qquad [15.16]$$

with the rate law

$$R = -\frac{1}{2}\frac{d[A]}{dt} = k[A]^2. \qquad [15.17]$$

Rearranging the last equality

$$-\frac{d[A]}{[A]^2} = 2k\,dt \qquad [15.18]$$

and integrating yields

$$\frac{1}{[A]} = \frac{1}{[A]_0} + 2kt. \qquad [15.19]$$

In analyzing data, one may plot the inverse concentration $[A]^{-1}$ against time t. If a straight line is obtained, the reaction is second order. The slope then equals $2k$, the intercept $[A]_0^{-1}$

For two different reactants, we have the process

$$A + B \rightarrow \text{products} \qquad [15.20]$$

with the rate law

$$R = -\frac{d[A]}{dt} = -\frac{d[B]}{dt} = k[A][B]. \qquad [15.21]$$

The two simultaneous equations can be easily reduced to a single differential equation. For, reaction (15.20) implies that [A] and [B] decrease at the same rate. We let x be the concentration units of each which have reacted at time t. Then one has

$$x = a - [A] = b - [B] \qquad [15.22]$$

and

$$\frac{dx}{dt} = -\frac{d[A]}{dt} = -\frac{d[B]}{dt}, \qquad [15.23]$$

where

$$a = [A]_0 \quad \text{and} \quad b = [B]_0. \qquad [15.24]$$

Substitute into equation (15.21),

$$\frac{dx}{dt} = k(a - x)(b - x), \qquad [15.25]$$

rearrange,

$$\frac{dx}{(a - x)(b - x)} = k\,dt, \qquad [15.26]$$

and expand into partial fractions,

$$\frac{dx}{(b - a)(a - x)} + \frac{dx}{(a - b)(b - x)} = k\,dt. \qquad [15.27]$$

Then integrate to get

$$\frac{1}{a - b}\left[\ln(a - x) - \ln(b - x)\right] - \frac{1}{a - b}\left[\ln a - \ln b\right] = kt \qquad [15.28]$$

or

$$\frac{1}{a-b}\ln\frac{b(a-x)}{a(b-x)}=kt, \qquad [15.29]$$

whence

$$\ln\frac{[\mathrm{A}]}{[\mathrm{B}]}=(a-b)kt+\ln\frac{a}{b}. \qquad [15.30]$$

In analyzing experimental data, one may plot ln ([A]/ [B]), or log ([A]/ [B]) against time t. If a straight line is obtained, the reaction is second order. The slope yields coefficient $(a - b)k$, or $(a - b)k \log e$, respectively.

But when the initial concentrations are equal, the differential equation reduces to (15.18) without the 2 factor. The solution is

$$\frac{1}{[\mathrm{A}]}=\frac{1}{[\mathrm{B}]}=\frac{1}{a}+kt. \qquad [15.31]$$

A plot of $|\mathrm{A}]^{-1}$ against t then yields a straight line with the slope k

As examples of second order reactions, we have various double decomposition reactions and displacement reactions.

15.6 Third Order Reactions

A reaction may be (a) third order in a single reactant, (b) second order in one reactant and first order in a second reactant, or (c) first order in three different reactants.

For the first type, we have the process

$$\mathrm{A}+\mathrm{A}+\mathrm{A}\rightarrow \text{products} \qquad [15.32]$$

with the rate law

$$R=-\frac{1}{3}\frac{\mathrm{d}[\mathrm{A}]}{\mathrm{d}t}=k[\mathrm{A}]^3. \qquad [15.33]$$

Rearranging the last equality

$$-\frac{\mathrm{d}[\mathrm{A}]}{[\mathrm{A}]^3}=3k\,\mathrm{d}t \qquad [15.34]$$

and integrating yields

$$\frac{1}{[\mathrm{A}]^2}=\frac{1}{[\mathrm{A}]_0^2}+6kt. \qquad [15.35]$$

Here one would plot the inverse squared concentration $[\mathrm{A}]^{-2}$ against time t. A straight line result would establish the third order rate law. The slope would equal $6k$, the intercept $[\mathrm{A}]_0^{-2}$

For the second type, we have the process

$$2\mathrm{A}+\mathrm{B}\rightarrow \text{products} \qquad [15.36]$$

with the rate law

$$R=-\frac{1}{2}\frac{\mathrm{d}[\mathrm{A}]}{\mathrm{d}t}=-\frac{\mathrm{d}[\mathrm{B}]}{\mathrm{d}t}=k[\mathrm{A}]^2[\mathrm{B}]. \qquad [15.37]$$

The two simultaneous equations can be easily reduced to a single differential equation. Indeed, reaction (15.36) implies that [A] decreases at twice the rate at which [B] decreases. Letting x be the concentration units of B which have reacted at time t, one has

$$x = \frac{1}{2}\left(a - [\mathrm{A}]\right) = b - [\mathrm{B}] \qquad [15.38]$$

where

$$a = [\mathrm{A}]_0 \quad \text{and} \quad b = [\mathrm{B}]_0. \qquad [15.39]$$

Substitute into line (15.37),

$$\frac{\mathrm{d}x}{\mathrm{d}t} = k(a - 2x)^2(b - x), \qquad [15.40]$$

rearrange

$$\frac{\mathrm{d}x}{(a - 2x)^2(b - x)} = k\,\mathrm{d}t, \qquad [15.41]$$

and expand into partial fractions. On integration, we obtain

$$\frac{1}{a - 2b}\left(\frac{1}{a} - \frac{1}{[\mathrm{A}]}\right) + \frac{1}{(a - 2b)^2}\ln\frac{[\mathrm{A}]b}{a[\mathrm{B}]} = kt. \qquad [15.42]$$

If a person makes the concentration of B much larger than the concentration of A initially, then

$$2b >> a \qquad [15.43]$$

and equation (15.42) reduces to the form

$$\frac{1}{[\mathrm{A}]} = \frac{1}{a} + 2bkt, \qquad [15.44]$$

which agrees with (15.19), The reaction is then pseudo second order.

On the other hand, if one sets the concentration of A equal to twice the concentration of B, then $a = 2b$ and equation (15.41) becomes

$$\frac{\mathrm{d}x}{(b - x)^3} = 4k\,\mathrm{d}t. \qquad [15.45]$$

Integrating yields

$$\frac{1}{(b - x)^2} - \frac{1}{b^2} = 2kt \qquad [15.46]$$

or

$$\frac{1}{[\mathrm{B}]^2} = \frac{1}{b^2} + 2kt. \qquad [15.47]$$

One may test the reaction to see if it is third order by starting with equivalent amounts of each reactant in the reaction vessel, The equivalent concentration of either reactant is determined as a function of time at a given temperature. Then the square of the reciprocal of this concentration is plotted against time. A straight line would establish the third order rate law. The slope would give $2k$, the intercept b^{-2}.

When a reaction is first order in three different reactants

$$A + B + C \rightarrow \text{products,} \tag{15.48}$$

the rate law is

$$R = -\frac{d[A]}{dt} = -\frac{d[B]}{dt} = -\frac{d[C]}{dt} = k[A][B][C]. \tag{15.49}$$

Reaction (15.48) implies that [A], [B], and [C] decrease at the same rate. Let x be the concentration units of A, B, or C reacted at time t. Then the three simultaneous equations reduce to the equation

$$\frac{dx}{dt} = k(a - x)(b - x)(c - x) \tag{15.50}$$

or

$$\frac{dx}{(a - x)(b - x)(c - x)} = k\ dt. \tag{15.51}$$

This can be integrated by expanding the left side into partial fractions.

If one starts with all three concentrations equal, with $a = b = c$, one has

$$\frac{dx}{(a - x)^3} = k\ dt. \tag{15.52}$$

Integrating then yields

$$\frac{1}{(a - x)^2} - \frac{1}{a^2} = 2kt. \tag{15.53}$$

or

$$\frac{1}{[A]^2} = \frac{1}{a^2} + 2kt. \tag{15.54}$$

A third order reaction often involves two steps. Two of the reactants combine to form an unstable complex. Some of these complexes are struck by the third reactant before they separate. This collision leads to formation of the products.

15.7 Determining the Differential Rate Law

A person can determine the order of a given reaction with respect to one or more reactants by measuring the rate at different concentrations at the given temperature and analyzing the results in a particular way.

Consider the reaction of a single reactant. If a is the initial concentration of the reactant and x equals the concentration units that have reacted at time t, the rate law may have the form

$$\frac{dx}{dt} = k(a - x)^n \tag{15.55}$$

in which n is the order and k the specific reaction rate. Taking the logarithm of both sides of equation (15.55) leads to

$$\log\frac{dx}{dt} = n\log(a - x) + \log k. \tag{15.56}$$

In practice, concentration $a - x$ is measured at various times. Then x is plotted against t and the slope dx/dt is measured at appropriate x's. The corresponding $(a - x)$'s are calculated and the log (dx/dt) is plotted against log $(a - x)$. If the reaction is of definite order, the points will define a straight line. The slope of the line gives the order n. See figure 15.1.

Now consider a reaction involving two or more reactants. One may start with equivalent concentrations of each. Then equivalent concentrations of the reactants will be present throughout the run and the rate law reduces to form (15.55) with n the overall order. A person can then proceed as previously described to determine n.

To find the order with respect to one of the reactants, an investigator introduces large excesses of all reactants except for this one. Then only the chosen reactant will exhibit a varying concentration and the above procedure will yield the order with respect to it, if indeed this number exists.

Example 15.3

Using a given temperature and initial pressure, a sample of acetaldebyde decomposed at 7.49 torr min^{-1} when 5.0 per cent had reacted and at 5.14 torr min^{-1} when 20.0 per cent had reacted. Estimate the order of the reaction.

With only two data points, we do not need to plot. Instead, put the numbers into formula (15.56),

$$\log 7.49 = n \log 95 + \log k,$$
$$\log 5.14 = n \log 80 + \log k,$$

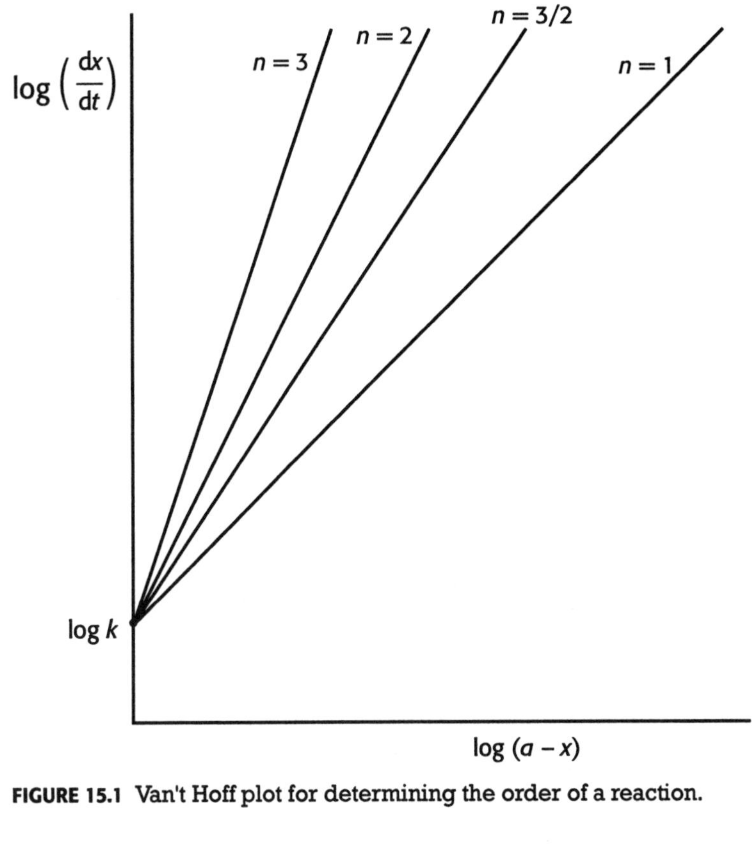

FIGURE 15.1 Van't Hoff plot for determining the order of a reaction.

and eliminate log k by subtraction. Thus we obtain

$$\log 7.49 - \log 5.14 = n \log 95 - n \log 80,$$

whence

$$n = \frac{\log(7.49/5.14)}{\log(95/80)} = 2.19$$

The reaction is probably second order.

15.8 Determining Order from an Integrated Form

A person can find the order of a given reaction with respect to one or more reactants from the variation of the extent of reaction with time and from the variation of time for a given extent of reaction with initial concentration.

Consider any reaction conducted under conditions such that the rate law

$$\frac{dx}{dt} = k(a - x)^n \qquad [15.57]$$

applies. Rearranging this equation

$$\frac{dx}{(a - x)^n} = k\ dt, \qquad [15.58]$$

and integrating with $n \neq 1$ yields

$$\frac{1}{n-1}\left[\frac{1}{(a-x)^{n-1}} - \frac{1}{a^{n-1}}\right] = kt. \qquad [15.59]$$

Into this formula, substitute the fraction unreacted,

$$\frac{a - x}{a} = \alpha, \qquad [15.60]$$

and rearrange to get

$$\alpha^{1-n} - 1 = (n-1)a^{n-1}kt = (n-1)\tau. \qquad [15.61]$$

To determine the order of a reaction, an investigator may plot observed α's against log t. On another sheet, he or she would plot against log τ theoretical α's calculated from formula (15.61) for n's that yield nearly the same shape of curve. Since

$$\log \tau = \log t + \log(a^{n-1}k), \qquad [15.62]$$

the experimental curve should match one of the theoretical curves with a shift along the log t axis, if the reaction is of definite order. The n used in plotting the matching curve yields the reaction order.

Alternatively, one may solve equation (15.61) for time t, obtaining

$$t = \frac{\alpha^{1-n} - 1}{(n-1)a^{n-1}k} = f(\alpha, n, k)a^{1-n}, \qquad [15.63]$$

whence

$$\log t_{1-\alpha} = (1-n)\log a + \log f_\alpha. \qquad [15.64]$$

For a reaction of order n at a given temperature T, expression f_α depends only on the fraction unreacted α at time $t_{1-\alpha}$.

In analyzing a given run, a person may plot the observed reactant concentration against time. Then one chooses evenly spaced concentrations along the curve as initial values. The time needed for each of these concentrations to be reduced by a chosen fraction 1 - α is measured. (For convenience, 1 - α may be chosen as 1/2.) Then log $t_{1-\alpha}$ is plotted against log a, where a symbolizes the chosen initial values. If the reaction is of definite order n, a straight line is obtained with the slope 1 - n.

Finally, equation (15.61) can be solved for k;

$$k = \frac{\alpha^{1-n} - 1}{(n-1)a^{n-1}t_{1-\alpha}}. \qquad [15.65]$$

With order n determined, one can employ this equation to calculate the specific reaction rate k from the data.

Example 15.4

At 518° C, the half life of acetaldebyde was 410 s when the initial pressure was 363 torr and 880 s when the initial pressure was 169 torr Calculate the reaction order.

Again, we have only two data points. So we put the numbers into formula (15.64),

$$\log 410 = (1-n)\log 363 + \log f,$$
$$\log 880 = (1-n)\log 169 + \log f,$$

and eliminate log f by subtraction. Thus we obtain

$$\log 880 - \log 410 = (n-1)\log 363 - (n-1)\log 169,$$

whence

$$n - 1 = \frac{\log(880/410)}{\log(363/169)} = 1.00$$

and

$$n = 2.00\ .$$

From the half lives, we conclude that the reaction is second order.

Example 15.5

How does formula (15.65) reduce when the reaction is second order and the fraction reacted is 1/2?

First solve for $t_{1-\alpha}$,

$$t_{1-\alpha} = \frac{\alpha^{1-n} - 1}{(n-1)ka^{n-1}}.$$

Then consider that n is 2 and set 1 - α equal to 1/2. We obtain

$$t_{1/2} = \frac{(1/2)^{-1} - 1}{(1)ka^1} = \frac{1}{k}\frac{1}{a}.$$

When a reaction is second order, a plot of the half life $t_{1/2}$ against the reciprocal of the chosen initial values should be a straight line with the slope $1/k$.

Example 15.6

Repeat the calculation in example 15.5 for a fraction reacted of 1/4.

Now α is 3/4. So substituting into the rearranged formula (15.65) leads to

$$t_{1/4} = \frac{(3/4)^{-1} - 1}{(1)ka^{1}} = \frac{1}{3k}\frac{1}{a}.$$

When a reaction is second order, a plot of the quarter life $t_{1/4}$ against the reciprocal of the chosen initial values should be a straight line with the slope $1/(3k)$. Alternatively, one may divide $1/a$ by $t_{1/4}$ for the chosen intervals. Approximate constancy would indicate that the reaction is second order. One would then proceed to average the results and divide by 3 to get k.

15.9 The Arrhenius Equation

The rate expression for a simple reaction consists of a rate constant k times a product of concentrations of various constituents raised to various powers. For most systems, the rate constant varies markedly with the temperature T. An approximate formulation of this dependence was made by Svante Arrhenius in 1889.

From rather extensive experimental data, Arrhenius found that

$$k = A\exp(-E_a / RT) \qquad [15.66]$$

whence

$$\ln k = \ln A - \frac{E_a}{RT} \qquad [15.67]$$

or

$$\log k = \log A - \frac{(\log e)E_a}{RT}. \qquad [15.68]$$

For a given reaction, a plot of $\ln k$, or of $\log k$, against $1/T$ yields an approximately straight line with the slope $-E_a/R$, or $-(\log e)\, E_a/R$, and intercept $\ln A$, or $\log A$, respectively. Here R is the gas constant.

The pre-exponential factor A is called the *frequency factor*. The energy E_a is called the *activation energy*. On comparing the Arrhenius equation with the Boltzmann distribution law, we see that E_a appears to be the energy the reactants must be given for any reaction to occur.

Indeed, we suppose that a reacting complex moves from a stable initial configuration to a stable final configuration over an energy barrier such as figure 15.2 shows. Furthermore, we have

$$\Delta H^0 = E_a(\text{forward}) - E_a(\text{reverse}), \qquad [15.69]$$

where ΔH^0 is the enthalpy of reaction. When the forward activation energy is less than the reverse activation energy, the reaction is exothermic. When the forward activation energy is greater than the reverse activation energy, the reaction is endothermic.

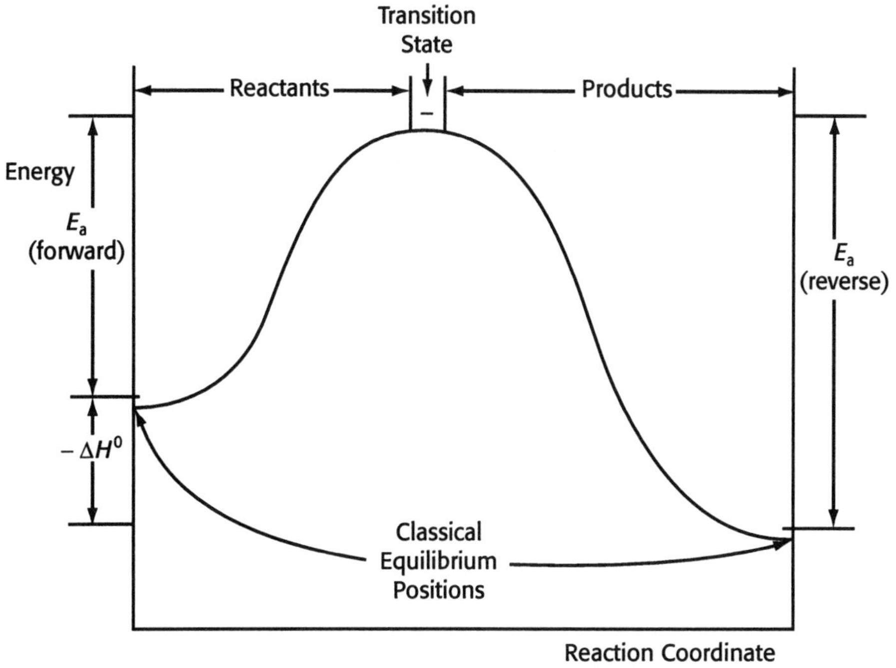

FIGURE 15.2 Variation of the effective potential energy along the reaction path between points of stable classical equilibrium.

Example 15.7

For the reaction

$$H_2 + I_2 \rightarrow 2HI,$$

rate constant k is 8.90×10^{-5} M^{-1} s^{-1} at 283° C and 5.04×10^{-3} M^{-1} s^{-1} at 356° C. Calculate the Arrhenius parameters.

Apply formula (15,67) to the two data points k_1, T_1 and k_2, T_2. Then eliminate $\ln A$ to form

$$\ln \frac{k_2}{k_1} = \frac{E_a}{R}\left(\frac{1}{T_1} - \frac{1}{T_2}\right) = \frac{E_a}{R}\frac{T_2 - T_1}{T_1 T_2}$$

whence

$$E_a = \frac{RT_1T_2 \ln\left(k_2 / k_1\right)}{T_2 - T_1}.$$

With the given numbers, we have

$$E_a = \frac{\left(8.3145 \text{ J K}^{-1} \text{ mol}^{-1}\right)\left(556.15 \text{ K}\right)\left(629.15\right)\ln\left(5.04 \times 10^{-3}\right) / \left(8.90 \times 10^{-5}\right)}{73 \text{ K}} = 160.9 \text{ kJ mol}^{-1}.$$

For the frequency factor, substitute into formula (15.67) rearranged,

$$\ln A = \ln 8.90 \times 10^{-5} + \frac{160.9 \times 10^3 \text{J mol}^{-1}}{\left(8.3145 \text{ kJ mol}^{-1}\right)\left(556.15 \text{ K}\right)} = 25.462,$$

and solve for

$$A = 1.14 \times 10^{11}\ \mathrm{M^{-1}s^{-1}}.$$

15.10 Reversible First-Order Reactions

In many systems, both forward and backward reactions need to be considered. For various isomerizations and rearrangements, these are first order. Consider a system in which both processes

$$\mathrm{A} \xrightarrow{k_1} \mathrm{B} \quad \text{and} \quad \mathrm{B} \xrightarrow{k_{-1}} \mathrm{A} \qquad [15.70]$$

are significant. Here k_1 and k_{-1} are the rate constants for the forward (A to B) and the backward (B to A) reactions. The rate law is

$$\frac{\mathrm{d}[\mathrm{A}]}{\mathrm{d}t} = -\frac{\mathrm{d}[\mathrm{B}]}{\mathrm{d}t} = -k_1[\mathrm{A}] + k_{-1}[\mathrm{B}]. \qquad [15.71]$$

Now let a be the equilibrium concentration of A, b the equilibrium concentration of B, at a given temperature. Also let x be the excess concentration units of A present at time t. Thus we set

$$[\mathrm{A}] = a + x, \qquad [\mathrm{B}] = b - x. \qquad [15.72]$$

Since a and b are constant at the given temperature,

$$\frac{\mathrm{d}[\mathrm{A}]}{\mathrm{d}t} = -\frac{\mathrm{d}[\mathrm{B}]}{\mathrm{d}t} = \frac{\mathrm{d}x}{\mathrm{d}t}, \qquad [15.73]$$

equation (15.71) reduces to

$$\frac{\mathrm{d}x}{\mathrm{d}t} = -k_1(a + x) + k_{-1}(b - x) = -k_1 a + k_{-1} b - (k_1 + k_{-1})x. \qquad [15.74]$$

But at equilibrium, the excess concentration of A is zero,

$$x = 0 \quad \text{and} \quad \frac{\mathrm{d}x}{\mathrm{d}t} = 0. \qquad [15.75]$$

As a consequence, we have

$$k_{-1} b = k_1 a \qquad [15.76]$$

and

$$\frac{b}{a} = \frac{k_1}{k_{-1}} = K_c. \qquad [15.77]$$

Parameter K_c is the equilibrium constant when the activity coefficients equal 1.

From (15.74) and (15.76), we obtain the differential equation

$$\frac{\mathrm{d}x}{x} = -(k_1 + k_{-1})\mathrm{d}t, \qquad [15.78]$$

which integrates to

$$\ln x = \ln x_0 - (k_1 + k_{-1})t \qquad [15.79]$$

or

$$x = x_0 e^{-(k_1 + k_{-1})t} = x_0 e^{-t/\tau}. \qquad [15.80]$$

The time constant or relaxation time for the reaction is

$$\tau = \frac{1}{k_1 + k_{-1}} \tag{15.81}$$

On comparing formulas (15.79) and (15.80) with (15.13) and (15.14), we see that initial concentration $[A]_0$ has been replaced with the initial excess concentration x_0 and rate constant k has been replaced with the sum $k_1 + k_{-1}$. The form of the integrated expression is unchanged.

15.11 First-Order Reaction Opposed by a Second-Order Reaction

Various decompositions are first order in the forward direction, second order in the backward direction. But when such a system is not far from equilibrium, the second-order reaction can be approximated as a first order reaction.

Consider a system in which both processes

$$A \xrightarrow{k_1} B + C \quad \text{and} \quad B + C \xrightarrow{k_{-1}} A \tag{15.82}$$

are significant. The rate law is

$$\frac{d[A]}{dt} = -\frac{d[B]}{dt} = -\frac{d[C]}{dt} = -k_1[A] + k_{-1}[B][C]. \tag{15.83}$$

Let a be the equilibrium concentration of A, b the equilibrium concentration of B, c the equilibrium concentration of C, at a given temperature. Also let x be the excess concentration units of A present at time t. Thus we set

$$[A] = a + x, \qquad [B] = b - x, \qquad [C] = c - x. \tag{15.84}$$

Since a, b, and c are constant in a given system at a given temperature

$$\frac{d[A]}{dt} = -\frac{d[B]}{dt} = -\frac{d[C]}{dt} = \frac{dx}{dt}, \tag{15.85}$$

equation (15.83) reduces to

$$\frac{dx}{dt} = -k_1(a + x) + k_{-1}(b - x)(c - x) = -k_1 a + k_{-1}bc - [k_1 + k_{-1}(b + c)]x + k_{-1}x^2. \tag{15.86}$$

At equilibrium, the excess concentration of A is zero,

$$x = 0 \quad \text{and} \quad dx/dt = 0 \tag{15.87}$$

With condition (15.86), we then have

$$k_{-1}bc = k_1 a \tag{15.88}$$

and

$$\frac{bc}{a} = \frac{k_1}{k_{-1}} = K_c. \tag{15.89}$$

Here K_c is the equilibrium constant.

When the system is close to equilibrium, $x^2 \cong 0$ and equation (15.86) reduces to

$$\frac{dx}{dt} = -[k_1 + k_{-1}(b + c)]x. \tag{15.90}$$

Integration yields

$$x = x_0 e^{-\left[k_1 + k_{-1}\left(b+c\right)\right]t} = x_0 e^{-t/\tau}. \qquad [15.91]$$

The time constant or relaxation time for the reaction is

$$\tau = \frac{1}{k_1 + k_{-1}\left(b+c\right)} = \frac{1}{k_{-1}\left(K_c + b + c\right)}. \qquad [15.92]$$

From the relaxation time, the equilibrium constant, and the equilibrium concentrations of B and C, a person can calculate k_{-1}. From k_{-1} and K_c, one can then calculate k_1.

15.12 Perturbation Experiments

In studying a fast reaction, one may perturb the system and then measure the relaxation time to the new equilibrium point. Knowing the equilibrium constant and the equilibrium concentrations at this point, one can then calculate the forward and backward rate constants.

Commonly, a person introduces a fast increase in temperature. This may be accomplished by suddenly releasing the charge on a large capacitor through a small volume of solution containing the reactants and products. Thus, 42 J of electrical energy may be discharged in 1 μs through 10 cm^3 of electrolytic solution. A rise of about 1° C would be produced. This would shift the equilibrium point following the van't Hoff equation (7.90).

The relaxation to the new equilibrium point may be followed by optical absorption. Some indicator may be added to enhance the effect. For ionic reactions, a person may follow the change in electrical conductance with time.

Alternatively, one may introduce a fast increase in pressure. Thus, a volume of compressed gas may be released into a chamber containing the solution. This would shift the equilibrium point following equation (7.96).

To investigate the properties of an excited electronic state, a person may subject the given system to a pulsed laser operating at the appropriate frequency. For weak acids, there may be a considerable change in the ionization constant. Relaxation to the new point of equilibrium would be followed by tracking the pH change.

Example 15.8

At 25° C, the process

$$H_2O\left(l\right) \rightleftarrows H^+\left(aq\right) + OH^-\left(aq\right)$$

relaxes in 37 μs. With

$$K_{\gamma c} = 1.002 \times 10^{-14}\ M^2,$$

calculate the rate constants for the forward and backward reactions.

The discussion in section 15.11 applies. For the equilibrium concentrations of H^+ (aq) and OH^- (aq), we have

$$b = c = \left(1.002 \times 10^{-14}\ M^2\right)^{1/2} = 1.001 \times 10^{-7}\ M.$$

The concentration of H_2O is

$$\left[H_2O\right] = \frac{1000\ g\ l^{-1}}{18.015\ g\ mol^{-1}} = 55.51\ M$$

So for expression K_c in formulas (15.89) and (15.92), we obtain

$$K_c = \frac{[H^+][OH^-]}{[H_2O]} = \frac{1.002 \times 10^{-14}\ \mathrm{M}^2}{55.51\ \mathrm{M}} = 1.805 \times 10^{-16}\ \mathrm{M}.$$

Solving equation (15.92) for k_{-1} and substituting in these numbers yields

$$k_{-1} = \frac{1}{\tau(K_c + b + c)} = \frac{1}{(37 \times 10^{-6}\ \mathrm{s})(2.002 \times 10^{-7}\ \mathrm{M})} = 1.35 \times 10^{11}\ \mathrm{M}^{-1}\ \mathrm{s}^{-1}.$$

Then solving (15.89) for k_{-1} gives us

$$k_1 = k_{-1}K_c = (1.35 \times 10^{11}\ \mathrm{M}^{-1}\ \mathrm{s}^{-1})(1.805 \times 10^{-16}\ \mathrm{M}) = 2.44 \times 10^{-5}\ \mathrm{s}^{-1}.$$

Note that the reaction between the hydrogen and the hydroxide ions is very fast. Parameter k_{-1} is approximately equal to the normal binary gas collision rate constant.

15.13 Reactions in High Pressure Gaseous and in Condensed Phases

In a low pressure gaseous solution, solvent molecules do not interfere appreciably with movement of solute molecules and collisions between two different species of solute molecules occur randomly in both space and time. But as the pressure and density are increased, a stage is reached at which the interference becomes appreciable. In a high pressure gaseous solution and in a condensed phase solution, each molecular pair that is formed tends to be caged in by neighboring solvent molecules. Then the two molecules may collide repeatedly before escaping from the cage and the collision rate is no longer random in time. Formation of a caged pair from two different solute molecules is called an *encounter*. The pair is called an *encounter complex*.

The rate at which two reactant molecules come together, on average, is governed by the phenomenological laws of diffusion and conduction. If the activation energy for chemical reaction is low, the encounter frequency governs the rate of reaction.

The processes taking place between reactants A and B are

$$\mathrm{A} + \mathrm{B} \underset{k_{-1}}{\overset{k_1}{\rightleftarrows}} \mathrm{AB} \xrightarrow{k_2} \text{products}. \qquad [15.93]$$

Here AB is the encounter complex. Reactions (15.93) have the rate laws

$$-\frac{d[\mathrm{A}]}{dt} = k_1[\mathrm{A}][\mathrm{B}] - k_{-1}[\mathrm{AB}], \qquad [15.94]$$

$$\frac{d[\mathrm{AB}]}{dt} = k_1[\mathrm{A}][\mathrm{B}] - k_{-1}[\mathrm{AB}] - k_2[\mathrm{AB}]. \qquad [15.95]$$

We expect the concentration of the intermediate AB to vary only slowly and to remain relatively small throughout the reaction. So we set

$$\frac{d[\mathrm{AB}]}{dt} \cong 0. \qquad [15.96]$$

Taking the rate of change of any intermediate as zero is known as the *steady-state approximation.*

Introducing condition (15.96) into equation (15.95) and solving for the concentration of AB yields

$$[\mathrm{AB}] = \frac{k_1[\mathrm{A}][\mathrm{B}]}{k_{-1} + k_2}. \qquad [15.97]$$

Substituting this result into equation (15.94) and reducing leads to the rate law

$$-\frac{\mathrm{d}[\mathrm{A}]}{\mathrm{d}t} = \frac{k_1 k_2}{k_{-1} + k_2}[\mathrm{A}][\mathrm{B}] = k[\mathrm{A}][\mathrm{B}]. \qquad [15.98]$$

In the last step, the coefficient of the concentration product is designated k

For a fast reaction, where

$$k_2 >> k_{-1}, \qquad [15.99]$$

parameter k_{-1} can be neglected in the denominator of k and we find that

$$k = k_1. \qquad [15.100]$$

The reaction rate then reduces to the encounter rate; the overall reaction is said to be *diffusion-limited*, For a slow reaction, where

$$k_2 << k_{-1}, \qquad [15.101]$$

parameter k_2 can be neglected in the denominator of k and we have

$$k = \frac{k_1}{k_{-1}} k_2 = Kk^2. \qquad [15.102]$$

Here K is the equilibrium constant for formation of the encounter complex from the reactants.

15.14 The Encounter Rate Constants

In high pressure gaseous and in condensed phases, solute molecules A and B come together to form the complex AB by a diffusion process. Furthermore, the dissociation of the complex into separate molecules A and B is a diffusion process. In the continuum approximation, these processes are governed by the formulas in section 5.13.

Consider reactant molecules A and B in a solution at a given temperature. Suppose that at a given time t the centers of A and B are distance r apart, as figure 15.3 shows. Let D_A be the diffusion coefficient for A, D_B the diffusion coefficient for B. Now, the concentration gradient governing the average flux density J_A of A molecules towards a B molecule is $\mathrm{d}[\mathrm{A}]_r/\mathrm{d}r$. Here $[\mathrm{A}]_r$ is the average concentration of A when the A-B separation is r.

Since both A and B molecules are diffusing, the effective diffusion coefficient for coordinate r is $D_A + D_B$ So from formula (5.104), we construct

$$J_{\mathrm{A},r} = -(D_{\mathrm{A}} + D_{\mathrm{B}})\frac{\mathrm{d}[\mathrm{A}]_r}{\mathrm{d}r} \qquad [15.103]$$

for the flux density of A at point r. The total flux towards the center of B, assuming spherical symmetry about this center and steady state conditions, is

$$-4\pi r^2 J_{\mathrm{A},r} = -4\pi R^2 J_{\mathrm{A},R}. \qquad [15.104]$$

Here R is the intercenter distance when the encounter has occurred, as figure 15.4 illustrates.

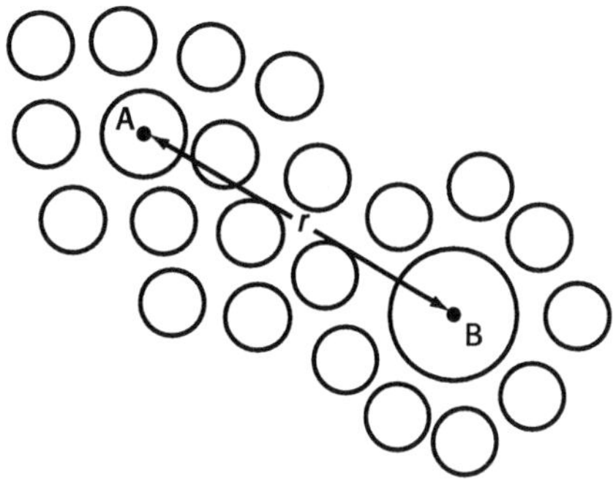

FIGURE 15.3 Reactant molecules A and B separated by solvent, each diffusing separately towards each other.

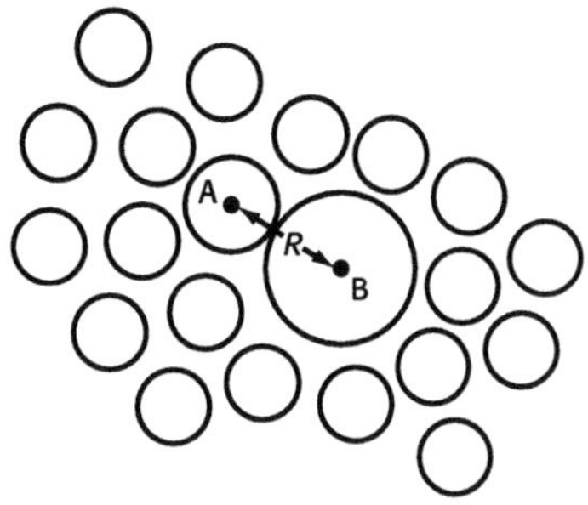

FIGURE 15.4 Reactant molecules A and B after diffusing together to form an encounter complex AB.

The total flux of A across the sphere of radius R about a B times the B concentration equals the encounter rate

$$k_1[\mathrm{A}][\mathrm{B}] = -4\pi R^2 J_{\mathrm{A},R}[\mathrm{B}] = -4\pi r^2 J_{\mathrm{A},r}[\mathrm{B}]. \qquad [15.105]$$

Introducing relation (15.103) into (15.105) and canceling [B] yields

$$k_1[\mathrm{A}] = 4\pi r^2 (D_\mathrm{A} + D_\mathrm{B}) \frac{\mathrm{d}[\mathrm{A}]_r}{\mathrm{d}r}. \qquad [15.106]$$

Then separating the variables and integrating,

$$k_1[\mathrm{A}] \int_R^\infty \frac{\mathrm{d}r}{r^2} = 4\pi (D_\mathrm{A} + D_\mathrm{B}) \int_{[\mathrm{A}]_R}^{[\mathrm{A}]} \mathrm{d}[\mathrm{A}]_r, \qquad [15.107]$$

leads to

$$k_1[\mathrm{A}] \frac{1}{R} = 4\pi (D_\mathrm{A} + D_\mathrm{B}) \left([\mathrm{A}] - [\mathrm{A}]_R\right). \qquad [15.108]$$

We consider that as the intercenter distance r decreases to R, the concentrations of A and B decrease to $[\mathrm{A}]_R$ and $[\mathrm{B}]_R$, while the concentration of AB increases from zero to $[\mathrm{AB}]_R$. Paralleling the negative concentration gradient - d $[\mathrm{A}]_r/\mathrm{d}r$ driving the association reaction is the negative concentration gradient - $\mathrm{d}[\mathrm{AB}]_r/\mathrm{d}r$ driving the dissociation reaction. Furthermore, in the latter, molecules A and B are diffusing apart. So we may take the diffusion coefficient to be $D_\mathrm{A} + D_\mathrm{B}$ as before.

For the flux density of AB at point r, we write

$$J_{\mathrm{AB},r} = -(D_\mathrm{A} + D_\mathrm{B}) \frac{\mathrm{d}[\mathrm{AB}]_r}{\mathrm{d}r} \qquad [15.109]$$

In the steady state, the total flux is

$$4\pi r^2 J_{AB,r} = 4\pi R^2 J_{AB,R}. \qquad [15.110]$$

The movement of AB out across the sphere of radius R times the concentration of AB equals the dissociation rate; thus

$$k_{-1}[AB]_R = 4\pi R^2 J_{AB,R}[AB]_R \qquad [15.111]$$

whence

$$k_{-1} = 4\pi R^2 J_{AB,R} = 4\pi r^2 J_{AB,r} = -4\pi r^2 (D_A + D_B)\frac{d[AB]_r}{dr}. \qquad [15.112]$$

Separating variables and integrating,

$$k_{-1}\int_R^{\infty}\frac{dr}{r^2} = -4\pi(D_A + D_B)\int_{[AB]_R}^{0} d[AB]_r, \qquad [15.113]$$

yields

$$k_{-1}\frac{1}{R} = 4\pi(D_A + D_B)[AB]_R. \qquad [15.114]$$

Within the sphere of radius R there is one encounter complex at the pertinent time. So if ΔV is the volume of this sphere, the numerical concentration is

$$[AB]_R = \frac{1}{\Delta V}, \qquad [15.115]$$

while the molar concentration is

$$[AB]_R = \frac{1}{N_A \Delta V}. \qquad [15.116]$$

In calculating the equilibrium constant for formation of the encounter complex, we may set $[A]_R$ in equation (15.108) equal to zero. Then we obtain

$$K = \frac{k_1}{k_{-1}} = N_A \Delta V. \qquad [15.117]$$

So far, we have considered only diffusive forces. But if A and B are charged, formation of the encounter complex involves a change in the potential energy $\phi(R)$. One then has to multiply the right side of (15.117) by the Boltzmann factor. If R is in centimeters, we then have

$$K = \frac{4}{3}\pi R^3 10^{-3} N_A e^{-\phi(R)/kT} \qquad [15.118]$$

for concentrations in moles per lifer (M).

15.15 Solution Reaction Rates

When A and B are not charged, the reaction to products may take place in both a low pressure gas phase and in solutions where the cage effect becomes significant. We expect that the activation energy for the reaction leading from encounter to products will not be altered much by the enveloping solvent. However, the encounter rate is.

Consider the sequence in line (15.93). Also suppose that the activation energy is small so that

$$[A]-[A]_R \cong [A]. \qquad [15.119]$$

Then formulas (15.100) and (15.108) reduce to

$$k = k_1 = 4\pi\left(D_A + D_B\right)R \equiv k_D. \qquad [15.120]$$

In the last symbol, the subscript D indicates that this k is the diffusion-limited rate constant.

As an approximation, introduce the Einstein-Stokes formula (5.113) for the diffusion coefficients,

$$D_A = \frac{kT}{6\pi\eta r_A}, \qquad D_B = \frac{kT}{6\pi\eta r_B}. \qquad [15.121]$$

Here η is the viscosity of the solvent while r_A and r_B are the radii of A and B. For simplicity, take

$$r_A \cong r_B = \frac{1}{2}R. \qquad [15.122]$$

Substituting into equation (15,120) now yields

$$k_D = 4\pi\left(\frac{4kT}{6\pi\eta R}\right)R = \frac{8kT}{3\eta}. \qquad [15.123]$$

This is the rate constant for [A] and [B] expressed numerically. Each concentration must be divided by Avogadro's number when we go to moles per unit volume. And the bimolecular rate constant has to be multiplied by Avogadro's number. Then we have

$$k_D = \frac{8R_gT}{3\eta}, \qquad [15.124]$$

where R_g is the molar gas constant $N_A k$.

When the activation energy is larger, concentration $[A]_R$ becomes appreciable with respect to concentration [A]. But if k_R is the rate constant for reaction during an encounter, we have

$$-\frac{d[A]}{dt} = k_R[A]_R[B] = k[A][B], \qquad [15.125]$$

whence

$$[A]_R = \frac{k}{k_R}[A]. \qquad [15.126]$$

Substituting (15.126) into formula (15.108), canceling [A], and introducing k_D for $4\pi(D_A + D_B)R$ yields

$$k_1 = k_D\left(1 - \frac{k}{k_R}\right). \qquad [15.127]$$

As an approximation, one may replace k_1 on the left with k. Then solving for k leads to

$$k = \frac{k_D}{1 + k_D / k_R}. \qquad [15.128]$$

When molecules A and B are identical, the preceding discussions count each reaction twice. Recall the similar statement in section 3.15. Then for a diffusion-limited process, we have

$$-\frac{1}{2}\frac{d[A]}{dt} = \frac{4R_gT}{3\eta}[A]^2 = k_D[A]^2. \quad [15.129]$$

Example 15.9

Calculate the diffusion-limited rate constant for the reaction between A and B in water at 25° C, where the viscosity is 0.891×10^{-3} kg m^{-1} s^{-1}.

Combine the parameters as formula (15.124) directs:

$$k_D = \frac{8R_gT}{3\eta} = \frac{8(8.3145 \text{ J K}^{-1} \text{ mol}^{-1})(298.15 \text{ K})(10^3 \text{ l m}^{-3})}{3(0.891\times10^{-3} \text{ kg m}^{-1} \text{ s}^{-1})} = 7.42\times10^9 \text{ M}^{-1} \text{ s}^{-1}.$$

Example 15.10

Calculate the zero-activation-energy rate constant for the gas phase reaction between A and B at 25° C. Consider that $r_A = r_B = 2.2 \times 10^{-10}$ m while $M_A = M_B = 15$ g mol^{-1}.

From formulas (3.99) and (3.102), we see that

$$k_{AB} = \sigma\bar{v}_{rel}N_A(10^3 \text{ l m}^{-3})$$

But with the given parameters, we have

$$\sigma = \pi(4.4\times10^{-10} \text{ m})^2 = 6.082\times10^{-19} \text{ m}^2$$

and

$$\bar{v}_{rel} = \left(\frac{16R_gT}{\pi M}\right)^{1/2} = \left(\frac{16(8.3145 \text{ J K}^{-1} \text{ mol}^{-1})(298.15 \text{ K})}{\pi(15\times10^{-3} \text{ kg mol}^{-1})}\right)^{1/2} = 917.4 \text{ m s}^{-1}.$$

So

$$k_{AB} = (6.082\times10^{-19} \text{ m}^2)(917.4 \text{ m s}^{-1})(6.0221\times10^{23} \text{ mol}^{-1})(10^3 \text{ l m}^{-3}) = 3.36\times10^{11} \text{ M}^{-1} \text{ s}^{-1}.$$

Dividing this result by that from example 15.9 yields

$$\frac{2.78\times10^{11} \text{ M}^{-1} \text{ s}^{-1}}{7.42\times10^9 \text{ M}^{-1} \text{ s}^{-1}} = 45.$$

In the aqueous phase, this reaction would be about 1/50 as fast as in the gas phase. Note that this ratio varies with the molecular radii and with the molecular masses.

15.16 Dependence of the Specific Reaction Rate on Activity Coefficients

For a reaction in which a sufficiently definite transition state exists, an alternative treatment of the density effect is feasible. With such a system, we recognize that the effective concentration of a participant equals its activity coefficient times its concentration. The rate coefficient is changed accordingly.

Consider the elementary reaction

$$A + \ldots \rightleftarrows M^{\ddagger} \rightleftarrows D + \ldots \qquad [15.130]$$

in which A, ... are reactants, D, ... products, and $M^{\ddagger}$ represents the transition state complex. The phenomenological rate law for the forward reaction is

$$R_1 = k_1'[A]\cdots \qquad [15.131]$$

where k_1' is the rate coefficient as determined by experiment. The prime has been added to indicate that this coefficient may vary with the activity coefficients. Similarly, the law for the reverse reaction is

$$R_2 = k_2'[D]\cdots. \qquad [15.132]$$

A person can consider $M^{\ddagger}$ as a reactant for production of products and for production of reactants. At *equilibrium*, symmetry prevails with respect to direction:

$$k^*[M^{\ddagger}] = R_1, \qquad [15.133]$$

$$k^*[M^{\ddagger}] = R_2; \qquad [15.134]$$

and

$$R_1 = R_2. \qquad [15.135]$$

This symmetry is expressed in the principle of *microscopic reversibility*: At equilibrium, each molecular process and the reverse of that process take place at the same rate.

Equations (15.131) and (15.133) yield the ratio

$$\frac{[M^{\ddagger}]}{[A]\cdots} = \frac{k_1'}{k^*}. \qquad [15.136]$$

But the equilibrium constant between the reactants and the complex is

$$K_1^* = \frac{\gamma^*[M^{\ddagger}]}{\gamma_A\cdots[A]\cdots}. \qquad [15.137]$$

Substituting (15.136) into (15.137) leads to the form

$$K_1^* = \frac{\gamma^* k_1'}{\gamma_A\cdots k^*}, \qquad [15.138]$$

which must be invariant at a given temperature.

We satisfy this invariance by keeping k^* constant and setting

$$k_1' = \frac{\gamma_A\cdots}{\gamma^*}k_1, \qquad [15.139]$$

where k_1 is the true rate constant for the forward reaction. Similarly, we have

$$k_2' = \frac{\gamma_D\cdots}{\gamma^*}k_2, \qquad [15.140]$$

where k_2 is the true rate constant for the backward reaction.

Note that the concentration per unit length along the reaction path is very small in the transition state. So the contribution of that degree of freedom to γ^* is the factor 1, When this is cancelled, we get the residue $\gamma^\ddagger$ Numerically, we have

$$\gamma^* = \gamma^\ddagger \tag{15.141}$$

and equations (15.139), (15.140) become

$$k_1' = \frac{\gamma_A \cdots}{\gamma^\ddagger} k_1, \tag{15.142}$$

$$k_2' = \frac{\gamma_D \cdots}{\gamma^\ddagger} k_2, \tag{15.143}$$

with k_1 and k_2 constant at a given temperature. In low density gas phase reactions of molecules, the activity coefficients equal 1 and they may be omitted. But for other systems, they need to be included in the formulas.

The construction here presumes that enough interaction occurs among the molecules so that effective equilibrium is maintained between the reactants and the transition state complexes moving to products. However, the interaction is considered to be small enough so that viscous effects do not appreciably slow the movement through the transition state. These effects will be considered in chapter 18.

Questions

15.1 How is the rate of the reaction

$$aA + bB \rightarrow cC + dD$$

defined to avoid ambiguity?

15.2 How may an experimenter follow the rate of a reaction?

15.3 Define specific reaction rate, order with respect to a constituent, overall order, elementary reaction, molecularity.

15 4 Integrate the rate laws for (a) zero order, (b) first order, (c) second order, (d) third order reactions.

15 5 How are these integrated forms used to determine the order of a reaction from experimental data?

15.6 When does the rate law have the form

$$\frac{dx}{dt} = k(a - x)^n \ ?$$

15.7 How may a person determine the derivative dx/dt?

15.8 What would a plot of log (dx/dt) against log $(a - x)$ tell one?

15.9 How may a plot of fraction of reactant unreacted against log t be used to determine the order?

15.10 How may one employ the times for various reactant concentrations to be reduced by a chosen fraction in determining the order?

15.11 How did Arrhenius represent the variation of the specific reaction rate with temperature?

15.12 Relate the activation energies for forward and backward directions for a given reaction.

15.13 Show that adding a backward first order reaction to a forward one does not alter the form of the integrated expression.

15.14 How can a person determine k_1 and k_{-1} for the reversible first order reaction?

15.15 A first-order reaction is opposed by a second- order reaction. Under what circumstances does this behave as a simple first-order reaction?

15.16 How can a person determine k_1 and k_{-1} for this system?

15.17 How may a rapidly reacting system be perturbed from its equilibrium state?
15.18 How may the relaxation of such a perturbed system be followed?
15.19 How do reactions in high pressure gaseous (supercritical) and in condensed phases differ from those in low pressure gaseous phases?
15.20 Construct the rate law for a reaction involving formation of an encounter complex.
15.21 How does this law reduce when the reaction of the complex is (a) fast, (b) slow?
15.22 How does Fick's law apply to the formation of an encounter complex in a solution?
15.23 Show how one obtains the equilibrium constant

$$K = \frac{4}{3}\pi R^3 10^{-3} N_A$$

for formation of the encounter complex.
15.24 How does one obtain the rate expression

$$k = 4\pi(D_A + D_B)R \ ?$$

15.25 What is the principle of microscopic reversibility?
15.26 Show how the specific reaction rate depends on pertinent activity coefficients.

Problems

15.1 How does the total pressure during the reaction

$$2\ NOBr(g) \rightarrow 2\ NO(g) + Br_2(g)$$

vary with the degree of dissociation at a given temperature? How is the partial pressure of NOBr related to the total pressure?

15.2 In the decomposition of N_2O_5 at 45° C, the partial pressures as seen in table 15.A, of N_2O_5 were obtained

TABLE 15.A

t, s	600	1800	3000	4200	5400	7200
$P_{N_2O_5}$, torr	69.2	39.8	22.3	12.6	7.1	2.8

Plot ln $P_{N_2O_5}$ against t to determine whether the reaction is first order. From the slope of the line, calculate the rate constant k.

15.3 In the condensation of acrolein with butadiene at 291.2° C, the results shown in table 15.B were obtained

TABLE 15.B

t, s	0	181	542	925	1374	1988
$P_{acrolein}$, torr	418.2	401.9	373.5	349.0	326.6	302.4
$P_{butadiene}$, torr	240.0	222.7	192.7	167.0	143.4	118.2

Plot ln $(P_{acrolein}/P_{butadiene})$ against t to determine whether the reaction is second order. Obtain the rate constant k from the slope.

15.4 The decomposition

$$CH_3CHO(g) \rightarrow CH_4(g) + CO(g)$$

was followed at a given temperature with a manometer. The partial pressures of acetaldebyde were calculated with the results seen in table 15.C

TABLE 15.C

t, s	0	42	105	242	480	840
P_{CH_3CHO}, torr	363	329	289	229	169	119

Plot $1/P_{CH_3CHO}$ against t to determine whether the reaction is second order. From the slope of the line, calculate the rate constant k.

15.5 Plot the data in problem 15.2 and at five or six well chosen pressures determine $dP_{N_2O_5}/dt$. Then plot ln $(-dP_{N_2O_5}/dt)$ against ln $P_{N_2O_5}$. From the slope of the best straight line through the points, obtain the order of the reaction.

15.6 Determine the order of the reaction in problem 15.4 by the method described in problem 15.5.

15.7 Solve equation (15.61) for ln α. Then determine the limit as n approaches 1. Compare the result with equation (15.13).

15.8 Calculate α at representative τ ranging from .01 to 10 for $n = 1$, $n = 2$, $n = 3$. Use equation (15.61) and the limiting equation constructed in problem 15.7. Then plot α against log τ for each integral order 1 through 3.

15.9 For each P_{CH_3CHO} in problem 15.4, calculate α. Plot this against log t. Then obtain the order n by comparing this curve with the three theoretical curves from problem 15.8.

15.10 From the plot of P_{CH_3CHO} against t in problem 15.6, determine the time to one-fourth reaction ($\alpha = 3/4$) from five or six well spaced partial pressures. Then test the reaction for second order status by the two methods outlined in example 15.6 and from each calculate the rate constant k.

15.11 For the decomposition of N_2O_5 at various temperatures, the rate, and constants seen in table 15.D were found

TABLE 15.D

T, K	273	298	308	318	328	338
k, s^{-1}	7.87×10^{-7}	3.46×10^{-5}	1.35×10^{-4}	4.98×10^{-4}	1.50×10^{-3}	4.87×10^{-3}

Plot ln k against $1/T$ and determine the Arrhenius parameters A and E_a.

15.12 Determine the relationship between the relaxation time and the rate constants for the equilibrium

$$A + B \rightleftarrows C + D$$

15.13 Calculate the second order rate constant for the reaction

$$I + I \rightarrow I_2$$

in benzene at 25 °C, where the viscosity is 0.601×10^{-3} kg m^{-1} s^{-1}

— — —

15.14 If the reaction

$$A \rightarrow B + \ldots$$

follows the differential rate law

$$-\frac{d[A]}{dt} = k[A][B],$$

what is the integrated rate law?

15.15 In the decomposition of azoisopropane at 270° C,

$$(CH_3)_2CHN = NCH(CH_3)_2 \rightarrow N_2 + C_6H_{14},$$

the total pressure P varied as we observe in table 15.E

TABLE 15.E

t, s	0	180	360	540	720	1020
P, torr	35.15	46.30	53.90	58.85	62.20	65.55

Plot log P_A against t to determine whether the reaction is first order. From the slope of the line, calculate the rate constant k.

15.16 In the dimerization of butadiene at 326° C,

$$2\ C_4H_6 \rightarrow C_8H_{12},$$

the total pressure P varied as in table 15.F

TABLE 15.F

t, s	0	367	731	1038	1751	2550	3652	5403	7140	10600
P, torr	632.0	606.6	584.2	567.3	535.4	509.3	482.8	453.3	432.8	405.3

Plot $1/P_A$ against t to determine whether the reaction is second order. From the slope of the line, calculate the rate constant k.

15.17 In the reaction between propionaldehyde and hydrocyanic acid in aqueous solution at 25° C, the concentrations varied as in table 15.G

TABLE 15.G

t, min	2.78	5.33	8.17	15.13	19.80	∞.
$[HCN]$, M	0.0990	0.0906	0.0830	0.0706	0.0653	0.0424
$[C_3H_7CHO]$, M	0.0566	0.0482	0.0406	0.0282	0.0229	0.0000

Plot ln ([A]/ [B]) against t to determine whether the reaction is second order. Calculate the rate constant k from the slope.

15.18 Plot the data in problem 15.15 and at five or six well chosen partial pressures determine dP_A/dt. Then plot ln ($-dP_A/dt$) against ln P_A. From the slope of the best straight line through the points, obtain the order of the reaction.

15.19 Determine the order of the reaction in problem 15.16 by the method described in problem 15.18,

15.20 Show that a reaction

$$A \rightarrow \text{products}$$

obeying the rate law

$$\frac{dx}{dt} = k(a - x)^n$$

acts as a first order reaction in the initial stages of a run. Obtain an expression for the pertinent first order rate constant.

15.21 For each P_A obtained in problem 15.15, calculate the fraction unreacted. Plot this against log t. Two cycle semilogarithmic graph paper may be used. Then obtain the order by comparing this curve with theoretical curves obtained as in problem 15.8.

15.22 For each P_A obtained in problem 15.16, calculate the fraction unreacted. Plot this against log *t* as in problem 15.21. Obtain the order by comparing this curve with appropriately plotted theoretical curves as before.

15.23 The time required for 0.310 of a sample of $(CH_3)_2O$ to decompose at 777 K for various initial concentrations as seen in table 15.H.

TABLE 15.H

a, 10^{-3} M	8.13	6.44	3.10	1.88
$t_{0.31}$, s	590	665	900	1140

Determine the order of the reaction and the corresponding rate constant.

15.24 The rate constant for

$$H_2 + I_2 \rightarrow 2\ HI$$

varies with temperature as in table 15.I

TABLE 15.I

T, K	599	629	647	666	683	700
k, 10^{-3} M^{-1} s^{-1}	0.54	2.5	5.2	14	25	64

Plot ln k against $1/T$ and determine the Arrhenius parameters A and E_a.

15.25 Consider equation (15.30) to hold and determine its limiting form as b approaches a.

15.26 Consider a diffusion controlled reaction obeying equation (15.124), Determine how its Arrhenius activation energy depends on $d\eta/dT$.

References

Books

Connors, K. A.: 1990, *Chemical Kinetics: The Study of Reaction Rates in Solution,* VCH Publishers, Inc., New York, pp. 1-186.

After a general introductory chapter, Connors covers phenomenological kinetics in chapters 2-4. Besides the various common rate laws, he discusses methods of measurement and data analysis. An introduction to Laplace transform theory and Monte Carlo methods is included. The emphasis is on reactions in condensed phases.

Steinfeld, J. I., Francisco, J. S., and Hase, W. L.: 1989, *Chemical Kinetics and Dynamics*, Prentice-Hall, Englewood Cliffs, NJ, pp. 1-177.

In chapters 1-4, the authors treat phenomenological kinetics, together with some mechanisms. The common rate laws, methods of analyzing data, and experimental techniques are covered in some detail. Mathematical methods explained include the use of matrices, the Laplace transform, and Monte Carlo techniques.

Articles

Barth, R.: 1992, "Mass Balance in the Physical Chemistry Curriculum: An Improved Approach to Chemical Kinetics," *J. Chem. Educ.* **69**, 622-623.

Berberan-Santos, M. N., and Martinho, J. M. G.: 1990, "The Integration of Kinetic Rate Equations by Matrix Methods," *J. Chem. Educ.* **67**, 375-379.

Bluestone, S., and Yan, K. Y.: 1995, "A Method to Find the Rate Constants in Chemical Kinetics of a Complex Reaction," *J. Chem. Educ.* **72**, 884-886.

Borderie, B., Lavagre, D., Levy, G., and Micheau, J. D.: 1990, "A Simple Method for Analyzing First-Order Kinetics," *J. Chem. Educ.* **67**, 459-460.

Clegg, R. M.: 1986, "Derivation of Diffusion-Controlled Chemical Rate Constants with the Help of Einstein's Original Derivation of the Diffusion Constant," *J. Chem. Educ.* **63**, 571- 574.

Green, M. E.: 1984, "Chemical Applications of Fluctuation Spectroscopy," *J. Chem. Educ.* **61**, 600-605.

Jianbin, H.: "The Further Use of Time-Lag Method for Mixed Second Order Kinetics," *J. Chem. Educ.* **66**, 723-724.

Kahley, M. J., and Novak, M.: 1996, "A Practical Procedure for Determining Rate Constants in Consecutive First-Order Systems," *J. Chem. Educ.* **73**, 359-364.

Laidler, K. J.: 1984, "The Development of the Arrhenius Equation," *J. Chem. Educ.* **61**, 494-498.

Levin, E., and Eberhart, J. G.: 1989, "Simplified Rate-Law Integration for Reactions that Are First-Order in Each of Two Reactants," *J. Chem. Educ.* **66**, 705.

Logan, S. R.: 1990, "The Kinetics of Isotopic Exchange Reactions," *J. Chem. Educ.* **67**, 371-373.

Marasinghe, P. A. B., and Wirth, L. M.: 1992, "A Graphical Solution of the Second-Reaction Rate Constant of a Two-Step Consecutive First-Order Reaction," *J. Chem. Educ.* **69**, 285-286.

Mata-Perez, F., and Perez-Benito, J. F.: 1987, "The Kinetic Rate Law for Autocatalytic Reactions," *J. Chem. Educ.* **64**, 925- 927.

Naqvi, K. R.: 1989, "Normal Modes—of Vibration and Relaxation," *J. Chem. Educ.* **66**, 703-705.

Pavlis, R. R.: 1997, "Kinetics without Steady State Approximations," *J. Chem. Educ.* **74**, 1139-1140.

Pladziewicz, J. R., Leaniak, J. S., and Abrahamson, A. J.: 1986, "Treatment of Kinetic Data for Opposing Second-Order and Mixed First and Second-Order Reactions," *J. Chem. Educ.* **63**, 850- 851.

Pogliani, L., and Terenzi, M.: 1992, "Matrix Formulation of Chemical Reaction Rates," *J. Chem. Educ.* **69**, 278-280.

Ramachandran, B. R., and Halpern, A. M.: 1996, "Chemical Kinetics in Real Time: Using the Differential Rate Law and Discovering the Reaction Orders," *J. Chem. Educ.* **73**, 686-689.

Salvador, F., Gonzales, J. L., and Tel, L. M.: 1984, "Non- Isothermic Chemical Kinetics in the Undergraduate Laboratory," *J. Chem. Educ.* **61**, 921-922.

Spencer, J. N.: 1992, "Competitive and Coupled Reactions," *J. Chem. Educ.* **69**. 281-284.

Strizhak, P. and Menzinger, M.: 1996, "Nonlinear Dynamics of the BZ Reaction: A Simple Experiment That Illustrates Limit Cycles, Chaos, Bifurcations, and Noise," *J. Chem. Educ.* **73**, 868-873.

Tan, X., and Lindenbaum, S.: 1994, "A Unified Equation for Chemical Kinetics," *J. Chem. Educ.* **71**, 566-567.

Trnhlar, D. G.: 1985, "Nearly Encounter-Controlled Reactions," *J. Chem. Educ.* **62**, 104-106.

16

Explanatory Mechanisms

16.1 Types of Elementary Reactions

IN STUDYING A REACTION, ONE IS CONCERNED not only with its stoichiometry and its rate law but also with the steps involved. The main steps, the principal contributing elementary reactions, constitute the chemical *mechanism* for the process.

To be significant, an elementary reaction must meet certain criteria. In absence of externally imposed radiation, the necessary excitation energy is supplied by one or more collisions. When a molecule is complex enough, it may store this energy for some time and then react unimolecularly:

$$A \rightarrow \text{products.} \tag{16.1}$$

In a low density gas phase, most collisions involve just two molecules. When these are reactant molecules and enough kinetic energy is converted to potential energy, reaction may proceed bimolecularly:

$$A + B \rightarrow \text{products.} \tag{16.2}$$

However, a significant number of collisions involve three molecules. When these are reactants and enough kinetic energy is converted to potential energy, reaction may proceed termolecularly:

$$A + B + C \rightarrow \text{products.} \tag{16.3}$$

A molecule with few vibrational degrees of freedom cannot store the energy for disruption of any bond, over an appreciable length of time. Then step (16.1) is replaced with

$$A + M \rightarrow \text{products} + M. \tag{16.4}$$

where M is some other molecule in the system. Furthermore, if reaction (16.2) involves a single product molecule with few vibrational degrees of freedom, some other molecule would be needed to carry away the excess energy. Then step (16.2) would be replaced with

$$A + B + M \rightarrow \text{products} + M. \tag{16.5}$$

In a high density gas phase and in a condensed phase, additional molecules are always near. Then one may not explicitly include the M's in the elementary reactions even though they may be significant.

Bimolecular and termolecular reactions generally require the reactant molecules to come together with a certain range of relative orientations. The more restrictive this range, the smaller Arrhenius parameter A is than that calculated from collision theory. As a consequence, an elementary reaction path involving one new bond being formed (or broken) at each stage is more probable than a path requiring two new bonds being formed (or broken) simultaneously.

Other conditions being comparable, an elementary reaction with a low activation energy is favored over an alternate one with a higher activation energy. Quantum effects may also limit the contribution of a reaction. Thus, a process in which the net spin changes is forbidden.

Example 16.1

In 1919, Perrin proposed that reactant molecules acquire their activation energy by absorbing electromagnetic radiation from the surrounding walls. What is wrong with this hypothesis?

The radiation hypothesis requires the confining walls to do work on the reacting system by inducing chemical change in the mixture. But both the walls and the mixture are at the same temperature *T*. To produce appreciable net chemical work, the source needs to be at a much higher temperature than the reaction mixture, according to the second law of thermodynamics.

On the other hand, radiation with a non-Maxwellian distribution of frequencies may contain considerable work. The energy in radiation limited to a very narrow band is practically pure work. As an example, we have the radiation produced by an infrared laser.

Such a laser operating at or near the frequency of a vibrational mode of a reactant can be very effective in exciting the reactant and causing unimolecular reaction.

16.2 Activation in Simple First-Order Reactions

A molecule can acquire the energy to react (a) in collisions or (b) on absorbing photons. In a normal first-order reaction, mechanism (a) predominates. But a collision is bimolecular. How does this process lead to first-order kinetics?

In 1922, Lindemann showed how. He proposed the following mechanism:

$$\mathrm{A} + \mathrm{M} \xrightarrow{k_1} \mathrm{A}^* + \mathrm{M}, \tag{16.6}$$

$$\mathrm{A}^* + \mathrm{M} \xrightarrow{k_{-1}} \mathrm{A} + \mathrm{M}, \tag{16.7}$$

$$\mathrm{A}^* \xrightarrow{k_2} \mathrm{B}. \tag{16.8}$$

Here M is any molecule that collides with A while A* is a reactant molecule with enough energy to react and B represents the products.

This mechanism presumes that the energizing of A and the de- energizing of A* are *single-step* processes. Thus, the pertinent collisions are considered strong, involving a large transfer of energy into A and out of A*. The energy is stored in the vibrational degrees of freedom. In process (16.8), enough energy becomes concentrated in the bond or bonds that are being broken and the movement to products occurs along the reaction coordinate.

Considering (16.6)—(16.8) as ordinary reaction processes leads to the rate expressions

$$-\frac{\mathrm{d}[\mathrm{A}]}{\mathrm{d}t} = k_1[\mathrm{A}][\mathrm{M}] - k_{-1}[\mathrm{A}^*][\mathrm{M}], \tag{16.9}$$

$$\frac{\mathrm{d}[\mathrm{A}^*]}{\mathrm{d}t} = k_1[\mathrm{A}][\mathrm{M}] - k_{-1}[\mathrm{A}^*][\mathrm{M}] - k_2[\mathrm{A}^*], \tag{16.10}$$

$$\frac{d[B]}{dt} = k_2[A^*]. \quad [16.11]$$

Since the concentration of A* should remain small, its rate of change is very small and we may set

$$\frac{d[A^*]}{dt} \cong 0. \quad [16.12]$$

With this steady-state approximation, equations (16.9)—(16.11) yield

$$-\frac{d[A]}{dt} = \frac{d[B]}{dt}. \quad [16.13]$$

Substituting (16.12) into (16.10) and solving for [A*] yields

$$[A^*] = \frac{k_1[A][M]}{k_{-1}[M] + k_2}. \quad [16.14]$$

Equations (16.9) and (16.11) now reduce to

$$-\frac{d[A]}{dt} = \frac{d[B]}{dt} = \frac{k_1 k_2[M]}{k_{-1}[M] + k_2}[A] = k[A]. \quad [16.15]$$

The reciprocal of the first-order coefficient k is

$$\frac{1}{k} = \frac{k_{-1}}{k_1 k_2} + \frac{1}{k_1[M]}. \quad [16.16]$$

Strictly, coefficients k_1 and k_{-1} depend on whether M is A, B, or an impurity C. If as an approximation, this effect is disregarded,

$$[M] \cong [A] + [B] + [C] = a, \quad [16.17]$$

the reaction is first-order at all pressures. Also equation (16.16) becomes

$$\frac{1}{k} = \frac{k_{-1}}{k_1 k_2} + \frac{1}{k_1 a}. \quad [16.18]$$

The theory predicts that a plot of $1/k$ against $1/a$ is a straight line. However, $1/k$ is found to fall below the straight line when $1/a$ is small. See figure 16.1. Refinements to the theory, allowing for the multiplicity of steps of excitation and de-excitation, lead to better results.

At high pressures, equation (16.15) gives us

$$k = \frac{k_1}{k_{-1}} k_2 = K k_2. \quad [16.19]$$

Here K is the equilibrium ratio

$$K = \frac{[A^*]}{[A]}. \quad [16.20]$$

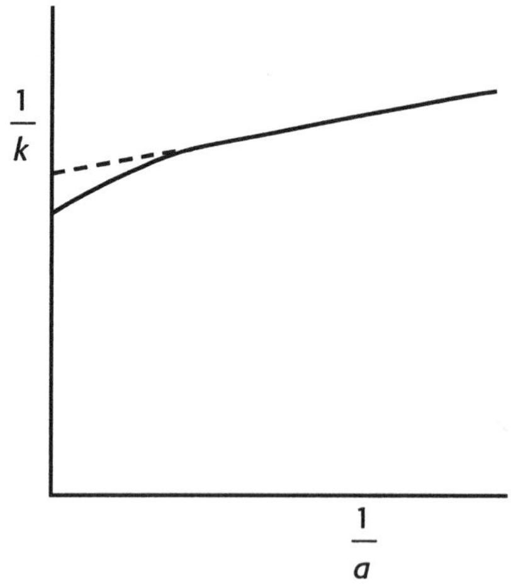

FIGURE 16.1 How the reciprocal of the first-order rate coefficient commonly varies with the reciprocal of the initial concentration.

At low pressures, the equation yields

$$k = k_1\left[\mathrm{M}\right]. \tag{16.21}$$

If only A were effective in exciting A molecules,

$$\left[\mathrm{M}\right] \cong \left[\mathrm{A}\right], \tag{16.22}$$

the reaction would become second order at low pressures.

16.3 Decomposition of Nitrogen Pentoxide

The fact that a reaction exhibits integral order does not ensure that it is simple. In this section, we will consider a complex first-order reaction; in the next, a complex second-order reaction.

The stoichiometric equation for the thermal decomposition of nitrogen pentoxide is

$$\begin{array}{c} \mathrm{N_2O_5} \rightarrow 2\mathrm{NO_2} + \frac{1}{2}\mathrm{O_2}. \\ \uparrow\downarrow \\ \mathrm{N_2O_4} \end{array} \tag{16.23}$$

The reaction proceeds at measurable rates throughout the temperature range 0-120° C and for pressures ranging from 5×10^{-4} torr to 700 torr. The reaction is homogeneous and first order, with the experimental rate constant

$$k = \left(3.21 \times 10^{13}\right)\exp\left(-102{,}500 / RT\right)\ \mathrm{s}^{-1} \tag{16.24}$$

in the gas phase. Similar results are found in the liquid phase with various solvents.

The rate coefficient k does fall off at pressures below 0.05 torr. However, all quasiunimolecular rate theories predict an appreciable decrease at much higher pressures. Note the discussion in section 16.2 and the data in problem 16.1.

In 1947, Ogg proposed a mechanism that resolves this difficulty. It involves the following elementary reactions:

$$\mathrm{N_2O_5} \xrightarrow{k_1} \mathrm{NO_2} + \mathrm{NO_3}, \tag{16.25}$$

$$\mathrm{NO_2} + \mathrm{NO_3} \xrightarrow{k_{-1}} \mathrm{N_2O_5}, \tag{16.26}$$

$$NO_2 + NO_3 \xrightarrow{k_2} NO_2 + NO + O_2, \qquad [16.27]$$

$$NO + NO_3 \xrightarrow{k_3} 2NO_2. \qquad [16.28]$$

Process (16.25) being a simple first-order reaction, coefficient k_1 varies as section 16.2 describes. Because k_1/k_{-1} is the equilibrium constant K for reactions (16.25) and (16.26), coefficient k_{-1} varies similarly. At very low pressures, formula (16.16) leads to

$$k_1 = k_1'\left[M\right] \qquad \text{and} \qquad k_{-1} = k_{-1}'\left[M\right], \qquad [16.29]$$

with k_1' and k_{-1}' constant.

Applying the rate laws for the elementary reactions, we find that

$$\frac{d\left[NO_3\right]}{dt} = k_1\left[N_2O_5\right] - k_{-1}\left[NO_2\right]\left[NO_3\right] - k_2\left[NO_2\right]\left[NO_3\right] - k_3\left[NO\right]\left[NO_2\right], \qquad [16.30]$$

$$\frac{d\left[NO\right]}{dt} = k_2\left[NO_2\right]\left[NO_3\right] - k_3\left[NO\right]\left[NO_3\right]. \qquad [16.31]$$

In a run at a given temperature, the concentrations of NO_3 and NO remain small and

$$\frac{d\left[NO_3\right]}{dt} \cong 0, \qquad \frac{d\left[NO\right]}{dt} \cong 0. \qquad [16.32]$$

With these steady-state approximations, equations (16.31) and (16.30) reduce to

$$k_2\left[NO_2\right]\left[NO_3\right] = k_3\left[NO\right]\left[NO_3\right] \qquad [16.33]$$

and

$$k_1\left[N_2O_5\right] - k_{-1}\left[NO_2\right]\left[NO_3\right] - 2k_2\left[NO_2\right]\left[NO_3\right] = 0. \qquad [16.34]$$

Solving (16.34) for $[NO_3]$ gives us

$$\left[NO_3\right] = \frac{k_1\left[N_2O_5\right]}{k_{-1}\left[NO_2\right] - 2k_2\left[NO_2\right]}. \qquad [16.35]$$

Each time a molecule of NO_2 reacts with one of NO_3 to produce NO_2 + NO + O_2, the NO formed reacts with another NO_3 to produce 2 NO_2. Since the two NO_3 molecules have come from two N_2O_5 molecules, the rate of decomposition of N_2O_5 is twice the rate of step (16.27):

$$-\frac{d\left[N_2O_5\right]}{dt} = 2k_2\left[NO_2\right]\left[NO_3\right]. \qquad [16.36]$$

Combining (16.35) and (16.36) leads to

$$-\frac{d\left[N_2O_5\right]}{dt} = \frac{2k_1k_2\left[N_2O_5\right]}{k_{-1} + 2k_2} = k\left[N_2O_5\right]. \qquad [16.37]$$

Thus, the reaction is first order even though the rate-determining step is bimolecular.

From (16.37), the first-order rate coefficient is

$$k = \frac{2k_1k_2}{k_{-1} + 2k_2} = \frac{2k_2}{k_{-1}/k_1 + 2k_2/k_1}. \qquad [16.38]$$

At low pressures, coefficients k_1 and k_{-1} are proportional to [M] following equations (16.29). Then formula (16.38) becomes

$$k = \frac{2k_2}{k_{-1}'/k_1' + 2k_2/k_1'[\mathrm{M}]}. \quad [16.39]$$

At extremely low pressures, the last term in the denominator predominates and we have

$$k = k_1'[\mathrm{M}]. \quad [16.40]$$

The rate has become proportional to the product [N_2O_5] [M]. However, heterogeneous reaction on the container walls competes significantly unless the vessel is appropriately large.

Above 0.05 torr, the last term in the denominators of (16.38) and (16.39) is negligible and

$$k = 2\frac{k_1}{k_{-1}}k_2 = 2Kk_2, \quad [16.41]$$

a constant. Thus, we have an explanation for the unusually low pressure at which the decrease in k becomes evident.

In a sequence of elementary reactions, reactive intermediates appear and disappear. In some mechanisms, all paths lead to stable molecules. This is true with NO_3 and NO in the Ogg mechanism. In other mechanisms, each intermediate is regenerated in a later step. Thus, a cycle of reactions exists, in each step of which one intermediate disappears and another one appears. The cycle is called a *chain*, each reactive intermediate a *chain carrier*.

16.4 Hydrogen-Iodine Reactions

Hydrogen reacts with iodine reversibly following the stoichiometric equation

$$H_2 + I_2 \rightleftarrows 2HI. \quad [16.42]$$

The reactions proceed at measurable rates over the temperature range 280 - 530° C. The forward reaction is homogeneous and second order with the approximate rate coefficient

$$k_f = (5.96 \times 10^{10})\exp(-161{,}000/RT)\mathrm{M}^{-1}\mathrm{s}^{-1}. \quad [16.43]$$

The backward reaction is also homogeneous and second order with the approximate rate coefficient

$$k_b = (1.4 \times 10^{11})\exp(-188{,}000/RT)\mathrm{M}^{-1}\mathrm{s}^{-1}. \quad [16.44]$$

But an empirical expression fitting the data points for the forward reaction yielded an activation energy with an abnormally high temperature coefficient. Furthermore, the logarithm of the equilibrium constant, ln K, calculated from the rate coefficients deviated considerably from a linear dependence on $1/T$. These results suggest that the reactions are not simple but complex.

If both the forward and reverse reactions were simple, the transition state would have the configuration depicted in figure 16.2. But making this complex from H_2 and I_2 along a single reaction path requires the formation of two partial H—I bonds simultaneously. For the reverse reaction, it requires the formation of a partial H—H and a partial I—I bond simultaneously. These coordinated actions are much less probable than the formation of a single partial bond.

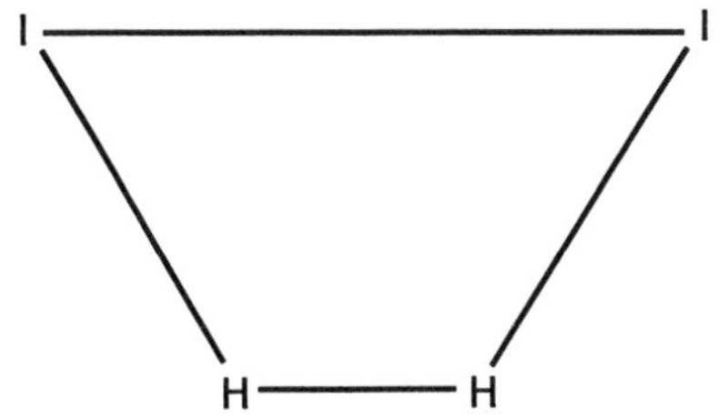

FIGURE 16.2 Transition state for reaction of molecular H_2 with molecular I_2 and for the reverse reaction.

Thus, one is led to consider the complex in figure 16.3. This is considerably lower in energy than the cis- iodine complex. Also, its formation from 2 HI involves a single action, formation of a partial H—H bond.

The trans-iodine complex would presumably break up into H_2I and I, and then into H_2 and another I. The iodine atoms would combine in a separate reaction. For the opposite direction, we have a termolecular reaction that can be broken down into two bimolecular elementary reactions.

Then, the mechanism for forward reaction (16.42) is

$$I_2 + M \xrightarrow{k_1} 2I + M, \qquad [16.45]$$

$$I + H_2 \underset{k_{-2}}{\overset{k_2}{\rightleftarrows}} H_2I, \qquad [16.46]$$

$$I + H_2I \xrightarrow{k_3} 2HI, \qquad [16.47]$$

$$2I + M \xrightarrow{k_{-1}} I_2 + M. \qquad [16.48]$$

Step (16.47) yields the rate law

$$\frac{d[HI]}{dt} = 2k_3[I][H_2I]. \qquad [16.49]$$

To an approximation, reactions (16.46) maintain an equilibrium concentration of the intermediate H_2I and

$$[H_2I] = \frac{k_2}{k_{-2}}[H_2][I]. \qquad [16.50]$$

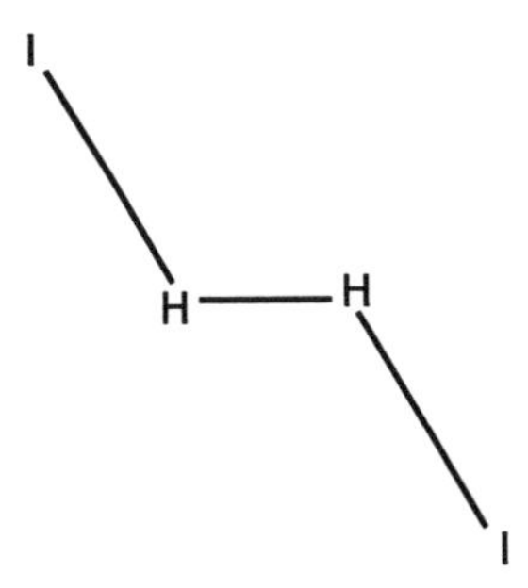

FIGURE 16.3 Alternate transition state for the hydrogen iodide decomposition.

Similarly, reactions (16.45) and (16.48) tend to maintain an equilibrium concentration of atomic I and

$$[\mathrm{I}]^2 = \frac{k_1}{k_{-1}}[\mathrm{I}_2]. \tag{16.51}$$

Substituting expressions (16.50) and (16.51) into (16.49) gives us

$$\frac{d[\mathrm{HI}]}{dt} = 2k_3 \frac{k_1}{k_{-1}} \frac{k_2}{k_{-2}}[\mathrm{H}_2][\mathrm{I}_2] = 2k_3 K_1 K_2 [\mathrm{H}_2][\mathrm{I}_2] = k[\mathrm{H}_2][\mathrm{I}_2], \tag{16.52}$$

a second-order rate law for forward reaction (16.42).

For the corresponding backward reaction, we have the elementary step

$$2\mathrm{HI} \xrightarrow{k_{-3}} \mathrm{I} + \mathrm{H}_2\mathrm{I}, \tag{16.53}$$

followed by the other steps listed. This yields the rate law

$$-\frac{d[\mathrm{HI}]}{dt} = 2k_{-3}[\mathrm{HI}]^2, \tag{16.54}$$

again, a second-order expression.

To explain the anomaly mentioned in the second paragraph of this section, Benson and Srinivasan proposed that a parallel chain-reaction path exists. This includes the elementary reactions

$$\mathrm{I}_2 + \mathrm{M} \xrightarrow{k_1} 2\mathrm{I} + \mathrm{M}, \tag{16.55}$$

$$\mathrm{I} + \mathrm{H}_2 \underset{k_{-4}}{\overset{k_4}{\rightleftarrows}} \mathrm{HI} + \mathrm{H}, \tag{16.56}$$

$$\mathrm{H} + \mathrm{I}_2 \xrightarrow{k_5} \mathrm{HI} + \mathrm{H}, \tag{16.57}$$

$$2\mathrm{I} + \mathrm{M} \xrightarrow{k_{-1}} \mathrm{I}_2 + \mathrm{M}. \tag{16.58}$$

Imposing the steady-state conditions, as we will see in section 16.5, then yields

$$\frac{d[\mathrm{HI}]}{dt} = \frac{2k_4(k_1/k_{-1})^{1/2}[\mathrm{H}_2][\mathrm{I}_2]^{1/2}}{1 + k_{-4}[\mathrm{HI}]/k_5[\mathrm{I}_2]}. \tag{16.59}$$

The complete rate for forward reaction (16.42) is the sum of expressions (16.52) and (16.59). Decreasing the initial iodine concentration reduces expression (16.52) more than expression (16.59). So it increases the relative contribution of the chain reaction. From the temperature dependence of the various parameters, one can also show that increasing the temperature increases the relative contribution of the chain reaction. At low temperatures and moderate pressures, the non-chain mechanism predominates; at high temperatures, the chain mechanism does.

16.5 Hydrogen-Bromine Reactions

Hydrogen reacts with bromine following the stoichiometric equation

$$\mathrm{H}_2 + \mathrm{Br}_2 \rightleftarrows 2\mathrm{HBr}. \tag{16.60}$$

The forward reaction proceeds at measurable rates over the temperature range 220- 340° C. It is apparently homogeneous. But it does not follow a simple rate law.

In 1906, Bodenstein and Lind reported that the initial rate is first order in H_2 and one-half order in Br_2. The equilibrium lies far to the right in (16.60). Nevertheless, the product HBr slows down, that is *inhibits*, the process. Bodenstein and Lind found that the inhibiting effect varies with the ratio [HBr]/ [Br_2]. Thus, they constructed the rate expression

$$\frac{d[HBr]}{dt} = \frac{k[H_2][Br_2]^{1/2}}{1 + k'[HBr]/[Br_2]}. \qquad [16.61]$$

This fit the experimental data when

$$k = \left(1.7 \times 10^{11}\right)\exp\left(-175,00/RT\right)M^{-1/2}s^{-1} \qquad [16.62]$$

and

$$k' = 10. \qquad [16.63]$$

In 1919 and 1920, Christiansen, Herzfeld, and Polanyi independently proposed a mechanism which we rewrite in the form

$$Br_2 + M \xrightarrow{k_1} 2Br + M, \qquad [16.64]$$

$$Br + H_2 \underset{k_{-2}}{\overset{k_2}{\rightleftarrows}} HBr + H, \qquad [16.65]$$

$$H + Br_2 \xrightarrow{k_3} HBr + Br, \qquad [16.66]$$

$$2Br + M \xrightarrow{k_{-1}} Br_2 + M. \qquad [16.67]$$

Here the initiation reaction is (16.64), the chain reactions are (16.65) and (16.66), the termination reaction is (16.67).

Introducing the rate laws for the elementary reactions gives us

$$\frac{d[H]}{dt} = k_2[Br][H_2] - k_3[H][Br_2] - k_{-2}[H][HBr], \qquad [16.68]$$

$$\frac{d[Br]}{dt} = 2k_1[Br_2][M] - k_2[Br][H_2] + k_3[H][Br_2] + k_{-2}[H][HBr] - 2k_{-1}[Br]^2[M]; \qquad [16.69]$$

$$\frac{d[HBr]}{dt} = k_2[Br][H_2] + k_3[H][Br_2] - k_{-2}[H][HBr]. \qquad [16.70]$$

In a run at a given temperature, the concentrations of H and Br remain small and

$$\frac{d[H]}{dt} \cong 0, \qquad \frac{d[Br]}{dt} \cong 0. \qquad [16.71]$$

With these steady-state approximations, equation (16.68) reduces to

$$0 = k_2[Br][H_2] - k_3[H][Br_2] - k_{-2}[H][HBr], \qquad [16.72]$$

whence

$$[\mathrm{H}] = \frac{k_2[\mathrm{H_2}][\mathrm{Br}]}{k_3[\mathrm{Br_2}] + k_{-2}[\mathrm{HBr}]}. \quad [16.73]$$

And equation (16.69) reduces to

$$0 = 2k_1[\mathrm{Br_2}][\mathrm{M}] - 2k_{-1}[\mathrm{Br}]^2[\mathrm{M}], \quad [16.74]$$

whence

$$[\mathrm{Br}] = \left(\frac{k_1}{k_{-1}}\right)^{1/2}[\mathrm{Br_2}]^{1/2}, \quad [16.75]$$

the equilibrium value.

But subtracting (16.72) from (16.70),

$$\frac{\mathrm{d}[\mathrm{HBr}]}{\mathrm{d}t} = 2k_3[\mathrm{H}][\mathrm{Br_2}], \quad [16.76]$$

then substituting in concentrations (16.73) and (16.75), leads to

$$\frac{\mathrm{d}[\mathrm{HBr}]}{\mathrm{d}t} = \frac{2k_3k_2(k_1/k_{-1})^{1/2}[\mathrm{H_2}][\mathrm{Br_2}]^{3/2}}{k_3[\mathrm{Br_2}] + k_{-2}[\mathrm{HBr}]} = \frac{2k_2(k_1/k_{-1})^{1/2}[\mathrm{H_2}][\mathrm{Br_2}]^{1/2}}{1 + k_{-2}[\mathrm{HBr}]/k_3[\mathrm{Br_2}]}. \quad [16.77]$$

Note that this result agrees with the empirical rate law (16.61). It can be integrated numerically to obtain the dependence of [HBr] on time t. Also, note that substituting iodine for bromine in the mechanism changes result (16.77) to formula (16.59).

Apparently, the non-chain mechanism analogous to that for iodine does not contribute appreciably in the thermal range. However, bromine can be dissociated with energetic photons. At room temperature, Sullivan found that the non-chain mechanism did then predominate.

16.6 Hydrogen-Chlorine Reactions

Hydrogen reacts with chlorine following the stoichiometric equation

$$\mathrm{H_2} + \mathrm{Cl_2} \rightarrow 2\mathrm{HCl}. \quad [16.78]$$

From what we know of the hydrogen-iodine and the hydrogen-bromine reactions, we consider the mechanism to be

$$\mathrm{Cl_2} + \mathrm{M} \xrightarrow{k_1} 2\mathrm{Cl} + \mathrm{M}, \quad [16.79]$$

$$\mathrm{Cl} + \mathrm{H_2} \underset{k_{-2}}{\overset{k_2}{\rightleftarrows}} \mathrm{HCl} + \mathrm{H}, \quad [16.80]$$

$$\mathrm{H} + \mathrm{Cl_2} \xrightarrow{k_3} \mathrm{HCl} + \mathrm{Cl}, \quad [16.81]$$

$$2\mathrm{Cl} + \mathrm{M} \xrightarrow{k_{-1}} \mathrm{Cl_2} + \mathrm{M}. \quad [16.82]$$

However, the hydrogen-chlorine reaction is very sensitive to the action of light, the presence of impurities, and surface effects that remove chlorine atoms.

The only atom - molecule reaction in the general mechanism with an appreciable energy barrier is the

$$X + H_2 \rightarrow HX + H \quad [16.83]$$

reaction. In table 16.1, the Arrhenius parameters for this reaction are compared.

We see that the $Cl + H_2$ reaction is much faster than the $Br + H_2$ and the $I + H_2$ reactions. As a consequence, forward reaction (16.80) competes much more effectively with chain-ending reaction (16.82) than in the analogous bromine and iodine processes. So the hydrogen-chlorine reaction exhibits very long chains. In the photochemical process, chains containing up to 10^6 cycles have been observed.

A common impurity that shortens the chain markedly is oxygen. Indeed, it reacts with both chain carriers. With the hydrogen atom, we have

$$H + O_2 + M \rightarrow HO_2 + M, \quad [16.84]$$

$$HO_2 \rightarrow \text{destruction at surface.} \quad [16.85]$$

And with the chlorine atom, we have

$$Cl + O_2 + M \rightarrow ClO_2 + M, \quad [16.86]$$

$$Cl + ClO_2 \rightarrow 2ClO, \quad [16.87]$$

$$2ClO \rightarrow Cl_2 + O_2 \quad [16.88]$$

16.7 Thermal Explosions

In general, reactants are separated from products by one or more energy barriers. As a result, the rate of a typical reaction increases markedly with temperature. When the reaction is exothermic, the rate of heat evolution tends to increase exponentially as the temperature rises. But the rate of heat transfer, from the reacting system to the walls, increases approximately linearly. When the heat is not removed fast enough, an explosive rise in reaction rate occurs.

In a given process, the rate of heat evolution may vary with temperature at reactant concentrations c_1, c_2, and c_3 as the curves in figure 16.4 indicate. But with the walls of the containing vessel at temperature T_0, the heat loss from the reacting system varies as the straight line shows, independent of concentration c.

When the reactants are introduced into the given vessel at concentration c_1 and temperature T_0, the rate of heat production is greater than the rate of heat loss until the curve for c_1 crosses the straight line at temperature T_s. This is a stable reactant temperature, since increasing the temperature results in a higher rate of heat loss than rate of heat production and return of the system to the crossing point.

TABLE 16.1 Arhenius Parameters for Halogen Atom — Hydrogen Molecule Reactions

Reaction	$A, M^{-1}s^{-1}$	E_a, kJ mol^{-1}
$Cl + H_2 \rightarrow HCl + H$	8.3×10^{10}	22.9
$Br + H_2 \rightarrow HBr + H$	6.3×10^{10}	74.1
$I + H_2 \rightarrow HI + H$	1.6×10^{11}	140.3

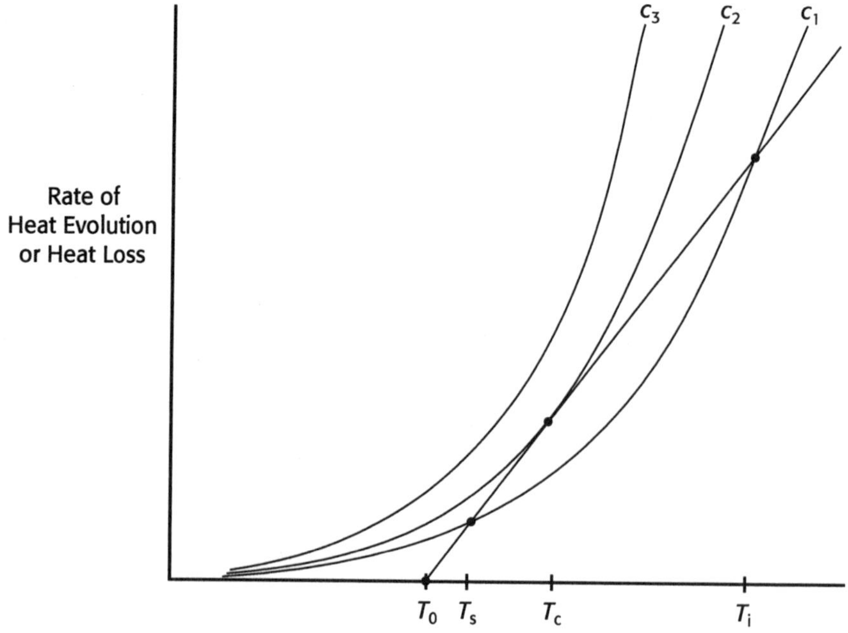

FIGURE 16.4 Heat evolution for a given amount of reactants at concentrations c_1, c_2, c_3 and heat loss to walls for a given vessel at temperature T_0.

But if the reacting material is heated to temperature T_i the system becomes unstable. Any rise beyond this point causes the rate of heat production to exceed the rate of heat loss and explosion results. So T_i is called the *ignition temperature* for the given system.

When the reactants are introduced into the vessel at concentration c_3 and temperature T_0, the rate of heat production exceeds the rate of heat loss at all temperatures and explosion ensues.

For a given reacting system, initial temperature, and vessel, there is a minimum concentration c_2 at which explosion occurs without added heating (ignition). This curve meets the line for heat loss tangentially at the *critical temperature* T_c.

The temperature dependence of the reaction rate of a given reacting mixture is given, to the Arrhenius approximation, by

$$R = f(c) A e^{-E/RT}. \qquad [16.89]$$

Function $f(c)$ depends on the mechanism. For an nth order reaction, it equals c^n. If H is the heat released per mole of reaction and V is the volume of the containing vessel, the rate at which heat is released in the whole vessel is

$$Q_R = HVR. \qquad [16.90]$$

The rate of heat loss to the walls is proportional to the temperature difference between the body of the system and the walls. If S is the surface area and h the coefficient of heat transfer, then the rate of heat loss is

$$Q_C = Sh(T - T_0). \qquad [16.91]$$

At the critical point, the rate of heat evolution equals the rate of heat loss:

$$HVf(c)Ae^{-E/RT} = Sh(T - T_0). \quad [16.92]$$

Also, the slope of the heat evolution curve equals the slope of the heat loss line:

$$\frac{dQ_R}{dT} = \frac{dQ_C}{dT} \quad [16.93]$$

or

$$HVf(c)Ae^{-E/RT}\frac{E}{RT^2} = Sh. \quad [16.94]$$

Now, divide (16.92) by (16.94):

$$\frac{RT^2}{E} = T - T_0. \quad [16.95]$$

Introduce the dimensionless variable,

$$\theta = \frac{T - T_0}{T_0} \quad \text{or} \quad \frac{T}{T_0} = \theta + 1, \quad [16.96]$$

so equation (16.95) becomes

$$\frac{RT_0}{E}(\theta + 1)^2 = \theta. \quad [16.97]$$

For chemical reactions, the fractional temperature rise θ is small with respect to 1. Then equation (16.97) yields the solution

$$\theta \cong \frac{RT_0}{E}. \quad [16.98]$$

The other root

$$\theta \cong \frac{E}{RT_0}, \quad [16.99]$$

is not physically significant.

From equations (16.96) and (16.98), we obtain

$$\frac{1}{T} = \frac{1}{T_0(1+\theta)} \cong \frac{1-\theta}{T_0} = \frac{1}{T_0} - \frac{R}{E}, \quad [16.100]$$

$$e^{E/RT} = \exp\left(\frac{E}{RT_0} - 1\right) = e^{-1}e^{E/RT_0}, \quad [16.101]$$

$$T - T_0 = \theta T_0 \cong \frac{RT_0^2}{E}. \quad [16.102]$$

Substituting into equation (16.92) and rearranging leads to

$$f(c) = \frac{ShRT_0^2}{HVAEe}e^{E/RT_0}. \quad [16.103]$$

The critical concentration comes from solving this equation for c.

For an nth order equation,

$$f(c) = c^n, \qquad [16.104]$$

one finds that

$$\ln c - \frac{2}{n}\ln T_0 = \frac{E}{nRT_0} + \frac{1}{n}\ln\frac{ShR}{HVAEe}. \qquad [16.105]$$

For $n = 2$ and $c = P/RT$, equation (16.105) has the form

$$\log\frac{P}{T} = \frac{A}{T} - B. \qquad [16.106]$$

The critical pressures P for many mixtures in a given vessel at temperature T satisfy this equation. Indeed, it does not discriminate between thermal and branched chain explosions.

In a *thermal explosion*, heat is produced in the reacting mixture faster than it can be conducted away. As a result, the temperature rises and speeds up the reaction until it becomes explosive. Warning is given of the instability by the way the temperature increases.

In a *branched chain explosion*, reactive intermediates multiply in one or more chain branching reactions. When conditions are adjusted, or they evolve after time, so that the chain breaking reactions do not destroy all the additional chain carriers produced, these accumulate. Beyond a certain point, the reaction becomes explosive. But before this point, there is no unusual temperature increase or pressure increase to give warning.

As an example of a branched chain process, we will consider the reaction between hydrogen and oxygen.

16.8 Hydrogen-Oxygen Reactions

Hydrogen reacts with oxygen following the stoichiometric equation

$$2H_2 + O_2 \rightarrow 2H_2O. \qquad [16.107]$$

The mechanism involves the reactive intermediates H, O, and OH, as we will see.

A representative mixture in an ordinary reaction vessel behaves as figure 16.5 shows. In a silica tube 5 cm in diameter, a stoichiometric mixture at 550° C reached the first limit at 1 torr pressure. This explosion pressure varied inversely with the diameter of the tube at the given composition, temperature, and wall material.

The product, water, acts as an inhibitor near the second limit. In its absence, the limit is practically independent of the tube diameter as long as this diameter is above 4 cm. Explosion then occurred in a 550° C stoichiometric mixture at about 100 torr.

Just outside the first and second explosion limits, the rate of the slow reaction is very small. But at the third limit, there is no such sharp transition; below the limit, the rate is high. At the limit, there is an induction period. This falls to zero as the system is taken higher. Furthermore, the explosion pressure varies roughly inversely with the diameter of the reaction tube. In a KCl coated vessel 5.8 cm in diameter, a mixture at 550° C reached the third limit at 560 torr.

The abrupt onset of explosion as the first and second limits are crossed from the slow reaction side indicates that a branched chain mechanism is involved. In this, the most important elementary reactions are initiation,

$$H_2 + M \rightarrow 2H + M, \qquad [16.108]$$

$$O_2 + M \rightarrow 2O + M, \qquad [16.109]$$

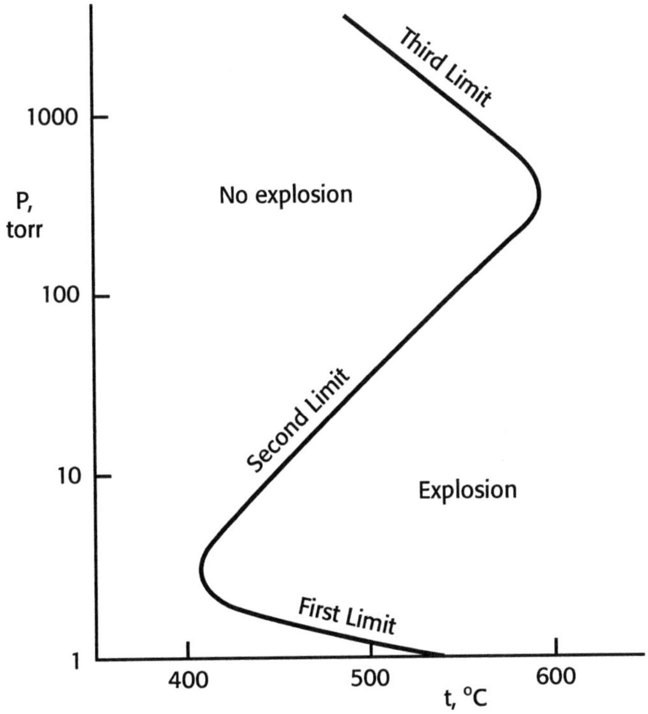

FIGURE 16.5 Curve separating the pressures and temperatures where explosion occurs from those where slow reaction occurs in a given hydrogen-oxygen mixture in a given reaction vessel.

chain branching,

$$H + O_2 \xrightarrow{k_2} OH + O, \tag{16.110}$$

$$O + H_2 \xrightarrow{k_3} OH + H, \tag{16.111}$$

chain propagation,

$$OH + H_2 \xrightarrow{k_1} H_2O + H, \tag{16.112}$$

chain breaking,

$$H + \text{wall} \xrightarrow{k_4} \frac{1}{2} H_2, \tag{16.113}$$

$$H + O_2 + M \xrightarrow{k_5} HO_2 + M. \tag{16.114}$$

Note that reaction (16.109) is followed by reaction (16.111), production of an H, and by reaction (16.112), production of another H. We link these together, letting the rate of production of H from the initiation reactions (16.108) and (16.109) be w_0. Furthermore, the fractions of reactions (16.111) and (16.112) so involved may be neglected.

Introducing the rate laws for the elementary reactions with these concessions gives us

$$\frac{d[H]}{dt} = w_0 - k_2[H][O_2] + k_3[O][H_2] + k_1[OH][H_2] - k_4[H] - k_5[H][O_2][M]. \tag{16.115}$$

$$\frac{d[OH]}{dt} = k_2[H][O_2] + k_3[O][H_2] - k_1[OH][H_2], \tag{16.116}$$

$$\frac{d[O]}{dt} = k_2[H][O_2] - k_3[O][H_2]. \tag{16.117}$$

From bond energy considerations, we expect reactions (16.111) and (16.112) to be fast compared to (16.110). So the concentrations of O and OH remain relatively small. As an approximation, we set

$$\frac{d[OH]}{dt} = 0, \qquad \frac{d[O]}{dt} = 0. \tag{16.118}$$

From (16.117), we then obtain

$$[O] = \frac{k_2[H][O_2]}{k_3[H_2]}. \tag{16.119}$$

Similarly from (16.116), we obtain

$$[OH] = \frac{k_2[H][O_2] + k_3[H_2]k_2[H][O_2]/k_3[H_2]}{k_1[H_2]} = \frac{2k_2[H][O_2]}{k_1[H_2]}. \tag{16.120}$$

Substituting into equation (16.115) gives us

$$\begin{aligned} \frac{d[H]}{dt} &= w_0 - k_2[H][O_2] + k_3\frac{k_2[H][O_2]}{k_3[H_2]}[H_2] + k_1[H_2]\frac{2k_2[H][O_2]}{k_1[H_2]} - k_4[H] - k_5[H][O_2][M] \\ &= w_0 + \left(2k_2[O_2] - k_4 - k_5[O_2][M]\right)[H]. \end{aligned} \tag{16.121}$$

When multiplied by [H], the term

$$2k_2[O_2] = f \tag{16.122}$$

represents the rate of chain branching while the term

$$k_4 + k_5[O_2][M] = g \tag{16.123}$$

represents the rate of chain breaking. Letting [H] = n then gives us

$$\frac{dn}{dt} = w_0 + (f - g)n \tag{16.124}$$

for the governing equation.

Integration of (16.124) with n = 0 at t = 0 yields

$$n = \frac{w_0}{g - f}\left[1 - e^{-(g-f)t}\right] = \frac{w_0}{f - g}\left[e^{(f-g)t} - 1\right]. \tag{16.125}$$

Below the first explosion limit, the oxygen molecule and the impinging molecule concentrations are very low and equation (16.123) reduces to

$$k_4 \cong g. \tag{16.126}$$

Expression (16.122) is also low,

$$g > f, \tag{16.127}$$

and the hydrogen atom concentration approaches $w_0/(g - f)$ as figure 16.6 shows. As the initial oxygen concentration increases, f increases until

$$g = f, \tag{16.128}$$

and the first explosion limit is reached. Between the first and the second explosion limits,

$$g < f, \tag{16.129}$$

the hydrogen atom concentration increases exponentially as figure 16.7 illustrates.

But increasing the initial pressure with composition constant increases the term $k_5[O_2][M]$ faster than the term $2k_2[O_2]$. A point is reached at which

$$g = f \tag{16.130}$$

again, the second explosion limit. Beyond this point,

$$g > f \tag{16.131}$$

and the sequence in figure 16.6 applies.

As the initial pressure is raised further, a point is reached at which the heat generated in the exothermic steps exceeds what can be removed by conduction and convection. Then thermal explosion ensues. This occurs at and above the third explosion limit.

At low pressures, the HO_2 radical presumably diffuses to the walls and is there transformed to H_2O, O_2, and H_2O_2. At higher pressures, additional elementary reactions become significant. Between the second and third limits, one needs to consider

$$HO_2 + H_2 \rightarrow H_2O_2 + H, \tag{16.132}$$

$$H_2O_2 + M \rightarrow 2OH + M, \tag{16.133}$$

and related reactions.

16.9 Decomposition of Ethane

The pyrolysis of ethane between 550 and 700° C produces ethylene and hydrogen with some methane, a little propane, propylene, other hydrocarbons, and carbon itself. The stoichiometric equation for the dehydrogenation is

$$C_2H_6 \rightarrow C_2H_4 + H_2. \tag{16.134}$$

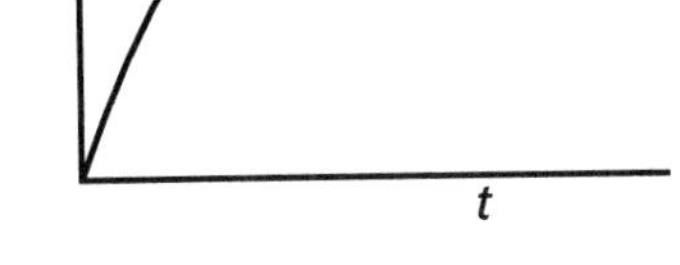

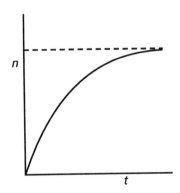

FIGURE 16.6 Variation of the hydrogen atom concentration with time below the first explosion limit and immediately above the second.

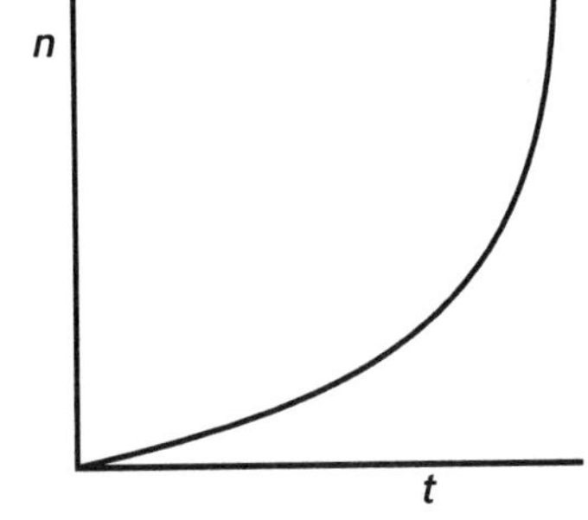

FIGURE 16.7 Variation of the hydrogen atom concentration with time between the first and the second explosion limits.

At 625° C, the equilibrium constant K is 0.05 atm, so the reverse reaction is extensive, particularly at higher pressures and temperatures.

The reaction proceeds homogeneously. But it is inhibited by compounds that react with radicals, such as NO and propylene. Furthermore, radicals have been detected by the mirror-removal process. A deposit of lead on the inner surface of a reaction tube is removed by decomposing ethane. Lead tetramethyl, $Pb(CH_3)_4$, is found among the products. Nevertheless, initial rates seem to fit a first-order law.

After Rice and Herzield, the most important elementary reactions include:

- for initiation,

$$C_2H_6 \xrightarrow{k_1} 2CH_3, \quad [16.135]$$

$$CH_3 + C_2H_6 \xrightarrow{k_2} CH_4 + C_2H_5, \quad [16.136]$$

- chain propagation,

$$C_2H_5 \xrightarrow{k_3} C_2H_4 + H, \quad [16.137]$$

$$H + C_2H_6 \xrightarrow{k_4} C_2H_5 + H_2, \quad [16.138]$$

- chain termination,

$$2C_2H_5 \xrightarrow{k_5} C_4H_{10}, \quad [16.139]$$

$$H + C_2H_5 \xrightarrow{k_6} C_2H_6. \quad [16.140]$$

The steady-state condition applies to CH_3, C_2H_5, and H. Furthermore, we consider that reaction (16.140) has a negligible effect on d [H]/dt. Also, the chain is long enough, on average, so that the overall rate is determined by the rate of reaction (16.138).

With these approximations, we find that

$$-\frac{d\left[C_2H_6\right]}{dt} = \frac{k_1^{1/2} k_3 \left[C_2H_6\right]^{1/2}}{\left(k_5 + k_3 k_6 / 2k_4\left[C_2H_6\right]\right)^{1/2}}. \quad [16.141]$$

At low C_2H_6 concentrations, parameter k_5 is negligible in the denominator and equation (16.141) reduces to a first-order law. At high concentrations, on the other hand, the second term in the denominator is negligible and the equation reduces to a one-half order law.

16.10 The Activity Coefficient Effect

For an elementary reaction proceeding freely through a sufficiently definite transition state, formula (15.142) is valid. In practice, the specific reaction rate is often symbolized by k, the ideal rate constant by k_0. Then the formula becomes

$$k = \frac{\gamma_A \cdots}{\gamma^{\ddagger}} k_0. \quad [16.142]$$

For a unimolecular reaction,

$$A \rightarrow M^{\ddagger} \rightarrow \text{products}, \quad [16.143]$$

the formula reduces to

$$k = \frac{\gamma_A}{\gamma^{\ddagger}} k_0. \quad [16.144]$$

For a bimolecular reaction,

$$A + B \rightarrow M^{\ddagger} \rightarrow \text{products}, \qquad [16.145]$$

equation (16.142) reduces to

$$k = \frac{\gamma_A \gamma_B}{\gamma^{\ddagger}} k_0. \qquad [16.146]$$

In low and moderate pressure gas phases, a person may consider the activity coefficients of molecules and uncharged radicals to be 1. Then the specific reaction rate k equals the rate constant k_0 to a good approximation. In the discussions so far in this chapter, we have made this identification.

In dilute solutions, the activity coefficients of neutral solutes may also not deviate much from 1. But the γ's of ions do. And this deviation needs to be taken into account.

For a unimolecular change in an ion, the charge on $M^{\ddagger}$ is the same as that on A. Furthermore, the effective diameter of $M^{\ddagger}$ is generally not much greater than that of A. So in dilute solutions, formula (8.97) leads to the simple result

$$\frac{\gamma_A}{\gamma^{\ddagger}} \cong 1 \qquad [16.147]$$

and

$$k \cong k_0. \qquad [16.148]$$

For a bimolecular elementary reaction, equation (16.146) yields

$$\log k = \log k_0 + \log \gamma_A + \log \gamma_B - \log \gamma^{\ddagger}. \qquad [16.149]$$

When A and B are charged and the solution is very dilute, the activity coefficients may be approximated by Debye-Hückel equation (8.96). We find that

$$\log \gamma_A + \log \gamma_B - \log \gamma^{\ddagger} = -A\sqrt{\mu}\left[z_A^2 + z_B^2 - \left(z_A + z_B\right)^2\right] = 2Az_A z_B\sqrt{\mu}. \qquad [16.150]$$

Equation (16.149) now becomes

$$\log k = \log k_0 + 2Az_A z_B\sqrt{\mu}. \qquad [16.151]$$

Here k is the specific reaction rate, k_0 is the ideal rate constant, parameter A may be calculated from the coefficient of $-z_j^2\sqrt{\mu}$ in (8.95), z_A and z_B are the units of charge on A and B, μ is the ionic strength. In aqueous solution at 25° C, formula (16.151) reduces to

$$\log = \log k_0 + 1.02 z_A z_B\sqrt{\mu}. \qquad [16.152]$$

A person may determine product $z_A z_B$ and the composition of the transition state complex from the limiting slope of a plot of log k against $\sqrt{\mu}$. Some examples appear in figure 16.8.

Example 16.2

Investigators found the reaction

$$S_2O_8^{2-} + 2I^- \rightarrow 2SO_4^{2-} + I_2$$

to be first order in persulfate ion and first order in iodide ion. At 25° C the second-order specific reaction rate k was found to equal 1.33 M^{-1} min^{-1} if initially the solution was 0.0100 M in KI and 0.00015 M in $K_2S_2O_8$. Calculate k for a run that starts with the solution 0.0020 M in KI and 0.00015 M in $K_2S_2O_8$.

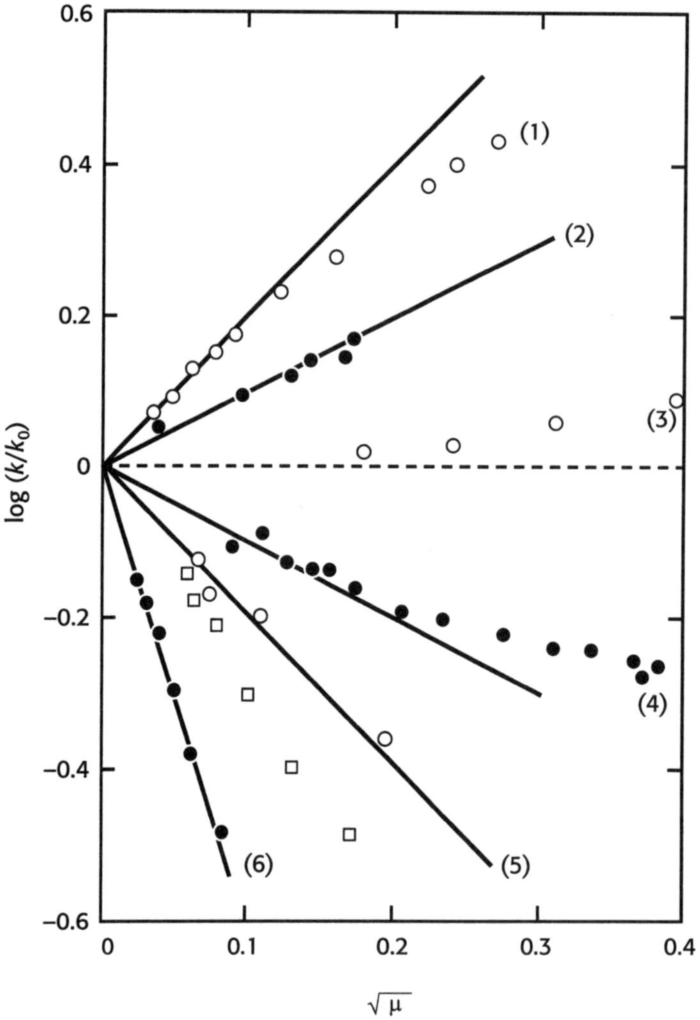

FIGURE 16.8 Plots of log (k/k_0) against $\sqrt{\mu}$ for aqueous solutions of the reactants (1) $BrCH_2COO^- + S_2O_3^{2-}$, (2) $e^- + NO_2^-$, (3) $H^+ + C_{12}H_{22}O_{11}$ (sucrose), (4) $H^+ + Br^- + H_2O_2$ (5) $OH^- + Co(NH_3)_5Br^{2+}$, (6) $Fe(H_2O)_6^{2+} + Co(C_2O_4)^{3-}$.

For the effective diameters of $S_2O_8^{2-}$ and I^-, table 8.5 has

$$a_{S_2O_8^{2-}} = 4\ \text{Å}, \qquad a_{I^-} = 3\ \text{Å}.$$

For the effective diameter of the transition state complex between $S_2O_8^{2-}$ and I^-, we estimate

$$a^{\ddagger} = 5\ \text{Å}.$$

The approximate ionic strength in the first solution is

$$\mu \cong 0.01\ \text{m}.$$

So from table 8.6, we obtain

$$\gamma_{S_2O_8^{2-}} = 0.661, \quad \gamma_{I^-} = 0.899, \qquad \gamma^{\ddagger} = 0.405.$$

Substituting these numbers into equation (16.146) leads to

$$1.33 = k_0 \frac{(0.661)(0.899)}{0.405}$$

whence

$$k_0 = 0.906 \text{ M}^{-1} \text{ min}^{-1}.$$

In the second solution, we have

$$\mu \cong 0.00245 \text{ m}.$$

Table 8.6 now gives us

$$\gamma_{S_2O_8^{2-}} = 0.804, \qquad \gamma_{I^-} = 0.946, \qquad \gamma^{\ddagger} = 0.616.$$

Inserting the k_0 and these activity coefficients into equation (16.146) yields

$$k = (0.906 \text{ M}^{-1} \text{ min}^{-1}) \frac{(0.804)(0.946)}{0.616} = 1.12 \text{ M}^{-1} \text{ min}^{-1}.$$

16.11 Conversion of Ammonium Cyanate to Urea

In studying a reaction, an investigator needs to determine whether the postulated reaction path is reasonable. The geometry of each transition state complex as well as its composition must be evaluated. One may then rule out various stoichiometrically possible mechanisms.

Consider the reaction of the ammonium ion with the cyanate ion to produce urea:

$$NH_4^+ + NCO^- \rightarrow (NH_2)_2CO. \tag{16.153}$$

This process has been extensively followed in aqueous and alcoholic solutions. In water, the rate is measurable between 30 and 70° C. The reaction is homogeneous, going practically to completion.

The rate law has the form

$$\frac{d[(NH_2)_2CO]}{dt} = k[NH_4^+][NCO^-]\frac{\gamma_+\gamma_-}{\gamma^{\ddagger}}, \tag{16.154}$$

where γ_+ is the activity coefficient of NH_4^+, γ_- the activity coefficient of NCO^-, $\gamma^{\ddagger}$ the activity coefficient of the neutral transition state complex, and k is the rate constant. This form suggests that reaction (16.153) is elementary.

But the ammonium ion probably complexes with the cyanate ion through a hydrogen bond:

$$\begin{array}{c} \text{H} \\ | \\ \text{H—N—H—N = C = O.} \\ | \\ \text{H} \end{array} \tag{16.155}$$

This transition state complex cannot rearrange readily to form urea. Instead, one expects it to dissociate into ammonia and isocyanic acid. A free ammonia molecule can then bind to the carbon and one of its hydrogen atoms can move to the other nitrogen, yielding urea.

Thus we are led to the mechanism

$$NH_4^+ + NCO^- \underset{k_{-1}}{\overset{k_1}{\rightleftarrows}} NH_3 + HNCO, \qquad [16.156]$$

$$NH_3 + HNCO \xrightarrow{k_2} (NH_2)_2CO, \qquad [16.157]$$

in which reaction (16.157) proceeds through the steps

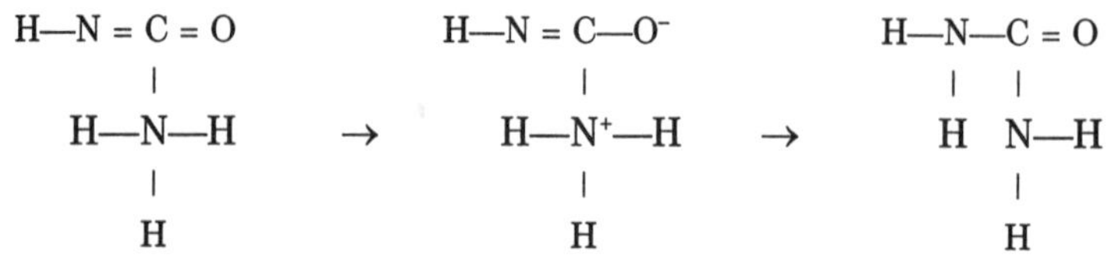

[16.158]

From (16.157), we have

$$\frac{d\left[(NH_2)_2CO\right]}{dt} = k_2[NH_3][HNCO]\frac{\gamma_{NH_3}\gamma_{HNCO}}{\gamma^{\ddagger}}. \qquad [16.159]$$

If k_2 is small with respect to k_{-1}, approximate equilibrium in reaction (16.156) is maintained and

$$\frac{[NH_3][HNCO]\gamma_{NH_3}\gamma_{HNCO}}{[NH_4^+][NCO^-]\gamma_+\gamma_-} \cong K. \qquad [16.160]$$

Combining (16.159) and (16.160) leads to the rate law

$$\frac{d\left[(NH_2)_2CO\right]}{dt} = k_2K[NH_4^+][NCO^-]\frac{\gamma_+\gamma_-}{\gamma^{\ddagger}}, \qquad [16.161]$$

which agrees with equation (16.154).

Note that the form of the empirical law does not justify the single-reaction mechanism (16.153). A similar statement applies to other ionic reactions.

An inert salt can affect the rate of an ionic reaction in two different ways:

(1) It can alter the specific reaction rate of the rate determining step by changing the activity coefficient ratio in the rate law.
(2) It can alter concentrations of reactants in the rate determining step by changing the equilibrium expression K_c that relates these concentrations to the given concentrations.

For the ammonium cyanate conversion, the second statement applies.

16.12 Nitration of Aromatic Compounds

An inert salt can also affect the rate of a reaction between molecules if the mechanism involves the production and consumption of ions.

As an example, consider the reaction

$$ArH + HNO_3 \rightarrow ArNO_2 + H_2O \qquad [16.162]$$

in which ArH is an aromatic compound. One finds that the usual orientation rules are followed. The more negative positions on the ring(s) are favored in the nitration. This fact suggests that the particle attacking the ring(s) is a positively charged particle, a cation.

Now, a nitric acid molecule has the structure

```
O       H
 \     /
   N—O
 /
O
```

[16.163]

This can react with a hydrogen ion from another acid molecule to produce the cation

```
O       H
 \     /
  N—⁺O
 /     \
O       H
```

[16.164]

Then the central N—O bond may break yielding NO_2^+ and H_2O.

The NO_2^+ cation is isoelectronic with CO_2. Its presence is indicated by a 1400 cm^{-1} band in the Raman spectrum of nitric acid, corresponding to the 1340 cm^{-1} band in CO_2. Furthermore, this band is strengthened on adding a strong acid to the nitric acid. So we consider NO_2^+ to be the positively charged particle that attacks the aromatic ring(s).

We are thus led to the mechanism

$$HNO_3 + HA \underset{k_{-1}}{\overset{k_1}{\rightleftarrows}} H_2NO_3^+ + A^-, \qquad [16.165]$$

$$H_2NO_3^+ \underset{k_{-2}}{\overset{k_2}{\rightleftarrows}} NO_2^+ + H_2O, \qquad [16.166]$$

$$NO_2^+ + ArH \xrightarrow{k_3} ArNO_2H^+, \qquad [16.167]$$

$$ArNO_2H^+ + A^- \xrightarrow{k_4} ArNO_2 + HA. \qquad [16.168]$$

When the aromatic compound reacts rapidly with NO_2^+, the rate of formation of this cation is relatively slow, reaction (16.166) is rate determining, and

$$\frac{d\left[ArNO_2\right]}{dt} = k_2\left[H_2NO_3^+\right]\frac{\gamma_{H_2NO_3^+}}{\gamma^{\ddagger}}. \qquad [16.169]$$

With (16.165) approximately at equilibrium, we have

$$\frac{\left[H_2NO_3^+\right]\left[A^-\right]}{\left[HNO_3\right]\left[HA\right]}\frac{\gamma_{H_2NO_3^+}\gamma_{A^-}}{\gamma_{HNO_3}\gamma_{HA}} = K_1 \qquad [16.170]$$

and equation (16.169) becomes

$$\frac{d\left[ArNO_2\right]}{dt} = k_2K_1\left[HNO_3\right]\frac{\left[HA\right]}{\left[A^-\right]}\frac{\gamma_{HNO_3}\gamma_{HA}}{\gamma_{A^-}\gamma^{\ddagger}}. \qquad [16.171]$$

Since the overall stoichiometry (16.162) does not involve HA or A^-, their concentrations remain constant during any given run and the reaction is first order in HNO_3, zero order in ArH. However, increasing [HA] speeds up the reaction while increasing [A^-] slows it down.

On the other hand, when the aromatic compound reacts slowly enough with NO_2^+, step (16.166) is also at approximate equilibrium. Then we have

$$\frac{[NO_2^+][H_2O][A^-]}{[HNO_3][HA]}\frac{\gamma_{NO_2^+}\gamma_{H_2O}\gamma_{A^-}}{\gamma_{HNO_3}\gamma_{HA}} = K. \qquad [16.172]$$

Reaction (16.167) is now rate determining and

$$\frac{d[ArNO_2]}{dt} = k_3[NO_2^+][ArH]\frac{\gamma_{NO_2^+}\gamma_{ArH}}{\gamma^{\ddagger}}. \qquad [16.173]$$

Combining equations (16.172) and (16.173) yields

$$\frac{d[ArNO_2]}{dt} = k_3K[ArH][HNO_3]\frac{[HA]\gamma_{ArH}\gamma_{HNO_3}\gamma_{HA}}{[H_2O][A^-]\gamma_{H_2O}\gamma_{A^-}\gamma^{\ddagger}}. \qquad [16.174]$$

The principal effect of an inert salt is on activity coefficients γ_{A^-} and $\gamma^{\ddagger}$. From (16.162), the concentrations of HA and A^- remain constant during a run. The H_2O produced may not increase its concentration significantly. Then the reaction is second order, first order in ArH and first order in HNO_3. Experiment confirms this result, as long as the aromatic compound is not too reactive.

Questions

16.1 What is a chemical mechanism? What kinds of elementary reactions may it contain?
16.2 How can a molecule acquire the energy to react by itself?
16.3 Show what rate law the Lindemann mechanism yields.
16.4 Why does the rate of a unimolecular reaction fall at low pressures?
16.5 Under what circumstances does a normally first- order reaction become second order?
16.6 Show what rate law the Ogg mechanism for N_2O_5 decomposition yields.
16,7 Why isn't the reaction between hydrogen and iodine a simple bimolecular process?
16.8 Show that the complex non-chain mechanism for the reaction between H_2 and I_2 yields a second-order rate law.
16.9 Formulate the parallel chain mechanism for the reaction between H_2 and I_2.
16.10 Use the data in table 16.1 to explain why this chain mechanism contributes appreciably only at high temperatures.
16.11 Show what rate law the chain mechanism for the hydrogen-bromine reaction yields.
16.12 How is the hydrogen-chlorine reaction analogous to the hydrogen-bromine reaction? How is it different?
16.13 Why is the hydrogen-chlorine reaction so sensitive to (a) the action of light, (b) the presence of impurities, (c) surface effects?
16.14 How does a thermal explosion develop?
16.15 What is a branched chain explosion?
16.16 Describe the mechanism for the hydrogen-oxygen reaction.
16.17 What determines (a) the first explosion limit, (b) the second explosion limit, (c) the third explosion limit?
16.18 What are the important elementary reactions in the pyrolysis of ethane?
16.19 When does the specific reaction rate vary with reactant and transition state concentrations at a given temperature?
16.20 How does the specific reaction rate vary when the reactants are ions?

16.21 May a reaction rate vary with ionic strength even when the rate determining step is a reaction between molecules. Explain.
16.22 How does ammonium cyanate react to form urea?
16.23 Describe the formation of the NO_2^+ ion in nitric acid solutions.
16.24 How does the NO_2^+ ion act on an aromatic compound ArH?

Problems

16.1 For the conversion of cyclopropane to propylene, the following first-order rate constants were obtained at 469.6° C as seen in table 16.A

TABLE 16.A

P, torr	760	388	210	110	51	26
$k \times 10^4$, s^{-1}	1.11	1.08	1.04	0.96	0.84	0.79

Plot $1/k$ against $1/P$ and note any deviation from linearity. Extrapolate the curve to infinite pressure and determine the limiting value of k.

16.2 Above 200° C the reaction

$$2ICl + H_2 \rightarrow I_2 + 2HCl$$

obeys the rate law

$$-\frac{d[H_2]}{dt} = k[ICl][H_2].$$

Suggest a mechanism.

16.3 Over the temperature range 25 - 600° C, ozone decomposes by the mechanism

$$O_3 + M \underset{k_{-1}}{\overset{k_1}{\rightleftharpoons}} O_2 + O + M,$$

$$O + O_3 \xrightarrow{k_2} 2O_2.$$

Formulate the overall rate law.

16.4 From the numbers in table 16.1, calculate the ratio of the specific reaction rates for

$$I + H_2 \xrightarrow{k_I} HI + H,$$
$$Br + H_2 \xrightarrow{k_{Br}} HBr + H,$$

at 300°.

16.5 Carry out the integration of equation (16.124).
16.6 What form does solution (16.125) assume when the rate of chain branching equals the rate of chain termination?
16.7 In the ethane decomposition of section 16.9, the activation energies for elementary reactions (16.135)—(16.140) are 359.8, 43.5, 165.3, 28.4, 0, and 0 kJ mol^{-1}, respectively. What is the activation energy for the first-order rate law?
16.8 Calculate the ratio k/k_0 for the reaction

$$NH_4^+ + NCO^- \rightarrow (NH_2)_2CO$$

when ammonium cyanate is dissolved in 0.095 m NaCl to give a solution that is initially 0.005 m in NH_4 and 0.005 m in NCO^- at 25° C. Use the activity coefficient tables.

— — —

16.9 The reaction

$$2NO_2 + O_3 \rightarrow N_2O_5 + O_2$$

follows the rate law

$$-\frac{d[NO_2]}{dt} = 2k[NO_2][O_3].$$

Formulate a mechanism.

16.10 Construct the rate law for the homogeneous para- to ortho-hydrogen conversion, which follows the mechanism

$$H_2 + M \leftrightarrows 2H + M,$$

$$H + p\text{-}H_2 \rightarrow o\text{-}H_2 + H.$$

16.11 The chlorination of carbon monoxide involves the following steps

$$Cl_2 + M \leftrightarrows 2Cl + M,$$

$$Cl + CO + M \leftrightarrows COCl + M,$$

$$COCl + Cl_2 \rightarrow COCl_2 + Cl.$$

First construct a rate law considering the first two reactions to be in equilibrium. Then introduce the steady state approximation and derive an improved rate law.

16.12 The decomposition of acetaldehyde involves the following steps:

initiation, $CH_3CHO \rightarrow CH_3 + CHO,$ (1)

chain I, $CH_3 + CH_3CHO \rightarrow CH_4 + CH_3CO,$ (2)

$CH_3CO + M \rightarrow CH_3 + CO + M,$ (3)

chain II, $CHO + M \rightarrow CO + H + M,$ (4)

$H + CH_3CHO \rightarrow H_2 + CH_3CO,$ (5)

termination, $2CH_3 \rightarrow C_2H_6.$ (6)

Most reaction occurs via chain I; both H_2 and C_2H_6 are minor products. With the steady state approximation, construct the appropriate rate law,

16.13 The decomposition of nitric acid vapor follows the mechanism

$$HNO_3 \leftrightarrows HO + NO_2, \quad (1)$$
$$HO + HNO_3 \rightarrow H_2O + NO_3, \quad (2)$$
$$NO_2 + NO_3 \rightarrow NO_2 + NO + O_2, \quad (3)$$
$$NO + NO_3 \rightarrow 2NO_2. \quad (4)$$

Construct the rate law.

16.14 Derive equation (16.141) from the mechanism for ethane decomposition.

16.15 For a substitution reaction of bromacetic acid at 25° C, the bimolecular specific reaction rates in table 16.B were found

TABLE 16.B

k, M^{-1} min^{-1}	0.288	0.301	0.334	0.371	0.429
μ, M	0.1×10^{-2}	0.2×10^{-2}	0.4×10^{-2}	1.0×10^{-2}	2.0×10^{-2}

Determine the product $z_A z_B$ of the charges on the reactants.

References

Books

Armstrong, D. A., and Holmes, J. L.: 1972, "Decomposition of Halides and Derivatives," in BT vol. 4, pp. 147-155.

In the first section of this reference, the thermal reactions of hydrogen with iodine, together with the reverses, are discussed critically. In the second section, the corresponding reactions of hydrogen with bromine are described in detail. In the third section, the hydrogen-chlorine reactions are similarly treated.

Bamford, C. H., and Tipper, C. F. H. (editors), *Comprehensive Chemical Kinetics*, continuing series, Elsevier Publishing Co., Amsterdam (abbreviated BT). The following sections are pertinent:

Benson, S. W.: 1960, *The Foundations of Chemical Kinetics*, McGraw-Hill Book Co., Inc., pp. 431-446.

A conduction model for thermal explosions is developed in considerable detail in this reference. This is followed by a description of branched chain explosions.

Dixon-Lewis, G., and Williams, D. J.: 1977, "The Oxidation of Hydrogen and Carbon Monoxide," in BT vol. 17, pp. 1-144.

The complex reactions of hydrogen with oxygen are described thoroughly, with many references to the journal literature.

Frost, A. A., and Pearson, R. G.: 1961, *Kinetics and Mechanism*, 2nd edn., John Wiley & Sons, Inc., New York, pp. 307- 315.

This section treats the ammonium cyanate-urea conversion with care.

Laidler, K. J., and Loucks, L. F.: 1972, "The Decomposition and Isomerization of Hydrocarbons," in BT vol 5, pp. 47-52.

Here the mechanism and the rate law for the pyrolysis of ethane are discussed.

Preston, K. F., and Cvetanovic, R. J.: 1972, "The Decomposition of Inorganic Oxides and Sulphides," in BT vol 4, pp 94-101.

In this part of the review, the decomposition of N_2O_5 is discussed in detail, with many references to the journal literature.

Steinfeld, J. I., Francisco, J. S., and Hase, W. L.: 1989, *Chemical Kinetics and Dynamics*, Prentice-Hall, Englewood Cliffs, NJ, pp. 164-168, 510-516.

In the first section cited, the effect of varying the ionic strength on reactions between ions is discussed. In the second section, a simplified mechanism for the hydrogen-oxygen reaction is presented and applied to an interpretation of the explosion limits.

Taylor, R.: 1972, "Kinetics of Electrophilic Aromatic Substitution," in BT vol. 13, pp. 1-40.

In these pages, the nitration of aromatic compounds by aqueous nitric acid alone, in the presence of other acids, and in the presence of organic solvents is surveyed in detail.

Articles

Avarino, J. M., and Pfartinez, E.: 1983, "Two-Body and Three-Body Atomic Recombination Reactions," *J. Chem. Educ.* **60**, 53-56.

Basza, G., Nagy, I. P., and Lengyei, I.: 1991, "The Nitric Acid/ Nitrous Acid and Ferrein/Rerriin System: A Reaction that Demonstrates Autocatalysis, Reversibility, Pseudo Orders, Chemical Waves, and Concentration Jump," *J. Chem. Educ.* **68**, 863- 868.

Brown, M. E., and Buchanan, K. J.: 1985, "Thermodynamically and Kinetically Controlled Products," *J. Chem. Educ.* **62**, 575- 578.

Gilman, J. J.: 1996, "Mechanochemistry," *Science* 274, October **4**, 65.

Griffiths, J. F.: 1985, "Thermokinetic Interactions in Simple Gaseous Reactions," *Annu. Rev. Phys. Chem.* **36**, 77-104.

Gupta, K. S., and Gupta, Y. K.: 1984, "Hydrogen-Ion Dependence of Reaction Rates and Mechanism," *J. Chem. Educ.* **61**, 972-978.

King, E. L.: 1986, "The Rate Laws for Reversible Reactions," *J. Chem. Educ.* **63**, 21-24.

Kiny, E. L.: 1991, "Reactant Fluxes in the Steady State," *J. Chem. Educ.* **68**, 897-901.

Laidler, K. J.: 1988, "Rate-Controlling Step: A Necessary or Useful Concept?" *J. Chem. Educ.* **65**, 250-254.

Loudon, &. M.: 1991, "Mechanistic Interpretation of pH—Rate Profiles," *J. Chem. Educ.* **68**, 973-984.

Miller, J. A., Kee, R. J., and Wistbrook, C. K.: 1990, "Chemical Kinetics and Combustion Modeling," *Annu. Rev. Phys. Chem.* **41**, 345-387.

Neumann, M. G.: 1986, "Chain Reaction Diayrams," *J. Chem. Educ.* **63**, 684.

Raines, R. T., and Hansen, D. E.: 1988, "An Intuitive Approach to Steady-State Kinetics," *J. Chem. Educ.* **65**, 757-759.

Tardy, D. C., and Cater, E. D.: 1983, "The Steady State and Equilibrium Assumptions in Chemical Kinetics," *J. Chem. Educ.* **60**, 109-111.

Temkin, O. N., and Bonchev, D. G.: 1992, "Application of Graph Theory to Chemical Kinetics Part 1. Kinetics of Complex Reactions," *J. Chem. Educ.* **69**, 544-550.

Viossat, V., and Ben-Aim, R. I.: 1993, "A Test of the Validity of Steady State and Equilibrium Approximations in Chemical Kinetics," *J. Chem. Educ.* **70**, 732-738.

Weston Jr., R. E.: 1988, "A Case Study in Chemical Kinetics: The OH + CO Reaction," *J. Chem. Educ.* **65**, 1062- 1066.

17

Statistical Thermodynamics

17.1 Earlier Discussions

IN CHAPTER 3, WE SAW HOW THE TRANSLATIONAL ENERGY of the molecules in an ideal gas determines the macroscopic pressure P and temperature T of the gas. And how at *equilibrium* at temperatures that are not too low, the molecular velocities and speeds are distributed. Furthermore, we saw how molecules are distributed on the average in a potential field. The results are summarized in the Boltzmann-Maxwell distribution law.

On small areas and in small regions, appreciable deviations from these results occur. Direct evidence comes from observation of the Brownian motion of colloidal particles. Also, in a resistor the random motion of the electrons causes fluctuations in a passing current that, when amplified, appear as noise.

In chapter 5, we related the entropy S of a system to the number of disjoint microstates contributing to a given macroscopic state. And we calculated the entropy change associated with a fluctuation. We found that appreciable changes would occur with appreciable probability only in very small systems.

In chapter 8, we used the Boltzmann distribution law to determine the structure of the ionic atmosphere about a given ion in an electrolytic solution.

In chapter 10, we constructed a partition function Z based on the Boltzmann distribution law. We inserted the quantized translational levels. Then with the continuum approximation, we obtained explicit expressions for Z_{tr}. Similarly, explicit expressions for the rotational partition function Z_{rot} and for the vibrational partition function Z_{vib} were obtained in chapters 11 and 12. We saw how these yielded formulas for the respective contributions to the internal energy.

We have still to relate other thermodynamic properties to the partition function for a system. Of particular interest are the expressions for the entropy, the Helmholtz energy, and the concentration equilibrium constant. We will make use of the latter in developing the transition state theory for rate constants.

We also need to determine how the partition function for an N particle system is related to the partition function for each contributing particle.

17.2 Relating Thermodynamic Properties to the Partition Function

Consider a macroscopic system in a particular macroscopic state. What individual molecules, radicals, ions are doing is not specified. So in a homogeneous region of appreciable size, at a temperature that is not too low, an enormous number of disjoint particle states are allowed. But over time, each of these states is visited with equal probability. So the symmetry assumed in the derivation of the Maxwell-Boltzmann distribution law prevails.

Indeed, a large system may be broken down into small regions in which the average distribution over energy levels is that described by this distribution law. Consider a typical such region. Let N be the number of molecules there. But if N_j is the number of molecules in the jth disjoint state, on the average, when the temperature there is T, if E_j. is the energy of this state, and if

$$\beta = \frac{1}{kT}, \qquad [17.1]$$

where k is the Boltzmann constant, then for the given region we have

$$N_j = \frac{Ne^{-\beta E_j}}{\sum e^{-\beta E_j}} = \frac{Ne^{-\beta E_j}}{Z}. \qquad [17.2]$$

The summation in the partition function

$$Z = \sum e^{-\beta E_j} \qquad [17.3]$$

extends over all allowed disjoint states.

Take the logarithm of equation (17.3) and differentiate the result with respect to β at constant volume to get

$$\left(\frac{\partial}{\partial\beta}\ln Z\right)_V = -\frac{\sum E_j e^{-\beta E_j}}{Z} = -\frac{\sum N_j E_j}{N} = -\langle E\rangle. \qquad [17.4]$$

The pointed brackets indicate that an average has been taken. In a region of appreciable size, fluctuations are insignificant and the average energy is identified with the internal energy:

$$E = -\left(\frac{\partial}{\partial\beta}\ln Z\right)_V. \qquad [17.5]$$

On the other hand, differentiation with respect to E_j at constant temperature yields

$$\left(\frac{\partial}{\partial E_j}\ln Z\right)_T = \frac{-\beta e^{-\beta E_j}}{Z} = -\frac{\beta N_j}{N}. \qquad [17.6]$$

We now have

$$\mathrm{d}\ln Z = \frac{\partial \ln Z}{\partial\beta}\mathrm{d}\beta + \sum\frac{\partial \ln Z}{\partial E_j}\mathrm{d}E_j = -\left[E\,\mathrm{d}\beta + \beta\sum\frac{N_j}{N}\mathrm{d}E_j\right], \qquad [17.7]$$

whence

$$\mathrm{d}(\ln Z + \beta E) = \beta\left(\mathrm{d}E - \sum\frac{N_j}{N}\mathrm{d}E_j\right). \qquad [17.8]$$

The shift in each energy level is effected by a change in external parameters λ_1, λ_2, ..., λ_n, . The corresponding energy change is the sum

$$\sum\frac{N_j}{N}\mathrm{d}E_j = \sum_j\frac{N_j}{N}\sum_k\frac{\partial E_j}{\partial\lambda_k}\mathrm{d}\lambda_k = \mathrm{d}w_{\mathrm{rev}} \qquad [17.9]$$

in which $\Sigma(N_j/N)(\partial E_j/\partial\lambda_k)$ acts as a force and $d\lambda_k$ as the corresponding displacement, Since

$$dE = dq + dw, \tag{17.10}$$

equation (17.8) becomes

$$d\left(\ln Z + \beta E\right) = \beta\, dq_{\text{rev}} = \beta\, T\, dS, \tag{17.11}$$

where S is the entropy.

Integrating (17.11) and setting the arbitrary constant equal to zero yields the relationship

$$\ln Z + \beta E = \beta\, TS. \tag{17.12}$$

Then introducing the Helmholtz energy gives us

$$A = E - TS = -\frac{1}{\beta}\ln Z = -kT\ln Z. \tag{17.13}$$

With equation (17.5), we obtain the statistical internal energy

$$E_{\text{stat}} = -\left(\frac{\partial \ln Z}{\partial \beta}\right)_V = kT^2\left(\frac{\partial \ln Z}{\partial T}\right)_V. \tag{17.14}$$

Note that this is relative to the zero for the energy levels. But the conventional internal energy includes the internal energy of the system at $T = 0$. For it, we have

$$E = kT^2\left(\frac{\partial \ln Z}{\partial T}\right)_V + E(0). \tag{17.15}$$

Applying formula (17.10) to a given substance yields

$$dE = T\, dS - P\, dV, \tag{17.16}$$

so that

$$dA = dE - T\, dS - S\, dT = -P\, dV - S\, dT. \tag{17.17}$$

Relationships (17.17), (17.13),and (17.15) lead to

$$P = -\left(\frac{\partial A}{\partial V}\right)_T = kT\left(\frac{\partial \ln Z}{\partial V}\right)_T \tag{17.18}$$

and

$$S = -\left(\frac{\partial A}{\partial T}\right)_V = k\left[\ln Z + T\left(\frac{\partial \ln Z}{\partial T}\right)_V\right] = k\ln Z + \frac{E - E(0)}{T}. \tag{17.19}$$

Example 17.1

Reduce equation (10.47) to a convenient form for calculating molecular translational partition functions.

Factor out the molecular mass, the absolute temperature, and the volume. Then introduce accepted values of the fundamental constants and carry out the indicated operations:

$$z_{\text{tr}} = \left(\frac{2\pi m_{\text{u}} k}{h^2}\right)^{3/2} M^{3/2}T^{3/2}V = \left[\frac{2\pi\left(1.660540\times10^{-27}\text{ kg u}^{-1}\right)\left(1.380658\times10^{-23}\text{ J K}^{-1}\right)}{\left(6.626076\times10^{-34}\right)^2(\text{J s})\left(\text{kg m}^2\text{ s}^{-1}\right)}\right]^{3/2} M^{3/2}T^{3/2}V$$

$$= \left(1.879333\times10^{26}\text{ u}^{-3/2}\text{ K}^{-3/2}\text{ m}^{-3}\right)M^{3/2}T^{3/2}V.$$

Here M is in atomic mass units, T in kelvins, and V in cubic meters.

Example 17.2

For a linear molecule, the rotational partition function is given by equation (11.54), Reduce this to a convenient form for calculations from spectroscopic data.

From the separations between successive maxima in the rotational spectrum, or in the vibration- rotation spectrum, one obtains the parameter $2B$ in reciprocal centimeters. So we employ (11.54) in the form

$$z_{\text{rot}} = \frac{kT}{\sigma B} = \frac{0.69504 \text{ cm}^{-1}\text{ K}^{-1}}{\sigma}\frac{T}{B},$$

in which the result from example 11.7 has been introduced.

Example 17.3

For a nonlinear molecule, the rotational partition function is given by equation (11.57), Reduce this to a convenient form for calculations from spectroscopic data,

The rotational parts of the vibration-rotation spectrum of the given molecule are fitted to the calculated levels as approximated in table 11.1, Thus, parameters A, B, and C are obtained in reciprocal centimeters. So we employ (11.57) in the form

$$z_{\text{rot}} = \frac{\pi^{1/2}k^{3/2}}{\sigma\left(ABC\right)^{1/2}}T^{3/2} = \frac{\pi^{1/2}\left(0.69504 \text{ cm}^{-1}\text{ K}^{-1}\right)^{3/2}}{\sigma\left(ABC\right)^{1/2}}T^{3/2} = \frac{\left(1.02704 \text{ cm}^{-3/2}\text{ K}^{-3/2}\right)T^{3/2}}{\sigma\left(ABC\right)^{1/2}}.$$

As before, σ is the symmetry number while A, B, and C are the rotational constants.

Example 17.4

For a vibrational mode, one form of the partition function appears in equation (12.53). Rewrite this employing the lowest energy level as the zero for energy. Also consider the Q line to be in reciprocal centimeters.

From the Q line in the vibrational and/or Raman spectra, one obtains the vibrational spacing $\hbar\omega$ for the mode in reciprocal centimeters. So referred to the lowest energy level as zero, we have

$$z_{\text{vib}} = \frac{1}{1-e^{-\hbar\omega/kT}} = \frac{1}{1-e^{-\hbar_0/\left(0.69504 \text{ cm}^{-1}\text{ K}^{-1}\right)T}}.$$

17.3 Relating System Partition Functions to Molecular Ones

In chapters 10, 11, and 12, we constructed partition functions for the translation, rotation, and vibration of a molecule. We now have to consider how the product in formula (11.47) must be treated to give the partition function for a thermodynamic system.

The energy of a system is additive; it equals the sum of the energies of the molecules making up the system. If ε_{jl} is the energy of the lth molecule-in its jth state, we have

$$E_i = \varepsilon_{j1} + \varepsilon_{k2} + \ldots + \varepsilon_{mN}. \qquad [17.20]$$

If the molecules are *distinguishable* and practically independent, the complete partition function factors:

$$\sum e^{-E_i/kT} = \sum e^{-\varepsilon_{j1}/kT} \sum e^{-\varepsilon_{k2}/kT} \ldots \sum e^{-\varepsilon_{mN}/kT}, \qquad [17.21]$$

or

$$Z = z_1 z_2 \ldots z_N. \qquad [17.22]$$

If in addition, the molecules are equivalent, their partition functions are equal and

$$Z = z^N. \qquad [17.23]$$

Equivalent molecules in a gas or fluid are not distinguishable. For N such molecules, we have $N!$ permutations that cannot be differentiated. Then the sum over states has to be divided by $N!$:

$$Z = \frac{z^N}{N!}. \qquad [17.24]$$

In the solid state, the different molecules are distinguished by the lattice position each occupies and formula (17.25) may be used.

However, this formula does not allow for the interaction among the molecules. This interaction becomes important in dense gases, liquids, and solids. It is allowed for by adding the factor

$$Z_{\text{con}} = \frac{1}{V^N} \int \exp\left[-\beta U\left(\mathbf{r}^N\right)\right] \mathrm{d}\mathbf{r}^N, \qquad [17.25]$$

where $U(\mathbf{r}^N)$ is the interaction potential energy and V the volume, This factor is known as the *configuration integral.*

Example 17.5

Evaluate the *gamma function*

$$\Gamma(N+1) = \int_0^\infty t^N e^{-t}\, dt$$

for $N = 0$ and $N = 1$

Carrying out the integration for $N = 0$ yields

$$\Gamma(1) = \int_0^\infty e^{-t}\, dt = -e^{-t}\Big|_0^\infty = 1.$$

Similarly for $N = 1$,

$$\Gamma(2) = \int_0^\infty te^{-t}\, dt = -te^{-t}\Big|_0^\infty + \int_0^\infty e^{-t}\, dt = 0 + 1 = 1.$$

Example 17.6

Construct a recurrence formula for the gamma function.

Integrate the defining integral by parts with $N > 0$:

$$\Gamma(N+1) = \int_0^\infty t^N e^{-t}\, dt = -t^N e^{-t}\Big|_0^\infty + N\int_0^\infty t^{N-1} e^{-t}\, dt = 0 + N\,\Gamma(N) = N\,\Gamma(N).$$

For factorial N, we similarly have

$$N! = N(N-1)!$$

But from example 17.5, one can set

$$\Gamma(1) = 0! \quad \text{and} \quad \Gamma(2) = 1!$$

So when N is a positive integer, we find that

$$N! = \int_0^\infty t^N e^{-t}\, \mathrm{d}t.$$

Example 17.7

Construct a closed form approximation for N!
In the gamma function integral, set

$$t^N e^{-t} = e^{g(t)},$$

so

$$g = N \ln t - t$$

and

$$\frac{\mathrm{d}g}{\mathrm{d}t} = \frac{N}{t} - 1,$$

$$\frac{\mathrm{d}^2 g}{\mathrm{d}t^2} = -\frac{N}{t^2}.$$

Then expand $g(t)$ about the point to $t_0 = N$, where

$$g = N \ln N - N, \qquad \frac{\mathrm{d}g}{\mathrm{d}t} = 0, \qquad \frac{\mathrm{d}^2 g}{\mathrm{d}t^2} = -\frac{1}{N},$$

in the Taylor series

$$g(t) = N \ln N - N - \frac{1}{2N}(t - t_0)^2 + \ldots$$

Neglect higher terms and substitute into the integral:

$$N! \cong e^{N \ln N - N} \int_0^\infty e^{-(t-t_0)^2/2N}\, \mathrm{d}t \cong \sqrt{2N}\, e^{N \ln N - N} \int_{-\infty}^\infty e^{-x^2}\, \mathrm{d}x = \sqrt{2\pi N}\left(\frac{N}{e}\right)^N.$$

This result is known as the *Stirling approximation*. Its logarithm has the form

$$\ln N! \cong N \ln N - N + \frac{1}{2}\ln(2\pi N).$$

When N is large, the last term may be neglected.

17.4 Ideal Translational Entropy

With many substances, the parameters for individual molecules can be accurately determined. However, the interaction potential $U(\mathbf{r}^N)$ and the dependent configuration integral are much less accurately known. So thermodynamic properties are calculated for the hypothetical ideal gas state. The experimental values then may be corrected to this state, as we saw in section 7.7.

For energies E and H, the distinction between formula (17.23) and (17.24) is not significant. But for entropy S and the functions A and G involving S, any indistinguishability must be taken into account.

Consider a mole of a pure substance in the ideal gas state at a given temperature T and pressure P. From equation (3.23), its internal energy is given by

$$E - E(0) = \frac{3}{2}NkT = \frac{3}{2}RT. \qquad [17.26]$$

Since the molecules are indistinguishable, the system partition function is related to the molecular one by the formula

$$Z = \frac{z^N}{N!}. \qquad [17.27]$$

Substitute these expressions into equation (17,19), introduce the truncated Stirling approximation, factor the partition function, and introduce the ideal gas law for V/N as follows:

$$\begin{aligned} S_{\text{tr}} &= Nk\ln z_{\text{tr}} - Nk\ln N + Nk + \frac{3}{2}Nk = R\left(\frac{5}{2} + \ln\frac{z_{\text{tr}}}{M^{3/2}T^{3/2}V} + \frac{3}{2}\ln M + \frac{3}{2}\ln T + \ln\frac{V}{N}\right) \\ &= R\left(\frac{5}{2} + \ln\frac{z_{\text{tr}}}{M^{3/2}T^{3/2}V} + \ln k - \ln P_0 + \frac{3}{2}\ln M + \frac{5}{2}\ln T - \ln P\right). \end{aligned} \qquad [17.28]$$

The term $\ln P_0$ lets us express P in bars; we merely set P_0 equal to 10^5 Pa bar^{-1}.

Introduce the other fundamental constants as in example 17.1 to get the formula

$$S_{\text{tr}} = R\left(-1.151693 + \frac{3}{2}\ln M + \frac{5}{2}\ln T - \ln P\right). \qquad [17.29]$$

This is known as the *Sackur-Tetrode equation.* As it stands, M is in atomic mass units, T in kelvins, P in bars.

Example 17.8

Construct a formula for the ideal rotational entropy of a system of identical linear molecules.

When B/kT is small, equation (11.51) or (11.52) holds. The corresponding internal energy is given by formula (11.60), For molecules, we have

$$\left[E - E(0)\right]_{\text{rot}} = NkT$$

and

$$Z_{\text{rot}} = z_{\text{rot}}{}^{N} = \left(\frac{kT}{\sigma B}\right)^N = \left(\frac{0.69504\ \text{cm}^{-1}\ \text{K}^{-1}}{\sigma}\frac{T}{B}\right)^N.$$

Substituting into equation (17.19) and letting N be the Avogadro number yields

$$S_{\text{rot}} = Nk\ln z_{\text{rot}} + Nk = R(1 + \ln 0.69504 + \ln T - \ln\sigma B) = R(0.63621 + \ln T - \ln\sigma B).$$

Here T is in kelvins, B in reciprocal centimeters, while σ is the symmetry number.

Example 17.9

Construct a formula for the ideal rotational entropy of a system of identical nonlinear molecules.

When $ABC/(kT)^3$ is small, equation (11.57) or (11,58) holds. The corresponding internal energy is given by formula (11.62), For N molecules, we have

$$\left[E - E(0)\right]_{\text{rot}} = \frac{3}{2}NkT$$

and

$$Z_{\text{rot}} = z_{\text{rot}}{}^{N} = \left(\frac{\pi^{1/2}k^{3/2}T^{3/2}}{\sigma(ABC)^{1/2}}\right)^{N} = \left[\frac{\left(1.02704\ \text{cm}^{-3/2}\ \text{K}^{-3/2}\right)T^{3/2}}{\sigma(\text{ABC})^{1/2}}\right]^{N}.$$

Substituting into equation (17.19) and letting N be the Avogadro number yields

$$S_{\text{rot}} = Nk\ln z_{\text{rot}} + \frac{3}{2}Nk = R\left[(3/2) + \ln 1.02704 + (3/2)\ln T - \ln\sigma(ABC)^{1/2}\right]$$
$$= R\left[1.52668 + (3/2)\ln T - \ln\sigma(ABC)^{1/2}\right].$$

Here T is in kelvins, A, B, C are in reciprocal centimeters, while σ is the symmetry number.

17.5 Energy and Entropy of a Vibrational Mode

At moderate temperatures, the level spacing $\hbar\omega$ for most molecular vibrational modes is large with respect to kT and approximation (12.54) is not valid. Then one has to calculate from the form in example 17.4.

Consider a vibrational mode for which the Q line is at $\bar{k}_0$. The molar partition function is

$$Z_{\text{vib}} = z_{\text{vib}}{}^{N} = \left(\frac{1}{1 - e^{-\beta\bar{k}_0}}\right)^{N}. \qquad [17.30]$$

Since

$$\ln Z_{\text{vib}} = -N\ln\left(1 - e^{-\beta\bar{k}_0}\right), \qquad [17.31]$$

formula (17.14) yields

$$E - E(0) = -\left(\frac{\partial \ln Z_{\text{vib}}}{\partial\beta}\right)_V = N\frac{\bar{k}_0 e^{-\beta\bar{k}_0}}{1 - e^{-\beta\bar{k}_0}} = \frac{N\bar{k}_0}{e^{\beta\bar{k}_0} - 1}. \qquad [17.32]$$

and

$$\frac{E - E(0)}{T} = Nk\frac{\bar{k}_0/kT}{e^{\beta\bar{k}_0} - 1} = R\frac{\beta\bar{k}_0}{e^{\beta\bar{k}_0} - 1}. \qquad [17.33]$$

Substituting into formula (17.19) leads to the result

$$S_{\text{vib}} = R\left[-\ln\left(1 - e^{-\beta\bar{k}_0}\right) + \frac{\beta\bar{k}_0}{e^{\beta\bar{k}_0} - 1}\right], \qquad [17.34]$$

in which

$$\beta = \frac{1}{kT} = \frac{1}{\left(0.69504\ \text{cm}^{-1}\ \text{K}^{-1}\right)T}. \qquad [17.35]$$

17.6 The Ideal Concentration Equilibrium Constant

The dependence of the Helmholtz energy for a given substance on its partition function enables a person to relate the equilibrium constant for a given reaction to spectroscopic and molecular parameters.

Consider a homogeneous system in which the reaction

$$aA + bB \rightarrow lL + mM \qquad [17.36]$$

takes place. Equation (7.116) gives the Helmholtz energy increase per mole unit of reaction for each concentration in moles per unit volume, When the mole unit is replaced by the number unit and each concentration is expressed as number of molecules of the given kind per unit volume, this equation becomes

$$\Delta A^0 = -kT \ln K_{c'}. \qquad [17.37]$$

And when Z in (17.13) is expressed as a number per *molecule* per *unit volume*, we have

$$A^0 = -kT \ln z'. \qquad [17.38]$$

For reaction (17.36), the standard Helmholtz energy change is

$$\Delta A^0 = lA_L^0 + mA_M^0 - aA_A^0 - bA_B^0. \qquad [17.39]$$

Combining relations (17.37), (17.38), (17.39) and solving for $K_{c'}$, yields

$$K_{c'} = \frac{z_L'^l z_M'^m}{z_A'^a z_B'^b}. \qquad [17.40]$$

Multiplying the molarity of a constituent by $N_A(10^3 \text{l m}^{-3})$ yields its number concentration in m^{-3}. With K_c the conventional concentration equilibrium constant, we have

$$K_{c'} = K_c\left[N_A\left(10^3 \text{ l m}^{-3}\right)\right]^{l+m-a-b}. \qquad [17.41]$$

In form (17.40), the implication is that all partition functions are calculated from the same zero of energy. But in the partition functions we will employ, the zero of energy is the lowest energy level for each molecule.

The change in this lowest energy per unit of reaction is

$$\Delta\varepsilon_0 = l\varepsilon_{0L} + m\varepsilon_{0M} - a\varepsilon_{0A} - b\varepsilon_{0B}. \qquad [17.40]$$

In terms of the conventional single molecule partition functions, constructed per *unit volume*, equation (17.40) yields

$$K_{c'} = \frac{z_L^l z_M^m}{z_A^a z_B^b} e^{-\Delta\varepsilon_0/kT} = \frac{z_L^l z_M^m}{z_A^a z_B^b} e^{-\Delta E_0/RT}. \qquad [17.43]$$

In the molarity concentration scheme, this becomes

$$K_c = \left[N_A\left(10^3 \text{ l m}^{-3}\right)\right]^{a+b-l-m} \frac{z_L^l z_M^m}{z_A^a z_B^b} e^{-\Delta E_0/RT}. \qquad [17.44]$$

For the reaction

$$A + B \leftrightarrows M, \qquad [17.45]$$

we have

$$K_{c'} = \frac{z_M}{z_A z_B} e^{-\Delta E_0/RT}, \qquad [17.46]$$

with the number concentration scheme. In the molarity scheme, we have

$$K_c = N_A\left(10^3 \text{ l m}^{-3}\right)\frac{z_M}{z_A z_B} e^{-\Delta E_0/RT}. \qquad [17.47]$$

Example 17.10

Construct a formula for the partition function of Na (g). The molecular mass M is 22.990 u, while the lowest electronic state is a doublet.

We consider that practically all the atoms are in their ground electronic state. Since this is a doublet, we have

$$z_{el} = 2.$$

Since there are no rotational or vibrational modes, we take

$$z_{rot} = z_{vib} = 1.$$

From the formula in example 17.1, we obtain

$$z_{tr} = \left(1.8793\times10^{26}\right)\left(22.990\right)^{3/2} T^{3/2} V\,\text{K}^{-3/2}\text{m}^{-3} = \left(2.0716\times10^{28}\right) T^{3/2} V\,\text{K}^{-3/2}\text{m}^{-3}.$$

Multiplying these factors yields

$$z_{Na} = z_{tr} z_{rot} z_{vib} z_{el} = \left(4.1432\times10^{28}\right) T^{3/2} V\,\text{K}^{-3/2}\text{m}^{-3}.$$

Example 17.11

Construct a formula for the partition function of Na_2 (g) The parameters for the molecule are $B = 0.1547\text{ cm}^{-1}$, $\tilde{k}_0 = 159.2\text{ cm}^{-1}$, $D = 70.4\text{ kJ mol}^{-1}$.

Since all electrons are paired in the ground state, we have

$$z_{el} = 1.$$

Since the molecule is linear, the rotational formula in example 17.2 applies:

$$z_{rot} = \frac{kT}{\sigma B} = \frac{0.69504}{2}\frac{T\,\text{K}^{-1}}{0.1547} = 2.24641\;T\,\text{K}^{-1}.$$

For the one vibrational mode, the formula in example 17.4 is employed:

$$z_{vib} = \frac{1}{1-e^{-\tilde{k}_0/kT}} = \frac{1}{1-e^{-(229.05\text{ K})/T}}.$$

For the three translational modes, we have

$$z_{tr} = \left(1.8793\times10^{26}\right)\left(45.98\right)^{3/2} T^{3/2}\;V\;\text{K}^{-3/2}\;\text{m}^{-3} = \left(5.8594\times10^{28}\right) T^{3/2}\;V\;\text{K}^{-3/2}\;\text{m}^{-3}.$$

Multiplying these four factors gives us z_{Na_2} Also note that

$$\frac{D}{RT} = \frac{70.4\times10^4\text{ K}}{8.31451\;T} = \frac{8.46713\times10^3}{T}\text{ K}.$$

Example 17.12

For the reaction

$$Na_2\text{ (g)} \rightleftarrows 2\text{ Na (g)}$$

formulate $K_{c'}$. Then calculate K_c and K_P at 1000 K.

Use the results from examples 17.10 and 17.11, with $V = 1$, in equation (17.43):

$$K_{c'} = \frac{z_{\text{Na}}^2}{z_{\text{Na}_2}} e^{-D/RT} = \frac{\left(4.1432 \times 10^{28}\right)^2 T^3 \left(1 - e^{-(229.05\text{ K})/T}\right)}{\left(5.8594 \times 10^{28}\right) T^{3/2} \left(2.24641\, T\right)} e^{-\left(8.4671\times 10^3\text{ K}\right)/T} \text{K}^{-1/2}\text{m}^{-3}$$

$$= \left(1.3042 \times 10^{28}\right) T^{1/2} \left(1 - e^{-(229.05\text{ K})/T}\right) e^{-\left(8.4671\times 10^3\text{ K}\right)/T} \text{K}^{-1/2}\text{m}^{-3}.$$

Setting $T = 1000$ K and converting to K_c then leads to

$$K_c = \frac{1.3042 \times 10^{28}}{6.0221 \times 10^{26}} (31.623)(0.20471)\left(2.1027 \times 10^{-4}\right) \text{M} = 2.95 \times 10^{-2} \text{ M}.$$

For K_P, we have

$$K_P = K_c \left(RT\right)^{\Delta n} = \left(2.948 \times 10^{-2} \text{ mol l}^{-1}\right)\left(0.083145 \text{ l bar K}^{-1} \text{ mol}^{-1}\right)\left(1000 \text{ K}\right) = 2.45 \text{ bar}$$

17.7 Basis for the Distribution Law

In statistical mechanics, one is concerned with how a small system of particles in a definite molecular state interacts with a sea of similar systems or particles. To the sea, one can apply standard thermodynamics. So it may be assigned a temperature T and an entropy S. This entropy is related to the number of available orthogonal states.

Let the definite state system consist of N_ν particles in volume V_ν. Let it be in thermal contact with a bath containing N_β particles at temperature T, volume V_β, and energy E_β. See figure 17.1.

Now, energy moves into and out of the small system, altering its state from time to time. We wonder what the statistical weight of each of the disjoint states of this system is. How is it related to the bath temperature and to the state energy?

When the small system is in energy state E_ν, the diversity is in the surrounding sea. So the total number of disjoint submicroscopic states is

$$W\left(E_\beta\right) = W\left(E - E_\nu\right), \qquad [17.48]$$

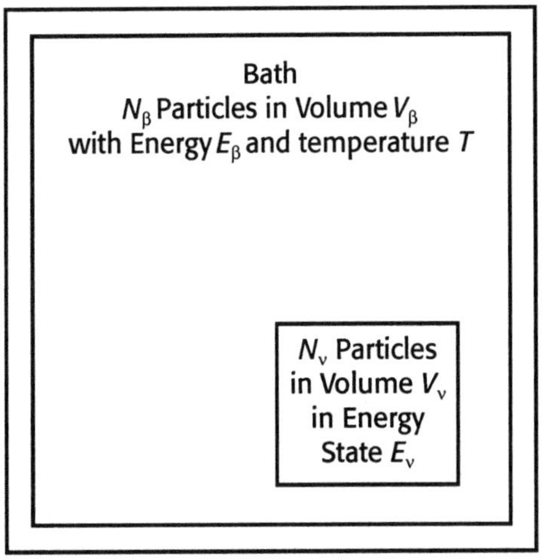

FIGURE 17.1 Set of N_ν particles in a definite energy state surrounded by a thermodynamic sea of N_β particles at temperature T.

where

$$E = E_\beta + E_\nu. \quad [17.49]$$

According to the statistical assumption, each disjoint state consistent with the imposed constraints is equally probable. So the statistical weight of this energy state is

$$w_\nu = aW(E - E_\nu) = a\exp[\ln W(E - E_\nu)]. \quad [17.50]$$

Let us expand the exponent using the Taylor formula

$$f(x) = f(a) + f'(a)(x-a) + f''(a)\frac{(x-a)^2}{2!} + \ldots. \quad [17.51]$$

We set

$$x = E - E_\nu, \qquad a = E, \qquad x - a = -E_\nu, \quad [17.52]$$

and obtain

$$\ln W(E - E_\nu) = \ln W(E) - \left(\frac{\partial \ln W(E)}{\partial E}\right)_V E_\nu + \ldots. \quad [17.53]$$

From thermodynamic relation (5.76), one has

$$\mathrm{d}S = \frac{1}{T}\,\mathrm{d}E + \frac{P}{T}\,\mathrm{d}V \quad [17.54]$$

and

$$\left(\frac{\partial S}{\partial E}\right)_V = \frac{1}{T}. \quad [17.55]$$

But entropy S is related to state number W by formula (5.55):

$$\frac{S}{k} = \ln W. \quad [17.56]$$

Differentiating equation (17.56) and introducing (17.55) yields

$$\left(\frac{\partial \ln W}{\partial E}\right)_V = \frac{1}{kT} = \beta. \quad [17.57]$$

Substituting result (17.57) into series (17.53) and neglecting the higher terms leads to the expression

$$\ln W(E - E_\nu) = \ln W(E) - \beta E_\nu, \quad [17.58]$$

which converts (17.50) to

$$w_\nu = a\exp[\ln W(E) - \beta E_\nu] = aW(E)e^{-\beta E_\nu} = \frac{e^{-\beta E_\nu}}{Z}. \quad [17.59]$$

Symbol Z has been introduced to represent $[aW(E)]^{-1}$. We now make w_ν equal the probability that the small system is in its νth state by setting

$$\sum w_\nu = 1. \quad [17.60]$$

Then we have

$$Z = \sum_\nu e^{-\beta E_\nu} = \sum_\nu e^{-E_\nu/kT}. \quad [17.61]$$

Expression Z is the partition function for the system.

For the setup in section 17.2, the probability is the ratio

$$w_\nu = \frac{N_\nu}{N}. \tag{17.62}$$

So result (17.59) agrees with formula (17.2),

17.8 Black-Body Radiation

The principle of microscopic reversibility implies that a surface will radiate electromagnetic waves with the same efficiency as it absorbs them. An ideal emitter is also an ideal absorber. Its surface is black at all wavelengths. But a small black surface is simulated by a hole with the same cross section, leading into a large cavity. So the radiation from each exhibits a similar distribution over wavelengths for a given source temperature.

Consider electromagnetic radiation in a rectangular volume. Let us impose periodic boundary conditions as in sections 10.6 and 10.8. In each translational state, there are two disjoint polarizations. So formula (10.57) applies. The number of disjoint modes with wavenumber less than $\bar{k}$ is

$$N = 2\left(\frac{4}{3}\pi\bar{k}^3\right). \tag{17.63}$$

Between $\bar{k}$ and $\bar{k} + \mathrm{d}\bar{k}$, the number of modes is

$$\mathrm{d}N = 8\pi\bar{k}^2\mathrm{d}\bar{k}\,. \tag{17.64}$$

The wavenumber is related to wavelength λ by the equation

$$\bar{k} = \frac{1}{\lambda}. \tag{17.65}$$

Differentiation yields

$$\mathrm{d}\bar{k} = -\frac{\mathrm{d}\lambda}{\lambda^2}. \tag{17.66}$$

By addition, the phase velocity c is given by

$$c = \lambda\nu, \tag{17.67}$$

where ν is the conventional frequency.

Substituting expressions (17.65) and (17.66) into formula (17.64) and reversing the sign gives

$$\mathrm{d}N = 8\pi\frac{\mathrm{d}\lambda}{\lambda^4} \tag{17.68}$$

for the number of modes between λ and $\lambda + \mathrm{d}\lambda$.

The electromagnetic field interacts with the walls as if the field was made up of photons. Furthermore, the energy of a photon with frequency ν, wavelength λ, is

$$\varepsilon = h\nu = \frac{hc}{\lambda}. \tag{17.69}$$

For a given cavity, a complete set of orthogonal (disjoint) energy states exists. These may be compounded from a complete set of orthogonal single photon states (the disjoint modes). All possible combinations of these must be employed in the partition function, no more, no less.

Let the allowed photon energies be ε_1, ε_2, ..., ε_j, . A particular cavity state arises with n_1 photons in the first state, n_2 photons in the second state,..., n_j photons in the jth state,... But the partition function is the sum over all allowed disjoint cavity states, over all n's for each ε. Thus, we have

$$Z = \sum_i e^{-\beta E_i} = \sum_i e^{-\beta\left[\ldots + n_j(i)\varepsilon_j + \ldots\right]} = \prod_j \sum_{n_j=0}^{\infty} e^{-\beta n_j \varepsilon_j} = \prod_j \frac{1}{1 - e^{-\beta\varepsilon_j}}, \tag{17.70}$$

whence

$$\ln Z = \sum_j -\ln\left(1 - e^{-\beta\varepsilon_j}\right). \tag{17.71}$$

The jth photon state is weighted by expression (17.59). So the average number of photons in this state is

$$\begin{aligned}\langle n_j \rangle &= \sum_i n_j(i) w_i = \sum_i n_j(i) e^{-\beta E_i} / Z = \sum_i n_j(i) e^{-\beta\left[\ldots + n_j(i)\varepsilon_j + \ldots\right]} / Z \\ &= \frac{\partial}{\partial(-\beta E_i)} \sum_i e^{-\beta \sum_k n_k(i)\varepsilon_k} / Z = \frac{\partial \ln Z}{\partial(-\beta E_i)}. \qquad (17.72)\end{aligned} \tag{17.72}$$

Substituting in expression (17.71) yields

$$\langle n_j \rangle = \frac{\partial}{\partial(-\beta E_i)} \sum_k -\ln\left(1 - e^{-\beta\varepsilon_k}\right) = \frac{e^{-\beta E_j}}{1 - e^{-\beta E_j}} = \frac{1}{e^{\beta E_j} - 1}. \tag{17.73}$$

The average number of photons of wavelength λ is

$$\langle n_\lambda \rangle = \frac{1}{e^{hc/\lambda kT} - 1}. \tag{17.74}$$

This is called the occupation number for λ. Multiplying the energy per photon by the occupation number and by the number of modes between λ and λ + dλ yields the energy density of radiation in this channel:

$$\mathrm{d}U = \frac{hc}{\lambda} \frac{1}{e^{hc/\lambda kT} - 1} \frac{8\pi}{\lambda^4} d\lambda = \frac{8\pi hc}{\lambda^5} \frac{1}{e^{hc/\lambda kT} - 1} d\lambda = \rho\, d\lambda. \tag{17.75}$$

The density over wavelength per unit volume is

$$\rho = \frac{8\pi hc}{\lambda^5} \frac{1}{e^{hc/\lambda kT} - 1}. \tag{17.76}$$

This is called the *Planck distribution.* It was first constructed by Max Planck in 1900,

Example 17.13

Calculate the constant products appearing in formula (17.76).

From the accepted values of the fundamental constants, we find that

$$8\pi hc = 8\pi\left(6.62608 \times 10^{-34}\ \mathrm{J\,s}\right)\left(2.99792 \times 10^{8}\ \mathrm{m\,s^{-1}}\right) = 4.99248 \times 10^{-24}\ \mathrm{J\,m}.$$

$$\frac{hc}{k} = \frac{\left(6.62608 \times 10^{-34}\ \mathrm{J\,s}\right)\left(2.99792 \times 10^{8}\ \mathrm{m\,s^{-1}}\right)}{1.38066 \times 10^{-23}\ \mathrm{J\,K^{-1}}} = 1.43877 \times 10^{-2}\ \mathrm{m\,K}.$$

Example 17.14

A spherical cavity 1.000 cm^3 in volume is heated to 2000 K. Calculate the electromagnetic energy in the wavelength range 500 to 525 nm in the cavity.

Since the average wavelength is 512.5 nm, we take

$$\frac{8\pi hc}{\lambda^5} = \frac{4.99248 \times 10^{-24}\ \text{J m}}{\left(512.5 \times 10^{-9}\ \text{m}\right)^5} = 1.412 \times 10^8\ \text{J m}^{-4}$$

and

$$e^{hc/\lambda kT} = \exp\frac{1.43877 \times 10^{-2}\ \text{m K}}{\left(512.5 \times 10^{-9}\ \text{m}\right)\left(2000\ \text{K}\right)} = 1.2477 \times 10^6.$$

In formula (17.76), these numbers yield the density over wavelength per unit volume

$$\rho = 1.412 \times 10^8 \frac{1}{1.2477 \times 10^6 - 1} = 1.132 \times 10^2\ \text{J m}^{-4}.$$

Multiply by $\Delta\lambda$ and the volume to get the energy:

$$E = \left(1.132 \times 10^2\ \text{J m}^{-4}\right)\left(25 \times 10^{-9}\ \text{m}\right)\left(1.000 \times 10^{-6}\ \text{m}^3\right) = 2.83 \times 10^{-12}\ \text{J}.$$

17.9 Allowance for Particle Transfer

The small system in a definite molecular state may interact with the surrounding bath through both energy and particle transfer. Additional expressions must then be introduced.

Let the definite state system consist of N_ν similar particles in volume V_ν, with energy E_ν. Let the interacting thermodynamic system contain N_β similar particles in volume V_β with internal energy E_β and temperature T. The total energy and the total number of particles are

$$E = E_\beta + E_\nu, \qquad N = E_\beta + N_\nu. \qquad [17.77]$$

All diversity is in the bath system. So the total number of disjoint submicroscopic states is

$$W\left(E_\beta, N_\beta\right) = W\left(E - E_\nu, N - N_\nu\right). \qquad [17.78]$$

The corresponding statistical weight becomes

$$w_\nu = aW = a \exp\left[\ln W\left(E - E_\nu, N - N_\nu\right)\right]. \qquad [17.79]$$

A Taylor expansion of the exponent is

$$\ln W\left(E - E_\nu, N - N_\nu\right) = \ln W\left(E, N\right) - \left(\frac{\partial \ln W\left(E, N\right)}{\partial E}\right)_{V,N} E_\nu - \left(\frac{\partial \ln W\left(E, N\right)}{\partial N}\right)_{E,V} N_\nu + \ldots. \qquad [17.80]$$

From thermodynamic relation (5.94) in the form

$$\mathrm{d}S = \frac{1}{T}\,\mathrm{d}E + \frac{P}{T}\,\mathrm{d}V - \frac{\mu}{T}\,\mathrm{d}N, \qquad [17.81]$$

one finds that

$$\left(\frac{\partial S}{\partial E}\right)_{V,N} = \frac{1}{T}, \qquad [17.82]$$

$$\left(\frac{\partial S}{\partial N}\right)_{E,V} = -\frac{\mu}{T}. \qquad [17.83]$$

Here μ is the chemical potential per particle and S the total entropy.

As before, the entropy is related to the state number by

$$\frac{S}{k} = \ln W, \qquad [17.84]$$

where k is the Boltzmann constant.

So we have

$$\left(\frac{\partial \ln W}{\partial E}\right)_{V,N} = \frac{1}{k}\left(\frac{\partial S}{\partial E}\right)_{V,N} = \frac{1}{kT} = \beta \qquad [17.85]$$

and

$$\left(\frac{\partial \ln W}{\partial N}\right)_{E,V} = \frac{1}{k}\left(\frac{\partial S}{\partial N}\right)_{E,V} = -\frac{\mu}{kT} = -\beta\mu. \qquad [17.86]$$

Substituting into expansion (17.80) and dropping the higher terms leads to the expression

$$\ln W\left(E - E_\nu, N - N_\nu\right) = \ln W\left(E, N\right) - \beta E_\nu + \beta\mu N_\nu, \qquad [17.87]$$

which converts equation (17.79) to

$$w_\nu = a \exp\left[\ln W\left(E, N\right) - \beta E_\nu + \beta\mu N_\nu\right] = aW\left(E, N\right)e^{-\beta\left(E_\nu - \mu N_\nu\right)} = \frac{e^{-\beta\left(E_\nu - \mu N_\nu\right)}}{\mathfrak{Z}}. \qquad [17.88]$$

We set

$$\mathfrak{Z} = \sum_\nu e^{-\beta\left(E_\nu - \mu N_\nu\right)} \qquad [17.89]$$

so that

$$\sum w_\nu = 1. \qquad [17.90]$$

Expression (17.89) is called the *grand partition function* for the system.

Formulas (17.88) and (17.89) differ from (17.59) and (17.61) in that E_ν - μN_ν has replaced E_ν. Since

$$E_\nu = \sum_k n_k\left(\nu\right)\varepsilon_{k'}, \qquad N_\nu = \sum_k n_k\left(\nu\right), \qquad [17.91]$$

weighting factor (17.88) may be rewritten as

$$w_\nu = \frac{e^{-\beta\sum_k\left(\varepsilon_k - \mu\right)n_k\left(\nu\right)}}{\mathfrak{Z}} \qquad [17.92]$$

with

$$\mathfrak{Z} = \sum_\nu e^{-\beta\sum_k\left(\varepsilon_k - \mu\right)n_k\left(\nu\right)}. \qquad [17.93]$$

The average number of particles in the jth state is given by

$$\langle n_j \rangle = \sum_\nu n_j(\nu) w_\nu = \frac{\sum_\nu n_j(\nu) e^{-\beta \sum_k (\varepsilon_k - \mu) n_k(\nu)}}{\mathfrak{Z}}. \tag{17.94}$$

But differentiating the logarithm of the grand partition function with respect to $-\beta\varepsilon_j$. yields

$$\frac{\partial \ln \mathfrak{Z}}{\partial(-\beta\varepsilon_j)} = \frac{1}{\mathfrak{Z}} \frac{\partial \mathfrak{Z}}{\partial(-\beta\varepsilon_j)} = \frac{\sum_\nu n_j(\nu) e^{-\beta \sum_k (\varepsilon_k - \mu) n_k(\nu)}}{\mathfrak{Z}}. \tag{17.95}$$

On comparing (17.94) with (17.95), we see that

$$\langle n_j \rangle = \frac{\partial \ln Z}{\partial(-\beta\varepsilon_j)}. \tag{17.96}$$

17.10 Fermi-Dirac Statistics

Consider a system in which all particles of a given kind are effectively independent of each other and of any other particles present. The eigenenergies ε_1, ε_2, ..., ε_j are then the same for all these particles. Also suppose that these particles obey the Pauli exclusion principle, so that n_j may be only 0 or 1, Such particles are called *fermions.*

The discussion in section 17.9 applies, with each n_k either 0 or 1. The grand partition function, in formula (17.89) or (17.93), reduces as follows:

$$\mathfrak{Z} = \sum_i e^{-\beta(E_i - \mu N_i)} = \prod_j \sum_{n_j=0}^{1} e^{-\beta(\varepsilon_j - \mu) n_j} = \prod_j \left[1 + e^{-\beta(\varepsilon_j - \mu)}\right], \tag{17.97}$$

So we have

$$\ln \mathfrak{Z} = \sum_i \ln\left[1 + e^{-\beta(\varepsilon_i - \mu)}\right]. \tag{17.98}$$

Substituting into formula (17.96) and carrying out the indicated differentiation with respect to $-\beta\varepsilon_j$ gives us the *Fermi-Dirac* distribution law

$$\langle n_j \rangle = \frac{e^{-\beta(\varepsilon_j - \mu)}}{1 + e^{-\beta(\varepsilon_j - \mu)}} = \frac{1}{e^{\beta(\varepsilon_j - \mu)} + 1}. \tag{17.99}$$

When eigenenergy ε_j equals the chemical potential μ, the occupation number $\langle n_j \rangle$ equals 1/2. Increasing ε_j causes $\langle n_j \rangle$ to fall; decreasing ε_j causes $\langle n_j \rangle$ to rise. At 0 K, the occupation number equals 1 below energy μ , 0 above. As the temperature is raised, the transition region smooths out and spreads over a broader and broader range.

Plots of formula (17.99) for representative temperatures appear in figures 17.2 - 17.4. The Fermi energy ε_f is the highest energy that would be occupied if all the lowest levels were filled and the other levels were empty. As the temperature of a metal is raised, the number density of free electrons falls slightly and ε_f falls following equation (10.59). The chemical potential μ is the energy of the level that would be half filled on the average. As the usage of higher levels increases with increasing temperature, the level that would be half full falls. We label the temperature at which the chemical potential reaches zero, T_0. Above this temperature, the chemical potential is negative.

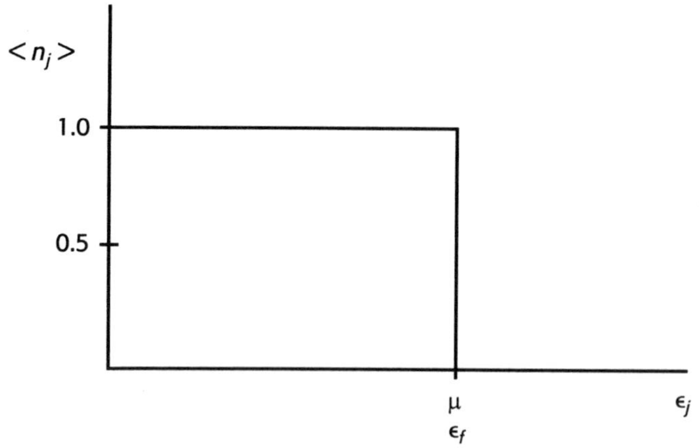

FIGURE 17.2 Fermi-Dirac occupation number $\langle n_j \rangle$ as a function of eigenenergy ϵ_j at 0 K.

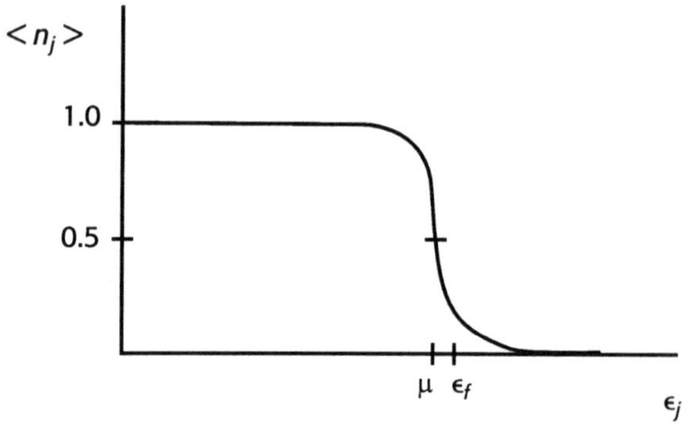

FIGURE 17.3 Fermi-Dirac occupation number as a function of the eigenenergy at a temperature very low with respect to T_0.

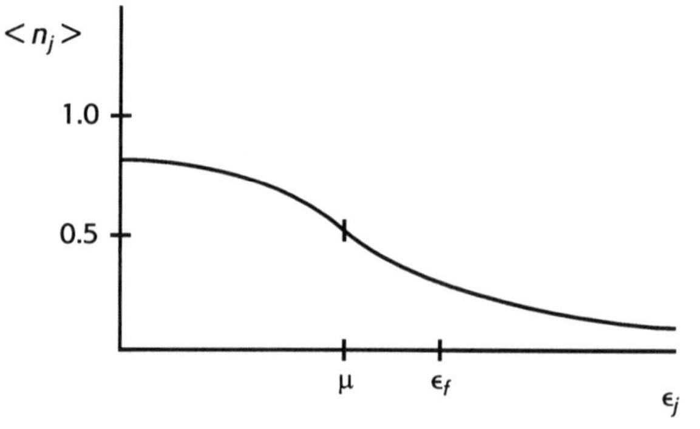

FIGURE 17.4 Fermi-Dirac occupation number as a function of the eigenenergy at a higher temperature, but one still below T_0.

Let us consider a system of equivalent fermions translating freely within a given volume. Let us also suppose that each carries one-half unit of spin. The number of orthogonal states with wavenumber between $\bar{k}$ and $\bar{k} + d\bar{k}$ per unit volume is given by formula (17.64)

$$dN = 8\pi\bar{k}^2 d\bar{k} \; . \tag{17.100}$$

But from the de Broglie equation,

$$\bar{k} = \frac{p}{h}, \tag{17.101}$$

and the kinetic energy formula,

$$\frac{p^2}{2m} = \varepsilon, \qquad [17.102]$$

we obtain

$$k = \frac{\left(2m\varepsilon\right)^{1/2}}{h} \qquad [17.103]$$

and

$$2k\,\mathrm{d}k = \frac{2m}{h^2}\,d\varepsilon. \qquad [17.104]$$

Substituting these relations into equation (17.100) gives us

$$\mathrm{d}N = 4\pi\frac{\left(2m\right)^{3/2}}{h^3}\varepsilon^{1/2}\,\mathrm{d}\varepsilon. \qquad [17.105]$$

Multiplying the occupation number $<n>$ by the number of disjoint states in interval $\mathrm{d}\varepsilon$ and integrating yields the number density

$$n = \int_0^\infty \rho\left(\varepsilon\right)\mathrm{d}\varepsilon = 4\pi\frac{\left(2m\right)^{3/2}}{h^3}\int_0^\infty \frac{\varepsilon^{1/2}\,\mathrm{d}\varepsilon}{e^{\beta\left(\varepsilon-\mu\right)}+1}. \qquad [17.106]$$

A plot of double density $\rho(\varepsilon)$ against ε at a temperature low with respect to T_0 appears in figure 17.5.

For the valence electrons in a metal, temperature T_0 is of the order 10^5 K. So in treating them at ordinary temperatures, one must employ Fermi-Dirac statistics. Furthermore, only the electrons around the Fermi level can then be readily excited. So these are the ones involved in the conduction of electricity.

Raising the temperature of a fermion system above T_0 causes $-\mu$ to increase much faster than kT. Above a certain temperature,

$$e^{\beta\left(\varepsilon_j-\mu\right)} >> 1, \qquad [17.107]$$

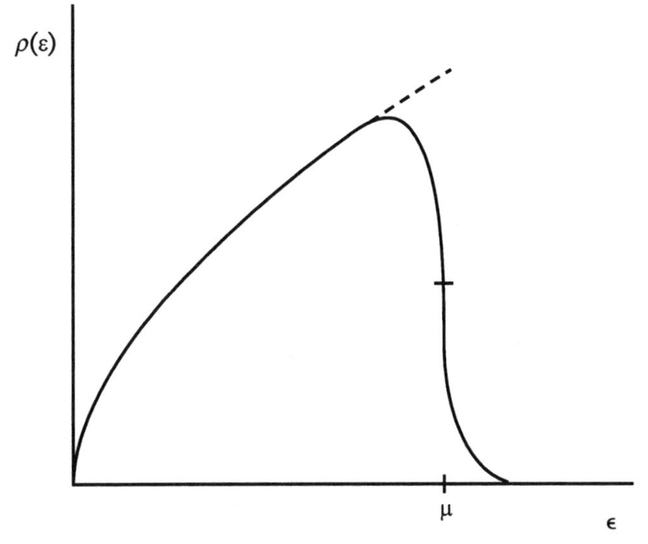

FIGURE 17.5 A Fermi-Dirac density over translational energy, per unit volume, as a function of the energy.

and formula (17.99) reduces to

$$\langle n_j \rangle \cong e^{-\beta(\varepsilon_j - \mu)}, \qquad [17.108]$$

a form of the Boltzmann distribution law.

Example 17.15

Evaluate the coefficient of the integral in formula (17.106) for a system of electrons. With the accepted values of the fundamental constants, we find that

$$\frac{4\pi(2m)^{3/2}}{h^3} = \frac{4\pi(2 \times 9.1094 \times 10^{-31}\ \text{kg})^{3/2}}{(6.6261 \times 10^{-34}\ \text{J s})^3} = 1.0622 \times 10^{56}\ \text{J}^{-3/2}\ \text{m}^{-3}.$$

17.11 Bose-Einstein Statistics

Again consider a system in which all particles of a given kind are effectively independent of each other and of any other particles present. Let the eigenenergies of each of these indistinguishable particles be $\varepsilon_1, \varepsilon_2, \ldots, \varepsilon_j, \ldots$. But now, suppose that n_j, the number in the jth state, can be any nonnegative integer. Such particles are called *bosons.*

The discussion in section 17.9 applies, with each ε_i allowing n_i to be 0, 1, 2,.... So in formula (17.93), each ε_i yields a geometric series:

$$\mathfrak{z} = \prod_i \sum_{n_i=0}^{\infty} e^{-\beta(\varepsilon_i - \mu)n_i} = \prod_i \frac{1}{1 - e^{-\beta(\varepsilon_i - \mu)}}. \qquad [17.109]$$

The logarithm of this grand partition function is the sum

$$\ln \mathfrak{z} = \sum_i -\ln\left[1 - e^{-\beta(\varepsilon_i - \mu)}\right]. \qquad [17.110]$$

Differentiating with respect to $-\beta\varepsilon_j$, as in equation (17.96), gives us the *Bose-Einstein distribution law*

$$\langle n_j \rangle = \frac{e^{-\beta(\varepsilon_i - \mu)}}{1 - e^{-\beta(\varepsilon_i - \mu)}} = \frac{1}{e^{\beta(\varepsilon_i - \mu)} - 1}. \qquad [17.111]$$

Note that this reduces to formula (17.73) when the chemical potential μ vanishes, as it does with photons. Photons are bosons.

Consider an ideal gas of spinless particles. Also employ the continuum approximation. Integral (10.57) now gives the number of orthogonal translational modes with wavenumber less than $\mathfrak{k}$:

$$N = \frac{4}{3}\pi \mathfrak{k}^3. \qquad [17.112]$$

So formula (17.105) is replaced with

$$dN = 2\pi \frac{(2m)^{3/2}}{h^3} \varepsilon^{1/2}\, d\varepsilon. \qquad [17.113]$$

Multiplying expression (17.111) by (17.113) and integrating over the energy levels yields the number density

$$\boldsymbol{n} = \int_0^\infty \rho(\varepsilon)\,\mathrm{d}\varepsilon = 2\pi\frac{(2m)^{3/2}}{h^3}\int_0^\infty \frac{\varepsilon^{1/2}\,\mathrm{d}\varepsilon}{e^{\beta(\varepsilon-\mu)}-1}. \qquad [17.114]$$

Let us order the eigenenergies in the ideal gas according to magnitude:

$$\varepsilon_{j+1} \geq \varepsilon_j. \qquad [17.115]$$

Under periodic boundary conditions, the lowest state has no translational energy:

$$\varepsilon_1 = 0. \qquad [17.116]$$

But for this level, equation (17.111) reduces to

$$\langle n_1 \rangle = \frac{1}{e^{-\mu/kT}-1}. \qquad [17.117]$$

Since this number cannot be negative, expression exp $(-\mu/kT)$ must be greater than 1 and the chemical potential μ must be negative. Raising the temperature excites particles and so decreases $<n_1>$. The denominator must then increase and $-\mu$ increases faster than kT.

Conversely, as the temperature is lowered, $-\mu$ decreases faster than kT. At a very low temperature, still above 0 K, $-\mu$ may reach 0. At this temperature T_0, the denominator of (17.117) reaches zero. Expression $<n_1>$ then becomes infinite. In any finite system, all particles would then condense into the ground state. This phenomenon is called the *Bose-Einstein condensation.*

The transition from normal fluid to superfluid in liquid ^{4}He, at 2.19 K, has been interpreted as such a condensation.

Considerably above temperature T_0, $-\mu$ is positive enough so that

$$e^{\beta(\varepsilon_j-\mu)} >> 1. \qquad [17.118]$$

Then formula (17.111) reduces to

$$\langle n_j \rangle \cong e^{-\beta(\varepsilon_j-\mu)}, \qquad [17.119]$$

a form of the Boltzmann distribution law.

Questions

17.1 What symmetry is presumed when the Maxwell-Boltzmann distribution law is employed?
17.2 What form does this law exhibit?
17.3 Show how the internal energy is related to the partition function.
17.4 How is the reversible work done on a system related to a change in the external parameters?
17.5 How is the entropy change related to the partition function?
17.6 Show how the Helmholtz energy is related to the partition function.
17.7 How does use of the partition function in these formulas imply use of the Boltzmann distribution law?
17.8 Relate the system partition function to the pertinent molecular ones.
17.9 What does introduction of the configuration integral accomplish?
17.10 Why are thermodynamic properties calculated for the hypothetical ideal gas state?
17.11 Can a person calculate the contribution of a particular molecular mode of motion to a thermodynamic function? Explain.

17.12 How is the concentration equilibrium constant related to the partition functions of the contributing molecules?
17.13 How is the pressure equilibrium constant related to the concentration one?
17.14 How may a person formulate the basic problem in statistical mechanics?
17.15 What is the statistical assumption? How does it apply to a definite energy state of a submicroscopic system?
17.16 Why does $e^{\ln x}$ equal x?
17.17 Justify the Taylor formula.
17.18 Show how temperature T is related to the entropy S of a system.
17.19 Evaluate the derivative $(\partial \ln W/\partial E)_V$.
17.20 Show how this derivative is employed in determining the statistical weight of the νth state of a molecular complex.
17.21 How is the radiation from a given surface related to absorption by the surface? What basic principle is involved?
17.22 Why can radiation from an ideal black surface be related to radiation within a cavity?
17.23 Why can we employ the Boltzmann distribution law in deriving the Planck distribution law? After all, photons are bosons.
17.24 Show how the chemical potential ν is related to the entropy S of a system..
17.25 Evaluate the derivative $(\partial \ln W/\partial N)_{E,\,V}$.
17.26 Then show how the result for statistical weight w_ν is modified to allow for particle transfer to or from the small system.
17.27 What is the grand partition function?
17.28 From this partition function, construct the Fermi- Dirac distribution law.
17.29 Describe how the Fermi-Dirac occupation number varies as the temperature is raised from 0 K.
17.30 What is the Fermi energy?
17.31 Explain how the chemical potential varies with temperature.
17.32 From the grand partition function, construct the Bose-Einstein distribution law.
17.33 Explain how Bose-Einstein condensation occurs.

Problems

17.1 Calculate the molecular translational partition function at 25° C and 1 cubic decimeter (1 l) for (a) H, (b) O, (c) H_2, (d) O_2, (e) H_2O.
17.2 Calculate the molecular rotational partition function at 25° C for (a) H_2, (b) O_2, (c) H_2O. For H_2 and O_2, rotational constant B equals 59.312 cm^{-1} and 1.4378 cm^{-1}, respectively. For H_2O, rotational constants A, B, C are 27.877 cm^{-1}, 14,512 cm^{-1} 9.285 cm^{-1}.
17.3 Calculate the molecular vibrational partition function at 298.15 K and at 1000 K for (a) H_2, (b) O_2, (c) H_2O. For H_2 and O_2, vibrational wave number $\tilde{k}_0$ is 4159 cm^{-1} and 1556 cm^{-1}, respectively. For H_2O, the vibrational wave numbers are 1595 cm^{-1}, 3652 cm^{-1}, and 3756 cm^{-1}.
17.4 Construct the molecular electronic partition function at 25° C for (a) H, (b) H_2, (c) O_2, (d) H_2O. The lowest electronic levels for these molecules are $^2S_{1/2}$, $^1\Sigma^+_g$, $^3\Sigma^-_g$, $^1A_1$, respectively.
17.5 The oxygen atom O has three low lying electronic levels, $^3P_2$ at 0 cm^{-1}, $^3P_1$ at 158.265 cm^{-1}, and $^3P_0$ at 226.977 cm^{-1}, with degeneracies 5, 3, and 1 respectively. Calculate its electronic partition function at 25° C.
17.6 Calculate the translational entropy of 1 mol of (a) H, (b) O, (c) H_2, (d) O_2, (e) H_2O at 25° C and 1 bar.
17.7 Calculate the rotational entropy of 1 mol of (a) H_2, (b) O_2, (c) H_2O at 25° C.
17.8 Calculate the vibrational entropy of 1 mol of (a) H_2, (b) O_2, (c) H_2O at 25° C.
17.9 Calculate the electronic entropy of 1 mol of (a) H, (b) H_2, (c) O_2, (d) H_2O at 25° C.
17.10 Calculate the electronic entropy of 1 mol of O at 25° C. Use the data in problem 17.5.
17.11 Construct a formula for the partition function of H (g).
17.12 Construct a formula for the partition function of H_2 (g)'
17.13 For the reaction

$$H_2\,(g) \leftrightarrows 2H\,(g),$$

the dissociation energy ΔE_0 is 36,116 cm^{-1} per molecule. Construct $K_{c'}$. Then calculate K_c and K_P at 1000 K.

17.14 The walls of a cavity 1.000 cm^3 in volume are maintained at 1000 K. Calculate the electromagnetic energy in the wavelength range 1000 to 1025 nm in the cavity.

17.15 Calculate the chemical potential at 25° C for the conduction electrons in Cu.

17.16 Calculate the temperature T_0 at which the chemical potential for the conduction electrons in Cu vanishes.

— — —

17.17 Calculate the molecular translational partition function at 25° C and 1 cubic decimeter (1) for (a) Cl, (b) HCl, (c) CH_4.

17.18 Calculate the molecular rotational partition function at 25° C for (a) HCl, (b) CH_4. The rotational constant B for HCl equals 10.588 cm^{-1}. The rotational constants A, B, C for CH_4 equal 5.252 cm^{-1}.

17.19 Calculate the molecular vibrational partition function at 298.15 K and at 1000 K for (a) HCl, (b) CH_4. For HCl, a fundamental vibration appears at 2889.6 cm^{-1}. For CH_4, a fundamental appears at 2914.2 cm^{-1}, a doubly degenerate fundamental at 1526 cm^{-1}, triply degenerate fundamentals at 3020.3 cm^{-1} and at 1306.2 cm^{-1}.

17.20 The chlorine atom Cl has two low lying electronic levels, $^2P_{3/2}$ at 0 cm and $^2P_{1/2}$ at 882.36 cm^{-1}, with degeneracies 4 and 2, respectively. Calculate its electronic partition function at 25° C.

17.21 Construct the molecular electronic partition function at 25° C for (a) HCl, (b) CH_4.

17.22 Calculate the translational entropy of 1 mol of (a) Cl, (b) HCl, (c) CH_4 at 25° C and 1 bar.

17.23 Calculate the rotational entropy of 1 mol of (a) HCl, (b) CH_4 at 25° C.

17.24 Calculate the vibrational entropy of 1 mol of (a) HCl, (b) CH_4 at 25° C.

17.25 Calculate the electronic entropy of 1 mol of Cl at 25° C.

17.26 Calculate the electronic entropy of 1 mol of (a) HCl, (b) CH_4 at 25° C.

17.27 Construct the factors in a formula for the partition function of Cl.

17.28 Construct the factors in a formula for the partition function of HCl.

17.29 For the reaction

$$\text{HCl (g)} \rightleftarrows \text{H (g)} + \text{Cl (g)},$$

the dissociation energy $\Delta\varepsilon$ is 35,743 cm^{-1}. Construct $K_{c'}$. Then calculate K_c and K_P at 1000 K.

17.30 Calculate the energy density in the wavelength range 650 to 655 nm inside a cavity 100 cm^3 in volume when the wall temperature is (a) 25° C, (b) 3000° C.

17.31 The wavelength of the emission maximum from a small hole in an electrically heated container varied with the temperature as in table 17.A.

TABLE 17.A

t, °C	1000	1500	2000	2500	3000	3500
λ_{max}, nm	2181	1600	1240	1035	878	763

From these data, calculate Planck's constant.

17.32 In a plasma, the concentration of electrons is 5.82×10^{18} m^{-3}. Calculate the electron temperature at which the chemical potential of the electrons would vanish. Can one apply Boltzmann statistics to these electrons?

References

Books

Chandler, D.: 1987, *Introduction to Modern Statistical Mechanics*, Oxford University Press, New York, pp, 3-118.

Chandler bases his derivation of the distribution laws on thermodynamic results. Thus, his discussion parallels the approach in this chapter. However, he does presume the reader has had previous exposure to the elementary principles.

Lavenda, B. H.: 1991, *Statistical Physics: A Probabilistic Approach*, John Wiley & Sons, Inc ., New York, pp. 1-358.

Lavenda adopts the historical approach to the subject. Thus, he expands on the reasoning followed by the original developers. Many questions and topics ignored by those in the mainstream are discussed.

Mayer, J. E. and Mayer, M. G.: 1940, *Statistical Mechanics*, John Wiley & Sons, Inc., New York, pp. 427-469.

Here we have a very useful mathematical appendix. Of particular interest are the derivations of the fundamental combinatorial formulas.

McQuarrie, D. A.: 1976, *Statistical Mechanics*, Harper & Row, New York, pp 1-193.

McQuarrie bases his development on the standard combinatorial formulas. To introduce the effects of constraints, he employs the method of Lagrange multipliers. Later in the book, McQuarrie presents a wealth of applications in sufficient detail.

Articles

Alberty, R. A.: 1988, "The Effect of a Catalyst on the Thermodynamic Properties and Partition Functions of a Group of Isomers," *J. Chem. Educ.* **65**, 409-413.

Alvarino, J. M., Veguillas, J., and Velasco, S.: 1989, "Equations of State, Collisional Energy Transfer, and Chemical Equilibrium in Gases," *J. Chem. Educ.* **66**, 139-141.

Compagner, A.: 1989, "Thermodynamics as the Continuum Limit of Statistical Mechanics," *Am. J. Phys.* **57**, 106-117.

David, C. W.: 1987, "A Tractable Model for Studying Solution Thermodynamics," *J. Chem. Educ.* **64**, 484-485.

Fowles, G. R.: 1994, "Time's Arrow: A Numerical Experiment," *Am. J. Phys.* **62**, 321-328.

Gill, S. J., Murphy, K. P., and Robert, C. H.: 1990, "Partition Function Formalism for Analyzing Calorimetric Experiments," *J. Chem. Educ.* **67**, 928-931.

Glasser, L.: 1989, "Order, Chaos, and All That!" *J. Chem. Educ.* **66**, 997-1001.

Grela, M. A.: 1990, "The Thermodynamic Functions of a Poscll-Teller Oscillator," *J. Chem. Educ.* **67**, 390-391.

Landsberg, P. T., Dunning-Davies, J., and Pollard, D.: 1994, "Entropy of a Column of Gas under Gravity," *Am. J. Phys.* **62**, 712- 717.

Larsen, R. D.: 1985, "The Planck Radiation Functions," *J. Chem. Educ.* **62**, 199-202.

Pimbley, J. M.: 1986, "Volume Exclusion Correction to the Ideal Gas with a Lattice Gas Model," *Am. J. Phys.* **54**, 54-57.

Prosper, H. B.: 1993, "Temperature Fluctuations in a Heat Bath," *Am. J. Phys.* **61**, 54-58.

Pyle, J. T.: 1985, "Planck's Radiation Law," *J. Chem. Educ.* **62**, 488-490.

Russell, D. K.: 1996, "The Boltzman Distribution," *J. Chem. Educ.* **73**, 299-300.

Smith, M. A.: 1993, "The Nature of Distribution Functions for Colliding Systems," *J. Chem. Educ.* **70**, 218-223.

Zitter, R. N., and Hilborn, R. S.: 1987, "Fermi-Dirac and Bose-Einstein Distributions and the Interaction of Light with Matter," *Am. J. Phys.* **55**, 522-524.

18

Reaction Rate Theory

18.1 The Transition State

IN CHAPTERS 15 AND 16, WE SAW THAT ELEMENTARY REACTIONS may be unimolecular, bimolecular, or termolecular. In a unimolecular step, a molecule with excess energy concentrates enough energy in the bond or bonds to be broken and movement to product occurs. In a bimolecular or termolecular step, the reactant molecules come together in an inelastic collision. When enough kinetic energy is transformed to potential energy, movement to products occurs. In most cases, these processes involve passage over a potential energy barrier, as figure 15.2 indicates.

An elementary reaction involves an array of nuclei and electrons going from a reactant configuration to a product configuration. According to the Born-Oppenheimer approximation, the nuclear and electronic motions are separable at any stage. So for each nuclear configuration, there is a definite electron energy which can be considered potential energy insofar as the nuclear motions are concerned. To this, one can add the potential energy of repulsion between the nuclei.

The sum of these, the apparent potential energy, can in principle be calculated and plotted against suitable internal coordinates. A person can then pick out the most probable path from the minimum for reactants to the minimum for products. Distance along this path, or a monotonic function thereof, is taken to be the *reaction coordinate*.

The apparent potential energy along this path varies as the curve in figure 18.1 indicates. The straight line segments represent the lowest energy level for reactants, transition state, and products, respectively. Along a coordinate orthogonal to the reaction coordinate, the potential energy varies as figure 18.2 suggests.

So the maximum in figure 18.1 is a saddle point whenever there is more than a single internal coordinate. When the reaction coordinate is within distance $\delta/2$ of this maximum, the system is considered to be in a *transition state*. Quantity δ is taken small enough so that the movement through the transition state is essentially a translation.

In developing transition state theory, Henry Eyring and independently M. G. Evans, M. Polyani assumed that the reactant molecules are distributed over their states following the Boltzmann distribution law. They also assumed that the movement through the transition state behaved as a classical translation. Thus, both reflection of higher energy systems and tunneling of lower energy ones are neglected.

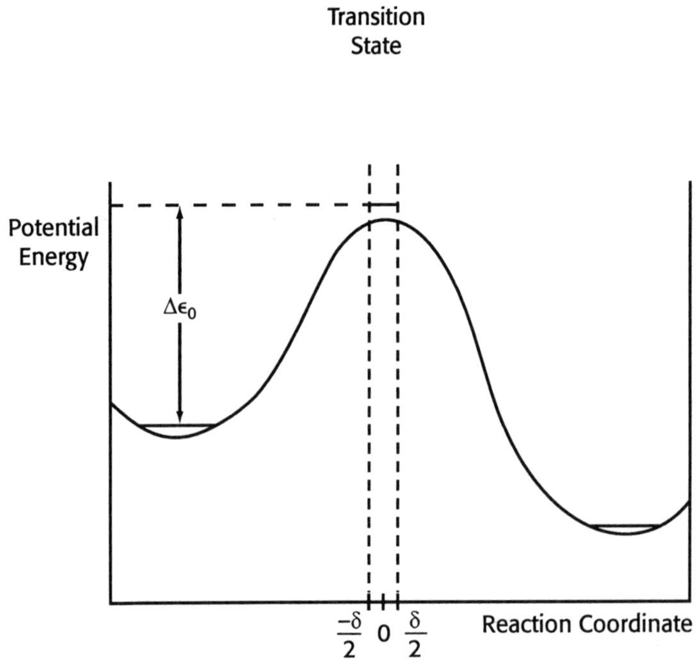

FIGURE 18.1 Apparent potential energy along the minimum energy path from the initial equilibrium configuration to the final one.

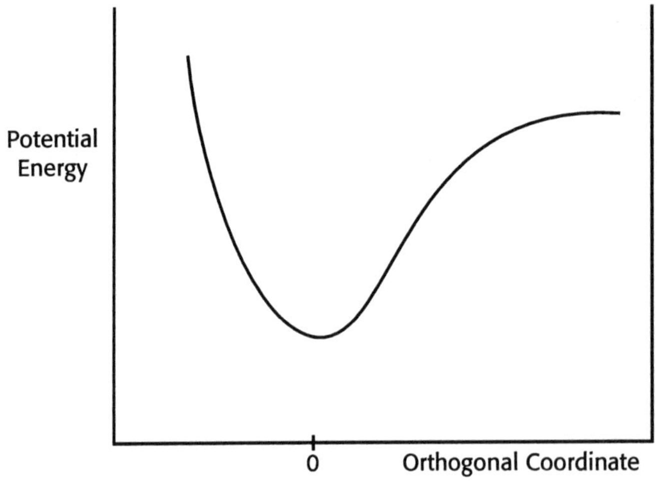

FIGURE 18.2 Apparent potential energy along a coordinate orthogonal to the reaction coordinate.

Example 18.1

Evaluate the integral

$$\int_0^\infty \mathrm{v} e^{-a\mathrm{v}^2}\, d\mathrm{v}.$$

Rearrange and identify the resulting integral as follows:

$$\int_0^\infty \mathrm{v} e^{-a\mathrm{v}^2}\, d\mathrm{v} = -\frac{1}{2a}\int_0^\infty e^{-a\mathrm{v}^2}\left(-2a\mathrm{v}\, d\mathrm{v}\right) = -\frac{1}{2a}\, e^{-a\mathrm{v}^2}\Big|_0^\infty = \frac{1}{2a}.$$

Example 18.2

Evaluate

$$\int_0^\infty e^{-a\mathrm{v}^2}\,d\mathrm{v}.$$

From example 10.5, this integral equals

$$\frac{1}{2}\left(\frac{\pi}{a}\right)^{1/2}.$$

18.2 The Eyring Equation

With the assumptions just noted, one can proceed to develop a formula for relating specific reaction rates to molecular parameters. These attributes may be embedded in partition functions for convenience.

Consider an elementary reaction

$$A + \ldots \rightarrow M^{\ddagger} \rightarrow D \qquad [18.1]$$

in which $M^{\ddagger}$ represents the transition state complex, A, ... are reactants, and D the product or products. There may be one, two, or three reactants. Let M_f and M_b represent the complexes moving forward and backward, respectively, through the transition zone.

At equilibrium, the number of complexes moving forward per unit volume equals the number moving backward per unit volume:

$$\left[M_f^{\ddagger}\right] = \left[M_b^{\ddagger}\right] = \frac{1}{2}\left[M^{\ddagger}\right]. \qquad [18.2]$$

But if the molecules are distributed over their states following the Boltzmann distribution law, the $M^{\ddagger}$s are in equilibrium with the reactants. If we set all activity coefficients equal to 1, we have

$$\frac{\left[M^{\ddagger}\right]}{\left[A\right]\cdots} = K^* \qquad [18.3]$$

and

$$\left[M_f^{\ddagger}\right] = \frac{1}{2}K^*\left[A\right]\ldots. \qquad [18.4]$$

Multiplying [A] is the concentration of each additional reactant in transformation (18.1).

The average speed of a forward moving complex is

$$\bar{\mathrm{v}} = \frac{\int_0^\infty \mathrm{v}e^{-\mu\mathrm{v}^2/2kT}\,d\mathrm{v}}{\int_0^\infty e^{-\mu\mathrm{v}^2/2kT}\,d\mathrm{v}} = \left(\frac{2kT}{\pi\mu}\right)^{1/2}. \qquad [18.5]$$

Here μ is the reduced mass for movement along the reaction coordinate. In the last step, the results summarized in examples 18.1 and 18.2 are used.

The time spent by an average complex in the transition zone is

$$\Delta t = \frac{\delta}{\bar{\mathrm{v}}}. \qquad [18.6]$$

So the time rate at which product is being formed per unit volume is

$$\frac{\mathrm{d}[\mathrm{D}]}{\mathrm{d}t}=\frac{1}{\Delta t}\left[\mathrm{M_f}^{\ddagger}\right]=\left(\frac{2kT}{\pi\mu}\right)^{1/2}\frac{K^*}{2\delta}[\mathrm{A}]\ldots. \qquad [18.7]$$

The equilibrium constant can be expressed in terms of the partition functions,

$$K^*=\frac{z^*}{z_{\mathrm{A}}\cdots}e^{-\Delta\varepsilon_0/kT}. \qquad [18.8]$$

Now the partition function for the reaction coordinate in the transition state is the translational one:

$$z_{\mathrm{tr}}^*=\left(2\pi\mu kT\right)^{1/2}\delta/h. \qquad [18.9]$$

This may be factored out:

$$z^*=z_{\mathrm{tr}}^*z^{\ddagger}. \qquad [18.10]$$

Combining (18.7), (18.8), (18.9), (18.10) leads to the result:

$$\frac{\mathrm{d}[\mathrm{D}]}{\mathrm{d}t}=\left(\frac{2kT}{\pi\mu}\right)^{1/2}\frac{\left(2\pi\mu kT\right)^{1/2}\delta}{h(2\delta)}\frac{z^{\ddagger}}{z_{\mathrm{A}}\cdots}e^{-\Delta E_0/RT}[\mathrm{A}]\cdots=\frac{kT}{h}\frac{z^{\ddagger}}{z_{\mathrm{A}}\cdots}e^{-\Delta\varepsilon_0/kT}[\mathrm{A}]\cdots=\frac{kT}{h}K^{\ddagger}[\mathrm{A}]\cdots. \qquad [18.11]$$

Here k is the Boltzmann constant, T the absolute temperature, h the Planck constant, $K^{\ddagger}$ the residue after z_{tr}* is factored out of the constant for equilibrium between the reactants and the transition state, $\Delta\varepsilon_0$ the energy of the lowest level for the transition state above the lowest level for the reactants. When $\Delta\varepsilon_0$ is multiplied by Avogadro's number, we get ΔE_0.

Note how parameter μ and distance δ have canceled out. Also note that the preexponential factor varies with temperature T, so it differs somewhat from the preexponential factor in the Arrhenius expression. Furthermore, $\Delta\varepsilon_0$ is not identical with the Arrhenius activation energy.

Example 18.3

Estimate the frequency factor A for a unimolecular reaction at 25° C.

The pre-exponential factor from formula (18.11) has the form

$$A=\frac{kT}{h}\frac{z^{\ddagger}}{z_{\mathrm{A}}},$$

in which z_{A} is the partition function per unit volume for the reactant and $z^{\ddagger}$ is the residual partition function per unit volume for the transition state. The translational factors for z_{A} and $z^{\ddagger}$ are equal; in the ratio they cancel. The moment(s) of inertia in the transition state is(are) somewhat larger than in the initial state. We neglect this change and let the rotational factors in z_{A} and z cancel. There is one less vibrational factor in z, that along the reaction coordinate. However, the vibrational factors are not much greater than 1 and the increase in the remaining ones may partially compensate for the one lost in z.

So we reach the approximation

$$A\cong\frac{kT}{h}$$

for the unimolecular reaction. At 25° C, we have

$$A = \frac{\left(1.3807 \times 10^{-23}\ \mathrm{J\,K^{-1}}\right)\left(298.15\ \mathrm{K}\right)}{6.6261 \times 10^{-34}\ \mathrm{J\,s}} = 6.21 \times 10^{12}\ \mathrm{s^{-1}}.$$

If the Boltzmann distribution is disturbed significantly by the reaction, the reaction may be abnormally slow. On the other hand, if the process is complex, the reaction may be abnormally fast.

Example 18.4

Formulate the specific reaction rate for atom A striking atom B as in section 3.15. Consider the electronic partition functions to equal 1.

From formula (10.47), the partition functions per unit volume for A and B are

$$z_{\mathrm{A}} = \frac{\left(2\pi m_{\mathrm{A}} kT\right)^{3/2}}{h^3}, \qquad z_{\mathrm{B}} = \frac{\left(2\pi m_{\mathrm{B}} kT\right)^{3/2}}{h^3}.$$

Similarly, the three-dimensional partition function per unit volume for the complex is

$$z_{\mathrm{tr}}^{\ddagger} = \frac{\left[2\pi\left(m_{\mathrm{A}} + m_{\mathrm{B}}\right)kT\right]^{3/2}}{h^3}.$$

From formula (11.54), the rotational partition function for the complex is

$$z_{\mathrm{rot}}^{\ddagger} = \frac{8\pi^2 \mu r^2 kT}{h^2}$$

where

$$\mu = \frac{m_{\mathrm{A}} m_{\mathrm{B}}}{m_{\mathrm{A}} + m_{\mathrm{B}}} \qquad \text{and} \qquad r = r_{\mathrm{A}} + r_{\mathrm{B}}.$$

Substituting into the coefficient of [AB] in (18.11) yields

$$k_{\mathrm{c}} = \frac{kT}{h} K^{\ddagger} = \frac{kT}{h} \frac{\left\{\left[2\pi\left(m_{\mathrm{A}} + m_{\mathrm{B}}\right)kT\right]^{3/2} / h^3\right\}\left(8\pi^2 \mu r^2 kT / h^2\right)}{\left[\left(2\pi m_{\mathrm{A}} kT\right)^{3/2} / h^3\right]\left[\left(2\pi m_{\mathrm{B}} kT\right)^{3/2} / h^3\right]} e^{-\Delta E_0 / RT} = \pi r^2 \left(\frac{8kT}{\pi\mu}\right)^{1/2} e^{-\Delta E_0 / RT}.$$

Note that this agrees with the coefficient in formula (3.101) when the energy barrier ΔE_0 vanishes,

Example 18.5

Atom A approaches a linear molecule B along its axis. A bond is formed with the struck atom while the atom at the other end of B is set free. Formulate the specific reaction rate. The translational partition functions per unit volume for A, B, and AB appear as in example 18.4. They introduce the factor

$$\frac{\left[2\pi\left(m_{\mathrm{A}} + m_{\mathrm{B}}\right)kT\right]^{3/2} / h^3}{\left[\left(2\pi m_{\mathrm{A}} kT\right)^{3/2} / h^3\right]\left[\left(2\pi m_{\mathrm{B}} kT\right)^{3/2} / h^3\right]} = \frac{h^3}{\left(2\pi\mu kT\right)^{3/2}}$$

into $K^{\ddagger}$. For the ratio of the rotational partition functions in $K^{\ddagger}$, we have

$$[8\pi^2 I^{\ddagger}kT/h^3] \,/\, [8\pi^2 I_B kT/\sigma_B h^3] = \sigma_B I^{\ddagger} / I_B$$

On forming the complex AB, two low frequency bending vibrational modes appear. They introduce the factor

$$z_{\text{vib}}^{\ddagger} = \frac{1}{\left(1 - e^{-\beta\hbar\omega_0}\right)^2}$$

into $K^{\ddagger}$. For the coefficient of [AB] in the rate expression, we obtain

$$\frac{kT}{h}K^{\ddagger} = \frac{kT}{h}\frac{h^3}{\left(2\pi\mu kT\right)^{3/2}}\frac{\sigma_B I^{\ddagger}}{I_B} z_{\text{vib}}^{\ddagger} e^{-\Delta E_0/RT} = \frac{h^2}{\left(2\pi\mu\right)^{3/2}\left(kT\right)^{1/2}}\frac{\sigma_B I^{\ddagger}}{I_B} z_{\text{vib}}^{\ddagger} e^{-\Delta E_0/RT}$$

Note how the additional structure in B has altered the expression k_c in the preceeding example. Since the new expression contains geometric constraints on how A must approach B, it generally yields a smaller specific reaction rate.

One might multiply k_c by a steric factor p that varies with the temperature T and thus get agreement.

18.3 The Kinetic Isotope Effect

In a reactant molecule, one may replace a mobile atom by its isotope. This change does not alter the electronic structure appreciably. So it does not significantly alter the potential energy surface on which the atom moves. At each stage along the reaction coordinate, the force constants remain essentially the same.

The change in nuclear mass does alter the frequencies of oscillation in the initial state and, as a consequence, the zero point energy there. In the transition state, the only oscillation appreciably affected is the one that has become a translation. This is encompassed in the preexponential factor. But the reduced mass for the translation cancels from this factor. So the isotope effect arises from the change in the energy of activation $\Delta\varepsilon_0$ caused by the shift in the initial zero point energy. Recall figure 18.1.

Consider the isotopic molecules R - H and R - D, each subject to an elementary reaction in which the indicated bond is broken. For simplicity, we treat molecular part R as a particle. The results obtained in chapter 12 will be employed.

Let f be the force constant for oscillation along the reaction coordinate in the initial state. For the reduced masses of R - H and R - D, we have

$$\mu_1 = \frac{m_H m_R}{m_H + m_R} \cong m_H, \qquad \mu_2 = \frac{m_D m_R}{m_D + m_R} \cong m_D, \tag{18.12}$$

where m_H is the mass of a hydrogen atom, m_D the mass of a deuterium atom, and m_R the mass of the molecular residue R. Setting the increase in the energy of activation equal to the decrease in the initial zero point energy yields

$$\delta\left(\Delta\varepsilon_0\right) = \frac{1}{2}\hbar\left(\omega_1 - \omega_2\right) = \frac{1}{2}\hbar f^{1/2}\left(\frac{1}{\mu_1^{1/2}} - \frac{1}{\mu_2^{1/2}}\right) = \frac{1}{2}\hbar\left(\frac{f}{\mu_1}\right)^{1/2}\left[1 - \left(\frac{\mu_1}{\mu_2}\right)^{1/2}\right] = \lambda. \tag{18.13}$$

The principal effect of the substitution is on motion along the reaction coordinate. We consider that the partition functions for the other internal modes are unchanged. Furthermore, the translational partition functions are little affected when the mass of R is relatively large, as we assume. So in the ratio of the rate constants, these partition functions cancel. We are left with

$$\frac{k_2}{k_1} = e^{-\delta(\Delta\varepsilon_0)/kT} = e^{-\lambda/kT}. \qquad [18.14]$$

If the reaction involves the simultaneous formation of a new bond, as in

$$\mathrm{A-H+B \rightarrow A\text{-}H\text{-}B \rightarrow A+H-B} \qquad [18.15]$$

or

$$\mathrm{A-D+B \rightarrow A\text{-}D\text{-}B \rightarrow A+D-B}, \qquad [18.16]$$

forming the transition state requires the introduction of two bending vibrations. The zero point energies for (18.15) and (18.16) in this state may differ appreciably. This difference would reduce quantity $\delta(\Delta\varepsilon_0)$ and reduce k_1/k_2, So generally, formula (18.14) yields a maximum for this ratio.

However, we have neglected quantum mechanical tunneling through the barrier. This becomes significant when the barrier is narrow and the initial force constant is low. Since the pertinent attenuation is less for the lighter isotope, the corresponding tunneling is greater. This acts to increase k_1/k_2.

Example 18.6

Estimate the maximum kinetic isotope effect when the H in a reactant molecule R - O - H is replaced with D at 25° C. The wave number for the O - H bond oscillation is 3300 $\mathrm{cm^{-1}}$.

With formula (18.13), we find that

$$\lambda = \frac{1}{2}k_1\left[1-\left(\frac{\mu_1}{\mu_2}\right)^{1/2}\right] = \frac{1}{2}\left(3300\ \mathrm{cm^{-1}}\right)\left[1-\left(\frac{1.007825}{2.0140}\right)^{1/2}\right] = \left(1650\ \mathrm{cm^{-1}}\right)\left(0.2926\right) = 482.8\ \mathrm{cm^{-1}}.$$

Substituting into equation (18.14) then yields

$$\frac{k_2}{k_1} = \exp\left[\left(-482.8\ \mathrm{cm^{-1}}\right)/\left(0.69504\ \mathrm{cm^{-1}\ K^{-1}}\right)\left(298.15\ \mathrm{K}\right)\right] = 0.097,$$

whence

$$\frac{k_1}{k_2} = 10.3.$$

Before deuteration, the reaction may be as much as 10.3 times as fast as after deuteration.

18.4 Static Nonideality

In the derivation of equation (18.11), we set all activity coefficients equal to 1, This is acceptable for the contribution from movement along the reaction coordinate through the transition state. But it is not generally acceptable for the other translational motions.

In a dense gas and in a condensed phase, the activity coefficients are usually not equal to 1, Then equation (18,3) is replaced with

$$\frac{\gamma^*\left[\mathrm{M}^\ddagger\right]}{\gamma_\mathrm{A}\left[\mathrm{A}\right]\cdots} = K^* \qquad [18.17]$$

or

$$\frac{[M^{\ddagger}]}{[A]\cdots} = \frac{\gamma_A \cdots}{\gamma^*} K^*. \qquad [18.18]$$

Coefficient γ^* arises from the static interactions between the activated complex and the surrounding molecules. So we set

$$\gamma^* = \gamma^{\ddagger}. \qquad [18.19]$$

In the derivation, K^* must be replaced with $(\gamma_A\ldots/\gamma^*)\,K^*$ and $K^{\ddagger}$ with $(\gamma_A\ldots/\gamma^{\ddagger})\,K^{\ddagger}$, For the specific reaction rate, we obtain

$$k = \frac{\gamma_A \cdots}{\gamma^{\ddagger}} \frac{k_B T}{h} K^{\ddagger} = \frac{\gamma_A \cdots}{\gamma^{\ddagger}} k_0. \qquad [18.20]$$

Note how this result agrees with equation (16.142). Also, note that the Boltzmann constant is here symbolized by k_B.

18.5 Transition State Recrossings

In a typical chemical conversion, the reacting complex moves along a potential energy valley up to and over a saddle point. The effective width of the valley and its height along the path vary continuously. These variations cause some of the de Broglie wave to be reflected at each stage.

Consider the reaction

$$A \rightarrow M^{\ddagger} \rightarrow B \qquad [18.21]$$

where A represents the reactants that combine to form the transition complex $M^{\ddagger}$ while B represents the products. Suppose that any complex that has traveled through the transition state has the probability w_f of being reflected back through, while any complex that has traveled back through the transition state has the probability w_b of being reflected forward through the transition state. The various recrossings described in figure 18.3 then occur.

Here a represents the concentration units of A entering the transition state directly from the initial state in unit time while b represents the concentration units of B coming directly from the final state in unit time. For simplicity, we neglect any variation of w_f or w_b with energy.

From the expressions in figure 18.3, the concentration units moving forward through the transition state in unit time are

$$\frac{d[B]_{tr}}{dt} = a\left(1 + w_f w_b + w_f^2 w_b^2 + \cdots\right) + b w_b\left(1 + w_f w_b + w_f^2 w_b^2 + \cdots\right) = \left(a + b w_b\right)\left(1 - w_f w_b\right)^{-1}. \qquad [18.22]$$

Similarly, the concentration units moving backward through the transition state in unit time are

$$\frac{d[A]_{tr}}{dt} = a w_f\left(1 + w_f w_b + w_f^2 w_b^2 + \cdots\right) + b\left(1 + w_f w_b + w_f^2 w_b^2 + \cdots\right) = \left(a w_f + b\right)\left(1 - w_f w_b\right)^{-1}. \qquad [18.23]$$

When the system is at equilibrium, rates (18.22) and (18.23) are equal:

$$\left(a + b w_b\right)\left(1 - w_f w_b\right)^{-1} = \left(a w_f + b\right)\left(1 - w_f w_b\right)^{-1}. \qquad [18.24]$$

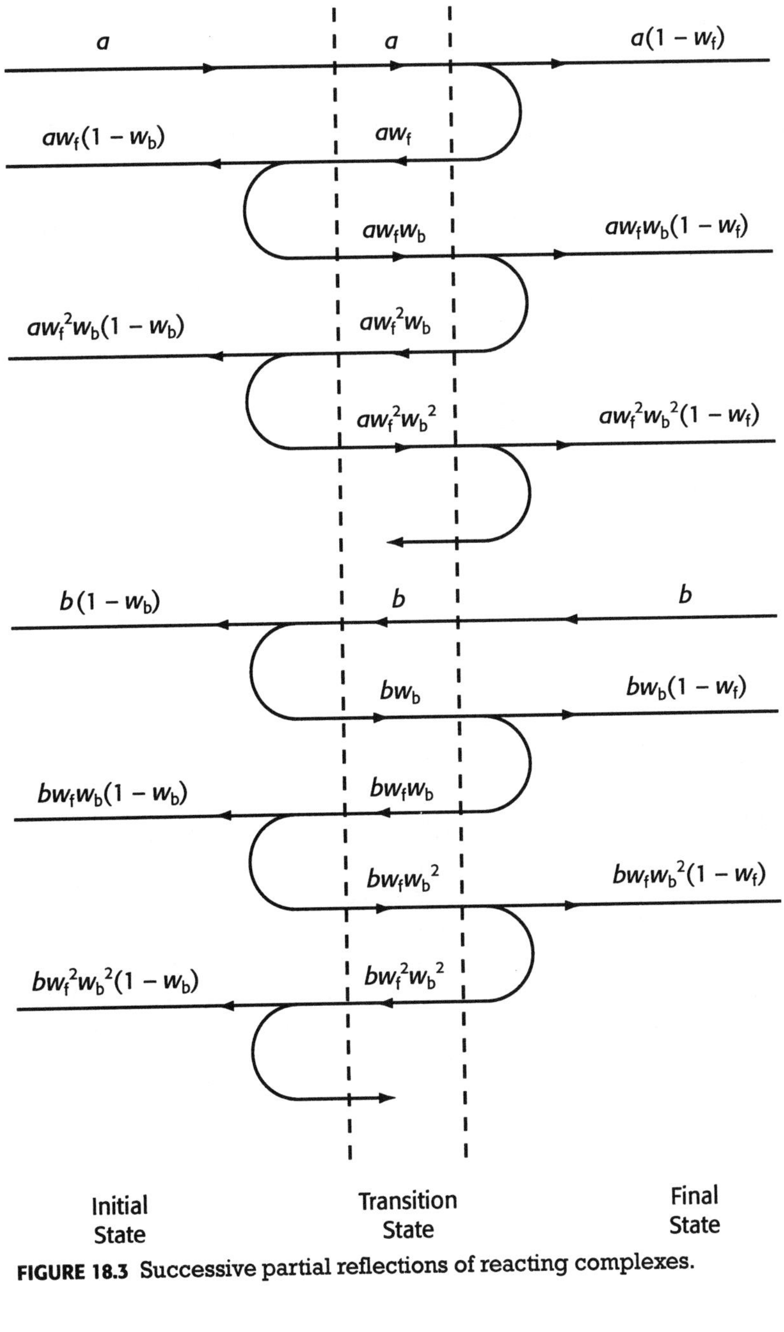

FIGURE 18.3 Successive partial reflections of reacting complexes.

Canceling the common factor and rearranging yields

$$b = a\left(1 - w_\mathrm{f}\right)\left(1 - w_\mathrm{b}\right)^{-1}. \tag{18.25}$$

Then substituting (18.25) into (18.22) gives the equilibrium rate

$$\frac{\mathrm{d}\left[\mathrm{B}\right]_\mathrm{tr}}{\mathrm{d}t} = a\left(1 + \frac{1 - w_\mathrm{f}}{1 - w_\mathrm{b}} w_\mathrm{b}\right)\left(1 - w_\mathrm{f} w_\mathrm{b}\right)^{-1} = \frac{a}{1 - w_\mathrm{b}}. \tag{18.26}$$

On the other hand, the net rate at which [B] is being produced from [A], whether the system is at equilibrium or not, is

$$\frac{d[B]}{dt} = a(1-w_f)\left(1 + w_f w_b + w_f^{\,2} w_b^{\,2} + \cdots\right) = a(1-w_f)(1-w_f w_b)^{-1}, \qquad [18.27]$$

according to figure 18.3.

Eyring equations (18.11) and (18.20) give the equilibrium $d[B]_{tr}/dt$. Chemical measurements yield the nonequilibrium $d[B]/dt$ .The ratio

$$\frac{d[B]/dt}{d[B]_{tr}/dt} = \kappa \qquad [18.28]$$

is called the *transmission coefficient*. Substituting (18.27) and (18.26) into (18.28) yields

$$\kappa = \frac{a(1-w_f)(1-w_f w_b)^{-1}}{a(1-w_b)^{-1}} = \frac{(1-w_f)(1-w_b)}{1-w_f w_b}. \qquad [18.29]$$

The corresponding specific reaction rate is

$$k = \kappa \frac{k_B T}{h} \frac{\gamma_A \cdots}{\gamma^{\ddagger}} K^{\ddagger}. \qquad [18.30]$$

18.6 Thermodynamic Considerations

In constructing factor k_BT/h in equation (18.11), we took z_{tr}* from the partition function for the transition state. As a consequence, expression $K^{\ddagger}$ is incomplete; it does not contain the partition function for motion along the reaction coordinate in its numerator.

Correspondingly, one can subtract from a thermodynamic function, the contribution from motion along the reaction coordinate. The result may be called a residual thermodynamic function. In figure 18.4, the standard residual Gibbs energy is plotted against the reaction coordinate for a particular process. The horizontal line is drawn at the standard Gibbs energy for the reactants. We note that

$$\Delta G^{\ddagger} = G^{\ddagger}_{TS} - G^{0}_{RE}. \qquad [18.31]$$

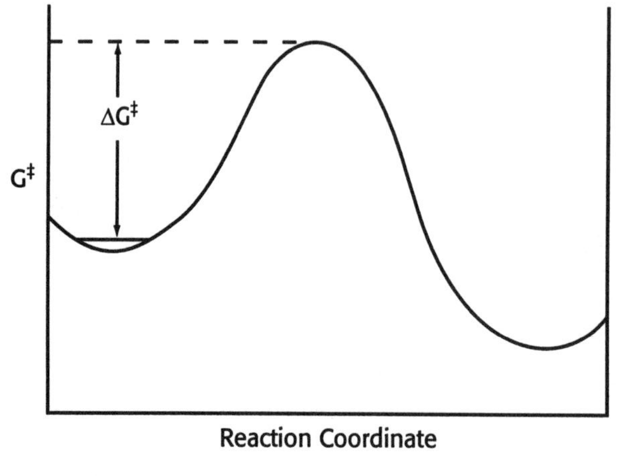

FIGURE 18.4 Residual standard Gibbs energy G‡ along the reaction coordinate.

Expression $\Delta G^{\ddagger}$ is called the *Gibbs energy of activation*, while $G_{TS}^{\ddagger}$ is the residual standard Gibbs energy in the transition state and G_{RE}^{0} is the standard Gibbs energy of the reactants.

From equation (7.55), we construct

$$\Delta G^{\ddagger} = -RT \ln K^{\ddagger}. \qquad [18.32]$$

Then solving for K and substituting the result into (18.30) yields

$$k = \kappa \frac{k_B T}{h} \frac{\gamma_A \cdots}{\gamma^{\ddagger}} e^{-\Delta G^{\ddagger}/RT}. \qquad [18.33]$$

The rate constant k comes out in $M^{-(n-1)}\ s^{-1}$ units when the standard state for each participant in $K^{\ddagger}$ is 1 M. Since

$$\Delta G^{\ddagger} = \Delta H^{\ddagger} - T\Delta S^{\ddagger}, \qquad [18.34]$$

where $\Delta H^{\ddagger}$ is the *enthalpy of activation* and $\Delta S^{\ddagger}$ the *entropy of activation*, we also have

$$k = \kappa \frac{k_B T}{h} \frac{\gamma_A \cdots}{\gamma^{\ddagger}} e^{\Delta S^{\ddagger}/R} e^{-\Delta H^{\ddagger}/RT}. \qquad [18.35]$$

18.7 Transition State Location

We have seen how recrossings reduce the reactive flux below the flux through the transition state. Presumably, the best location for this state is the one that makes this difference as small as possible. Variational theory allows us to find this location.

Consider a reaction for which the classical theory yields the specific rate

$$k = \frac{\gamma_A \cdots}{\gamma^{\ddagger}} \frac{k_B T}{h} K^{\ddagger}. \qquad [18.36]$$

Suppose that the transition state is at position q along the reaction coordinate. Also suppose that the variation of the potential energy within the state itself can be neglected.

In principle, one can make different choices of q and calculate the corresponding k. Those choices away from the best one will involve more recrossings and so a greater k.

As a consequence, the best position is the point where

$$\left(\frac{\partial k}{\partial q^{\ddagger}}\right)_T = 0. \qquad [18.37]$$

But there we have

$$\left(\frac{\partial \ln K^{\ddagger}}{\partial q^{\ddagger}}\right)_T = 0. \qquad [18.38]$$

and

$$\left(\frac{\partial \Delta G^{\ddagger}}{\partial q^{\ddagger}}\right)_T = 0. \qquad [18.39]$$

Condition (18.39) is met when the transition state is located where $G^{\ddagger}$ is a maximum along the reaction coordinate. In general, this point differs from the saddle point on the potential energy surface.

For common reactions of or between molecules, the difference may not be significant. But when the energy increment $\Delta\varepsilon_0$ in formula (18.11) is small, the difference may be appreciable. And when there is no potential energy barrier, the only barrier is the Gibbs energy one. Then criterion (18.39) is essential.

The latter condition occurs when a bond is formed without another one being broken. As examples, we have

$$\mathrm{H} + \mathrm{CH_3} \rightarrow \mathrm{CH_4}, \quad [18.40]$$

$$\mathrm{H} + \mathrm{C_2H_5} \rightarrow \mathrm{C_2H_6}. \quad [18.41]$$

18.8 Viscous Nonideality

As a reacting complex moves along the reaction coordinate through the transition state, it may be retarded by viscous forces. In high density gas phases and in condensed phases, such forces become appreciable and reduce the rate of reaction.

Consider a reaction with a definite initial state, transition state, and final state. Let a reaction coordinate be established. Let us also suppose that the final state is not occupied. Presumably, the forward reaction is not affected by the reverse reaction, so this is not a limitation on the result. The net velocity of reacting complexes then rises from zero in the middle of the initial state to a maximum at the end of the transition zone. The viscous forces act on this velocity. Furthermore, movement towards the final state is driven by a net thermodynamic force.

Let distance along the reaction coordinate from the neutral point be x while the reduced mass for movement of the reacting complex along this coordinate is μ. Then the viscous retarding force acting on an average complex at position x may be expressed as

$$f_1 = -\mu\beta\dot{x}, \quad [18.42]$$

where $\dot{x}$ is the average velocity of the complex when it is at position x. The force due to the variation of the apparent potential energy, acting on an average complex, is

$$f_2 = \mu\omega_0^2 x, \quad [18.43]$$

while the effective thermodynamic force is μA.

Applying Newton's second law to a representative complex yields

$$\mu\ddot{x} - f_1 - f_2 = \mu A, \quad [18.44]$$

where $\ddot{x}$ is the acceleration d^2x/dt^2. Introducing the expressions for the forces and canceling the mass leads to the differential equation

$$\ddot{x} + \beta\dot{x} - \omega_0^2 x = A. \quad [18.45]$$

At random points throughout the transition zone, complexes are reflected. To the approximation that the effect is continuous and uniform over x in this zone, the average velocity is

$$\dot{x} = \omega(x + B) \quad [18.46]$$

with ω and B constant. Equation (18.45) now reduces to

$$\left(\omega^2 + \beta\omega - \omega_0^2\right)x + \omega^2 B + \beta\omega B - A = 0. \quad [18.47]$$

Since this equation must hold at the crest where $x = 0$, we have

$$\omega^2 B + \beta\omega B - A = 0. \quad [18.48]$$

Substituting (18.48) into (18.47), canceling x, and solving the resulting equation for ω leads to

$$\omega = -\frac{\beta}{2} + \left(\frac{\beta^2}{4} + \omega_0^{\,2}\right)^{1/2} = \left\{\left[\left(\frac{\beta}{2\omega_0}\right)^2 + 1\right]^{1/2} - \frac{\beta}{2\omega_0}\right\}\omega_0. \qquad [18.49]$$

At the beginning of the transition zone, where $x = -\delta/2$, approximate equilibrium with reactants prevails. So there, the average velocity $\dot{x}$ equals zero and B in (18.46) must equal $\delta/2$. At the end of the reaction zone, where $x = \delta/2$, equation (18.46) now yields

$$\dot{x} = \omega\delta. \qquad [18.50]$$

We expect the specific reaction rate to be proportional to this velocity. So it is proportional to ω.

When there is no viscous retardation, parameter β is zero, ω equals ω_0, and the specific reaction rate is given by formula (18.30). So in the presence of the retardation, the rate is multiplied by the factor in curly brackets in formula (18.49). We set

$$\left[\left(\frac{\beta}{2\omega_0}\right)^2 + 1\right]^{1/2} - \frac{\beta}{2\omega_0} = \kappa^{\mathrm{KR}} \qquad [18.51]$$

and call it the *Kramers factor* in honor of its originator H. A. Kramers. The specific reaction rate now assumes the form

$$k = \kappa^{\mathrm{KR}}\kappa\frac{\gamma_{\mathrm{A}}\cdots}{\gamma^{\ddagger}}\frac{k_{\mathrm{B}}T}{h}K^{\ddagger} \qquad [18.52]$$

At low densities, the viscous forces are relatively low,

$$\beta << 2\omega_0, \qquad [18.53]$$

and equation (18.51) reduces to

$$\kappa^{\mathrm{KR}} = 1. \qquad [18.54]$$

At high densities, the viscous forces are relatively high,

$$\beta >> 2\omega_0, \qquad [18.55]$$

and the radical in (18.51) expands as follows:

$$\left[\left(\frac{\beta}{2\omega_0}\right)^2 + 1\right]^{1/2} = \frac{\beta}{2\omega_0} + \left(\frac{1}{2}\right)\left(\frac{2\omega_0}{\beta}\right) + \;\ldots\;. \qquad [18.56]$$

The succeeding terms are relatively small. Neglecting these, we find that

$$\kappa^{\mathrm{KR}} = \frac{\beta}{2\omega_0} + \frac{\omega_0}{\beta} - \frac{\beta}{2\omega_0} = \frac{\omega_0}{\beta}. \qquad [18.57]$$

A plot of the Kramers factor against the ratio β/ω_0 appears in figure 18.5.

Note that β is directly proportional to the viscosity of the medium for a given reacting complex. A small variation of ω_0 with solvent is also observed. The section between limit (18.54) and the region where approximation (18.57) is valid is called the *Kramers turnover* region.

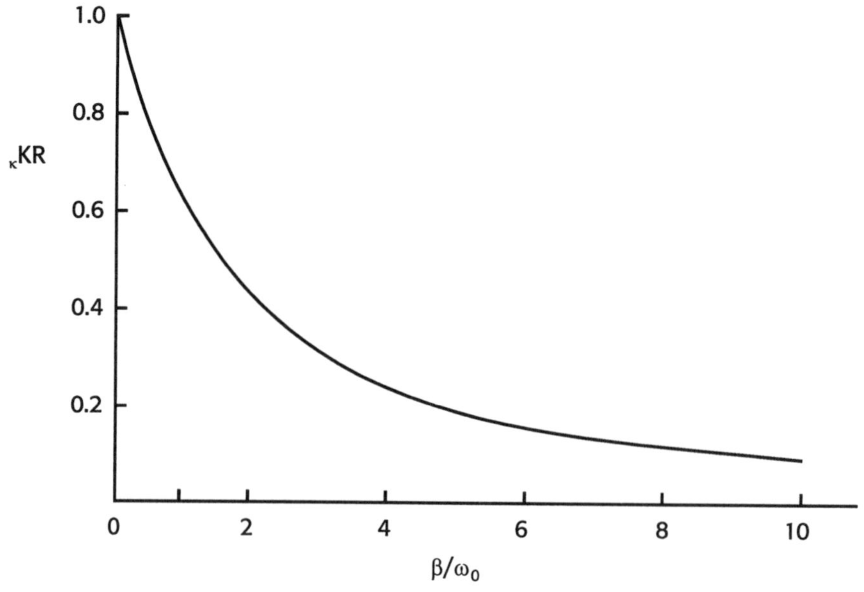

FIGURE 18.5 Variation of the Kramers transmission coefficient with the frictional parameter β/ω_0.

Example 18.7

When the friction coefficient $\mu\beta$ is large, movement to the transition state may be described as a diffusion process. Relate the Kramers coefficient κ^{KR} to the pertinent diffusion coefficient.

From the Stokes law, equation (5.106), the retarding force on the average reacting complex is

$$f = -6\pi\eta r\dot{x}.$$

Set the coefficient of x in (18.42) equal to the coefficient here and introduce relationship (5.113) to construct

$$\mu\beta = 6\pi\eta r = \frac{k_B T}{D}.$$

Then substitute the resulting β into formula (18.57) to get

$$\kappa^{KR} = \frac{\mu\omega_0 D}{k_B T}.$$

18.9 Variation of the Rate Coefficients with Temperature

Transition state theory enables one to describe the temperature dependence of the specific reaction rates in considerable detail.

For simplicity, let us consider reactions described by formulas (18.11). Thus, we suppose that each pertinent activity coefficient and transmission coefficient equals 1 to a good approximation,

Note that $\Delta\epsilon_0$ equals the zero point energy in the transition state minus that in the initial state. Number ΔE_0 is this difference multiplied by Avogadro's number. For a given process, ΔE_0 does not vary with temperature. However, one must consider the variations in the partition functions as well as in the $k_B T/h$ factor.

For a unimolecular reaction at ordinary temperatures, the approximations in example 18.3 apply. The resulting specific reaction rate has the form

$$k = BTe^{-\Delta E_0/RT} \qquad [18.58]$$

with

$$B \cong \frac{k_B}{h}. \qquad [18.59]$$

For a bimolecular reaction between atoms, the arguments in example 18.4 apply. The resulting specific reaction rate has the form

$$k = BT^{1/2}e^{-\Delta E_0/RT} \qquad [18.60]$$

with

$$B = \left(\frac{8\pi k_B}{\mu}\right)^{1/2} r^2. \qquad [18.61]$$

For a bimolecular reaction between an atom and a linear molecule, proceeding along the axis, the discussion in example 18.5 applies. The specific reaction rate then has the form

$$k = BT^{-1/2}e^{-\Delta E_0/RT} \qquad [18.62]$$

with a more complicated, but still constant, B. When the transition state is nonlinear, the specific reaction rate reduces to the Arrhenius form

$$k = Be^{-\Delta E_0/RT} \qquad [18.63]$$

with the stated approximations.

For a bimolecular reaction between two linear molecules, with a linear transition state, our approximations yield the specific reaction rate

$$k = BT^{-3/2}e^{-\Delta E_0/RT} \qquad [18.64]$$

with B constant. When the transition state is nonlinear, we have the approximate form

$$k = BT^{-1}e^{-\Delta E_0/RT} \qquad [18.65]$$

with B constant.

For a bimolecular reaction between two nonlinear molecules, we similarly find the temperature dependence

$$k = BT^{-2}e^{-\Delta E_0/RT}. \qquad [18.66]$$

18.10 Further Relationships

In determining how the transition state for a given reaction is related to the reactant state, one may make use of the temperature dependence of its rate coefficient k and of various standard thermodynamic formulas.

In practice, the temperature variation is often represented by fitting experimental data for the given reaction to the Arrhenius equation

$$k = Ae^{-E_a/RT} \qquad [18.67]$$

with T the absolute temperature. But in view of final expression (18.11), it may also be fitted to

$$\frac{k}{T} = Be^{-E_b/RT}. \qquad [18.68]$$

Differentiating the logarithm of equation (18.67) yields

$$\frac{\mathrm{d}}{\mathrm{d}T}\ln k = \frac{E_\mathrm{a}}{RT^2}, \tag{18.69}$$

while differentiating the logarithm of equation (18.68) gives us

$$\frac{\mathrm{d}}{\mathrm{d}T}\ln\frac{k}{T} = \frac{E_\mathrm{b}}{RT^2}. \tag{18.70}$$

From final expression (18.11), we find that

$$k = \frac{k_\mathrm{B}T}{h}K^{\ddagger} \tag{18.71}$$

and

$$\ln k = \ln\left(k_\mathrm{B}/h\right) + \ln T + \ln K^{\ddagger}. \tag{18.72}$$

The residual equilibrium constant $K^{\ddagger}$ is a concentration constant. To its temperature dependence, formula (7.93) applies.

Differentiating equation (18.72) and introducing (7.93) leads to

$$\frac{\mathrm{d}}{\mathrm{d}T}\ln k = \frac{1}{T} + \frac{\mathrm{d}}{\mathrm{d}T}\ln K^{\ddagger} = \frac{1}{T} + \frac{\Delta E^{\ddagger}}{RT^2}. \tag{18.73}$$

Combining (18.69), (18.73), and solving for the Arrhenius activation energy yields

$$E_\mathrm{a} = RT + \Delta E^{\ddagger} = RT + \Delta H^{\ddagger} - P\Delta V^{\ddagger} \tag{18.74}$$

where $\Delta V^{\ddagger}$ is the standard volume change on activation. Proceeding similarly with the formula

$$\frac{k}{T} = \frac{k_\mathrm{B}}{h}K^{\ddagger} \tag{18.75}$$

leads to

$$E_\mathrm{b} = \Delta E^{\ddagger} = \Delta H^{\ddagger} - P\Delta V^{\ddagger}. \tag{18.76}$$

Thus, we have

$$E_\mathrm{a} = E_\mathrm{b} + RT. \tag{18.77}$$

In the approximation that the reactants are gaseous and ideal, one has

$$P\Delta V^{\ddagger} = \left(1 - x\right)RT \tag{18.78}$$

where x is the molecularity of the reaction.

But with the transmission coefficient and the activity coefficients equal to 1, formula (18.35) reduces to

$$k = \frac{k_\mathrm{B}T}{h}e^{\Delta S^{\ddagger}/R}e^{-\Delta H^{\ddagger}/RT}. \tag{18.79}$$

Introducing formula (18.74) for $\Delta H^{\ddagger}$ and (18.78) for $P\Delta V^{\ddagger}$ now gives us

$$k = e^{x}\frac{k_\mathrm{B}T}{h}e^{\Delta S^{\ddagger}/R}e^{-E_\mathrm{a}/RT}. \tag{18.80}$$

With formula (18.77), we also have

$$k = e^{x-1}\frac{k_\mathrm{B}T}{h}e^{\Delta S^{\ddagger}/R}e^{-E_\mathrm{b}/RT}. \tag{18.81}$$

Here each standard state involves the given constituent at unit concentration (1 M).

To change to unit partial pressure as standard state for each constituent, we convert the residual equilibrium constant by the formula

$$K_c^{\ddagger} = K_P^{\ddagger}\left(RT\right)^{-\Delta n} = K_P^{\ddagger}\left(RT\right)^{x-1} \qquad [18.82]$$

based on equation (7.64). Then equation (18.71) becomes

$$k = \frac{k_B T}{h} K_P^{\ddagger}\left(RT\right)^{x-1} \qquad [18.83]$$

while (18.80) becomes

$$k = e^{x}\frac{k_B T}{h}\left(RT\right)^{x-1} e^{\Delta S_P^{\ddagger}/R} e^{-E_a/RT} \qquad [18.84]$$

and (18.81) becomes

$$k = e^{x-1}\frac{k_B T}{h}\left(RT\right)^{x-1} e^{\Delta S_P^{\ddagger}/R} e^{-E_b/RT}. \qquad [18.85]$$

Here x is the molecularity.

In condensed phases, the standard volume change on activation, $\Delta V^{\ddagger}$, is small and may be neglected. In place of (18.80), we obtain

$$k = e\frac{k_B T}{h} e^{\Delta S^{\ddagger}/R} e^{-E_a/RT}, \qquad [18.86]$$

while in place of (18,81), we have

$$k = \frac{k_B T}{h} e^{\Delta S^{\ddagger}/R} e^{-E_b/RT}. \qquad [18.87]$$

Example 18.8

For the chair- chair conformational inversion of cyclohexane, the specific reaction rate is given by the formula

$$k = 1.76 \times 10^{13} e^{-(46.86 \text{ kJ mol}^{-1})/RT} \text{ s}^{-1}$$

over the temperature range -116 .7° to -24 .0° C. Calculate the thermodynamic properties $\Delta S^{\ddagger}$, $\Delta H^{\ddagger}$, $\Delta G^{\ddagger}$ about -70.35° C.

Since the reaction is occurring in a condensed phase, equation (18.86) applies. On comparing this with the given expression, we see that

$$e\frac{k_B T}{h} e^{\Delta S^{\ddagger}/R} = 1.76 \times 10^{13} \text{ s}^{-1} \qquad \text{at} \qquad 202.50 \text{ K}.$$

But since

$$e\frac{k_B T}{h} = e\frac{1.38066 \times 10^{-23} \text{ J K}^{-1}}{6.62608 \times 10^{-34} \text{ J s}} 202.50 \text{ K} = 1.147 \times 10^{13} \text{ s}^{-1},$$

we find that

$$e^{\Delta S^{\ddagger}/R} = \frac{1.76 \times 10^{13} \text{ s}^{-1}}{1.147 \times 10^{13} \text{ s}^{-1}} = 1.5345$$

whence

$$\Delta S^{\ddagger} = \left(0.4282\right)\left(8.31451\right) \text{ J K}^{-1} \text{ mol}^{-1} = 3.56 \text{ J K}^{-1} \text{ mol}^{-1}.$$

We also see that

$$E_a = 46.86 \text{ kJ mol}^{-1}.$$

For the condensed phase, $\Delta V^{\ddagger} \cong 0$ and equation (18.74) is replaced with

$$E_a = RT + \Delta H^{\ddagger}.$$

So here

$$\Delta H^{\ddagger} = \left[46{,}860 - (8.31451)(202.50)\right] \text{J mol}^{-1} = 45{,}180 \text{ J mol}^{-1}.$$

Furthermore, we have

$$\Delta G^{\ddagger} = \Delta H^{\ddagger} - T\Delta S^{\ddagger} = \left[45{,}180 - (202.50)(3.56)\right] \text{J mol}^{-1} = 44{,}460 \text{ J mol}^{-1}.$$

In the Lindemann mechanism, converting an average reactant molecule A into the average one with enough energy to react A* involves a considerable increase in entropy. Then proceeding to the transition state requires movement of some of the excess energy into the reaction coordinate. This involves a decrease in entropy. Generally, the former is greater than the latter. Then the overall change, going from A to A*, is positive, as here.

Questions

18.1 What is the kinematics of an elementary reaction?

18.2 How does the Born-Oppenheimer approximation allow one to treat this behavior?

18.3 What is a reaction coordinate? Is it a coordinate for a normal mode?

18.4 How does the apparent potential energy vary along the reaction coordinate?

18.5 How is the transition state defined?

18.6 How is the average forward speed through the transition state calculated?

18.7 Construct a formula for the mean time spent by a forward moving complex in the transition state.

18.8 Construct a formula for the concentration of forward moving complexes in the transition state.

18.9 How is the equilibrium constant K^* expressed in terms of the partition functions?

18.10 How are these formulas combined to produce the Eyring equation

$$k = \frac{k_B T}{h} K^{\ddagger}?$$

18.11 Summarize the approximations made in deriving this form for the specific reaction rate.

18.12 Show when the pre-exponential factor in the Arrhenius equation equals $k_B T/h$.

18.13 Describe the kinetic isotope effect.

18.14 How does the change in the mass of a participating nucleus act to change the specific reaction rate?

18.15 How is the Eyring equation modified to allow for nonunitary activity coefficients?

18.16 Relate the transmission coefficient κ to the reflection probabilities w_f and w_b.

18.17 When is tunneling through the energy barrier significant?

18.18 What is a residual thermodynamic function? How may it be determined?

18.19 Describe the behavior of the standard residual Gibbs energy along the reaction coordinate.

18.20 Why is consideration of this behavior indispensable in treating some elementary reactions? Cite an example.

18.21 How may a variational procedure allow one to determine the location of an appropriate transition state.

18.22 When may a viscous retarding force be significant for a reaction?

18.23 Formulate an expression for this retarding force.

18.24 Formulate an expression for the thermodynamic force driving the reacting complex forward.

18.25 From these expressions, construct the Kramers factor.

18.26 How may the Kramers factor be related to a pertinent diffusion coefficient?

18.27 Summarize how the specific reaction rate may vary with temperature.
18.28 How does the entropy of activation appear in the specific reaction rate?
18.29 How is a positive $\Delta S^{\ddagger}$ interpreted?

Problems

18.1 From example 18.3, the specific reaction rate of a first order reaction has the form

$$k = BTe^{-E/RT}.$$

Take the data from problem 15.11, plot ln (k/T) against $1/T$, and determine B and E. Calculate BT at 306 K.

18.2 Fit the data of problem 15.24 to the form

$$k = BT^{-1}e^{-E/RT}.$$

From the slope of the appropriate straight line, determine the energy barrier height E. Also determine factor B and the product BT^{-1} at 650 K.

18.3 Calculate the specific rate of reactive collisions between F atoms and H_2 molecules at 25° C. Consider the pertinent radii of F and H_2 to be 1.22 Å and 0.76 Å. Energy ΔE_0 is 6.57 kJ mol^{-1}.

18.4 The reaction

$$F + H_2 \rightarrow HF + H$$

proceeds through a linear transition state. The moment of inertia of H_2 is 0.277 u $Å^2$; that of the complex F-H-H, 7.433 u $Å^2$, The doubly degenerate bending wave number of the complex is about 400 cm^{-1} . Calculate the specific reaction rate, at 25° C.

18.5 Calculate the maximum kinetic isotope effect for replacing the H in reactant molecule

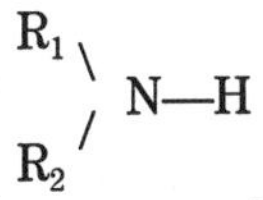

with D (a) at 25° C, (b) at 100° C. The wave number for the N-H bond is 3100 cm^{-1}.

18.6 The reaction between $S_2O_8^{=}$ and I^- is first order in each ion. If in a solution 0.024 M in KCl, 0.00015 M in $K_2S_2O_8$ and 0.0005 M in KI, the second-order rate constant k is 1.70 M^{-1} min^{-1} at 25° C, what is k in a solution 0.0020 M in KI and 0.00015 M in $K_2S_2O_8$ at the same temperature?

18.7 If variation of the potential energy along a reaction path and a changing valley width cause 0.160 part of a typical forward moving de Broglie wave and 0.090 part of a typical reverse moving de Broglie wave to be returned to the transition zone after passing through, what does the transmission coefficient equal?

18.8 Determine the β/ω_0 at which the Kramers transmission coefficient equals 0.45.

18.9 The reaction

$$Br + Br + M \rightarrow Br_2 + M$$

was followed in a high pressure argon bath. At very high pressures, the second order rate coefficient varied with the diffusion coefficient as in table 18.A.

TABLE 18.A

D, 10^{-3} cm^2 s^{-1}	0.116	0.076	0.057	0.043
k, 10^{12} cm^3 mol^{-1} s^{-1}	17.0	12.5	9.0	7.6

Show how these results support the Kramers theory.

— — —

18.10 For the decomposition of a bans azoalkane in an organic solvent, the rate constants seen in table 18.B were found, at the given Celsius temperatures.
Determine the Arrhenius parameters A and E_a.

18.11 From the data in problem 18.10, determine the parameters B and E in the form

$$k = BTe^{-E/RT}.$$

TABLE 18.B

t, °C	114.55	119.16	123.83	128.51	133.33
$k \times 10^4$, s^{-1}	1.02	1.65	2.68	4.66	8.44

18.12 Decomposition of NO_2 to NO and O_2 is second order with a rate constant varying with temperature as in table 18.C.

TABLE 18.C

T, K	592	603	627	652	656
k, $M^{-1}\,s^{-1}$	0.522	0.755	1.700	4.020	5.030

Determine the Arrhenius parameters A and E_a.

18.13 Fit the data of problem 18.12 to the form

$$k = BT^{-2}e^{-E/RT}.$$

Thus, determine the parameters B and E.

18.14 Calculate the specific rate of reactive collisions between the HC radical and the N_2 molecule at 25° C. Energy ΔE_0 is 2.21 kJ mol^{-1} while the pertinent radii of HC and N_2 are 0.75 Å and 0.70 Å.

18.15 The reactants and the transition state for the reaction

$$HC + N_2 \rightleftarrows HCN + N$$

have the properties shown in table 18.D.

TABLE 18.D

n	HC	N_2	HCN_2
M, g mol^{-1}	13.019	28.013	41.032
I, 10^{-40} g cm^2	1.935	13.998	73.2
$\tilde{\nu}$, cm^{-1}	2733	2330	3130, 2102, 1252, 1170, 564, 401
z_{elec}	2	1	2
E_0, kJ mol^{-1}			2.21

Calculate the specific reaction rate at 25° C.

18.16 Calculate the maximum kinetic isotope effect for replacing the H in reactant molecule R_3C-H with D (a) at 25° C, (b) at 100° C. The wave number for the C-H bond is 2800 cm^{-1}.

18.17 For the decomposition of N_2O_5, the specific reaction rate is given by the formula

$$k = 3.79 \times 10^{10} Te^{-\left(99.9 \text{ kJ mol}^{-1}\right)/RT} \text{ s}^{-1}.$$

Calculate the thermodynamic properties as $\Delta S^{\ddagger}$, $\Delta H^{\ddagger}$, $\Delta G^{\ddagger}$ at 306 K.

18.18 For the reaction between hydrogen and iodine, the specific reaction rate fits the formula

$$k = 1.15 \times 10^{11}\, e^{-\left(164.8 \text{ kJ mol}^{-1}\right)/RT} \text{ M}^{-1}\text{ s}^{-1}$$

in the temperature range 600 K to 700 K. Calculate the thermodynamic properties $\Delta S^{\ddagger}$, $\Delta H^{\ddagger}$, $\Delta G^{\ddagger}$ at 650 K.

References

Books

Connors, K. A.: 1990, *Chemical Kinetics: The Study of Reaction Rates in Solution*, VCH Publishers, Inc., New York, pp 187-243.

In his chapter 5, Connors presents an introduction to collision theory, potential energy surfaces, and transition state theory. His treatment of the first two fields is useful. However, his derivation of the Eyring equation is defective.

Steinfeld, J. I., Francisco, J. S., and Hase, W. L.: 1989, *Chemical Kinetics and Dynamics*, Prentice-Hall, Englewood Cliffs, NJ, pp. 209-245, 308-341, 402-414.

In chapter 7, various approximate interatomic potentials are discussed. For polyatomic complexes, internal coordinates and the normal mode concept are introduced. Then various potential energy surfaces leading from a given reactant configuration to the corresponding product configuration are described. Of these, the most accurate are computed by ab initio methods. On these, one can pick out the most favorable reaction path.

In chapter 10, transition state theory is developed in considerable detail. The various approximations and limitations in the theory are pointed out.

In the third reference, in chapter 12, an introduction to Kramers' theory appears.

Articles

Andres, J., Moliner, V., and Silla, E.: 1994, "Comparison of Several Semiempirical and ab Initio Methods for Transition State Structure Characteristics. Addition of CO_2 to $CH_3NHCONH_2$," *J. Phys. Chem.* **98**, 3664-3668.

Bauer, S. H., and Wilcox Jr., C. F.: 1995, "What's in a Name-Transition State or Critical Transition Structure?" *J. Chem Educ.* **72**, 13-16.

Castano, R., de Juan, J., and Martinez, E.: 1983, "The Calculation of Potential Energy Curves of Diatomic Molecules: The RKR Method," *J. Chem Educ.* **60**, 91-93.

Clary, D. C.: 1990, "Fast Chemical Reactions: Theory Challenges Experiment," *Annu. Rev. Phys. Chem.* **41**, 51-90.

Fernandez, G. M., Sordo, J. A., and Sordo, T. L.: 1988, "Analysis of Potential Energy Surfaces," *J. Chem. Educ.* **65**, 565- 667.

Green Jr., W. H., Moore, C. B., and Polik, W. F.: 1992, "Transition States and Rate Constants for Unimolecular Reactions," *Annu. Rev. Phys. Chem.* **43**, 591-625.

Hamann, S. D., and le Noble, W. J.: 1984, "The Estimation of Activation Parameters: Corrections and Incorrections," *J. Chem Educ.* **61**, 658-660.

Lehman, J. J., and Goldstein, E.: 1996, "The Potential Energy Surface of ClF_3," *J. Chem Educ.* **73**, 1096-1098.

Lehmann, K. K., Scoles, G., and Pate, B. H.: 1994, "Intramolecular Dynamics from Eigenstate-Resolved Infrared Spectra," *Annu. Rev. Phys. Chem.* **45**, 241-274.

Neumark, D. M.: 1992, "Transition State Spectroscopy of Bimolecular Chemical Reactions," *Annu. Rev. Phys. Chem.* **43**, 153-176.

Pilling, M. J.: 1996, "Radical-Radical Reactions," *Annu. Rev. Phys. Chem.* **47**, 81-108.

Rayez, J. C., and Forst, W.: 1989, "Statistical Calculation of Unimolecular Rate Constant," *J. Chem. Educ.* **66**, 311-313.

Reid, S. A., and Reisler, H.: 1996, "Experimental Studies of Resonances in Unimolecular Decomposition," *Annu. Rev. Phys. Chem.* **47**, 495-525.

Sathymaurthy, N., and Joseph, T.: 1984, "Potential Energy Surface and Molecular Reaction Dynamics," *J. Chem. Educ.* **61**, 968- 971.

Schatz, G. C.: 1988, "Quantum Effects in Gas Phase Bimolecular Chemical Reactions," *Annu. Rev. Phys. Chem.* **39**, 317- 340.

Schroedar, J., and Troe, J.: 1987, "Elementary Reactions in the Gas-Liquid Transition Range," *Annu. Rev. Phys. Chem.* **38**, 163- 190.

Truhlar, D. G.: 1984, "Variational Transition State Theory," *Annu. Rev. Phys. Chem.* **35**, 159-189.

Tsao, J. Y.: 1989, "Transition-State Theory for Quantum and Classical Particle Escape from a Finite Square Well," *Am. J. Phys.* **57**, 269-274.

Weston Jr., R. E., and Flynn, G. W.: 1992, "Relaxation of Molecules with Chemically Significant Amounts of Vibrational Energy: The Dawn of the Quantum State Resolved Era," *Annu. Rev. Phys. Chem.* **43**, 559-589.

19

Photochemistry

19.1 Pertinent Wave Properties

THE CLASSICAL BEHAVIOR OF ELECTRIC AND MAGNETIC fields is described by Maxwell's equations. For any region where the permittivity ε and permeability μ are fixed, the equations are linear. Then in any wave, each frequency ν is independent. By the way, specifying a frequency implies that the wave, when not diverging, is sinusoidal in space and time.

The wavelength λ of a sinusoidal wave is the distance between successive points at the same phase. And since the frequency is the number of such points passing a given position in unit time, the phase velocity c equals the product

$$c = \lambda \nu. \tag{19.1}$$

The wave number $\tilde{\nu}$ is the number of wavelengths per unit length,

$$\tilde{\nu} = \frac{1}{\lambda}. \tag{19.2}$$

Each component frequency interacts with matter as if it were composed of particles called photons. The energy of N photons is

$$E = Nh\nu = N\hbar\omega, \tag{19.3}$$

where h is Planck's constant, $\hbar$ is $h/2\pi$, and ω is the angular frequency. One mole of photons is called an *einstein*.

The *intensity* I of a wave reaching a certain small cross section may be given by

(a) the number of photons falling on the cross section per unit area per unit time,
(b) the energy falling on the cross section per unit area per unit time,
(c) a number proportional to either of these, or
(d) a property varying monotonically with either of these (such as the amplitude of the electric intensity or the amplitude of the magnetic intensity at the given position).

Common detectors employ the excitation of electrons. Thus, they yield a number proportional to the energy per unit cross sectional area per unit time. In the following discussions, symbol I represents such a quantity.

The *transmittance* of a given sample is the ratio I/I_0 where I_0 is the incident intensity and I the transmitted intensity. At low and moderate intensities, this is found to be a property of the sample, its composition and its relative configuration.

Example 19.1

At 460 nm a blue filter transmits 72.7% of the light and a yellow filter 40.7%. What is the transmittance at the same wavelength of the two filters in combination?

First, consider the light to pass through the blue filter and then the yellow filter. Let I_0 be the initial intensity, I_1 the intensity after passing through the blue filter, and I_2 the intensity after passing through the yellow filter. Since the transmittance of the first filter is 0.727, we have

$$\frac{I_1}{I_0} = 0.727.$$

And since the transmittance of the second filter is 0.407, we have

$$\frac{I_2}{I_1} = 0.407.$$

Multiplying these yields

$$\frac{I_2}{I_0} = \frac{I_2}{I_1}\frac{I_1}{I_0} = (0.407)(0.727) = 0.296.$$

Since multiplication is commutative, interchanging the two filters does not alter the result.

19.2 Pertinent Particle Properties

Whenever an electromagnetic wave interacts with matter, it does so in a discrete manner. It loses or gains energy and momentum in bundles or quanta. Thus, it acts as if it were composed of particles. These are called photons.

On measuring the energy ε in a unit of interaction, one finds that

$$\varepsilon = h\nu = hc\tilde{\nu} = \frac{hc}{\lambda}. \qquad [19.4]$$

Here h is Planck's constant, ν the conventional frequency, c the phase velocity, $\tilde{\nu}$ the wave number, and λ the wavelength. The energies of the photons at various borderline wavelengths in the electromagnetic spectrum are listed in table 19.1.

One considers photons to travel at the speed that energy is carried. This would equal the group velocity of the electromagnetic wave at the pertinent frequency ν.

In free space, the group velocity equals the phase velocity c. But this equals the fundamental speed of Einstein relativity theory. Since this theory implies that the mass of a particle varies with its speed v as

$$m = \frac{m_0}{\sqrt{1 - v^2 / c^2}}, \qquad [19.5]$$

the rest mass of a photon must be zero.

By the Einstein theory, the energy associated with relativistic mass m is

$$E = mc^2. \qquad [19.6]$$

One may then define the mass of a photon as

$$m = \frac{h\nu}{c^2} = \frac{h}{\lambda c} \qquad [19.7]$$

TABLE 19.1 Photon Energies at Representative Wavelengths of the Electromagnetic Spectrum

Description	Vacuum Wavelenth λ	Vacuum Wave Number $\tilde{\nu}$	Individual Photon Energy ε	Molar Photon Energy E
γ Rays	1.00×10^{-11} m	1.00×10^{9} cm^{-1}	1.99×10^{-14} J	1.20×10^{10} J
X Rays	1.00×10^{-8} m	1.00×10^{6} cm^{-1}	1.99×10^{-17} J	1.20×10^{7} J
Vacuum UV	2.00×10^{-7} m	5.00×10^{4} cm^{-1}	9.93×10^{-19} J	5.98×10^{5} J
Near UV	3.80×10^{-7} m	2.63×10^{4} cm^{-1}	5.23×10^{-19} J	3.15×10^{5} J
Visible	7.80×10^{-7} m	1.28×10^{4} cm^{-1}	2.55×10^{-19} J	1.53×10^{5} J
Near IR	2.50×10^{-6} m	4.00×10^{3} cm^{-1}	7.95×10^{-20} J	4.49×10^{4} J
Mid IR	5.00×10^{-5} m	2.00×10^{2} cm^{-1}	3.97×10^{-21} J	2.39×10^{3} J
Far IR	1.00×10^{-3} m	1.00×10^{1} cm^{-1}	1.99×10^{-22} J	1.20×10^{2} J
Microwaves	1.00×10^{-1} m	1.00×10^{-1} cm^{-1}	1.99×10^{-24} J	1.20 J
Radio Waves				

and its momentum as

$$p = mc = \frac{h}{\lambda} = h\tilde{\nu}. \qquad [19.8]$$

Note that overall (19.8) is a form of the de Broglie equation. From equations (19.6) and (19.8), we also have

$$E = pc. \qquad [19.9]$$

When a target particle absorbs a photon, both energy and momentum are conserved. But in the infrared, visible, and ultraviolet regions of the spectrum, the recoil energy of a molecule is very small, relatively, and can be neglected.

In the Compton effect, where the photon is in the X-ray region and the target particle is an electron, this recoil energy has to be taken into account.

Example 19.2

Let a photon be absorbed by a particle of mass M initially at rest and determine what fraction of the photon energy appears as recoil kinetic energy.

Here the momentum p of the photon ends up as momentum of the excited particle. Let E_{re} be the recoil energy while E is the photon energy.

From formula (19.9), we have

$$p = \frac{E}{c}.$$

But the recoil energy is

$$E_{re} = \frac{p^2}{2M} = \frac{E^2}{2Mc^2}$$

whence

$$\frac{E_{re}}{E} = \frac{E}{2Mc^2}.$$

Example 19.3

Consider the target particle to be an electron and calculate the photon energy for which the recoil energy would be 1.00×10^{-3} times this photon energy.

When the target particle is an electron, the denominator in the final formula in example 19.2 is

$$2Mc^2 = 1.0220 \text{ MeV} = 1.0220 \times 10^6 \text{ eV}.$$

Then for

$$E_{re} / E = 1.00 \times 10^{-3},$$

we have

$$E = \left(1.00 \times 10^{-3}\right)\left(1.022 \times 10^6 \text{ eV}\right) = 1.02 \times 10^3 \text{ eV} = \left(1.02 \times 10^3 \text{ eV}\right)\left(1.60 \times 10^{-19} \text{ J eV}^{-1}\right) == 1.63 \times 10^{-16} \text{ J}.$$

This is in the X-ray range.

19.3 Phenomenology of Absorption

Since light interacts with matter as if both the matter and light were composed of particles, the absorption process is discrete rather than continuous. But for macroscopic layers, where the number of both kinds of particles is large, the continuum approximation suffices.

Consider a beam of electromagnetic radiation traveling in the x direction through a homogeneous material. And at this stage, let us suppose that the wave is practically monochromatic and that only one substance is absorbing the radiation.

On passing through a thin layer of thickness Δx, the intensity decreases by an amount proportional to the number of photons reaching the layer per unit time and to the number of absorbing molecules in the layer. For unit cross section, we have

$$-\Delta I = \kappa I\left[\mathrm{B}\right]\Delta x. \qquad [19.10]$$

Here κ is the constant of proportionality while [B] is the concentration of the absorbing molecules.

In the continuum approximation, we rewrite equation (19.10) as

$$-\frac{\mathrm{d}I}{I} = \kappa\left[\mathrm{B}\right]\mathrm{d}x. \qquad [19.11]$$

Integrating (19.11) over absorbing length l yields

$$I = I_0 e^{-\kappa[\mathrm{B}]l}. \qquad [19.12]$$

If we set

$$\kappa = \varepsilon \ln 10, \qquad [19.13]$$

then equation (19.12) becomes

$$I = I_0 10^{-\varepsilon[\mathrm{B}]l}. \qquad [19.14]$$

By definition, the negative decimal logarithm of I/I_0 is the *absorbance* A. From (19.14), we find that

$$A = -\log\frac{I}{I_0} = \varepsilon\left[\mathrm{B}\right]l. \qquad [19.15]$$

Ratio I/I_0 is called the transmittance of the solution, ε the *molar absorption coefficient*, while [B] is the concentration of the absorbing substance and l the length of the absorbing path.

In a mixture, each absorbing substance acts independently. So each one contributes independently to the space rate of absorption of the light. In place of

$$-\left(\frac{\partial I}{\partial x}\right)_{[\mathrm{B}]} = \kappa\left[\mathrm{B}\right]I, \qquad [19.16]$$

we have

$$-\left(\frac{\partial I}{\partial x}\right)_{\mathrm{all}[\mathrm{B}_i]\mathrm{'s}} = \left(\sum \kappa_i\left[\mathrm{B}_i\right]\right)I. \qquad [19.17]$$

And in place of (19.15), we have

$$A = -\log\frac{I}{I_0} = \left(\sum \varepsilon_i\left[\mathrm{B}_i\right]\right)l = \sum A_i. \qquad [19.18]$$

The total absorbance equals the *sum* of the absorbencies of the constituents.

Example 19.4

The transmittance of a 0.0500 M C_6H_6 solution through a cell 1.00 mm in thickness at 256 nm was 0.160. Calculate the absorbance A and the molar absorption coefficient ε.

Using the definition of absorbance A in equation (19.15), we find that

$$A = -\log\frac{I}{I_0} = -\log 0.160 = 0.796.$$

Then the second equality in (19.15) yields

$$\varepsilon = -\frac{\log I/I_0}{[\mathrm{B}]l} = \frac{0.796}{(0.0500\ \mathrm{M})(0.100\ \mathrm{cm})} = 159\ \mathrm{M}^{-1}\ \mathrm{cm}^{-1}.$$

Example 19.5

Evaluate the product $N_A hc$ that occurs in the formula

$$E = N_A h\frac{c}{\lambda}.$$

Substitute accepted values of the fundamental constants into the expression and carry out the indicated operations:

$$N_A hc = (6.0221\times10^{23}\ \mathrm{mol}^{-1})(6.6261\times10^{-34}\ \mathrm{J\ s})(2.99792\times10^{8}\ \mathrm{m\ s}^{-1}) = 1.19627\times10^{-1}\ \mathrm{J\ m\ mol}^{-1}.$$

19.4 Absorption Intensities

In the energy spectrum, rotational transitions occur in the far infrared, vibrational transitions in the near infrared, electronic transitions in the visible and ultraviolet regions. As a consequence, a given vibrational transition is accompanied by various rotational transitions, And a given electronic transition is accompanied by various vibrational and rotational transitions.

With the ever present line broadening, a given vibrational or electronic transition occurs over a band of wavelengths. So its intensity is measured by an integral of the absorption coefficient over the band.

In a smoothed-out representation, absorption coefficient ε varies with wave number $\tilde{\nu}$ following a Gaussian curve:

$$\varepsilon = \varepsilon_{\max}e^{-a(\tilde{\nu}-\tilde{\nu}_{\max})^2} = \varepsilon_{\max}e^{-ax^2}. \qquad [19.19]$$

Here we have set

$$\tilde{\nu} - \tilde{\nu}_{\max} = x. \qquad [19.20]$$

By experiment, one can determine the maximum in the absorption coefficient $\varepsilon_{\max}$. Also, one can measure the width at half maximum $\Delta\tilde{\nu}_{1/2}$.

In our approximation, the absorption curve is symmetric. So at

$$x = \frac{1}{2}\Delta\tilde{\nu}_{1/2}, \qquad [19.21]$$

we have

$$ax^2 = \ln 2. \qquad [19.22]$$

Solving for a then yields

$$a = \frac{\ln 2}{\left(\frac{1}{2}\Delta\tilde{\nu}_{1/2}\right)^2}. \qquad [19.23]$$

The integral of the absorption coefficient over the band is

$$\int_{\text{band}} \varepsilon \, d\tilde{\nu} \cong \int_{-\infty}^{\infty} \varepsilon_{\text{max}} e^{-ax^2} \, dx = \varepsilon_{\text{max}} \left(\frac{\pi}{a}\right)^{1/2} = \varepsilon_{\text{max}} \left(\frac{\pi}{\ln 2}\right)^{1/2} \frac{\Delta\tilde{\nu}_{1/2}}{2} = 1.0645 \varepsilon_{\text{max}} \Delta\tilde{\nu}_{1/2}. \qquad [19.24]$$

A strong electronic absorption band may present an ε_{max} of 10^4 to 10^5 M^{-1} cm^{-1} and a $\Delta\tilde{\nu}_{1/2}$ of 1000 to 5000 cm^{-1}. A weak band may yield an ε_{max} of 10 M^{-1} cm^{-1} and a $\Delta\tilde{\nu}_{1/2}$ of 100 cm^{-1}. A forbidden band may appear with ε_{max} equaling 10^{-4} to 10^{-3} M^{-1} cm^{-1}.

A standard electronic system consists of an electron in the potential field

$$V = \frac{1}{2} k r^2 \qquad [19.25]$$

with parameter k chosen so that

$$\left(\frac{k}{\mu}\right)^{1/2} = \omega = 2\pi\nu. \qquad [19.26]$$

Here ν is the average frequency for the absorption band, μ the mass of the electron, and r the distance from the center of the field. The *oscillator strength* f is defined as the ratio of the intensity for the given band to that for the corresponding standard electronic system.

By calculation, one can show that

$$f = \left(4.32 \times 10^{-9} \text{ M cm}^2\right) \int_{\text{band}} \varepsilon \, d\tilde{\nu}. \qquad [19.27]$$

Example 19.6

A particular solute yielded a band with a peak absorbency of 44,000 M^{-1} cm^{-1} at 30,000 cm^{-1} and a width at half maximum of 5000 cm^{-1}. Calculate its oscillator strength.

In the approximation that the band is Gaussian, formula (19.24) applies:

$$\int \varepsilon \, d\tilde{\nu} = (1.0645)\left(44{,}000 \text{ M}^{-1} \text{ cm}^{-1}\right)\left(5000 \text{ cm}^{-1}\right) = 2.34 \times 10^8 \text{ M}^{-1} \text{ cm}^{-2}.$$

Substituting this result into equation (19.27) yields

$$f = \left(4.32 \times 10^{-9} \text{ M cm}^2\right)\left(2.34 \times 10^8 \text{ M}^{-1} \text{ cm}^{-2}\right) = 1.01.$$

19.5 Electronic Transitions

One can consider that emission or absorption involves creation or annihilation of an electromagnetic wave segment. A molecule effects this through a redistribution of its electric charge, a process measured by the associated dipole moment.

If ψ_{i} is the initial state function and ψ_{f} the final state function of the molecule, the *transition dipole moment* is

$$\vec{\mu}_{\text{if}} = \int \psi_{\text{i}}^* \vec{\mu} \psi_{\text{f}} \, d^3\mathbf{r}, \qquad \vec{\mu} = -e\mathbf{r}. \qquad [19.28]$$

Here e is the charge on an electron and $\mathbf{r}$ is the radius vector coordinate for the electron. One can show that the oscillator strength for this transition is related to the moment by the equation

$$f = \left(1.4094 \times 10^{42}\ \mathrm{s\ C^{-2}\ m^{-2}}\right)\nu\left|\vec{\mu}_{\mathrm{if}}\right|^2. \quad [19.29]$$

For transitions among vibrational levels, the discussion in section 12,7 applies. The dissociation energy for a bond may be obtained as example 12.10 describes. However, the results of a Birge-Sponer extrapolation are not very accurate. But direct transition from the ground level to a dissociated state, without electronic excitation, is forbidden.

In the Born-Oppenheimer approximation, the electronic motion separates from the vibrational motion. Then the state function factors into an electronic factor and a vibrational factor:

$$\psi = \psi_\varepsilon \psi_\mathrm{v}. \quad [19.30]$$

The transition dipole moment for the change ε, $v \rightarrow \varepsilon'$, v' is

$$\vec{\mu}_{\mathrm{if}} = -e\int \psi_\varepsilon^* \mathbf{r}\, \psi_{\varepsilon'} \mathrm{d}^3\mathbf{r}_\varepsilon \int \psi_\mathrm{v}^* \psi_{\mathrm{v}'}\, \mathrm{d}R_{\mathrm{nuc}}. \quad [19.31]$$

Here $\mathbf{r}_\varepsilon$ is the pertinent electron coordinate, R_{nuc} the appropriate nuclear generalized coordinate. The last integral represents the overlap between the initial and final vibrational state functions:

$$S_{vv'} = \int \psi_v^* \psi_{v'} \mathrm{d}R_{\mathrm{nuc}}. \quad [19.32]$$

At low temperatures, most of the absorbing molecules are in their lowest vibrational state. Transitions may then occur to vibrational levels in a higher electronic state for which the first integral in (19.31) does not vanish. The most intense transition occurs to the v' for which $S_{vv'}$ is a maximum. For the potential energy surfaces in figure 19.1, this occurs when $v' = 8$. Weaker transitions to the neighboring vibrational states also occur.

Usually the minimum in the potential lies at a larger bond distance in the excited electronic state. Then some of the transitions carry the system above the dissociation limit,

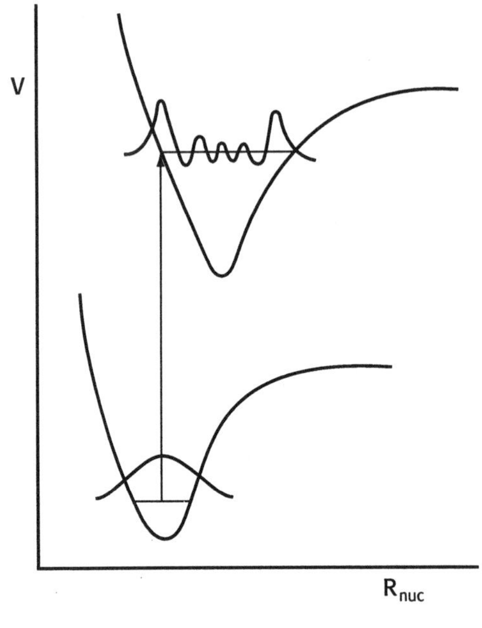

FIGURE 19.1 Electronic transition from the ground state on one potential energy curve to the eighth excited state on a higher curve. For this transition, the overlap between the vibrational eigenfunctions (sketched in) is the greatest.

where the two parts of the molecule fly apart. The resulting translational levels form a continuum. See figure 19.2.

Generally there are several important excited electronic states, each with its potential energy curve. When one of these crosses the excited bound-state curve, as figure 19.3 shows, transition to the unbound-state curve occurs near the crossing point. As a result, there is a blurring of the level or levels about this point.

When crossing is not a problem, the dissociation energies for both the ground and the excited electronic states can be determined from the dissociation limit. Let us employ the notation in figure 19.4. The wave number for the onset of the continuum, as figure 19.2 shows, is labeled $\tilde{\nu}_{\text{limit}}$. We see that

$$\tilde{\nu}_{\text{limit}} = \tilde{\nu}_0 + D_0{}' = D_0{}'' + \Delta\tilde{\nu}_{\text{atomic}}. \qquad [19.33]$$

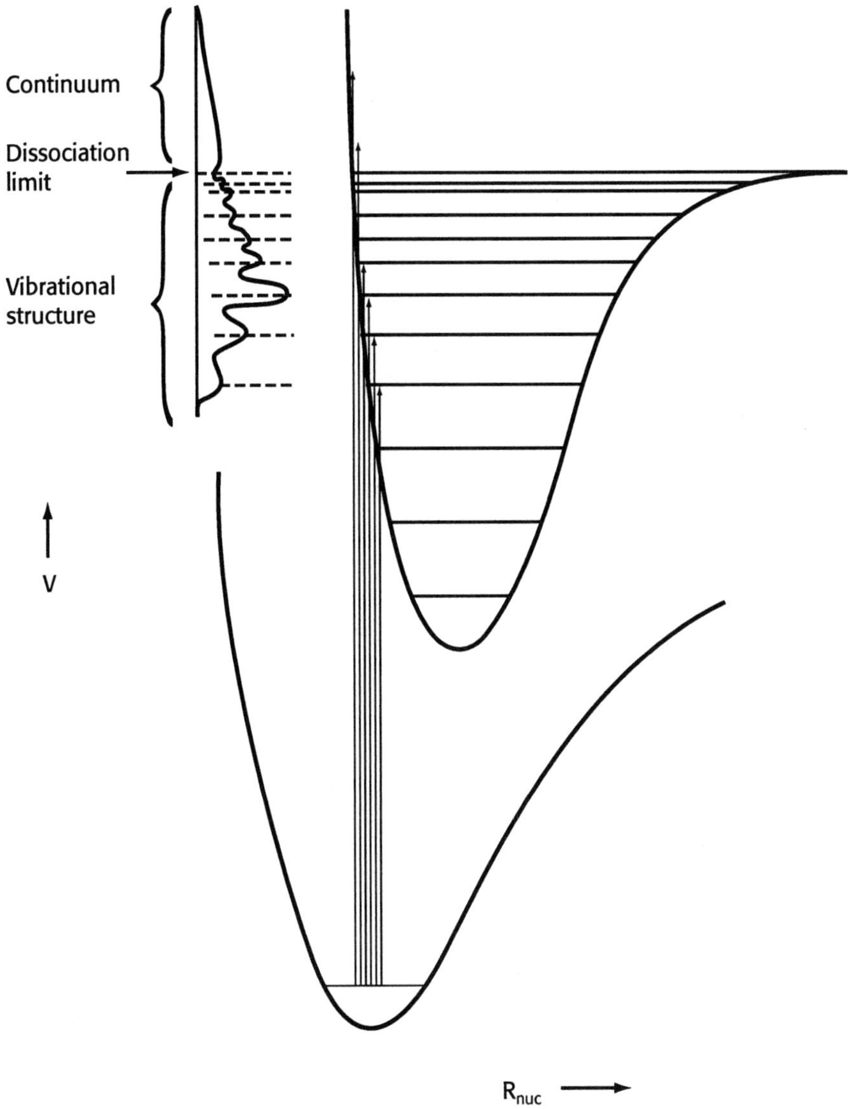

FIGURE 19.2 Transitions from the ground state on one potential energy curve to vibrational and translational (continuum) levels on a higher potential energy curve.

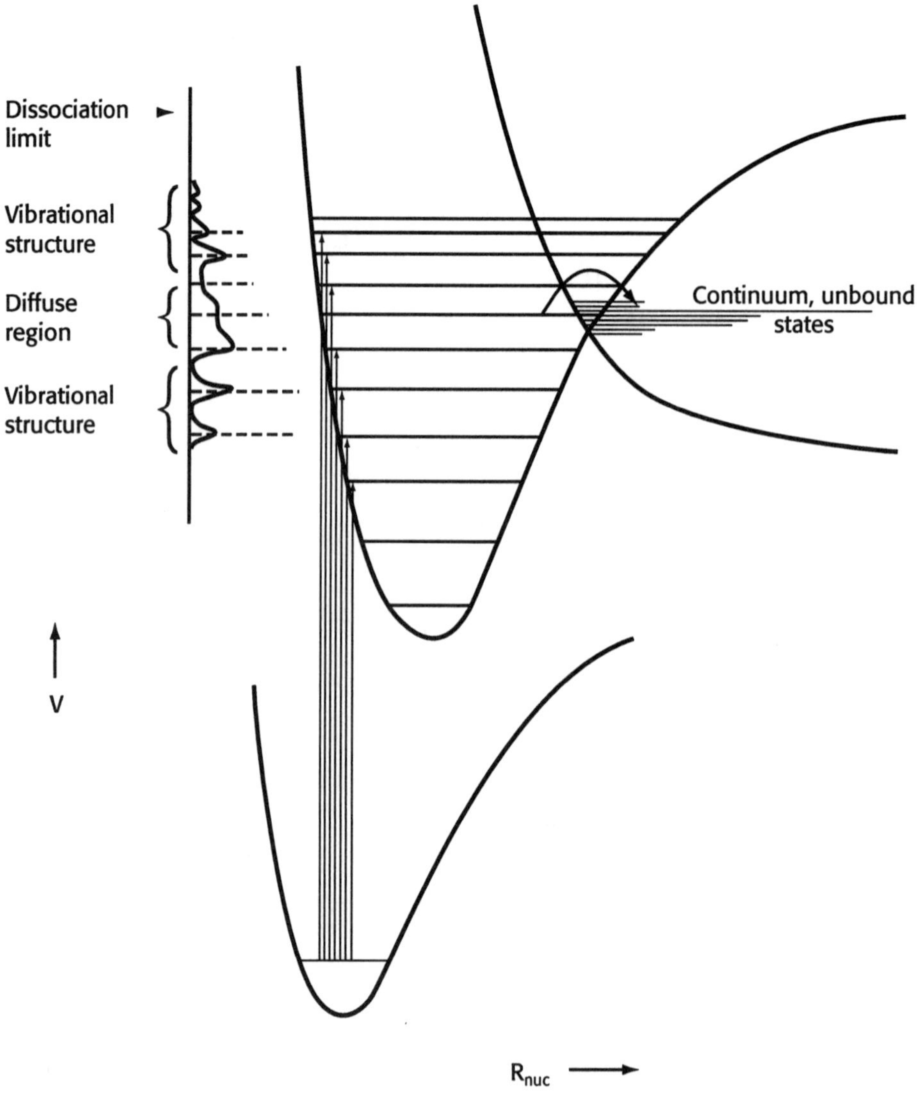

FIGURE 19.3 How dissociation may occur around a point where the potential for an unbound state crosses that for an excited bound state.

The states of the atoms (or molecular fragments) produced from each curve are identified. Their energy difference in wave numbers is labeled $\Delta\tilde{\nu}_{atomic}$. Also, energy $\tilde{\nu}_0$ is obtained from the vibrational spectrum. Then both D_0'' and D_0' can be calculated from the observed $\tilde{\nu}_{limit}$.

19.6 Reaction Possibilities

We have seen how the molecules of a substance may be dissociated on absorbing electromagnetic radiation. The molecular fragments produced include atoms and radicals. In a given composite system, some of these may recombine. Others may combine with reactant molecules to form intermediates and/or products. The intermediates would

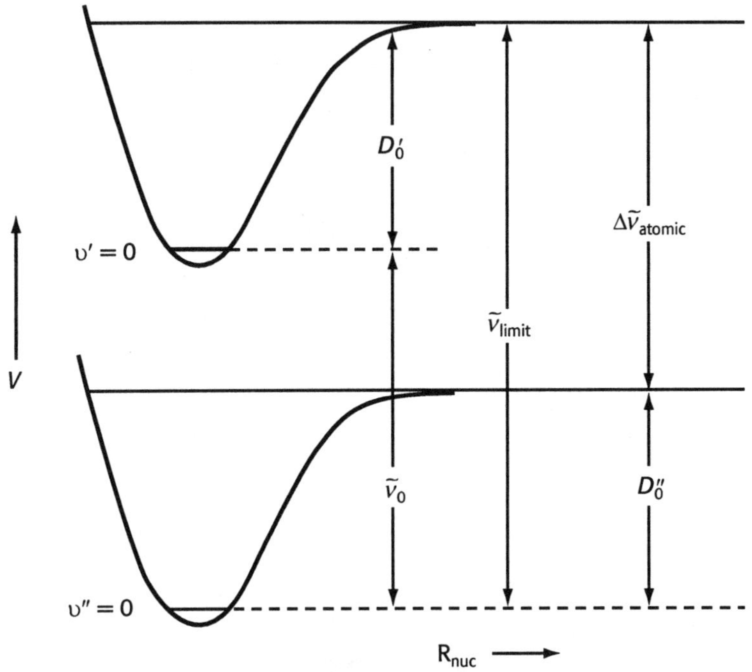

FIGURE 19.4 Breakdown of the dissociation-limit energy into the dissociation energy in the ground state and the atomic (or fragment) energy difference, and into the dissociation energy in the excited state and the energy difference between the ground levels.

react further to yield products. Still others may involve the reactant(s) in a chain process. Consequently, the amount of reaction produced per photon absorbed (or per einstein absorbed) varies widely from system to system.

In the laboratory, a person can determine the energy absorbed from an incident beam by a reacting system. The absorbed energy E is related to the number N of photons absorbed and the einsteins n of photons absorbed by the formula

$$E = Nh\nu = Nhc\tilde{\nu} = nN_A hc\tilde{\nu}. \qquad [19.34]$$

Here h is the Planck constant, ν the conventional frequency, c the speed of light, $\tilde{\nu}$ the wave number, N_A the Avogadro number.

One can also determine the number of moles reacting as a result of the measured absorption. When conditions are fixed, the amount of reaction is proportional to the number of photons absorbed, The ratio

$$\phi = \frac{n_{re}}{n_{ph}}, \qquad [19.35]$$

where n_{re} equals the moles of reaction produced while n_{ph} equals the moles of photons absorbed, is called the *quantum yield*. Some experimental quantum yields are listed in table 19.2.

19.7 The Hydrogen-Iodine Reaction

In the thermal reaction of hydrogen with iodine, the reactive intermediates include I and H. The concentration of H becomes appreciable and mechanism (16.55) - (16.58)

TABLE 19.2 Quantum Yields in Gas Phase Reactions

Stoichiometric Equation	Wavelength, nm	Quantum Yield
$2NH_3 \rightarrow N_2 + 3H_2$	210.0	0.2
$2NO_2 \rightarrow 2NO + O_2$	405.0	0.7
$2H_2O_2 \rightarrow 2H_2O + O_2$	253.7	1.7
$2HI \rightarrow H_2 + I_2$	207.0 - 282.0	2.0
$2HBr \rightarrow H_2 + Br_2$	207.0 - 253.0	2.0
$nC_2H_2 \rightarrow (C_2H_2)_n$	215.0	9
$CO + Cl_2 \rightarrow COCl_2$	400.0 - 436.0	10^3
$CH_4 + Cl_2 \rightarrow CH_3Cl + HCl$	253.6 - 436.0	10^4
$H_2 + Cl_2 \rightarrow 2HCl$	400.0	10^5

contributes significantly only at the higher temperatures. At the lower temperatures, mechanism (16.45) - (16.48) predominates.

Below 550 K, the thermal production of iodine atoms becomes very slow. But then, one can replace this with photochemical production of the iodine atoms.

The most accessible excited electronic potential curve appears as figure 19,2 indicates. In the corresponding absorption spectrum, a vibrational structure extends from 548.3 nm to 499.1 nm. From 499.1 nm on, the spectrum is continuous. Considerably higher in energy are nonbonding molecular states. Transitions to these yield only continuous spectra.

Throughout the continuous spectrum regions, the reaction

$$I_2 + h\nu \xrightarrow{I_a} I + I \qquad [19.36]$$

proceeds, With the low lying potential curve, the dissociation yields one iodine atom in the $^2P_{3/2}$ level and one in the $^2P_{1/2}$ level. The $^2P_{1/2}$ level lies 7603 cm^{-1} above the $^2P_{3/2}$ level.

Reaction (19.36) is followed by

$$I + H_2 + M \rightarrow H_2I + M, \qquad [19.37]$$

$$I + H_2I \rightarrow 2HI. \qquad [19.38]$$

In the approximation that the reverse of (19.37) and reaction (16.48) can be neglected, two HI molecules are produced for each photon absorbed.

If I_a is the local absorbed light intensity per unit path length, expressed as einsteins per unit volume per unit time, then the local rate of reaction is

$$\frac{d[HI]}{dt} = 2I_a. \qquad [19.39]$$

This may be averaged over the reaction space to yield

$$\left(\frac{d[HI]}{dt}\right)_{av} = 2\bar{I}_a. \qquad [19.40]$$

For simplicity, the averaging may be understood and equation (19.40) written in form (19.39).

19.8 The Hydrogen-Bromine Reaction

In the thermal reaction of hydrogen with bromine, reactive intermediates Br and H play the key role; the chain mechanism (16.64) - (16.67) predominates. But below 490 K, thermal production of bromine atoms becomes very slow. Nevertheless, this step can be replaced with photochemical production of the bromine atoms.

In the absorption spectrum of bromine vapor, dissociation first appears at 510.6 nm. Beyond this limit, the reaction

$$Br_2 + h\nu \xrightarrow{I_a} Br + Br \qquad [19.41]$$

goes on. As with iodine, both $^2P_{3/2}$ and $^2P_{1/2}$ atoms are produced. Here the $^2P_{1/2}$ level lies 3685 cm^{-1} above the $^2P_{3/2}$ level.

Between 490 K and 330 K, reaction (19.41) is followed by mechanism (16.65) - (16.67). Furthermore, some Br atoms diffuse to the walls and recombine there:

$$Br + wall \rightarrow Br(wall). \qquad [19.42]$$

This elementary reaction is first order in Br, while (16.67) is second order.

In the neighborhood of room temperature, reactions (16.65) and (16.66) are too slow to explain the rate at which HBr is produced. Then reaction (19.41) is followed by

$$Br + H_2 + M \underset{k_{-2}}{\overset{k_2}{\rightleftarrows}} H_2Br + M, \qquad [19.43]$$

$$Br + H_2Br \xrightarrow{k_3} 2HBr, \qquad [19.44]$$

as in the hydrogen-iodine reaction. But also using up bromine atoms at a considerable rate are the reactions

$$2Br + Br_2 \xrightarrow{k_4} 2Br_2, \qquad [19.45]$$

$$2Br + H_2 \xrightarrow{k_5} H_2 + Br_2. \qquad [19.46]$$

As an approximation, we consider that the concentration of Br is determined by reactions (19.41), (19.45), and (19.46). We also introduce the steady-state approximation. Then we find that

$$\frac{d[HBr]}{dt} = \frac{2k_3KI_a[H_2]}{k_4[Br_2] + k_5[H_2]} \qquad [19.47]$$

where

$$K = \frac{k_2}{k_{-2}}. \qquad [19.48]$$

19.9 The Hydrogen-Chlorine Reaction

In the chain mechanism for the reaction of a halogen with hydrogen, elementary reaction (16.83) is hindered by the energy barrier listed in table 16.1. This decreases markedly on going from iodine to bromine to chlorine. As a consequence, the role played by the chain mechanism increases markedly in that order. So this mechanism governs the photochemical reaction of chlorine with hydrogen, even at low temperatures.

From the absorption spectrum of chlorine, the dissociation limit appears at 478.6 nm. Beyond this limit, the reaction

$$Cl_2 + h\nu \xrightarrow{I_a} Cl + Cl \quad [19.49]$$

proceeds. In the continuum regions, both $^2P_{3/2}$ and $^2P_{1/2}$ atoms appear. The $^2P_{1/2}$ level is 881 cm^{-1} above the $^2P_{3/2}$ level.

Reaction (19.49) is followed by the chain process

$$Cl + H_2 \underset{k_{-2}}{\overset{k_2}{\rightleftarrows}} HCl + H, \quad [19.50]$$

$$H + Cl_2 \xrightarrow{k_3} HCl + Cl. \quad [19.51]$$

Terminating the chain is the reaction

$$Cl + Cl \xrightarrow{k_4} Cl_2. \quad [19.52]$$

With the steady-state approximation, this mechanism yields the rate law

$$\frac{d[HCl]}{dt} = \frac{2k_2k_3(1/k_4)^{1/2}[H_2][Cl_2]I_a^{1/2}}{k_3[Cl_2] + k_{-2}[HCl]}. \quad [19.53]$$

Over a wide range of concentrations, the experimental rate is nearly proportional to the square root of the light intensity at 405.0 nm. Furthermore, the quantum yield of HCl fits the expression

$$\phi = \frac{2.83 \times 10^3 [H_2][Cl_2]}{([Cl_2] + 1.7[HCl])I_a^{0.4}}. \quad [19.54]$$

From (19.53), one would expect the power of I_a in the denominator to be 0.5 rather than 0.4. Otherwise, formulas (19.53) and (19.54) are consistent.

In the presence of oxygen, the reaction

$$H + O_2 + M \rightarrow HO_2 + M \quad [19.55]$$

acts to end the chain. The HO_2 presumably diffuses to a wall and is there destroyed. Since reactions (19.49) and (19.50) must proceed before reaction (19.55), some HCl is produced regardless of how much O_2 is present.

Other inhibitors react with the chlorine atom:

$$Cl + X(+M) \rightarrow ClX(+M). \quad [19.56]$$

Molecule M is needed if X is not complex enough.

Examples of X include ammonia and organic nitrogen compounds, chlorine dioxide, nitric oxide, and ozone. Note that reaction (19.56) can follow (19.49) to the exclusion of the other elementary reactions. The production of HCl is then stopped until the inhibitor is used up. The time during which this occurs is called an *induction period*.

19.10 Photosensitization

The continuum limit in the absorption spectrum of a reactant may lie much higher than the energy needed for dissociation. But then, a substance may be found which will absorb a lower energy photon and pass the energy on to the reactant in a collision, thus causing dissociation. The atom(s) and/or radical(s) produced may then act as reactive intermediate(s) for further reaction. The process is called *photosensitization*; the substance that absorbs the photon, a *sensitizer*.

As an example, the hydrogen molecule does not absorb appreciably at wavelengths longer than 84.9 nm. Bat a 276.9 nm photon carries enough energy to dissociate an H_2 molecule.

There is an intense mercury line at 253.7 nm. Irradiation of mercury vapor with this line from a mercury lamp leads to the reaction

$$\mathrm{Hg}\left({}^1\mathrm{S}_0\right) + h\nu \rightarrow \mathrm{Hg}\left({}^3\mathrm{P}_1\right). \quad [19.57]$$

The excited mercury atom carries enough energy to dissociate a hydrogen molecule.

Let us rewrite reaction (19.57) in the form

$$\mathrm{Hg} + h\nu \rightarrow \mathrm{Hg}^*. \quad [19.58]$$

The excited mercury atom reacts with a hydrogen molecule in two ways:

$$\mathrm{Hg}^* + \mathrm{H}_2 \rightarrow \mathrm{Hg} + 2\mathrm{H}, \quad [19.59]$$

$$\mathrm{Hg}^* + \mathrm{H}_2 \rightarrow \mathrm{HgH} + \mathrm{H}. \quad [19.60]$$

Each hydrogen atom may act as a reactive intermediate for another reaction. Thus when carbon monoxide is also present, the following reactions ensue:

$$\mathrm{H} + \mathrm{CO} \rightarrow \mathrm{HCO}, \quad [19.61]$$

$$\mathrm{HCO} + \mathrm{H}_2 \rightarrow \mathrm{HCHO} + \mathrm{H}, \quad [19.62]$$

$$\mathrm{HCO} + \mathrm{HCO} \rightarrow \mathrm{HCHO} + \mathrm{CO}. \quad [19.63]$$

The atoms that have been employed as sensitizers include Xe, Hg, Zn, Cd, Na. The molecules include SO_2 and $C_6H_5COC_6H_5$. The ions include UO_2^{++}.

Indeed when uranyl oxalate solution is irradiated with 435.0 nm light, the following reactions proceed:

$$\mathrm{UO_2}^{++} + h\nu \rightarrow \left(\mathrm{UO_2}^{++}\right)^*, \quad [19.64]$$

$$\left(\mathrm{UO_2}^{++}\right)^* + \left(\mathrm{COOH}\right)_2 \rightarrow \mathrm{UO_2}^{++} + \mathrm{CO_2} + \mathrm{CO} + \mathrm{H_2O}. \quad [19.65]$$

The observed quantum yield is 0.58.

19.11 Chemiluminescence

In a complex process, one or more of the elementary reactions may produce an excited atom or molecule that loses its excess energy by radiating a photon. Such emission, in excess of the black-body rate, is called *chemiluminescence*. As examples, we have the cold light of a firefly, the glow of certain bacteria, the slow oxidation of phosphorus, the brilliant light of some flames.

Indeed, in the oxidation of carbon monoxide, much more light is emitted than would be if the system were a black body. On studying the emission spectrum, one finds that most of the light comes from a transition of O_2. Furthermore, one photon is produced for about every 125 molecules of CO_2 formed.

19.12 The Carbon Monoxide-Oxygen Reaction

Carbon monoxide reacts with oxygen following the stoichiometric equation

$$2CO + O_2 \rightarrow 2CO_2. \qquad [19.66]$$

At low temperatures, the reaction proceeds as a steady- state thermal reaction. But as the temperature is raised, the reaction becomes explosive. Furthermore, the explosive region is separated from the slow reaction region by a curve like that in figure 16.5.

In a stoichiometric mixture, the second limit appears at 53 torr when the temperature is 687° C and at 213 torr when the temperature is 740° C. Small amounts of water or hydrogen increase the rate of reaction and shift the explosion limits by providing additional chain carriers.

The important elementary reactions probably include for initiation

$$CO + O_2 \xrightarrow{k_1} CO_2 + O, \qquad [19.67]$$

chain propagation,

$$O + O_2 + M \xrightarrow{k_2} O_3 + M, \qquad [19.68]$$

$$O + CO + M \underset{k_{-3}}{\overset{k_3}{\rightleftarrows}} CO_2^* + M, \qquad [19.69]$$

chain branching,

$$O_3 + CO \xrightarrow{k_4} CO_2 + 2O, \qquad [19.70]$$

$$CO_2^* + O_2 \xrightarrow{k_5} CO_2 + 2O, \qquad [19.71]$$

chain breaking,

$$O_3 + CO + M \xrightarrow{k_6} CO_2 + O_2 + M, \qquad [19.72]$$

$$CO_2^* + O_2 \xrightarrow{k_7} CO_2 + O_2^*, \qquad [19.73]$$

$$O_2^* \xrightarrow{k_8} O_2 + h\nu. \qquad [19.74]$$

In the absence of impurities, reaction (19.67) is the most likely chain starting reaction. Reaction (19.68) has been studied separately; it has a very small activation energy. The carbon dioxide produced in reaction (19.69) is electronically excited because of quantum restrictions. Reaction (19.70) is endothermic by 47.7 kJ mole^{-1}, so its activation energy is considerable. Reaction (19.71) is relatively slow because the concentration of CO_2* is low. Reaction (19.72) explains the second explosion limit, an upper limit. As the total pressure is increased with temperature and composition constant, its rate increases faster than the rate of the lower order branching reactions. Reactions (19.73) and (19.74) explain the emission of photons. Direct radiation by CO_2* is forbidden.

Carbon dioxide molecules are produced in reactions (19.67), (19.70), (19.71), (19.72), and (19.73), while photons are produced only in reaction (19.74) following (19.73), Since the ratio of CO_2 molecules to photons produced is roughly 125, the amount of reaction (19.73) is relatively small. The rate of reaction (19.71) is comparable, so it may also be neglected as a source of CO_2.

Furthermore, the chain branching reactions overshadow reaction (19.67). We are left with reactions (19.70) and (19.72), At the *second explosion limit*, the rate of chain carrier production equals the rate of chain carrier destruction; thus

$$k_4[O_3][CO] = k_6[O_3][CO][M] \quad [19.75]$$

whence

$$k_4 = k_6[M]. \quad [19.76]$$

Since CO, O_2, and any added substance X all act as molecule M, expression $k_6[M]$ is the sum of the corresponding terms:

$$k_6[M] = k_{CO}[CO] + k_{O_2}[O_2] + k_X[X]. \quad [19.77]$$

Substituting into equation (19.76) yields

$$k_4 = k_{CO}[CO] + k_{O_2}[O_2] + k_X[X]. \quad [19.78]$$

Experimental results on the second explosion limit support this equation.

Questions

19.1 Summarize the wave properties of a traveling electromagnetic field.
19.2 How may the intensity of the pertinent wave be defined?
19.3 What effect may be employed in measuring this intensity?
19.4 What is the transmittance of a medium through which the wave travels?
19.5 Does this transmittance vary with the wavelength or wave number employed?
19.6 What is a photon?
19.7 How are a photon's energy and momentum related to properties of the pertinent wave?
19.8 What law governs the absorption of photons by macroscopically homogeneous material?
19.9 How is this law altered by the continuum approximation?
19.10 Integrate the resulting equation.
19.11 Define absorbance, molar absorption coefficient.
19.12 How do the absorbencies of different constituents combine in a solution? Explain.
19.13 Why do vibrational and electronic transitions occur over a band of wavelengths?
19.14 How may a formula for the integral of ε over a band be constructed?
19.15 What is the oscillator strength of a solute?
19.16 What is the best measure of a band's intensity?
19.17 Describe the transition dipole moment of a molecule. How does this moment determine the corresponding oscillator strength?
19.18 What rule governs the intensity of an electronic transition?
19.19 Explain what photon energies cause dissociation of a diatomic molecule.
19.20 How can photochemical dissociation lead to reaction?
19.21 What is quantum yield? May this vary with the wavelength absorbed?
19.22 Outline the mechanism of the photochemical reaction of hydrogen with iodine.
19.23 Construct the corresponding rate law.
19.24 Why does the local absorbed light intensity per unit path length vary along the path?
19.25 How may one describe the net effect throughout the reacting volume?
19.26 Outline the mechanism of the photochemical reaction of hydrogen with bromine.
19.27 Construct the corresponding rate law.
19.28 Describe the photochemical reaction of hydrogen with chlorine.
19.29 What is (a) photosensitization, (b) chemiluminescence?
19.30 Outline the mechanism for the reaction of carbon monoxide with oxygen.
19.31 Where does chemiluminescence appear in the sequence of reactions?
19.32 What does the ratio of CO_2 molecules to photons produced tell us?
19.33 What reactions explain the second explosion limit?

Problems

19.1 When a hydrogen atom absorbs a 10.21 eV photon, what fraction of this energy appears as recoil energy?

19.2 If the transmittance of a given solution is 11.8 percent in a cell 5.00 cm thick, what is its transmittance when it is studied in a 1.00 cm cell at the same wavelength?

19.3 Use the molar absorption coefficient obtained in example 19.4 to calculate the transmittance of a 0.0340 M C_6H_6 solution through a cell 2.60 mm thick at wavelength 256 nm.

19.4 The complex between ferric and tiron ions dissociates following the equation

$$FeR_3^{3-} \rightleftharpoons Fe^{3+} + 3R^{=}.$$

The absorbance of a solution in which the initial concentration of Fe^{3+} was 6×10^{-5} M and that of $R^{=}$ 18×10^{-5} M was 0.25. When the initial concentration of Fe^{3+} was 6×10^{-5} M while a large excess of $R^{=}$ was added, the absorbance was 0.35. All measurements were made in the same cell at 480 nm where only FeR^{3-} absorbs appreciably. Calculate the degree of dissociation of the complex in the first solution, α, and the equilibrium constant K_c.

19.5 The absorption band in example 19.4 is centered about 256 nm with a width at half maximum of 4000 cm^{-1}. Calculate its oscillator strength.

19.6 With the quantum yield from table 19.2, calculate the joules of 220.0 nm radiation needed to decompose 0.400 mol HBr.

19.7 The absorption spectrum of iodine vapor becomes continuous at 499.1 nm. But photodissociation of I_2 in this region yields one $^2P_{3/2}$ atom and one $^2P_{1/2}$ atom. If the excess energy of the latter is 7603 cm^{-1}, what is ΔE for the dissociation of I_2 into two ground state atoms?

19.8 In the photolysis of HI with 253.7 nm wavelength light, absorption of 307 J energy decomposed 1.300×10^{-3} mol HI. Calculate the corresponding quantum yield.

19.9 The photolysis of hydrogen iodide follows the mechanism:

$$\begin{aligned} HI + h\nu &\xrightarrow{I_a} H + I, \\ H + HI &\xrightarrow{k_2} H_2 + I, \\ I + I + M &\xrightarrow{k_3} I_2 + M, \\ H + I_2 &\xrightarrow{k_4} HI + I. \end{aligned}$$

Construct the rate law.

— — —

19.10 If the transmittance of 0.100 M A is 0.862 in a given absorption cell at a given wavelength, what is the transmittance of 0.500 M A under the same conditions?

19.11 Absorption of radiation by solid germanium abruptly becomes large when the wavelength is lowered below 1.8 μm. At the same time, the conductivity rises sharply. How much energy is needed to excite an electron from the valence band across the energy gap to the conduction band of Ge?

19.12 An investigator kept the initial thiocyanate ion concentration at 0.000050 M while increasing the bismuth ion concentration in the equilibrium

$$BiSCN^{++} \rightleftharpoons Bi^{3+} + SCN^{-}.$$

The absorbance at a given wavelength, of the solution, in a cell 1.00 cm thick approached 0.286. What was the absorption coefficient of the $BiSCN^{++}$?

19.13 When the initial bismuth concentration in problem 19.12 was 0.500 M, the absorbance was 0. 240. Calculate equilibrium constant K_c for the dissociation of $BiSCN^{++}$.

19.14 Azoethane has a peak absorbency of 9.6 M cm^{-1} at 28,000 cm^{-1} with a width at half maximum of 4900 cm^{-1}. Calculate its oscillator strength.

19.15 A cell was filled with HI gas and illuminated with 253.7 nm radiation for 15 min. Then all gases except H_2 were condensed out. If the hydrogen left filled 2.5 cm^3 at 25° C and 15 torr pressure, how much radiation was absorbed?

19.16 (a) If the quantum yield in the photodecomposition of gaseous NOCl into NO and Cl_2 is 2, what is a possible mechanism? (b) What additional reaction would lower the quantum yield?

19.17 Derive formula (19.53).

19.18 For the stoichiometric reaction

$$2COCl_2 + O_2 \rightarrow 2CO_2 + 2Cl_2$$

the photochemical rate law is

$$\frac{d[CO_2]}{dt} = \frac{2I_a}{1 + k'[Cl_2]/[O_2]}.$$

Construct a possible mechanism.

References

Books

Atkins, P. W.: 1990, *Physical Chemistry*, 4th edn, W. H. Freeman and Co., New York, pp. 500-534.

In chapter 17, Atkins deals with various electronic transitions in an introductory manner.

Bamford, C. H., and Tipper, C. F. H. (editors), *Comprehensive Chemical Kinetics*, continuing series, Elsevier Publishing Co., Amsterdam (abbreviated BT). The following sections are pertinent:

Bauman, R. P.: 1962, *Absorption Spectroscopy*, John Wiley & Sons, Inc., New York, pp. 364-433.

In this book, Bauman emphasizes the practical side of the subject. Chapter IX starts with a statement of the continuum form of the law of absorption, the Bouguer-Beer law. This is followed by a wide range of applications to chemical systems.

Burton, C. S., and Noyes, W. A., Jr.: 1969, "Effect of Low Energy Radiation," in BT vol. 3, pp. 1-66.

The authors survey the primary effects of electromagnetic radiation on atoms, diatomic molecules, and polyatomic molecules from the kinetic chemists' standpoint.

Carrington, T., and Garvin, D.: 1969, "The Chemical Production of Excited States," in BT vol. 3, pp. 107-181.

The authors discuss the production of excited forms of gaseous molecules through chemical reaction. These may lose their excess energy by radiation or by inelastic collision.

Dixon-Lewis, G., and Williams, D. J.: 1977, "The Oxidation of Hydrogen and Carbon Monoxide," in BT vol. 17, pp. 174-222.

The combustion of carbon monoxide is less well defined than that of hydrogen. Nevertheless, much is known and summarized here. Many references to the journal literature are listed.

White, J. M.: 1972, "Other Reactions Involving Halogen, Nitrogen, and Sulfur Compounds," in BT vol. 6, pp. 201-224.

The photochemical hydrogen-iodine, hydrogen-bromine, and hydrogen-chlorine reactions are discussed critically, with many references to the journal literature.

Articles

Adamson, A. W.: 1983, "Properties of Excited States," *J. Chem. Educ.* **60**, 797-802.

Ahmed, F.: 1987, "A Good Example of the Franck-Condon Principle," *J. Chem. Educ.* **64**, 427-428.

Alvarez J., G., and Roca R. M., S.: 1996, ".A Simple Algorithm for the Distinction of Reaction Mechanisms Based on the Measurement of Additive Properties," *J. Chem. Educ.* **73**, 214-216.

Berberau-Santos, M. N.: 1990, "Beer's Law Revisited," *J. Chem. Educ.* **67**, 757.

Bruneau, E., Lavabre, D., Levy, G., and Micheau, J. C.: 1992, "Quantitative Analysis of Continuous-Variation Plots with a Comparison of Several Methods: Spectrophotometric Study of Organic and Inorganic 1:1 Stoichiometry Complexes," *J. Chem. Educ.* **69**, 833- 837.

Dunbrack Jr., R. T..: 1986, "Calculation of Franck-Condon Factors for Undergraduate Quantum Chemistry," *J. Chem. Educ.* **63**, 953-955.

Fleming, G. R.: 1986, "Subpicosecond Spectroscopy," *Annu. Rev. Phys. Chem.* **37**, 81-104.

Grossman, W. E. L.: 1993, "The Optical Characteristics and Production of Diffraction Gratings," *J. Chem. Educ.* **70**, 741-748.

Harris, A. L., Brown, J. K., and Harris, C. B.: 198~3, "The Nature of Simple Photodissociation Reactions in Liquids on Ultrafast Time Scales," *Annu. Rev. Phys. Chem.* **39**, 341-366.

Hirota, E., and Kawaguchi, K.: 1985, "High Resolution Infrared Studies of Molecular Dynamics," *Annu. Rev. Phys. Chem.* **36**, 53-76.

Horvath, O., and Papp, S.: 1988, "Complex Equilibria Changing in Photochemical Reaction: Computerized Evaluation and Simulation," *J. Chem. Educ.* **65**, 1102-1105.

Kovalenko, L. J., and Leone, S. R.: 1988, "Innovative Laser Techniques in Chemical Kinetics," *J. Chem. Educ.* **65**, 681-687.

Lee, S. -Y.: 1985, "Potential Energy Surfaces and Effects on Electronic and Raman Spectra," *J. Chem. Educ.* **62**, 561- 566.

Logan, S. R.: 1990, "Spatial Inhomogeneity Effects in Photochemical Kinetics," *J. Chem. Educ.* **67**, 872-875.

Lykos, P.: 1992, "The Beer-Lambert Law Revisited: A Development without Calculus," *J. Chem. Educ.* **69**, 730-732.

Martin, X. B.: 1997, "Disadvantages of Double Reciprocal Plots," *J. Chem. Educ.* **74**, 1238-1240.

McDevitt, J. T.: 1984, "Photoelectrochemical Solar Cells," *J. Chem. Educ.* **61**, 217-221.

Meyer, G. J.: 1997, "Efficient Light-to-Electrical Energy Conversion: Nanocrystalline TiO_2 Films Modified with Inorganic Sensitizers," *J. Chem. Educ.* **74**, 652-656.

Porter, O. B.: 1983, "Introduction to Inorganic Photochemistry: Principles and Methods," *J. Chem. Educ.* **60**, 785- 790.

Reisler, H., and Wittig, C.: 1986, "Photo-Initiated Unimolecular Reactions," *Annu. Rev. Phys. Chem.* **37**, 307-349.

Schanze, K. S., and Schmehl, R. H.: 1997, "Applications of Inorganic Photochemistry in the Chemical and Biological Sciences— Contemporary Developments," *J. Chem. Educ.* **74**, 633-635.

Seltzer, M. D.: 1995, "Interpretation of the Emission Spectra of Trivalent Chromium-Doped Garnet Crystals Using Tanabe-Sugano Diagrams," *J. Chem. Educ.* **72**, 886-888.

Sheehy, B., and Di Mauro, L. F.: 1996, "Atomic and Molecular Dynamics in Intense Optical Fields," *Annu. Rev. Phys. Chem.* **47**, 463- 494.

Sturm, J. E.: 1990, "Grid of Expressions Related to the Einstein Coefficients," *J. Chem. Educ.* **67**, 32-33.

Tyler, D. R.: 1997, "Organometallic Photochemistry: Basic Principles and Applications to Materials Chemistry," *J. Chem. Educ.* **74**, 668-672.

Yan, S. G., Lyon, L. A., Lemon, B. I., Preiskorn, J. S., and Hupp, J. T.: 1997, "Energy Conversion Chemistry: Mechanisms of Charge Transfer at Metal-Oxide Semiconductor/Solution Interfaces," *J. Chem. Educ.* **74**, 657-662.

Zilio, S. C., and Bagnato, V. S.: 1989, "Radiative Forces on Neutral Atoms—A Classical Treatment," *Am. J. Phys.* **57**, 471-474.

Glossary

Anisotropic Having the property under consideration vary with direction from a typical point in the given material

Base For numbers, a reference number, such as 10 or e.

For vectors, a set of reference unit vectors, such as **i**, **j**, and **k**.

For functions, a set of reference normalized functions, such as the s, p, and d valence orbitals for a given atom.

For operators, a set of reference operators that combine linearly to yield any pertinent operator.

Coefficient A multiplying numerical factor.

Continuum Approximation Replacing discrete changes by continuous changes in a calculation. Thus replacing a relationship involving increments by one involving the corresponding differentials.

Cosine Ratio of side adjacent the pertinent angle of a right triangle to the hypotenuse. Thus it represents the projection of the hypotenuse in the direction of this side divided by the magnitude of the hypotenuse.

Density Mass per unit volume about a point or within a specified volume. May also apply to a concentration in entities per unit volume, as with electron density, molecule density, or ion density.

Differential An infinitesimally small quantity. When it is an infinitesimal change in a function of the independent variables, it is exact.

Dipole Moment Displacement of the center of the pertinent positive charge from the center of the pertinent negative charge. Measured by the magnitude of the positive charge times the displacement of its center from that of the negative charge.

Disjoint Completely independent, with no overlap. Applies to sets with no common elements.

Distribution Law Describes how a property varies with the pertinent independent variables. May be expressed as a probability density or frequency.

e The base of the natural system of logarithms. Thus, the limit of $(1 + 1/n)^n$ as n increases without limit.

Electric Current A flow of charged particles producing a net transport of charge. In a metal, the valence electrons flow. In an electrolyte, both cations and anions travel. In a plasma, some ion flow accompanies the electron flow.

Empirical Based on experiment rather than on theory.

Energy A conserved quantity that can be transferred as work or heat. For a mechanical system, it is subdivided into kinetic energy and potential energy.

Exponential Variation Rate of change in the dependent variable proportional to minus (or plus) its value at a given time.

Generalized Coordinate A variable expressing an aspect of the configuration of a system. Its changes coordinate with a generalized force to make work.

Gradient Spatial rate of change of a property. It is a vector since such a rate varies not only with position but also with the direction chosen.

Isotropic Having the same pertinent properties in all directions from a typical point in the given material.

Lattice A regular arrangement of structural units, such as atoms, ions, molecules, in space.

Linear For a scale, the distance from the origin is made proportional to the pertinent variable.

For a variable dependent on another, a plot with linear scales appears as a straight line.

Logarithm Of a number, the power to which the base must be raised to give the number.

Logarithmic For a scale, the distance from the origin is made proportional to the logarithm of the pertinent variable.

For a variable dependent on another variable, the dependent one varies directly with the logarithm of the independent one.

Matrix A rectangular array of numbers obeying the rule for matrix multiplication.

Matrix Element For matrix **A**, number A_{jk} where ordinal j identifies the row, k identifies the column, where it would appear in the array. The subscript for the row (or column) may be omitted when there is only one row (or column).

Matrix Multiplication For **C** = **AB**, we have $C_{jl} = \sum_k A_{jk}B_{kl}$.

Operator A mathematical entity that acts to convert various functions of the independent variables to different functions of these variables.

Orbital The function describing a possible state, apart from spin, for an electron in an atom or molecule.

Phase For a material, a given state of aggregation.

For a function, a particular stage in its variation.

Sine Ratio of side opposite the pertinent angle of a right triangle to the hypotenuse.

Singularity A point where the function under consideration, or one of its derivatives, becomes infinite or not uniquely defined.

Sinusoidal Expressible as an amplitude times the sine of an independent variable plus or minus a phase angle (a constant).

Space The three dimensional continuum in which all physical and chemical processes take place.

Spectrum A record of a distribution or the distribution itself. Thus, it may apply to the intensity of radiation over wavelength, frequency, or energy.

Spin The intrinsic angular momentum of a particle, leading to quantization of its magnetic moment.

State Function A complex function describing the quantum mechanical state of a system. Often called a wave function even though nothing identifiable is waving.

Submicroscopic Below the range of an optical or ultraviolet microscope. At the molecular or formula unit level.

Vector An entity that can be represented by a directed line segment. Thus it possesses both a magnitude and a direction.

Vector Addition Performed by adding the corresponding line segments. On an initial point, place the tail of the first vector. Then place the tail of the second one on the head of the first, and so on. The resultant runs from the initial point to the position of the final head.

Appendix 1
Fundamental Units

Physical Quantity	*Unit*	*Definition*
Time t	second s	A second is the duration of 9 192 631 770 cycles of the radiation from the transition between two hyperfine levels of the ground state of ^{133}Cs (originally, 1/86 400 of the mean solar day).
Length l	meter m	A meter is the distance traveled by light in free space during 1/299 792 458 second (originally, 10^{-7} of a meridean quadrant of the earth).
Mass m	kilogram kg	A kilogram is the mass of a cylinder of platinum-iridium alloy kept in a vault at Sevres, France by the International Bureau of Weights and Measures (originally the mass of one cubic centimeter of water at 4° C).
Electrical current I	ampere A	An ampere is the current which if maintained in two straight parallel conductors one meter apart in free space lead to a force between the conductors of 2×10^{-7} N m^{-1}, where end effects are negligible. In practice, a current balance, exploiting the force between current carrying coils, is employed.
Temperature T	kelvin K	A kelvin is 1/273.16 of the temperature of the triple point of water (originally, 1/100 of the distance from the freezing point to the normal boiling point of water as measured with a mercury-in glass thermometer).
Chemical amount of substance n	mole mol	A mole is the amount of substance that contains the same number of elementary units as there are in 12.0000 g of pure ^{12}C.

Appendix 2
Derived Units

Physical Quantity	*Unit*	*Definition*
Force $\mathcal{F}$	newton N	kg m s^{-2}
Work w	joule J	N m
Power $\mathcal{P}$	watt W	J s^{-1}
Pressure P	pascal Pa	N m^{-2}
Electric charge Q	coulomb C	A s
Electric potential $\mathcal{E}$	volt V	J C^{-1}
Electric resistance R	ohm Ω	V A^{-1}
Electric capacitance C	farad F	C V^{-1}
Frequency ν	hertz Hz	cycles s^{-1}
Magnetic flux density B	tesla T	N A^{-1} m^{-1}

Appendix 3
Physical Constants

Quantity	*Symbol*	*Value*
Speed of light	c	$2.997\,924\,58 \times 10^{8}$ m s^{-1}
Planck constant	h	$6.626\,075 \times 10^{-34}$ J s
Planck constant reduced	$\hbar = h/2\pi$	$1.054\,572\,7 \times 10^{-34}$ J s
Elementary charge	e	$1.602\,177\,3 \times 10^{-19}$ C
Electron mass	m_e	$9.109\,390 \times 10^{-31}$ kg
Proton mass	m_p	$1.672\,623 \times 10^{-27}$ kg
Neutron mass	m_n	$1.674\,929 \times 10^{-27}$ kg
Atomic mass unit	u	$1.660\,540 \times 10^{-27}$ kg
Avogadro number	N_A	$6.022\,137 \times 10^{23}$ mol^{-1}
Gas constant	R	$8.314\,51$ J K^{-1} mol^{-1}
Boltzmann constant	k_B	$1.380\,66 \times 10^{-23}$ J K^{-1}
Permittivity of free space	ε_0	$(4\pi c^2 \times 10^{-7})^{-1}$ A^{2}s^{2}N^{-1}m^{-2} = $8.854\,197\,817 \times 10^{-12}$ A^{2}s^{2}N^{-1}m^{-2}
Permeability of free space	μ_0	$4\pi \times 10^{-7}$ N A^{-2} = $12.566\,370\,614 \times 10^{-7}$ N A^{-2}

References

Books

Mills, I., Cvitas, T., Homann, Kl, Kallay, N., and Kuchitsu, K.: 1993, *Quantities, Units and Symbols in Physical Chemistry*, 2nd edn., Blackwell Scientific Publications, London, pp. 3-149.

This book presents the recommendations of the Commission on Physicochemical Symbols, Terminology, and Units of the International Union of Pure and Applied Chemistry, Physical Chemistry Division. Properties of particles, elements, and nuclides are listed together with the 1986 revision of the fundamental physical constants.

Articles

Particle Data Group: 1996, "Review of Particle Properties," *Phys. Rev.* D **54**, 65-76.

In the section cited, physical constants, SI units, atomic and nuclear properties of materials, and structures of elements are tabulated.

Appendix 4

Primitive Character Vectors and Basis Functions for Important Symmetry Groups

The symbol in the upper left corner of each table of the set identifies the symmetry group described. Each letter under it identifies a primitive symmetry species of the group. The numbers in a row are the components of the corresponding character vector. The heading for each column of numbers labels the pertinent class by a typical operation and states the number of covering operations therein. The functions and rotation operators in a row belong to the primitive symmetry species for the row. They form bases for the corresponding irreducible representation of the group.

C_1	I				
A	1	x, y, z	R_x, R_y, R_z	x^2, y^2, z^2	xy, yz, zx

S_1	I	σ_h				
A'	1	1	x, y	R_z	x^2, y^2, z^2	xy
A''	1	-1	z	R_x, R_y		yz, zx

S_2	I	i				
A_g	1	1		R_x, R_y, R_z	x^2, y^2, z^2	xy, yz, zx
A_u	1	-1	x, y, z			

S_4	I	S_4	C_2	S_4^3			
A	1	1	1	1		R_z	$x^2 + y^2, z^2$
B	1	-1	1	-1	z		$x^2 - y^2, xy$
C	1	i	-1	$-i$	$x - iy$	$R_x + iR_y$	$(x + iy)z$
D	1	$-i$	-1	i	$x + iy$	$R_x - iR_y$	$(x - iy)z$

S_6	I	C_3	C_3^2	i	S_6^5	S_6			
A_g	1	1	1	1	1	1		R_z	$x^2 + y^2, z^2$
C_g	1	ω	ω^2	1	ω	ω^2		$R_x - iR_y$	$(x + iy)^2, (x - iy)z$
D_g	1	ω^2	ω	1	ω^2	ω		$R_x + iR_y$	$(x - iy)^2, (x + iy)z$
A_u	1	1	1	-1	-1	-1	z		
C_u	1	ω	ω^2	-1	$-\omega$	$-\omega^2$	$x - iy$		
D_u	1	ω^2	ω	-1	$-\omega^2$	$-\omega$	$x + iy$		

$\omega = \exp(2\pi i/3)$

C_2	I	C_2			
A	1	1	z	R_z	x^2, y^2, z^2, xy
B	1	-1	x, y	R_x, R_y	xz, yz

C_3	I	C_3	C_3^2			
A	1	1	1	z	R_z	$x^2 + y^2, z^2$
C	1	ω	ω^2	$x - iy$	$R_x - iR_y$	$(x + iy)^2, (x - iy)z$
D	1	ω^2	ω	$x + iy$	$R_x + iR_y$	$(x - iy)^2, (x + iy)z$

$\omega = \exp(2\pi i/3)$

C_4	I	C_4	C_2	C_4^2			
A	1	1	1	1	z	R_z	$x^2 + y^2, z^2$
B	1	-1	1	-1			$x^2 - y^2, xy$
C	1	i	-1	$-i$	$x - iy$	$R_x - iR_y$	$(x - iy)z$
D	1	$-i$	-1	i	$x + iy$	$R_x + iR_y$	$(x + iy)z$

C_5	I	C_5	C_5^2	C_5^3	C_5^4			
A	1	1	1	1	1	z	R_z	x^2+y^2, z^2
C_1	1	ω	ω^2	ω^3	ω^4	$x - iy$	$R_x - R_y$	$(x-iy)z$
D_1	1	ω^4	ω^3	ω^2	ω	$x + iy$	$R_x + R_y$	$(x+iy)z$
C_2	1	ω^2	ω^4	ω	ω^3			$(x+iy)^2$
D_2	1	ω^3	ω	ω^4	ω^2			$(x-iy)^2$

$\omega = \exp(2\pi i/5)$

C_6	I	C_6	C_3	C_2	C_3^2	C_6^5			
A	1	1	1	1	1	1	z	R_z	x^2+y^2, z^2
B	1	-1	1	-1	1	-1			
C_1	1	$-\omega^2$	ω	-1	ω^2	$-\omega$	$x - iy$	$R_x - iR_y$	$(x-iy)z$
D_1	1	$-\omega$	ω^2	-1	ω	$-\omega^2$	$x + iy$	$R_x + iR_y$	$(x+iy)z$
C_2	1	ω	ω^2	1	ω	ω^2			$(x-iy)^2$
D_2	1	ω^2	ω	1	ω^2	ω			$(x+iy)^2$

$\omega = \exp(2\pi i/3)$

C_{2v}	I	C_2	$\sigma_v(zx)$	$\sigma_v(yz)$			
A_1	1	1	1	1	z		x^2, y^2, z^2
A_2	1	1	-1	-1		R_z	xy
B_1	1	-1	1	-1	x	R_y	xz
B_2	1	-1	-1	1	y	R_x	yz

C_{3v}	I	$2C_3$	$3\sigma_v$			
A_1	1	1	1	z		x^2+y^2, z^2
A_2	1	1	-1		R_z	
E	2	-1	0	(x, y)	(R_x, R_y)	$(x^2-y^2, xy), (xz, yz)$

C_{4v}	I	$2C_4$	C_2	$2\sigma_v$	$2\sigma_d$			
A_1	1	1	1	1	1	z		$x^2 + y^2, z^2$
A_2	1	1	1	-1	-1		R_z	
B_1	1	-1	1	1	-1			$x^2 - y^2$
B_2	1	-1	1	-1	1			xy
E	2	0	-2	0	0	(x, y)	(R_x, R_y)	(xz, yz)

C_{5v}	I	$2C_5$	$2C_5^2$	$5\sigma_v$			
A_1	1	1	1	1	z		$x^2 + y^2, z^2$
A_2	1	1	1	-1		R_z	
E_1	2	2 cos 72°	2 cos 144°	0	(x, y)	(R_x, R_y)	(xz, yz)
E_2	2	2 cos 144°	2 cos 72°	0			$(x^2 - y^2, xy)$

C_{6v}	I	$2C_6$	$2C_3$	C_2	$3\sigma_v$	$3\sigma_d$			
A_1	1	1	1	1	1	1	z		$x^2 + y^2, z^2$
A_2	1	1	1	1	-1	-1		R_z	
B_1	1	-1	1	-1	1	-1			
B_2	1	-1	1	-1	-1	1			
E_1	2	1	-1	-2	0	0	(x, y)	(R_x, R_y)	(xz, yz)
E_2	2	-1	-1	2	0	0			$(x^2 - y^2, xy)$

C_{2h}	I	C_2	i	σ_h			
A_g	1	1	1	1		R_z	x^2, y^2, z^2, xy
B_g	1	-1	1	-1		R_x, R_y	xz, yz
A_u	1	1	-1	-1	z		
B_u	1	-1	-1	1	x, y		

C_{3h}	I	C_3	C_3^2	σ_h	S_3	S_3^5			
A'	1	1	1	1	1	1		R_z	$x^2 + y^2, z^2$
C'	1	ω	ω^2	1	ω	ω^2	$x - iy$		$(x + iy)^2$
D'	1	ω^2	ω	1	ω^2	ω	$x + iy$		$(x - iy)^2$
A''	1	1	1	-1	-1	-1	z		
C''	1	ω	ω^2	-1	$-\omega$	$-\omega^2$		$R_x - iR_y$	$(x - iy)z$
D''	1	ω^2	ω	-1	$-\omega^2$	$-\omega$		$R_x + iR_y$	$(x + iy)z$

$\omega = \exp(2\pi i/3)$

C_{4h}	I	C_4	C_2	C_4^3	i	S_4^3	σ_h	S_4			
A_g	1	1	1	1	1	1	1	1		R_z	$x^2 + y^2, z^2$
B_g	1	-1	1	-1	1	-1	1	-1			$x^2 - y^2, xy$
C_g	1	i	-1	$-i$	1	i	-1	$-i$		$R_x - iR_y$	$(x - iy)z$
D_g	1	$-i$	-1	i	1	$-i$	-1	i		$R_x + iR_y$	$(x + iy)z$
A_u	1	1	1	1	-1	-1	-1	-1	z		
B_u	1	-1	1	-1	-1	1	-1	1			
C_u	1	i	-1	$-i$	-1	$-i$	1	i	$x - iy$		
D_u	1	$-i$	-1	i	-1	i	1	$-i$	$x + iy$		

D_2	I	$C_2(z)$	$C_2(y)$	$C_2(x)$			
A	1	1	1	1			x^2, y^2, z^2
B_1	1	1	-1	-1	z	R_z	xy
B_2	1	-1	1	-1	y	R_y	zx
B_3	1	-1	-1	1	x	R_x	yz

D_3	I	$2C_3$	$3C_2$			
A_1	1	1	1		$x^2 + y^2, z^2$	
A_2	1	1	-1	z	R_z	
E	2	-1	0	(x, y)	(R_x, R_y)	$(x^2 - y^2, xy), (xz, yz)$

D_4	I	$2C_4$	C_2	$2C_2'$	$2C_2''$			
A_1	1	1	1	1	1			$x^2 + y^2, z^2$
A_2	1	1	1	-1	-1	z	R_z	
B_1	1	-1	1	1	-1			$x^2 - y^2$
B_2	1	-1	1	-1	1			xy
E	2	0	-2	0	0	(x, y)	(R_x, R_y)	(xz, yz)

D_5	I	$2C_5$	$2C_5^2$	$5C_2$			
A_1	1	1	1	1			$x^2 + y^2, z^2$
A_2	1	1	1	-1	z	R_z	
E_1	2	2 cos 72°	2 cos 144°	0	(x, y)	(R_x, R_y)	(xz, yz)
E_2	2	2 cos 144°	2 cos 72°	0			$(x^2 - y^2, xy)$

D_6	I	$2C_6$	$2C_3$	C_2	$3C_2'$	$3C_2''$			
A_1	1	1	1	1	1	1			$x^2 + y^2, z^2$
A_2	1	1	1	1	-1	-1	z	R_z	
B_1	1	-1	1	-1	1	-1			
B_2	1	-1	1	-1	-1	1			
E_1	2	1	-1	-2	0	0	(x, y)	(R_x, R_y)	(xz, yz)
E_2	2	-1	-1	2	0	0			$(x^2 - y^2, xy)$

D_{2h}	I	$C_2(z)$	$C_2(y)$	$C_2(x)$	i	$\sigma(xy)$	$\sigma(zx)$	$\sigma(yz)$			
A_g	1	1	1	1	1	1	1	1			x^2, y^2, z^2
B_{1g}	1	1	-1	-1	1	1	-1	-1		R_z	xy
B_{2g}	1	-1	1	-1	1	-1	1	-1		R_y	zx
B_{3g}	1	-1	-1	1	1	-1	-1	1		R_x	yz
A_u	1	1	1	1	-1	-1	-1	-1			
B_{1u}	1	1	-1	-1	-1	-1	1	1	z		
B_{2u}	1	-1	1	-1	-1	1	-1	1	y		
B_{3u}	1	-1	-1	1	-1	1	1	-1	x		

D_{3h}	I	$2C_3$	$3C_2$	σ_h	$2S_3$	$3\sigma_v$			
A'_1	1	1	1	1	1	1			x^2+y^2, z^2
A'_2	1	1	-1	1	1	-1		R_z	
E'	2	-1	0	2	-1	0	(x, y)		
A''_1	1	1	1	-1	-1	-1			
A''_2	1	1	-1	-1	-1	1	z		
E''	2	-1	0	-2	1	0		(R_x, R_y)	(xz, yz)

D_{4h}	I	$2C_4$	C_2	$2C_2'$	$2C_2''$	i	$2S_4$	σ_h	$2\sigma_v$	$2\sigma_d$			
A_{1g}	1	1	1	1	1	1	1	1	1	1			x^2+y^2, z^2
A_{2g}	1	1	1	-1	-1	1	1	1	-1	-1		R_z	
B_{1g}	1	-1	1	1	-1	1	-1	1	1	-1			x^2-y^2
B_{2g}	1	-1	1	-1	1	1	-1	1	-1	1			xy
E_g	2	0	-2	0	0	2	0	-2	0	0		(R_x, R_y)	(xz, yz)
A_{1u}	1	1	1	1	1	-1	-1	-1	-1	-1			
A_{2u}	1	1	1	-1	-1	-1	-1	-1	1	1	z		
B_{1u}	1	-1	1	1	-1	-1	1	-1	-1	1			
B_{2u}	1	-1	1	-1	1	-1	1	-1	1	-1			
E_u	2	0	-2	0	0	-2	0	2	0	0	(x, y)		

D_{2d}	I	$2S_4$	C_2	$2C_2'$	$2\sigma_d$			
A_1	1	1	1	1	1			x^2+y^2, z^2
A_2	1	1	1	-1	-1		R_z	
B_1	1	-1	1	1	-1			x^2-y^2
B_2	1	-1	1	-1	1	z		
E	2	0	-2	0	0	(x, y)	(R_x, R_y)	(xz, yz)

D_{3d}	I	$2C_3$	$3C_2$	i	$2S_6$	$3\sigma_d$			
A_{1g}	1	1	1	1	1	1			$x^2 + y^2, z^2$
A_{2g}	1	1	-1	1	1	-1		R_z	
E_g	2	-1	0	2	-1	0		(R_x, R_y)	$(x^2 - y^2, xy), (xz, yz)$
A_{1u}	1	1	1	-1	-1	-1			
A_{2u}	1	1	-1	-1	-1	1	z		
E_u	2	-1	0	-2	1	0	(x, y)		

D_{4d}	I	$2S_8$	$2C_4$	$2S_8^3$	C_2	$4C_2'$	$4\sigma_d$			
A_1	1	1	1	1	1	1	1			$x^2 + y^2, z^2$
A_2	1	1	1	1	1	-1	-1		R_z	
B_1	1	-1	1	-1	1	1	-1			
B_2	1	-1	1	-1	1	-1	1	z		
E_1	2	$\sqrt{2}$	0	$-\sqrt{2}$	-2	0	0	(x, y)		
E_2	2	0	-2	0	2	0	0			$(x^2 - y^2, xy)$
E_3	2	$-\sqrt{2}$	0	$\sqrt{2}$	-2	0	0		(R_x, R_y)	(xz, yz)

T	I	$4C_3$	$4C_3^2$	$3C_2$			
A	1	1	1	1			$x^2 + y^2 + z^2$
C	1	ω	ω^2	1			$x^2 + \omega^2 y^2 + \omega z^2$
D	1	ω^2	ω	1			$x^2 + \omega y^2 + \omega^2 z^2$
F	3	0	0	-1	(x, y, z)	(R_x, R_y, R_z)	(xy, yz, zx)

$\omega = \exp(2\pi i/3)$

T_d	I	$8C_3$	$3C_2$	$6S_4$	$6\sigma_d$			
A_1	1	1	1	1	1			$x^2 + y^2 + z^2$
A_2	1	1	1	-1	-1			
E	2	-1	2	0	0			$(2z^2 - x^2 - y^2, x^2 - y^2)$
F_1	3	0	-1	1	-1		(R_x, R_y, R_z)	
F_2	3	0	-1	-1	1	(x, y, z)		(yz, zx, xy)

O	I	$8C_3$	$3C_2$	$6C_4$	$6C_2'$	
A_1	1	1	1	1	1	$x^2 + y^2 + z^2$
A_2	1	1	1	-1	-1	
E	2	-1	2	0	0	$(2z^2 - x^2 - y^2, x^2 - y^2)$
F_1	3	0	-1	1	-1	(x, y, z) (R_x, R_y, R_z)
F_2	3	0	-1	-1	1	(yz, zx, xy)

I	I	$12C_5$	$12C_5^2$	$20C_3$	$15C_2$	
A	1	1	1	1	1	$x^2 + y^2 + z^2$
F_1	3	-2 cos 144°	-2 cos 72°	0	-1	(x, y, z) (R_x, R_y, R_z)
F_2	3	-2 cos 72°	-2 cos 144°	0	-1	
G	4	-1	-1	1	0	
H	5	0	0	-1	1	$(2z^2 - x^2 - y^2, x^2 - y^2, yz, zx, xy)$

Index

A

D

F

G

H

T

Z